VOLUME EIGHTY THREE

CURRENT TOPICS IN MEMBRANES

Membrane Transporters in the Pathogenesis of Cardiovascular and Lung Disorders

CURRENT TOPICS IN MEMBRANES, VOLUME 83

Series Editor

IRENA LEVITAN
Division of Pulmonary and Critical Care Medicine
University of Illinois at Chicago
Chicago, IL, United States

VOLUME EIGHTY THREE

CURRENT TOPICS IN MEMBRANES

Membrane Transporters in the Pathogenesis of Cardiovascular and Lung Disorders

Edited by

SERGEI N. ORLOV

National Research Tomsk State University;
Siberian State Medical University, Tomsk;
Faculty of Biology, Lomonosov Moscow State University, Moscow, Russia

ELSEVIER

ACADEMIC PRESS

An imprint of Elsevier

Academic Press is an imprint of Elsevier
50 Hampshire Street, 5th Floor, Cambridge, MA 02139, United States
525 B Street, Suite 1650, San Diego, CA 92101, United States
The Boulevard, Langford Lane, Kidlington, Oxford OX5 1GB, United Kingdom
125 London Wall, London, EC2Y 5AS, United Kingdom

First edition 2019

ISBN: 978-0-12-817764-8
ISSN: 1063-5823

For information on all Academic Press publications
visit our website at https://www.elsevier.com/books-and-journals

Publisher: Zoe Kruze
Acquisition Editor: Ashlie Jackman
Editorial Project Manager: Peter Llewellyn
Production Project Manager: Abdulla Sait
Cover Designer: Alan Studholme

Typeset by SPi Global, India

Contents

Contributors

Dennis A. Ausiello
Center for Assessment for Technology and Continuous Health, MGH, Boston, MA, United States

Alexei Y. Bagrov
National Institute on Aging, NIH, Baltimore, MD, United States; Sechenov Institute of Evolutionary Physiology and Biochemistry, St. Petersburg, Russia

Anastasia Bakulina
Novosibirsk State University, Novosibirsk, Russia

Ajay Bjargava
The Francis Crick Institute, London, United Kingdom

Francis Boudreault
Centre de recherche, Centre hospitalier de l'Université de Montréal (CRCHUM), Montréal, QC, Canada

Elena V. Bouzinova
Department of Biomedicine, Health, Aarhus University, Aarhus, Denmark

Maria Castañeda-Bueno
Molecular Physiology Unit, Department of Nephrology and Mineral Metabolism, Instituto Nacional de Ciencias Médicas y Nutrición Salvador Zubirán, Mexico City, Mexico

Alexander V. Chibalin
Department of Molecular Medicine and Surgery, Integrative Physiology, Karolinska Institutet, Stockholm, Sweden; National Research Tomsk State University, Tomsk, Russia

Christopher Cutler
Biology Department, Georgia Southern University, Statesboro, GA, United States

Paola de los Heros
Unidad de Investigación Traslacional UNAM-INCICH, División de Investigación, Facultad de Medicina, Universidad Nacional Autónoma de México, México City, Mexico

Phat Vinh Dip
Nanyang Technological University, School of Biological Sciences, Singapore, Republic of Singapore

Nickolai O. Dulin
Department of Medicine, The University of Chicago, Chicago, IL, United States

Igor V. Emelyanov
Almazov Federal Heart, Blood and Endocrinology Centre, St. Petersburg, Russia

Olga V. Fedorova
National Institute on Aging, NIH, Baltimore, MD, United States

Kishio Furuya
Mechanobiology Laboratory, Nagoya University, Graduate School of Medicine, Nagoya, Japan

Gerardo Gamba
Molecular Physiology Unit, Department of Nephrology and Mineral Metabolism, Instituto Nacional de Ciencias Médicas y Nutrición Salvador Zubirán; Instituto de Investigaciones Biomédicas, Universidad Nacional Autónoma de México, Mexico City; Tecnológico de Monterrey, Escuela de Medicina y Ciencias de la Salud, Monterrey, NL, Mexico

Gerhard Grüber
Nanyang Technological University, School of Biological Sciences, Singapore, Republic of Singapore

Ryszard Grygorczyk
Department of Medicine, Université de Montréal; Centre de recherche, Centre hospitalier de l'Université de Montréal (CRCHUM), Montréal, QC, Canada

Svetlana V. Gusakova
Siberian State Medical University, Tomsk, Russia

Hiroyuki Hosokawa
Center for Systems Biology, Program in Membrane Biology, MGH, Boston, MA, United States

Leonid V. Kapilevich
National Research Tomsk State University, Tomsk, Russia

Elizaveta A. Klimanova
Faculty of Biology, Lomonosov Moscow State University, Moscow; National Research Tomsk State University, Tomsk, Russia

Alexandra O. Konradi
Almazov Federal Heart, Blood and Endocrinology Centre, St. Petersburg, Russia

Edward G. Lakatta
National Institute on Aging, NIH, Baltimore, MD, United States

Hongtao Liu
Division of Cell Signaling, National Institute for Physiological Sciences (NIPS), Okazaki, Japan

Olga D. Lopina
Faculty of Biology, Lomonosov Moscow State University, Moscow, Russia

Anna D. Manis
Department of Physiology, Medical College of Wisconsin, Milwaukee, WI, United States

Vladimir Marshansky
Neuro-Horizon Pharma, Sharon, MA, United States

Vladimir V. Matchkov
Department of Biomedicine, Health, Aarhus University, Aarhus, Denmark

Maria Merkulova
Center for Systems Biology, Program in Membrane Biology, MGH, Boston, MA, United States

Erika Moreno
Department of Nephrology and Mineral Metabolism, Instituto Nacional de Ciencias Médicas y Nutrición Salvador Zubirán, México City, Mexico

Shin-ichiro Mori
Division of Cell Signaling, National Institute for Physiological Sciences (NIPS), Okazaki, Japan

Shigeru Morishima
Division of Cell Signaling, National Institute for Physiological Sciences (NIPS), Okazaki, Japan

Adrián Rafael Murillo-de-Ozores
Molecular Physiology Unit, Department of Nephrology and Mineral Metabolism, Instituto Nacional de Ciencias Médicas y Nutrición Salvador Zubirán; Facultad de Medicina, Universidad Nacional Autónoma de México, Mexico City, Mexico

Tomohiro Numata
Department of Physiology, School of Medicine, Fukuoka University, Fukuoka, Japan

Yasunobu Okada
Department of Physiology and Systems Bioscience, Kyoto Prefectural University of Medicine, Kyoto; Division of Cell Signaling, National Institute for Physiological Sciences (NIPS), Okazaki, Japan

Sergei N. Orlov
National Research Tomsk State University; Siberian State Medical University, Tomsk; Faculty of Biology, Lomonosov Moscow State University, Moscow, Russia

Sergej Pirkmajer
Institute of Pathophysiology, Faculty of Medicine, University of Ljubljana, Ljubljana, Slovenia

Consuelo Plata
Department of Nephrology and Mineral Metabolism, Instituto Nacional de Ciencias Médicas y Nutrición Salvador Zubirán, México City, Mexico

Olga Ponomarchuk
Centre de recherche, Centre hospitalier de l'Université de Montréal (CRCHUM), Montréal, QC, Canada; Faculty of Biology, M.V. Lomonosov Moscow State University, Moscow, Russia

Ravshan Z. Sabirov
Laboratory of Molecular Physiology, Institute of Biophysics and Biochemistry, National University of Uzbekistan, Tashkent, Uzbekistan

Kaori Sato-Numata
Japan Society for the Promotion of Science, Tokyo, Japan

Svetlana V. Sidorenko
Faculty of Biology, Lomonosov Moscow State University, Moscow; National Research Tomsk State University, Tomsk, Russia

Larisa V. Smolyaninova
National Research Tomsk State University, Tomsk; Faculty of Biology, Lomonosov Moscow State University, Moscow, Russia

Masahiro Sokabe
Mechanobiology Laboratory, Nagoya University, Graduate School of Medicine, Nagoya, Japan

Denisha Spires
Department of Physiology, Medical College of Wisconsin, Milwaukee, WI, United States

Christian Staehr
Department of Biomedicine, Health, Aarhus University, Aarhus, Denmark

Alexander Staruschenko
Department of Physiology, Medical College of Wisconsin; Clement J. Zablocki VA Medical Center, Milwaukee, WI, United States

Ju Jing Tan
Department of Medicine, Université de Montréal; Centre de recherche, Centre hospitalier de l'Université de Montréal (CRCHUM), Montréal, QC, Canada

Youg Raj Thaker
Nanyang Technological University, School of Biological Sciences, Singapore, Republic of Singapore

James R. Tonra
BeyondSpring Pharmaceuticals, New York, NY, United States

Norma Vázquez
Department of Nephrology and Mineral Metabolism, Instituto Nacional de Ciencias Médicas y Nutrición Salvador Zubirán; Molecular Physiology Unit, Instituto de Investigaciones Biomédicas, Universidad Nacional Autónoma de México, México City, Mexico

Antonio Vega-Mateos
Department of Nephrology and Mineral Metabolism, Instituto Nacional de Ciencias Médicas y Nutrición Salvador Zubirán, México City, Mexico

Fanxing Zeng
Department of Exercise Physiology, Beijing Sport University, Beijing, China

Lin Zhang
Department of Biomedicine, Health, Aarhus University, Aarhus, Denmark; Department of Exercise Physiology, Beijing Sport University, Beijing, China

Preface

In the human genome, almost 6% of genes encode membrane ion transporters, whereas 20% of functionally characterized genes are involved in membrane delivery of ion transporters and the regulation of their functional activity. Ion transporters are critical for normal physiologic function including maintenance of resting and action electrical membrane potentials, cell volume regulation, intracellular accumulation of sugars, nucleotides, and other small molecular weight organic compounds, cell proliferation, differentiation, and death. Since initial observations of abnormal ion transport in spontaneously hypertensive rats, single point mutations of anion channel CFTR in patients with cystic fibrosis, and mutations of sodium channel ENaC and chloride-coupled cations carriers NKCC2 and NCC in monogenous forms of hypertension and hypotension, numerous research teams demonstrated an implication of abnormal ion transport in the pathogenesis of cardiovascular and lung disorders. The current volume highlights a number of recent findings on the molecular structure and function of membrane ion transporters and their relation to cardiovascular and lung pathology.

In 2 years of the discovery of Na^+,K^+-ATPase, Jens Christian Skou reported that cardiotonic steroids suppress enzyme activity. More recently, it was shown that side by side with dissipation of transmembrane gradients of monovalent cations low doses of these compounds trigger diverse signaling pathways independently of Na^+,K^+-ATPase inhibition. These findings are consistent with dual effect of cardiotonic steroids on tissue fibrosis considered in the first two chapters. Indeed, Dr. Bagrov et al. show that the low doses of endogenous cardiotonic steroids contribute to the cardiovascular fibrosis, whereas our team observed that at high concentrations these compounds suppress lung myofibroblast transformation via inhibition of the Na^+,K^+-ATPase and elevation of the intracellular $[Na^+]/[K^+]$ ratio. Features of Na^+,K^+-ATPase function in vascular smooth muscle cells in the context of the crosstalk of Na^+_i,K^+_i-dependent, Na^+/Ca^{2+} exchange-mediated, and Na^+_i,K^+_i-independent and Src- and phosphatidylinositol-4,5-bisphosphate 3 kinase-mediated signaling are further considered by Dr. Matchkov and coworkers (Chapter 6). In the Chapter 10, Drs. Pirkmajer and Chibalin employed evolutionary approach to demonstrate key developments contributed to diversification and integration of Na^+-K^+-ATPase and the endocrine system functions including vesicular traffic of GLUT4 and aquaporin 2.

Extracellular ATP and other nucleotides are important autocrine/paracrine mediators associated with pathogenesis of acute respiratory distress syndrome and ventilator-induced lung injury. In Chapter 3, Dr. Grygorczyk and coworkers provide the current state of knowledge of the mechanisms of ATP release in the lung overview of the recent technical advances in studying cellular ATP release in vitro and ex vivo. In Chapter 4, Dr. Marshansky's team using a combination of homology modeling and NMR analysis generated the structural model of N-terminal cytosolic domain of a2 subunit of V-ATPase. Their data assume that cardiovascular hypertension and acute lung injury/adult respiratory distress syndrome may be effectively treated by drugs targeting these proteins.

Recent studies demonstrated that elevation of $[Na^+]_i/[K^+]_i$ ratio is one of the major signals triggering transcriptomic changes. Dr. Klimanova et al. identified ubiquitous and cell type-specific $Na^+{}_i$,$K^+{}_i$-sensitive genes in rat aorta smooth muscle and endothelial cells as well as in rat cerebellar granule cells and mouse C2C12 myoblasts. These data presented in Chapter 5 strongly indicate that this novel mechanism of excitation–transcription coupling contributes to the development of cardiovascular and lung disorders and complications such as hypoxia accompanied by dissipation of transmembrane gradient of monovalent cations.

The thiazide-sensitive Na^+–Cl^- cotransporter (NCC) is the major pathway for salt reabsorption in the distal convoluted tubule and a target for thiazide-type diuretics widely employed for the treatment of essential hypertension. In Chapter 7, Dr. Gamba et al. present a review on NCC structure–function relationships with the emphasis on the main functional, pharmacological, and regulatory properties of the spiny dogfish kidney NCC in the comparison with other NCC orthologous. In Chapter 9, the same research team considers the molecular mechanisms by which extracellular potassium modulates the activity of the renal NaCl cotransporter, thus providing a beneficial effect of high-potassium diet in the treatment of hypertension and its cardiovascular and renal complications. Alterations in renal ion transport, triggered by pathophysiological conditions in diabetes, can exacerbate hypertension, accelerate renal injury, and are integral to the development of diabetic kidney disease. The chapter of Dr. Staruschenko and coworkers aims to gather current knowledge of epithelial sodium channels ENaC and other renal ion channels and carriers with altered expression and activity under diabetic conditions.

In many types of cells, apoptosis starts with cell shrinkage caused by activation of the volume-sensitive outwardly rectifying anion channel (VSOR).

Necrotic cell death exhibits persistent cell swelling, which is coupled to impaired VSOR function. Both VSOR and Maxi-Cl channels are also involved in ischemia- and ischemia reperfusion-induced apoptotic or necrotic cell death in cerebral and myocardial cells. Data presented in Chapter 8 by Dr. Okada and coworkers show that cardiac CFTR and probably other anion channels exert a protective action against necrotic myocardial infarction. Cardiac Maxi-Cl activity may participate in protection against ischemia(-reperfusion) injury by mediating ATP release.

We firmly believe that integration of these individual concepts will inspire new investigations to eventually improve our overall understanding of the role of membrane ion transporters in the cardiac, vascular, and lung functions with a final goal of the development of new therapeutic approaches for the treatment of cardiovascular and thoracic diseases.

Sergei N. Orlov

CHAPTER ONE

Acute salt loading and cardiotonic steroids in resistant hypertension

Igor V. Emelyanov[a], Alexandra O. Konradi[a], Edward G. Lakatta[b], Olga V. Fedorova[b], Alexei Y. Bagrov[b,c,*]

[a]Almazov Federal Heart, Blood and Endocrinology Centre, St. Petersburg, Russia
[b]National Institute on Aging, NIH, Baltimore, MD, United States
[c]Sechenov Institute of Evolutionary Physiology and Biochemistry, St. Petersburg, Russia
*Corresponding author: e-mail address: aybagrov@gmail.com

Contents

Abstract

The study addresses the association of marinobufagenin (MBG), a natriuretic and vasoconstrictor steroid, and Na/K-ATPase (NKA) activity with pressor response to salt-loading and arterial stiffness in resistant hypertension (RH). Thirty-four patients (18 males and 16 females; 56 ± 8 years) with RH on a combined (lisnopril/amlodipine/hydrochlorothiazide) therapy and 11 healthy age-matched normotensive subjects (7 males and 4 females; 54 ± 2 years) were enrolled in this study. Salt-loading was performed via intravenous infusion of 1000 mL saline (0.9% NaCl) for 1 h. Arterial stiffness was measured by Sphygmocor Px device with a calculation of pulse-wave velocity (PWV). Activity of NKA was measured in erythrocytes. We demonstrated that plasma levels of MBG and magnitude of NaCl-induced MBG-dependent NKA inhibition are associated with PWV, and that this association has gender- and age-specific fashion in RH patients.

Current Topics in Membranes, Volume 83
ISSN 1063-5823
https://doi.org/10.1016/bs.ctm.2019.01.005

Abbreviations

BP blood pressure
CO cardiac output
CS cardiotonic steroids
DBP diastolic blood pressure
MBG marinobufagenin
NKA Na/K-ATPase
PWV pulse-wave velocity
RH resistant hypertension
SBP systolic blood pressure

1. Introduction

Excessive salt ingestion plays an important role in pathogenesis of hypertension (Folkow, 1982; Weinberger, Miller, Luft, Grim, & Fineberg, 1986). One of major factors mediating compromised ability of the kidneys to excrete sodium is Na/K-ATPase (NKA) and its endogenous inhibitors, cardiotonic steroids (CS) including marinobufagenin (MBG) (Bagrov, Shapiro, & Fedorova, 2009). In the kidney, NKA and MBG represent major sodium transporting mechanism (Fedorova & Bagrov, 1997; Hamlyn & Blaustein, 1986). According to the "concept of a natriuretic hormone," the primary role of CTS is to promote natriuresis via inhibition of sodium reabsorption in the renal tubules (Hamlyn & Blaustein, 1986). CS promote natriuresis via inhibition of the NKA in the proximal tubules and thick ascending limb (TAL) of Henle's loop (Fedorova, Kashkin, Zakharova, Lakatta, & Bagrov, 2012). In addition to the effect of CS per se, a number of substances (natriuretic peptides, dopamine, calcitonin, etc.) exhibit natriuretic effect via modulation of NKA activity by cGMP in the TAL of Henle's loop (Aperia et al., 1994; Bailly, 2000; Florkowski, Richards, Espiner, Yandle, & Frampton, 1994). This segment of nephron, therefore, represents a unique site for the interaction of various natriuretic compounds. Only about 15%–25% of filtered sodium is reabsorbed in TAL; the importance of this segment of nephron in sodium metabolism could be pictured by the fact that a reduction in salt reabsorption by TAL could determine pathogenesis of Bartter's disorder (Hebert, 2003). In addition, CS exhibit growth-promoting and pro-fibrotic effects via NKA-dependent signaling in cardiovascular tissues (Fedorova, Emelianov, et al., 2015; Khalaf et al., 2018; Xie & Askari, 2002).

Despite significant progress in the development of antihypertensive therapies, hypertension remains a major determinant of cardiovascular mortality worldwide (Zanchetti, Grassi, & Mancia, 2009), and the achievement of blood pressure (BP) target level remains to be poor on the population level. About 10%–12% of hypertensive patients could be characterized as truly drug-resistant, i.e., when target BP cannot be achieved with a three-drug combination therapy, including a diuretic, given in adequate doses (Mancia et al., 2007). Mechanisms responsible for drug resistance are complex and are not fully understood. One of major predictors of drug resistance includes an uncontrolled salt and liquid intake (Folkow, 1982). Previously we studied CS in two human NaCl-loading studies, SALT and SARAH (Anderson et al., 2008; Fedorova, Lakatta, Bagrov, & Melander, 2015). Data from both SALT and SARAH studies suggest that MBG is an important determinant of BP, and that its reaction to salt and its relation to salt-sensitivity of BP differ by age and gender (Anderson et al., 2008; Fedorova, Lakatta, et al., 2015). While in older females (SALT study), a relative failure in MBG production may cause volume expansion and contribute to hypertension, in males aging is associated with lower baseline MBG levels, but with a greater responsiveness of MBG to HS (SARAH study) (Anderson et al., 2008; Fedorova, Lakatta, et al., 2015).

The major goal of this study was to investigate whether NaCl-induced MBG response in resistant hypertension (RH) is associated with salt-sensitivity of BP and with vascular stiffness. We hypothesized that (i) levels of MBG and magnitude of NaCl-induced MBG-dependent NKA inhibition would be associated with pulse-wave velocity (PWV), and (ii) that this association will exhibit gender and age specificity.

2. Methods

2.1 General

We examined 21 patients with true RH (11 males and 10 females, mean age 56 ± 8 years) from the Outpatient Resistant Hypertension Centre of the Almazov Federal Heart, Blood and Endocrinology Centre (St. Petersburg, Russia). The study was approved by local Ethic Committee and by Medstar Research Institute, and the informed consent was obtained from all participants. The inclusion criteria were: essential hypertension with a duration over 5 years, signed informed consent, stable three-drug full dose conventional antihypertensive therapy (lisnopril/amlodipine/hydrochlorothiazide) with office BP greater than 140/90 mmHg. All patients were carefully examined to exclude secondary causes of hypertension including obstructive sleep apnea; ambulatory BP monitoring was performed to exclude the "white coat" effect. Patients with concomitant coronary artery disease, previous stroke or transient ischemic attacks, renal insufficiency, severe concomitant diseases, known poor treatment adherence, drug abuse, body mass index (BMI) over 35 kg/m^2 were excluded. For 2 weeks before the test, patents were prescribed standard three-drug antihypertensive therapy in rational combinations: ACE inhibitor lisinopril 20 mg per day, calcium channel blocker amlodipine 10 mg per day and diuretic hydrochlorothiazide 25 mg per day. The diagnosis of RH was reconfirmed after 2 weeks. The control group included 11 sex- and age-matched subjects (7 males and 4 females, mean age 54 ± 2 years).

2.2 Hemodynamic measurements

Heart rate was monitored by electrocardiography and BP was measured by finger plethysmography (Finometer; TNO, Amsterdam, Netherlands) of the right index or middle finger intermittently recalibrated against oscillometry. Beat-to-beat arterial BP was noninvasively and continuously measured by a fully automated Finometer® device (Finapres Medical Systems, TNO-BMI, Netherlands), which uses a volume clamp technique and has been well-validated against intra-arterial BP measurements. Systolic BP (SBP) and diastolic BP (DBP) were measured on a beat-by-beat basis. The beat-by-beat stroke volume was derived from the blood pressure waveform using Beatscope® software (1.1a, TNO-BMI, Amsterdam, Netherlands), a program that employs the Modelflow technique computing cardiac output (CO) by simulating a nonlinear three-element Windkessel model of aortic input impedance.

2.3 Salt-loading protocol

The salt-loading test was performed in the morning in a quiet room with a stable air temperature 19–22°C. Patients fasted 12 ± 2 h before the test, and morning medications were not taken. Test was started after 30 min of rest in a supine position with an i.v. cubital catheter and with an empty urine bladder. All measurement equipment was also applied before the test. The test was divided into three parts. First, 60 min of baseline rest during which hemodynamic parameters were monitored and blood samples obtained at baseline and after 60 min of rest. Second, after 60 min of rest the infusion of 1000 mL of 0.9% NaCl was performed for 60 min (16 mL/min) followed by continuous BP and hemodynamic monitoring. Finally, the blood samples were obtained after the infusion peak, and the hemodynamic parameters were registered for an additional 60 min as a recovery period followed by the blood sampling at the end. Salt-sensitivity of BP was defined as SBP increase during the test 20 mmHg or more and/or DBP 10 mmHg or more (Luft, Weinberger, Grim, & Fineberg, 1986). Activity of NKA in erythrocytes was measured after the ex vivo incubation of the red blood cells with monoclonal anti-MBG mAb or with vehicle as described below.

2.3.1 *Erythrocyte Na/K-ATPase*

Erythrocytes were washed three times in an isotonic medium (145 mmol/L NaCl in 20 mmol/L Tris buffer; pH 7.6 at 4°C). Erythrocytes were preincubated with Tween-20 (0.5%) in sucrose (250 mmol/L) and Tris buffer (20 mmol/L; pH 7.4, $t = 37°C$) for 30 min, and were incubated for 30 min in the medium (mmol/L): Na 100, K 10, $MgCl_2$ 3, EDTA 0.5, Tris 50, ATP 2 (pH 7.4, $t = 37°C$) in the final dilution 1:40. The reaction was stopped by the addition of trichloroacetic acid to final concentration 7%. Total ATPase activity was measured by the production of inorganic phosphate (P_i), and NKA activity was estimated as the difference between ATPase activity in the presence and in the absence of 5 mmol/L ouabain. Whole blood was preincubated for 1 h with a vehicle or 3E9 anti-MBG monoclonal antibody.

2.3.2 *MBG immunoassay*

Plasma samples were extracted on Sep-Pak C-18 cartridges (Waters, Milford, MA, USA), and MBG competitive fluoroimmunoassay based on a murine monoclonal anti-MBG 4G4 antibody was performed as described recently. The cross-reactivity of 4G4 anti-MBG antibody is: MBG—100%, ouabain—0.005%, digoxin—0.03%, digitoxin <0.001%, bufalin—0.08%, cinobufagin—0.07%, cinobufatalin—40%, prednisone, spironolactone, aldosterone, proscillaridin, and progesterone <0.001%.

2.4 Statistical analyses

Data analysis was performed by Statistica for Windows, version 7.0 (StatSoft). All data are presented as means ± SEM. Normality of data distribution was assessed using Kolmogorov–Smirnov test. Nonparametric statistical methods were used when variables did not normally distribute. Hemodynamic data at different experimental stages were assessed by one-way or two-way analysis of variance (ANOVA) for repeated measures or by the Friedman test for nonparametric distributions, followed by the Bonferroni correction. The correlations were assessed using linear regression and Pearson correlation analysis.

3. Results

We studied 34 patients (18 males and 16 females; 56 ± 8 years) with RH, in which BP remained over 140/80 on a combined therapy. NaCl loading was performed via intravenous infusion of 1000 mL saline for 1 h. Control group and patient characteristics are presented in Table 1. Table 2 compares hemodynamic changes in patients and in the control group during salt loading. BP and total peripheral resistance increased in both groups, although in the hypertensive group it exceeded those in controls. As presented in Fig. 1, NaCl loading resulted in the pressor response, increase in plasma levels of MBG and in a twofold inhibition of erythrocyte NKA which was blocked by anti-MBG antibody. Because anti-MBG mAb reversed NaCl-induced NKA inhibition, MBG is responsible for this NKA-inhibitory effect of NaCl loading. As presented in Table 2, plasma levels of MBG have increased in salt-loaded subjects, which was associated with inhibition of erythrocyte NKA.

Table 1 Clinical and demographic characteristics.

Parameter	Control subjects ($n = 11$)	Patients with RH ($n = 34$)
Age, years	54 ± 2	56 ± 8
Body mass index, kg/m^2	28.5 ± 0.4	32.2 ± 5.4*
Gender, males/females	7/4	18/16
24 h systolic BP, mmHg	131.7 ± 1.8	149.5 ± 5.8**
24 h diastolic BP, mmHg	82.0 ± 1.5	91.1 ± 5.2**
Heart rate, beats/min	70.2 ± 1.4	72.5 ± 5.7
PWV, m/s	7.1 ± 0.5	7.9 ± 1.8*
Serum creatinine, μmol/L	75.5 ± 2.2	85.8 ± 9.1**
Glomerular filtration rate, mL/min	94.6 ± 4.3	76.7 ± 14.9**
Plasma sodium, mmol/L	140.7 ± 1.4	141.6 ± 4.9
Fasting glucose, mmol/L	5.55 ± 0.21	6.07 ± 0.66*
Total cholesterol, mmol/L	5.31 ± 0.16	6.11 ± 0.53**

Data are means ± SEM, *$P < 0.05$; **$P < 0.01$, RH vs. controls.

Table 2 Effects of acute salt loading.

Parameter	Control subjects ($n = 11$)	Patients with RH ($n = 34$)
SBP baseline, mmHg	133 ± 3	156 ± 3***
DBP baseline, mmHg	82 ± 2	90 ± 2***
CO baseline, mL/min	5.9 ± 0.2	5.5 ± 0.3
Plasma MBG baseline, nmol/L	0.14 ± 0.02	0.15 ± 0.01
NKA baseline, μmol P_i/mL/h	3.18 ± 0.45	2.2 ± 0.1*
Total peripheral resistance, baseline, din × s × sm^{-5}	1087 ± 18	1414 ± 28*
SBP final, mmHg	145 ± 4	180 ± 5***
DBP final, mmHg	88 ± 3	102 ± 3***
CO final, mL/min	6.4 ± 0.3	6.0 ± 0.4
Plasma MBG final, nmol/L	0.19 ± 0.01	0.23 ± 0.01
NKA final μmol P_i/mL/h	2.41 ± 0.44	1.6 ± 0.01**
Total peripheral resistance, final, din × s × sm^{-5}	1345 ± 24	2025 ± 27**

Data are means ± SEM. 2-way ANOVA, *$P < 0.05$, **$P < 0.01$, ***$P < 0.001$, RH vs. controls. *baseline*, before salt-loading; *CO*, cardiac output; *DBP*, diastolic blood pressure; *final*, after salt-loading; *RH*, resistant hypertension; *SBP*, systolic blood pressure.

Fig. 2 presents data on correlations of the magnitude of NaCl-induced NKA inhibition in the hypertensive group. While there were no association between the magnitude of NaCl-induced NKA inhibition and magnitude of BP response (A–C), delta NKA exhibited a positive linear correlation with age (D–F), and in male, but not in the female subjects, delta NKA positively correlated with PWV (G–I). Thus, in patients with RH magnitude of MBG-induced NKA inhibition associates not with the salt-sensitivity of BP per se, but with age of patients and with PVW which characterizes vascular stiffness, a major feature of cardiovascular aging.

In accordance with the above finding, in male, but not in female subjects, PVW exhibited positive linear correlation with both NaCl-induced renal MBG response, and with its physiological effect, i.e., natriuretic response (Fig. 3). Importantly, a magnitude of renal MBG response did not correlate with salt-sensitivity of BP.

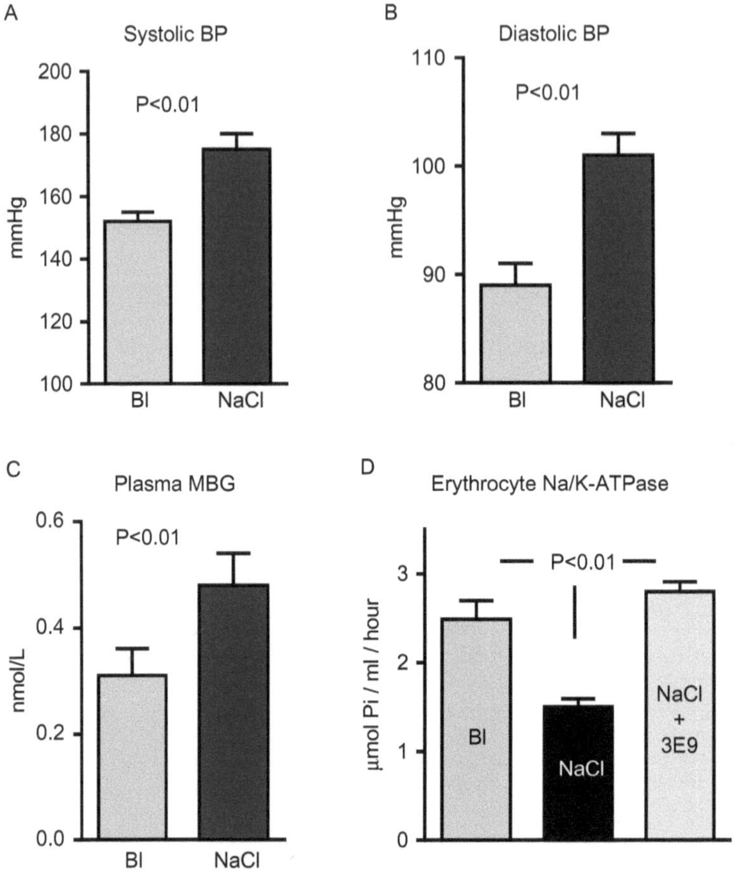

Fig. 1 Effect of acute NaCl loading on systolic (A) and diastolic (B) BP, on plasma MBG concentration (C) and on erythrocyte Na/K-ATPase (D) in patients with resistant hypertension. "NaCl + 3E9"—ex vivo effect of anti-MBG mAb on Na/K-ATPase in erythrocytes obtained following NaCl loading. *Bl*, baseline. *t*-test and 1-way ANOVA.

4. Discussion

In the present study we demonstrate that in patients with RH on a high NaCl, MBG response do not correlate with BP response, but rather is related to PWV. Our present findings are in agreement with the growing body of evidence that, with respect to pathogenesis of hypertension, the effects of CS on growth and collagen synthesis (vascular stiffness) may be more important than the effects of CS on BP. However, it appears that CS responses on the background of established hypertension are not

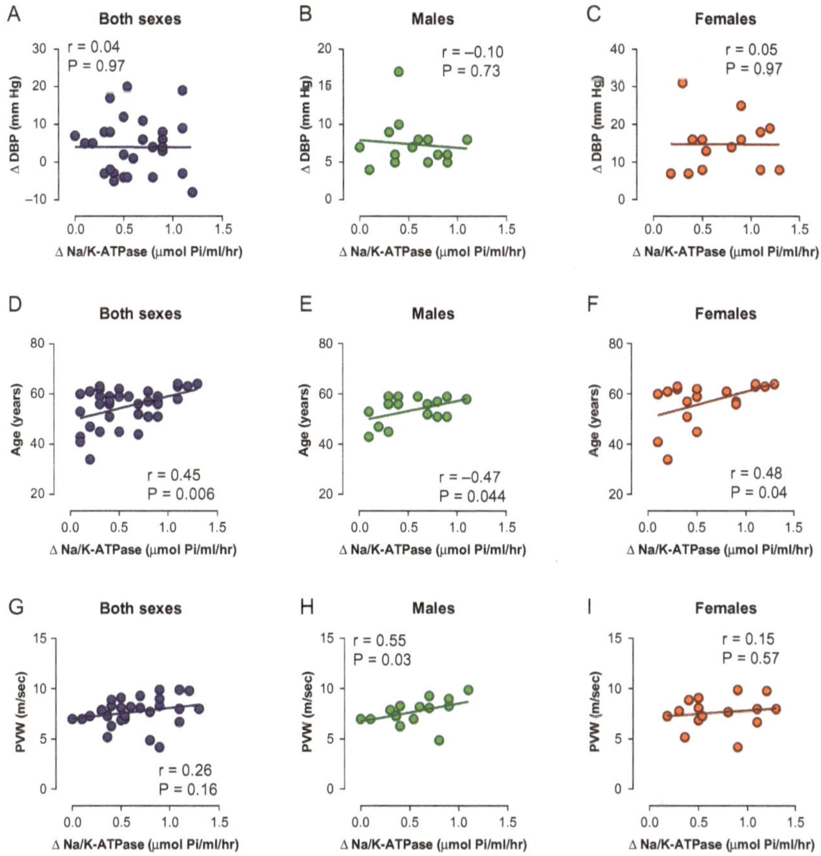

Fig. 2 Correlations of NaCl-loading-induced inhibition of erythrocyte Na/K-ATPase. Relationship between MBG-dependent inhibition of Na/K-ATPase and magnitude of NaCl-induced change of diastolic BP (A–C), age of the patients (D–F), and PWV (G–I).

anymore associated with the salt-sensitivity of BP, but rather are implicated in cardiovascular remodeling (Fig. 4). Analysis of present data suggests that involvement of CS in the response to NaCl challenge is altered with age, and that there is a gender difference in the response of CS to the salt-loading.

Data from the previous studies (Anderson et al., 2008; Fedorova, Lakatta, et al., 2015) suggest that MBG is an important determinant of BP and that its reaction to salt and its relation to salt-sensitivity of BP differ by age and gender. While in older females, a relative failure in MBG production may cause volume expansion and contribute to hypertension, in males, aging is associated with lower baseline MBG levels, but a greater responsiveness of

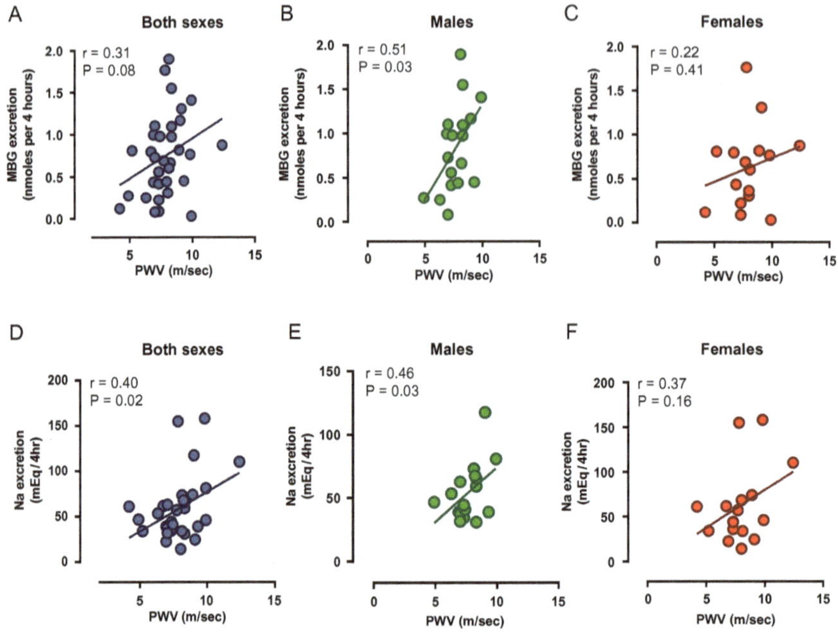

Fig. 3 Pearson correlation of NaCl-induced MBG response and PWV (A–C), and relationship between natriuretic response to the saline loading and PWV (D–F) in patients with resistant hypertension.

MBG to administration of high salt diet (Anderson et al., 2008; Fedorova, Lakatta, et al., 2015). Subsequently, independent positive associations of SBP with urine MBG/Na ratio were found in black women, which supported idea that reduced MBG-mediated Na excretion can contribute to adverse hemodynamics (Strauss, Smith, et al., 2018). More recently it was shown that at a young age heightened endogenous MBG levels may contribute to large artery stiffness in women via pressure-independent mechanisms, increasing their risk for future cardiovascular disease (Strauss, Smith, Wei, Fedorova, & Schutte, 2018). These human observations were supported by the experimental data showing that high-salt induced aortic stiffness in normotensive animals occurred in the presence of elevated MBG (Grigorova et al., 2018). Interestingly, a decrease in a salt consumption restored aortic elasticity and diminished the risk of cardiovascular disease by reduction of production of the pro-fibrotic factor MBG (Grigorova et al., 2018).

Two signaling pathways were shown to modulate pro-fibrotic effects of MBG (Fig. 4). Pro-fibrotic effects may be initiated by MBG is

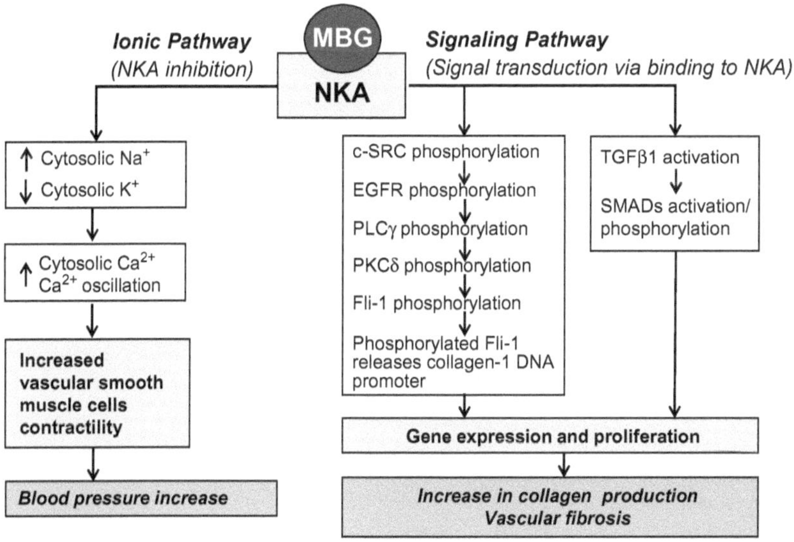

Fig. 4 The schematic presentation of the possible molecular mechanisms of the implication of MBG in blood pressure regulation via ionic pathway (Na^+/K^+-mediated signaling) and stimulation of vascular fibrosis via Na^+/K^+-independent signaling. *c-SRC*, proto-oncogene tyrosine-protein kinase; *EGFR*, epidermal growth factor receptor; *Fli-1*, Friend leukemia integration 1 transcription factor, a negative regulator of collagen-1 production; *MBG*, marinobufagenin; *NKA*, Na/K-ATPase; *PKCδ*, protein kinase C delta; *PLCγ*, phospholipase C gamma; *SMADs*, mothers against DPP homologs; *TGFβ*, transforming growth factor beta.

SMAD-dependent TGFβ-1 signaling which underlies vascular fibrosis in salt-induced aortic stiffness in normotensive rats (Grigorova et al., 2018). Another system, inhibition of Fli1, a nuclear transcription factor and a member of ETS family is implicated in MBG-induced fibrosis (Fedorova, Emelianov, et al., 2015). Fli1 acts as a negative regulator of collagen-1 synthesis and it competes with another transcription factor, ETS-1, to maintain a balance between stimulation and repression of *Col1a2* gene (Trojanowska, 2010). Preeclampsia is associated with the elevated levels of BP and MBG and fibrosis of umbilical arteries, and by a dramatic reduction in the levels of Fli1 (Fedorova et al., 2018). In a group of patients with mild hypertension in the presence of NaCl reduction levels of MBG also reduced and they were negatively related to aortic PWV (Jablonski et al., 2013). Recently in a small group of RH patients, an addition of spironolactone to the therapy was associated with a drop in PWV and with an increase in erythrocyte NKA activity (Fedorova, Emelianov, et al., 2015). In summary, in the present study it was demonstrated that MBG plays an important role in the regulation of vascular stiffness, and the further clinical studies will provide additional information related to the metabolic pathways activated by MBG.

Acknowledgments

Authors are grateful to Elena Frolova and Anton Bzhelyanskiy for technical support. This study was supported by Intramural Research Program, National Institute on Aging, National Institutes of Health (O.V.F., E.G.L., A.Y.B.), by a grant from Russian Federal Agency for Scientific Organizations (Mechanisms of the formation of physiological functions in phylogenesis and ontogenesis. Endogenous and exogenous regulation) (A.Y.B.), by OÜ Haakon, Narva, Estonia, and by a grant from Russian Federation No. 14.740.11.0928 (I.V.E., A.O.K.).

References

Anderson, D. E., Fedorova, O. V., Morrell, C. H., Longo, D. L., Kashkin, V. A., Metzler, J. D., et al. (2008). Endogenous sodium pump inhibitors and age-associated increases in salt sensitivity of blood pressure in normotensives. *American Journal of Physiology. Regulatory, Integrative and Comparative Physiology*, *294*, R1248–R1254.

Aperia, A., Holtback, U., Syren, M. L., Svensson, L. B., Fryckstedt, J., & Greengard, P. (1994). Activation/deactivation of renal Na(+),K(+)-ATPase: A final common pathway for regulation of natriuresis. *The FASEB Journal*, *8*, 436–469.

Bagrov, A. Y., Shapiro, J. I., & Fedorova, O. V. (2009). Endogenous cardiotonic steroids: Physiology, pharmacology, and novel therapeutic targets. *Pharmacological Reviews*, *61*, 9–38.

Bailly, C. (2000). Effect of luminal atrial natriuretic peptide on chloride reabsorption in mouse cortical thick ascending limb: Inhibition by endothelin. *Journal of the American Society of Nephrology*, *11*, 1791–1797.

Fedorova, O. V., & Bagrov, A. Y. (1997). Inhibition of Na/K ATPase from rat aorta by two endogenous Na/K pump inhibitors, ouabain and marinobufagenin. Evidence of interaction with different alpha-subunit isoforms. *American Journal of Hypertension*, *10*, 929–935.

Fedorova, O. V., Emelianov, I. V., Bagrov, K. A., Grigorova, Y. N., Wei, W., Juhasz, O., et al. (2015). Marinobufagenin-induced vascular fibrosis is a likely target for mineralocorticoid antagonists. *Journal of Hypertension*, *33*, 1602–1610.

Fedorova, O. V., Ishkaraeva, V. V., Grigorova, Y. N., Reznik, V. A., Kolodkin, N. I., Zazerskaya, I. E., et al. (2018). Antibody to marinobufagenin reverses placenta-induced fibrosis of umbilical arteries in preeclampsia. *International Journal of Molecular Sciences*, *19*(8), pii: E2377.

Fedorova, O. V., Kashkin, V. A., Zakharova, I. O., Lakatta, E. G., & Bagrov, A. Y. (2012). Age-associated increase in salt sensitivity is accompanied by a shift in the atrial natriuretic peptide modulation of the effect of marinobufagenin on renal and vascular sodium pump. *Journal of Hypertension*, *30*, 1817–1826.

Fedorova, O. V., Lakatta, E. G., Bagrov, A. Y., & Melander, O. (2015). Plasma level of the endogenous sodium pump ligand marinobufagenin is related to the salt-sensitivity in men. *Journal of Hypertension*, *33*, 534–541.

Florkowski, C. M., Richards, A. M., Espiner, E. A., Yandle, T. G., & Frampton, C. (1994). Renal, endocrine, and hemodynamic interactions of atrial and brain natriuretic peptides in normal men. *The American Journal of Physiology*, *266*, R1244–R1250.

Folkow, B. (1982). Physiological aspects of primary hypertension. *Physiological Reviews*, *62*(2), 347–504, 1982.

Grigorova, Y. N., Wei, W., Petrashevskaya, N., Zernetkina, V., Juhasz, O., Fenner, R., et al. (2018). Dietary sodium restriction reduces arterial stiffness, vascular TGF-β-dependent fibrosis and marinobufagenin in young normotensive rats. *International Journal of Molecular Sciences*, *19*(10), pii: E3168.

Hamlyn, J. M., & Blaustein, M. P. (1986). Sodium chloride, extracellular fluid volume, and blood pressure regulation. *The American Journal of Physiology*, *251*(4 Pt 2), F563–F575.

Hebert, S. C. (2003). Bartter syndrome. *Current Opinion in Nephrology and Hypertension*, *12*, 527–532.

Jablonski, K. L., Fedorova, O. V., Racine, M. L., Geolfos, C. J., Gates, P. E., Chonchol, M., et al. (2013). Dietary sodium restriction and association with urinary marinobufagenin, blood pressure, and aortic stiffness. *Clinical Journal of the American Society of Nephrology*, *8*, 1952–1959.

Khalaf, F. K., Dube, P., Mohamed, A., Tian, J., Malhotra, D., Haller, S. T., et al. (2018). Cardiotonic steroids and the sodium trade balance: New insights into trade-off mechanisms mediated by the Na^+/K^+-ATPase. *International Journal of Molecular Sciences*, *19*, pii: E2576.

Luft, F. C., Weinberger, M. H., Grim, C. E., & Fineberg, N. S. (1986). Effects of volume expansion and contraction on potassium homeostasis in normal and hypertensive humans. *Journal of the American College of Nutrition*, *5*, 357–369.

Mancia, G., De Backer, G., Dominiczak, A., Cifkova, R., Fagard, R., Germano, G., et al. (2007). Guidelines for the management of arterial hypertension: The task force for the management of arterial hypertension of the European Society of Hypertension (ESH) and of the European Society of Cardiology (ESC). *Journal of Hypertension*, *25*(2007), 1105–1187.

Strauss, M., Smith, W., Kruger, R., Wei, W., Fedorova, O. V., & Schutte, A. E. (2018). Marinobufagenin and left ventricular mass in young adults: The African-PREDICT study. *European Journal of Preventive Cardiology*, *25*, 1587–1595.

Strauss, M., Smith, W., Wei, W., Fedorova, O. V., & Schutte, A. E. (2018). Marinobufagenin is related to elevated central and 24-h systolic blood pressures in young black women: The African-PREDICT study. *Hypertension Research*, *41*, 183–192.

Trojanowska, M. (2010). Cellular and molecular aspects of vascular dysfunction in systemic sclerosis. Cellular and molecular aspects of vascular dysfunction in systemic sclerosis. *Nature Reviews Rheumatology*, *6*, 453–460.

Weinberger, M. H., Miller, J. Z., Luft, F. C., Grim, C. E., & Fineberg, N. S. (1986). Definitions and characteristics of sodium sensitivity and blood pressure resistance. *Hypertension*, *8*, II127–II134.

Xie, Z., & Askari, A. (2002). Na(+)/K(+)-ATPase as a signal transducer. *European Journal of Biochemistry*, *269*, 2434–2439.

Zanchetti, A., Grassi, G., & Mancia, G. (2009). When should antihypertensive drug treatment be initiated and to what level should systolic blood pressure be lowered? A critical reappraisal. *Journal of Hypertension*, *27*, 923–934.

Further reading

Blaustein, M. P., Ashida, T., Goldman, W. F., Wier, W. G., & Hamlyn, J. M. (1986). Sodium/calcium exchange in vascular smooth muscle: A link between sodium metabolism and hypertension. *Annals of the New York Academy of Sciences*, *488*, 199–216.

CHAPTER TWO

Control of lung myofibroblast transformation by monovalent ion transporters

Nickolai O. Dulin[a,*], Larisa V. Smolyaninova[b,c], Sergei N. Orlov[b,c,d,*]

[a]Department of Medicine, The University of Chicago, Chicago, IL, United States
[b]National Research Tomsk State University, Tomsk, Russia
[c]Faculty of Biology, Lomonosov Moscow State University, Moscow, Russia
[d]Siberian State Medical University, Tomsk, Russia
*Corresponding authors: e-mail address: ndulin@medicine.bsd.uchicago.edu; sergeinorlov@yandex.ru

Contents

Abstract

Myofibroblast differentiation is a critical process in the pathogenesis of tissue fibrosis. We focus our mini-review on recent data showing an implication of monovalent ion transporters in fibroblast to myofibroblast transformation of human lung fibroblasts (HLF). In cultured HLF, cardiotonic steroids (CTS) known as potent inhibitors of Na^+, K^+-ATPase suppress myofibroblast differentiation in parallel with up- and down-regulated expression of cyclooxygenase-2 (COX-2) and TGF-β receptor subunit TGFBR2, respectively. K^+-free medium mimics antifibrotic action of CTS indicating a key role of elevated intracellular $[Na^+]_i/[K^+]_i$ ratio. Augmented expression of COX-2 is abolished by inhibition of Na^+/Ca^{2+} exchanger. Side-by-side with CTS acting via elevation of the $[Na^+]_i/[K^+]_i$ ratio fibroblast to myofibroblast transformation is also suppressed by potent inhibitors of Ca^{2+}-activated chloride channels tannic acid and K^+,Cl^- cotransporter DIOA. The relative impact of $Ca^{2+}{}_i$-mediated and -independent signaling triggered

ISSN 1063-5823
https://doi.org/10.1016/bs.ctm.2019.01.002

by elevated $[Na^+]_i/[K^+]_i$ ratio and altered intracellular anion handling in transcriptomic changes involved in myofibroblast differentiation should be examined further.

Abbreviations

COX-2 inducible isoform of cyclooxygenase
CTS cardiotonic steroids
HLF human lung fibroblasts
IPF idiopathic pulmonary fibrosis
TGF-β transforming growth factor-β
TGFBR1 and TGFBR2 TGF-β receptor type I and type II, respectively

1. Introduction

From the pathophysiological point of view, fibrosis is a disorder of abnormal wound healing, in which the initial trigger to the fibrotic response is an injury, followed by an exuberant, non-resolving wound-healing response (Hardie, Glasser, & Hagood, 2009; Noble & Homer, 2005). Fibrosis is a characteristic feature of many chronic diseases that result in end-stage organ failure. In the developed world, fibrotic-driven diseases such as diabetic nephropathy, cardiac fibrosis, vascular fibrosis, liver cirrhosis and pulmonary fibrosis are the major cause of morbidity and mortality, accounting for 45% of all deaths (Wynn, 2008). The common pathological hallmark of fibrotic disorders is the presence of collagen-secreting fibroblasts sharing biochemical similarities with smooth muscle cells that are termed myofibroblasts. These cells are induced to differentiate by tissue-specific profibrotic factors released in the local inflammatory microenvironment, such as transforming growth factor-β (TGF-β), connective tissue growth factors (CTGF), angiotensin II, and endothelin-1 (Harvey, Montezano, Lopes, & Touyz, 2016; Wynn, 2008). Despite rapidly accumulated knowledge on the molecular mechanisms associated with myofibroblast differentiation, effective and organ-selective therapies for prevention of fibrotic disease are currently unavailable.

Idiopathic pulmonary fibrosis (IPF) is a progressive, fatal disease characterized by parenchymal fibrosis and structural distortion of the lungs. Age-adjusted mortality due to pulmonary fibrosis is increasing (Olson et al., 2003), and it poses a vexing clinical challenge given the lack of proven effective therapy. Similar to other fibrotic disorders injury of alveolar epithelial cells results in the elaboration of a fibrinous provisional matrix and activation of several pro-inflammatory, pro-coagulant, and pro-fibrotic mediators, of which TGF-β1 is the most established. Indeed, TGF-β1 has been localized to areas of fibrosis in both experimental pulmonary fibrosis and human disease (Broekelmann, Limper, Colby, & McDonald, 1991; Phan & Kunkel, 1992; Santana, Saxena, Noble, Gold, & Marshall, 1995). Lung-targeted overexpression of TGF-β1 results in the development of lung fibrosis in animals (Kang, Cho, Lee, Homer, & Elias, 2007; Sime, Xing, Graham, Csaky, & Gauldie, 1997). Conversely, inhibition of TGF-β1 via soluble TGF-β receptors can inhibit in vivo fibrogenesis (Giri, Hude, & Hollinger, 1993; Wang et al., 1999). Therefore, it is conceivable that inhibiting TGF-β 1 signaling could be an approach for treating fibrotic disease.

We focus our mini-review on recent data showing that monovalent ion transporters affect myofibroblast differentiation in response to TGF-β. Data on the action of ion transport inhibitors in in vivo models of tissue fibrosis are considered in recent comprehensive reviews (Khalaf et al., 2018; Orlov, La, Smolyaninova, & Dulin, 2017) and are out of scope of our current mini-review.

2. TGF-β-signaling involved in myofibroblast differentiation

Fibroblasts, under stimulation by TGF-β1, differentiate to myofibroblasts characterized by de novo expression of cytoskeletal and contractile proteins, modified focal adhesion complexes (Hinz, 2006) and components of the extracellular matrix (Phan, 2014). Several cytoskeletal and smooth muscle proteins are expressed in myofibroblasts, including smooth muscle α-actin, the most established marker for myofibroblast differentiation (Evans, Kelley, Krill, Low, & Adler, 1983; Hinz et al., 2007; Phan, 1996; Tomasek, Gabbiani, Hinz, Chaponnier, & Brown, 2002). Functionally, this phenotypic switch is thought to increase the ability of the myofibroblast to attach to the remodeling matrix and ultimately facilitate wound contraction during wound healing. However, induction of myofibroblast phenotype is also associated with an increased resistance to apoptosis (Thannickai & Horowitz, 2006; Zhang & Phan, 1999) and secretion of extracellular matrix proteins (collagen isoforms, cellular fibronectin, etc.) and pro-fibrotic factors (connective tissue growth factor [CTGF], insulin-like growth factor [IGF-1], etc.) (Leask & Abraham, 2004), thus perpetuating ongoing tissue remodeling and fibrosis. Myofibroblasts are invariably found in histologic sections of human lung specimens from patients with pulmonary fibrosis and are thought to be a critical pathogenic mechanism responsible for the progressive nature of IPF.

Therefore, understanding cellular and molecular mechanisms which induce and/or maintain TGF-β-dependent myofibroblast differentiation are of high importance for understanding the pathogenesis of the disease and for identifying potentially targetable molecules that are critical for myofibroblast differentiation.

TGF-β signals through a receptor serine/threonine kinase complex consisting of TGF-β receptor type I (TGFBR1) and TGF-β receptor type II (TGFBR2). Upon binding of TGF-β to the receptor complex, serine-threonine kinase TGFβR2 activates TGFβR1 via phosphorylation of its TTSGSGSG motif, which in turn phosphorylates receptor-activated Smad (R-Smad) transcription factors—Smad2 and Smad3. This activation mechanism might be blocked via dephosphorylation of TGFBR1 by serine-threonine protein phosphatases PP1C and PP2A. Phosphorylated Smad2/3 then bind to the common Smad mediator, Smad4. The Smad complex translocates to the nucleus and preferentially binds to Smad binding elements (SBE) at the promoters of target genes to initiate Smad-dependent gene transcription. For more details, see comprehensive reviews (Feng & Derynck, 2005; Huang & Chen, 2012; Leask & Abraham, 2004). More recently, we reported that TGF-β-induced differentiation of human lung fibroblasts (HLF) is also accompanied by activation of serum response factor (SRF) transcription factor. Importantly, downregulation of SRF expression by the siRNA approach or suppression of its activity by the pharmacological inhibitor CCG-1423 blocked TGFβ1-induced smooth muscle α-actin expression, indicating the requirement of SRF in this process (Sandbo, Kregel, Taurin, Bhorade, & Dulin, 2009). We also found that SRF activation by TGFβ1 is mediated by (*i*) Smad-dependent activation of Ras homolog gene family member A (RhoA), (*ii*) stress fiber formation, triggered by activation of Rho-associated protein kinase (ROCK), (*iii*) nuclear accumulation of megakaryoblastic leukemia 1(MKL1), a coactivator of SRF, and (*iv*) induction of transcription of SRF-dependent genes (Sandbo et al., 2011).

3. Na^+,K^+-ATPase as the only known receptor of cardiotonic steroids

Na^+,K^+-ATPase is an integral plasma membrane protein consisting of ~110 kDa α-subunit, ~35 kDa β-subunit and in most cells studied so far 8 kDa γ-subunit. It was shown that ATP hydrolysis leads to Asp369 phosphorylation within the α-subunit, which provides the E_1–E_2 conformational transition and electrogenic ion transport ($3Na^+$ vs $2K^+$) at a baseline rate of 60–80 phosphorylation–dephosphorylation cycles per second. In addition to the ubiquitous α1-isoform, three other α-subunits are expressed in a tissue-specific manner with high abundance in neuronal cells (α3 and α2), heart and skeletal muscle (α2), and testis (α4). Four isoforms of β-subunits are highly glycosylated and are obligatory for enzymatic activity. The functional significance of seven γ-subunits sharing a Pro-Phe-X-Tyr-Asp motif (FXYD) remains poorly understood. For more details, see reviews Blanco (2015) and Scheiner-Bobis (2002).

Cardiotonic steroids (CTS) are known as compounds sharing a common structure formed by a steroid nucleus with a lactone ring at C-17 and a hydroxyl group at C-14. The 5-membered and 6-membered lactone rings are the most significant feature of cardenolides and bufadienolides, isolated from plants and amphibians, respectively. In 1938, Wood and Moe reported that treatment with cardenolides caused the gain of Na^+ and loss of K^+ in the canine ventricular musculature (Wood & Moe, 1938). Fifteen years later, Schatzmann demonstrated that in human erythrocytes these compounds inhibit energy-dependent accumulation of $[K^+]_o$ and extrusion of $[Na^+]_i$ (Schatzmann, 1953). Finally, 2 years after the discovery of Mg^{2+}-dependent (Na^++K^+)-stimulated adenosine triphosphatase (Na/K-ATPase) (Skou, 1957), Jens Skou reported that CTS completely suppressed the enzyme's activity (Skou, 1960). The mechanism of Na^+,K^+-ATPase inhibition by CTS has been mainly explored with ouabain extracted from *Strophanthus gratus* and possessing much higher water solubility compared to other CTS. Lingrel and coworkers showed that transmembrane regions of the α-subunit H1, H5, and H7 and its H1-H2, H5-H6, and H7-H8 extracellular loops affect the affinity of Na^+,K^+-ATPase for ouabain (Lingrel, 2010).

As predicted, exposure to CTS affects numerous cellular functions related to maintenance of the transmembrane gradient of monovalent cations, such as electrical membrane potential (E_m), cell volume, transepithelial movement of salt and osmotically obliged water, Na^+/H^+ and $Na^+(K^+)/Ca^{2+}$ exchange, and symport of Na^+ with inorganic phosphate, glucose, amino acids, nucleotides, etc. During the last two decades, it has been shown that along with the above-mentioned canonical $[Na^+]_i$,$[K^+]_i$-, E_m-, and cell volume-mediated cellular responses, CTS can affect gene expression, membrane trafficking, cell adhesion, proliferation and death (Xie & Askari, 2002; Orlov, Akimova, & Hamet, 2005; Aperia, 2007; Liu & Xie, 2010; Riganti et al., 2011; Orlov, Klimanova, et al., 2017; Orlov, La, et al., 2017). The recent data showing cell type-specific impact of CTS-induced $[Na^+]_i$,$[K^+]_i$-mediated and -independent signaling pathways in cell survival and TGF-β-dependent myofibroblast differentiation are discussed below.

4. Cardiotonic steroids suppress TGF-β-induced myofibroblast differentiation via elevation of the $[Na^+]_i/[K^+]_i$ ratio

Unlike the ubiquitous impact of CTS on Na^+_i,K^+_i-dependent cell functions, their actions on cell survival and differentiation are tissue- and species-specific. For instance, ouabain triggers massive death of renal epithelial cells from the Madin-Darby canine kidney (MDCK) (Bolivar et al., 1987; Ledbetter, Young, & Wright, 1986), endothelial cells from the pig aorta (Orlov et al., 2004), human prostatic smooth muscle cells (Chueh, Guh, Chen, Lai, & Teng, 2001), human prostate adenocarcinoma cells (McConkey, Lin, Nutt, Ozel, & Newman, 2000), human monocytes (Kurosawa, Tani, Nishimura, Numazawa, & Yoshida, 2001), human erythroleukemia cell line (HEL) (Perne et al., 2009), human neuroblastoma cell line SH-SY5Y (Kulikov, Eva, Kirch, Boldyrev, & Scheiner-Bobis, 2007), and human neuronal precursor NT2 cells (Hennion, el-Masri, Huff, & el-Mailakh, 2002; Rosen, Glukmann, Feldman, Fridman, & Lichtstein, 2004). In striking contrast, we did not detect any impact of 48-h exposure to high concentrations of ouabain on survival of human lung fibroblasts (La, Reed, Koltsova, et al., 2016), rat vascular smooth muscle cells (Orlov, Thorin-Trescases, Kotelevtsev, Tremblay, & Hamet, 1999), rat endothelial cells and rat astrocytes (Akimova et al., 2015; Akimova, Mongin, Hamet, & Orlov, 2006). Moreover, ouabain inhibited apoptosis in rat vascular smooth muscle cells transfected with E1A adenoviral protein and subjected to serum-deprivation (Orlov et al., 1999).

Suppression of apoptosis in CTS-treated cells can be mediated by membrane depolarization, accumulation of Na^+ or loss of K^+. In addition, conformational transition of the Na^+,K^+-ATPase α-subunit, triggered by interaction with CTS, may be sufficient per se to generate the antiapoptotic signal. To examine the role of above-listed upstream signals in the suppression of apoptosis, we treated VSMC with ouabain in control and high-K^+ medium. Sustained depolarization in high-K^+ medium did not affect apoptosis. In contrast, dissipation of the transmembrane gradient of monovalent cations sharply diminished the effect of ouabain on Na^+_i and K^+_i content and completely abolished its antiapoptotic action (Orlov et al., 1999). Later on, the antiapoptotic action of ouabain and K^+-depleted medium was detected in other cell types (Isaev et al., 2000; Trevisi, Visentin, Cusinato, Pighin, & Luciani, 2004; Zhou et al., 2001).

Several research teams reported that TGF-β-induced myofibroblast differentiation is accompanied by activation of both pro- and antiapoptotic signaling. Thus, it was well-documented that in TGF-β inhibits proliferation of cultured hepatocytes and epithelial cell lines of different origin via cell cycle arrest and accumulation of apoptotic cells (for review, see (Song & Shi, 2018)) It was also shown that during the process of myofibroblast differentiation TGF-β-treated cells acquire resistance to apoptosis (Thannickai & Horowitz, 2006; Zhang & Phan, 1999). Keeping in mind the above-mentioned data on cell type-specific impact of CTS on apoptosis we investigated their actions on myofibroblast differentiation in TGF-β-treated HLF.

We initially observed that accumulation of smooth muscle-specific α-actin (SM-α-actin), fibronectin, and collagen-1 in primary-cultured human lung fibroblasts was inhibited by ouabain and another clinically relevant CTS—digoxin (Fig. 1A and B). To examine the role of dissipation of transmembrane gradient of monovalent cations in antifibrotic action of CTS we investigated the dose-dependent action of ouabain on intracellular Na^+ and K^+ content and employed K^+-free medium as an alternative approach for Na^+,K^+-ATPase inhibition. We found that inhibition of myofibroblast differentiation by ouabain correlated with the loss of K^+ and gain of Na^+

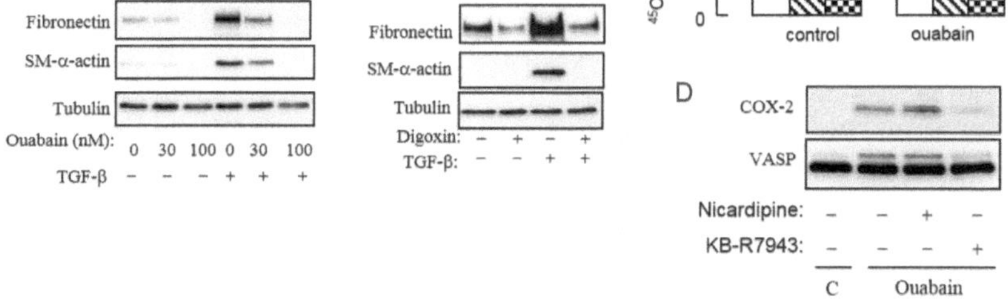

Fig. 1 CTS as antifibrotic compounds. Ouabain (A) and digoxin (B) augment COX-2 expression and inhibit myofibroblast differentiation in cultured human lung fibroblasts triggered by TGF-β. The action of ouabain, nicardipine and KB-R7943 on 45Ca influx (C), expression of COX-2 and VASP phosphorylation (D) in cultured human lung fibroblasts (La, Reed, Koltsova, et al., 2016; La, Reed, Smolyaninova, et al., 2016).

and was also observed in K^+-free medium (La, Reed, Koltsova, et al., 2016; La, Reed, Smolyaninova, et al., 2016). Viewed collectively, these data strongly suggest that cardiac glycosides inhibit myofibroblast differentiation via a signaling pathway mediated by suppression of the Na^+,K^+-ATPase activity and elevation of the $[Na^+]_i/[K^+]_i$ ratio.

It is well-documented that tumor progression is accompanied by the activation of stromal fibroblasts into myofibroblasts, often termed cancer-associated fibroblasts (CAFs) contributing to the pathogenesis of cancer (Polanska & Orimo, 2013; Ronnov-Jensen, Petersen, & Bissell, 1996). Through a high throughput screening of a chemical library for fibronectin expression as an initial CAF marker in TGF-β-treated human prostate fibroblasts, Coleman and coworkers identified seven cardiac glycosides as the most potent CAF blocking agents among 1700 other compounds (Coleman, Gray, Stephens, Scott, & Cardelli, 2016). They showed that CTS such as digoxin, strophanthin, and lanatoside inhibited TGF-β-induced fibronectin expression with IC_{50} of 30, 40, and 110nM, respectively. Importantly, all tested compounds suppressed myofibroblast differentiation via Smad-mediated signaling pathway without any impact on cell survival. It is well documented that the affinity of CTS for the mouse and rat α1-Na^+/K^+ ATPase is 10^3-fold less than that for the human protein. Considering this, Coleman et al. tested the efficacy of digoxin in the J2 mouse fibroblast cell line. They found that 240nM digoxin had no impact on the increment of fibronectin content in TGF-β-treated J2 cells (Coleman et al., 2016). These results are consistent with our conclusions that CTS suppress TGF-β-induced myofibroblast differentiation by inhibiting α1-Na^+/K^+-ATPase and elevation of the $[Na^+]_i/[K^+]_i$ ratio.

4.1 Upregulation of COX-2 expression

Numerous studies demonstrated that the expression of an inducible isoform of cyclooxygenase (COX-2) is decreased in lungs of patients with pulmonary fibrosis (Bauman et al., 2010; Coward, Watts, Feghali-Bostwick, Knox, & Pang, 2009; Gabasa et al., 2013; Petkova, Clelland, Ronan, Lewis, & Knox, 2003; Vancheri et al., 2000; Wilborn et al., 1995). This phenomenon may be caused by signaling triggered by augmented stiffness of fibrotic tissue (Liu et al., 2010). It was also shown that in animal models of IPF both pharmacological and genetic suppression of COX-2 worsens fibrosis (Bonner et al., 2002; Giri & Hyde, 1987; Hodges et al., 2004; Leindecker Stumm, Wettlaufer, Janckar, & Peters-Golden, 2011) (Keerthisingam et al., 2001) and fibrosis-associated lung function decline (Card et al., 2007). Prostaglandin E_2 (PGE_2) is a major lipid mediator produced by COX-2. In cultured lung fibroblasts, PGE_2 inhibits TGF β and endothelin-1-induced expression of smooth muscle α-actin and collagen through prostanoid receptor 2 (EP2) and elevation of intracellular cAMP content (Garrison et al., 2013; Kolodsick et al., 2003). More recently, Wettlaufer and co-workers reported that PGE_2 reversed the effect of TGF-β1 on the expression of ~50% of differentially expressed genes in CCL210 human lung fibroblasts cell line (Wettlaufer, Scott, McEachin, Peters-Golden, & Huang, 2016).

Our group demonstrated that COX-2 mRNA is upregulated in ouabain-treated vascular smooth muscle cells from rat aorta, human umbilical vein endothelial cells, and HeLa cancer cell line (Koltsova et al., 2012). Using K^+-free medium as an alternative approach for Na^+,K^+-ATPase inhibition, we observed that augmented expression of COX-2 in ouabain-treated HLF cells is triggered by elevation of the $[Na^+]_i/[K^+]_i$ ratio (Koltsova et al., 2012). In vascular smooth muscle and HeLa cells, ouabain treatment resulted in profound expression of immediate response gene c-Fos. In these cells, time-dependent elevation of c-Fos mRNA correlates with the gain of $[Na^+]_i$ (Haloui et al., 2007; Taurin et al., 2002). These results strongly suggest that $[Na^+]_i$ augmentation rather than loss of $[K^+]_i$ plays a key role in transcriptomic changes triggered by CTS. The possible molecular origin of $[Na^+]_i$ sensors involved in transcription regulation are considered elsewhere (Orlov & Hamet, 2015).

Data presented above allowed us to suggest that inhibition of myofibroblast differentiation by CTS in TGF-β-treated HLF is mediated by augmented COX-2 expression. Indeed, we found that treatment of human lung fibroblasts with 100 nM ouabain and digoxin resulted in profound and sustained expression of COX-2 mRNA and protein (Fig. 1A and B), which was associated with activation of PKA as assessed by phosphorylation-dependent electrophoretic mobility shift of the PKA substrate, vasodilator-stimulated phosphoprotein (VASP) (Fig. 1D). Importantly, both inhibitor of COX-2, NS-398, and a knockdown of COX-2 mRNA abolished ouabain-induced PKA activation (La, Reed, Koltsova, et al., 2016; La, Reed, Smolyaninova, et al., 2016).

Does COX-2 overexpression mediate the inhibition of myofibroblast differentiation by CTS? To answer this question we used COX-2 siRNA and COX-2 inhibitor NS-398. Apparently, neither COX-2 siRNA nor NS-398 affected the ability of ouabain to inhibit TGF-β-induced myofibroblast differentiation (La, Reed, Koltsova, et al., 2016; La, Reed, Smolyaninova, et al., 2016). These data strongly suggest that overexpression of COX-2 alone is not sufficient for suppression of TGF-β-induced myofibroblast differentiation by CTS.

4.2 Downregulation of TGFBR2 expression

Keeping in mind the lack of any actions of COX-2 inhibitors on the suppression of TGF-β-induced myofibroblast differentiation by CTS, we focused on upstream intermediates of TGF-β-induced signaling (La, Reed, Koltsova, et al., 2016; La, Reed, Smolyaninova, et al., 2016). We found that treatment of HLFs with ouabain resulted in a sustained downregulation of TGFβR2 mRNA and protein, which was accompanied by inhibition of TGF-β-induced Smad2 phosphorylation and Smad-dependent gene transcription (Fig. 2A). Importantly, downregulation of TGFBR2 was observed in HLF

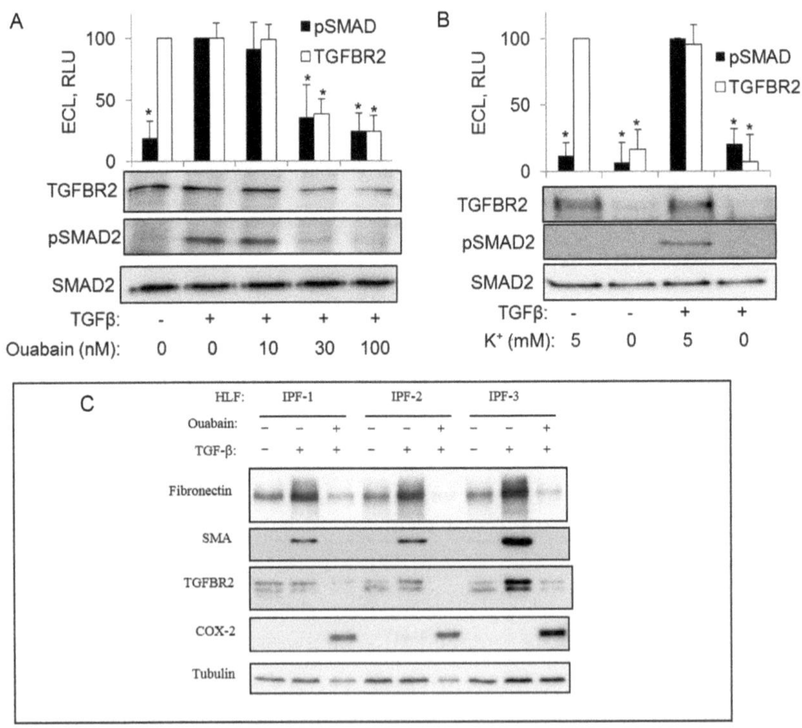

Fig. 2 Dose-dependent action of ouabain (A) and (C) and K+-free medium (B) on TGFBR2 expression, SMAD2 phosphorylation and expression of COX-2 and markers of myofibroblast differentiation (fibronectin and α-smooth muscle actin, SMA) in cultured fibroblasts isolated from control lungs (A) and (B) and lungs of three patients with idiopathic pulmonary fibrosis (IPF-1–IPG-3) (C) (La, Reed, Koltsova, et al., 2016; La, Reed, Smolyaninova, et al., 2016).

cultured from fibrotic lungs of patients with IPF (Fig. 2C). It should be underlined that these effects of ouabain were mimicked by inhibition of Na/K-ATPase through removal of extracellular K^+ (Fig. 2B) thus indicating a key role of $[Na^+]_i/[K^+]_i$-mediated signaling.

5. Role of $Ca^{2+}{}_i$-mediated signaling pathways

It has been shown that in several types of cells agonist-induced COX-2 expression is mediated by elevated $[Ca^{2+}]_i$ (Nakao et al., 2001; Zhu et al., 2002). We found that exposure of HLF to 100 nM ouabain increased the rate of ^{45}Ca influx by ~50% (La, Reed, Koltsova, et al., 2016; La, Reed, Smolyaninova, et al., 2016). These differences were preserved in the presence of an inhibitor of voltage-gated L-type Ca^{2+} channels, nicardipine, but were abolished by Na^+/Ca^{2+} exchanger inhibitor, compound KB-R4943 (Fig. 1C). Importantly, KB-R4943, but not nicardipine, sharply decreased COX-2 expression and PKA activation in response to ouabain (Fig. 1D). Importantly, the involvement of Na^+/Ca^{2+} exchanger in COX-2 expression seems to be cell type-specific phenomenon. Thus, in rat vascular smooth muscle cells, Na^+,K^+-ATPase inhibition by ouabain or in K^+-free medium increased the rate of $^{45}Ca^{2+}$ influx by 2-fold, and this increment was abolished by nicardipine, but it was resistant to KB-R7943. In these cells, COX-2 expression triggered by ouabain was diminished by nicardipine, calmodulin antagonists and compound KN-93, i.e., a potent inhibitor of Ca^{2+}/calmodulin-dependent protein kinase II (CaMKII) suggesting an implication of Ca^{2+}/calmodulin-CaMKII-mediated signaling pathway (Smolyaninova, Koltsova, Sidorenko, & Orlov, 2017).

As a working hypothesis we propose that Na^+,K^+-ATPase inhibition in CTS-treated HLF increases COX-2 expression via elevation of $[Na^+]_i$, activation of $Na^+{}_i/Ca^{2+}{}_o$ exchanger, elevation of $[Ca^{2+}]_i$, that, in turn, triggers $[Ca^{2+}]_i$-mediated mechanism of excitation-transcription coupling (Fig. 3) documented in numerous studies (for reviews, see (Gundersen, 2011; Ma, Groth, Wheeler, Barrett, & Tsien, 2011)). We did not examined the action of Ca^{2+} chelators and inhibitors of Na^+/Ca^{2+} exchanger on TGFBR2 expression. Thus, the relative impact of $Na^+{}_i$-mediated, $Ca^{2+}{}_i$-mediated and -independent mechanisms of excitation transcription coupling documented in recent studies (Koltsova et al., 2012; Koltsova, Tremblay, Hamet, & Orlov, 2015; Sidorenko et al., 2018; Smolyaninova et al., 2017) in downregulation of TGFBR2 transcription seen in CTS-treated HLF remains unknown.

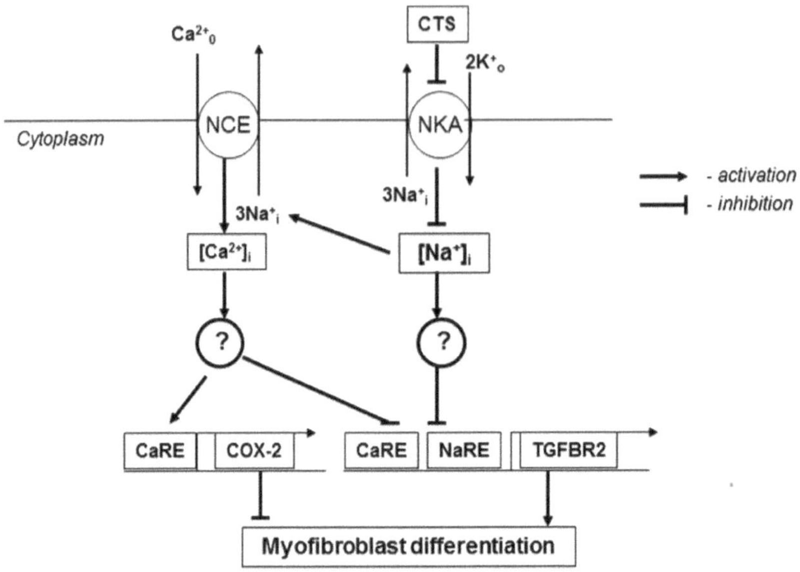

Fig. 3 Cross-talk of intracellular Na^+ and Ca^{2+} in the regulation of COX-2 and TGFB2R expression. ?—unknown steps of intracellular signaling; CaRE and NaRE—Ca^{2+} and Na^+-response elements of transcription regulation, respectively. For other abbreviations, see text.

6. Evidence for pro-fibrotic actions of endogenous cardiotonic steroids

During the last two decades ouabain, marinobufagenin and several other CTS have been identified in mammalian tissues and extracellular fluids and are defined as endogenous CTS (for review, see (Bagrov, Shapiro, & Fedorova, 2009; Orlov et al., 2005; Riganti et al., 2011; Schoner & Scheiner-Bobis, 2007)). Evidence for pro-fibrotic actions of endogenous CTS are based on two major observations. *First*, in several experimental models, treatment with antimarinobufagenin monoclonal antibodies prevented fibrotic changes in the heart, kidney and vasculature (for more details, see (Bagrov et al., 2009; Grigorova et al., 2016; Haller et al., 2014; Khalaf et al., 2018; Xie, Shapiro, & Shapiro, 2014). *Second*, in the absence of TGF-β the addition of low doses of CTS increased myofibroblasts accumulation in cultured rat heart fibroblasts (Elkareh et al., 2007), in rat cardiac fibroblasts (Drummond et al., 2016; El-Okdi et al., 2008), human skin fibroblasts (El-Okdi et al., 2008), rat aortic explants and isolated smooth muscle cells (Fedorova et al., 2015).

It was shown that an increment of collagen content in cardiac fibroblasts treated by low doses of marinobufagenin was abolished in the presence of tyrosine kinase inhibitor herbimycin, inhibitor of EGFR AG1478, Src antagonist PP2 (El-Okdi et al., 2008), and pNaKtide, a specific peptide inhibitor of Src (Leask & Abraham, 2004). Similar results were obtained in human skin fibroblasts, thus indicating an impact of EGFR/Src/tyrosine kinase-mediated signaling (El-Okdi et al., 2008). Using dermal fibroblasts, it was shown that the transcription factor Friend leukemia integration-1 (Fli-1) might be considered as a negative regulator of collagen synthesis (Czuwara-Ladykowska, Shirasaki, Jackers, Watson, & Trojanowska, 2001). A strong inverse relationship between the expressions of Fli-1 and procollagen was also demonstrated in primary culture of rat cardiac fibroblasts. Importantly, in these cells, 24-h exposure to 10 nM marinobufagenin attenuated Fli-1 content by 30%–40%, whereas transfection of a Fli-1 expression vector prevented increased procollagen-1 expression induced by marinobufagenin (Elkareh et al., 2009). The inverse effect of marinobufagenin on Fli-1 and collagen content was also revealed in rat aortic explants and in primary culture of vascular smooth muscle cells from rat aorta. In these experiments, the actions of marinobufagenin were completely abolished by mineralocorticosteroid antagonist canrenone (Fedorova et al., 2015), suggesting the crosstalk of mineralocorticoid receptors, Fli-1-, and an EGFR/Src-mediated pathway in collagen accumulation triggered by marinobufagenin in cardiac fibroblasts.

Do low doses of CTS trigger fibrosis via above-mentioned signaling pathways independently of Na^+,K^+-ATPase inhibition and elevation of the $[Na^+]_i/[K^+]_i$ ratio? This question becomes important because with a few exceptions (Bagrov et al., 1998; Bauer et al., 2005) the concentration of endogenous CTS in plasma and urine does not exceed 1 nM (Khalaf et al., 2018), i.e., 1–2 orders of magnitude less then IC50 of all isoforms of the Na+,K+-ATPase studied so far. To analyze this issue answer this question Elkareh et al. compared dose-dependent action of MBG and ouabain on ^{86}Rb uptake and collagen accumulation in rat cardiac fibroblasts. They observed that the low pro-fibrotic doses of CTS had no impact on ^{86}Rb uptake, and its half maximal inhibition was observed at concentration of MBG and ouabain $\sim 10^{-6}$M and 10^{-4}M, respectively (Elkareh et al., 2007). It should be noted, however, that incubation times selected for estimation of effects of CTS on collagen content (24 h) and ^{86}Rb uptake (10 min) were sharply different. This is important because of the slow actions of CTS at low concentrations documented in human lymphocytes (Segel & Lichtman, 1980) and endothelial cells (Tverskoi et al., 2016). Indeed, the

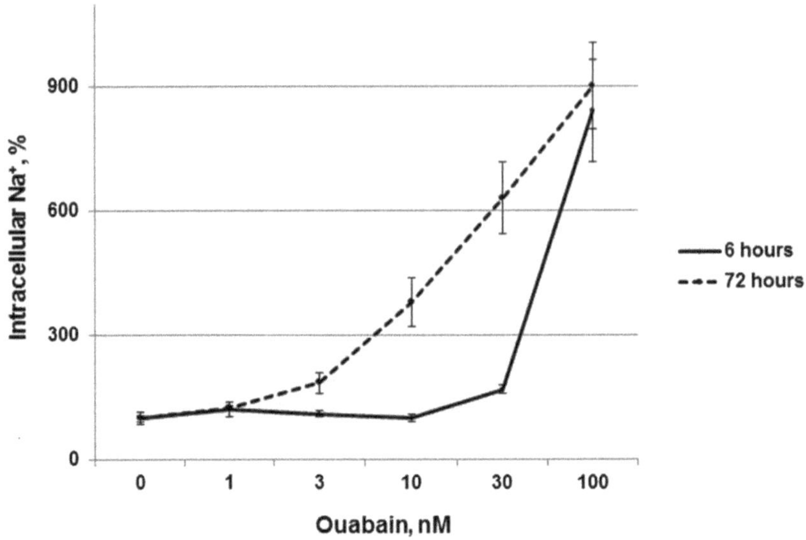

Fig. 4 Dose-dependent action of ouabain on intracellular Na^+ content in human lung fibroblasts. Cells were treated with ouabain for 6 or 72 h. Intracellular Na^+ content was measured as a steady-state distribution of $^{22}Na^+$ (La, Reed, Koltsova, et al., 2016; La, Reed, Smolyaninova, et al., 2016); its values obtained in the absence of ouabain and 73 ± 5 nmol/mg protein was taken as 100%. Means $\pm$ S.E. for experiment performed in triplicates are shown.

half-maximal elevation of $[Na^+]_i$ was detected in HUVEC within 6 h of treatment with 100 nM ouabain, whereas within 24 and 48 h the same increment was detected at ouabain concentrations of 10 and 3 nM, respectively (Tverskoi et al., 2016). Similar results were obtained in the comparative analysis of dose-dependent action of ouabain on $[Na^+]_i$ in HLF treated with this compound for 6 and 72 h (Fig. 4). Along with slow kinetics of interaction of ouabain with α1-Na^+,K^+-ATPase observed in early studies (Segel & Lichtman, 1980), this phenomenon might be caused by slow elevation of $[Na^+]_i$ and accumulation of Na^+,K^+-ATPase in P-E_2Na_3 conformation having high affinity for CTS.

In recent experiments, we demonstrated that elevation of the $[Na^+]_i/[K^+]_i$ ratio is an obligatory step for transcriptomic changes in ouabain- and marinobufagenin-treated human endothelial cells (Klimanova et al., 2017). These data strongly suggest that pro-fibrotic action of low doses of CTS noted in above-cited studies is not mediated by de novo gene expression, but contradict to studies showing that 24-h treatment of primary cultures of rat cardiac fibroblasts with 1 nM marinobufagenin or 100 nM ouabain resulted in ~5-fold attenuation of miR-29b-3p content, whereas transfection of miR-29b-3p inhibited CTS-induced collagen synthesis (Leask & Abraham, 2004). Using genetically engineering $\alpha 1^{+/-}$ mice it was shown that suppression of miR-29b-3p expression by low doses of ouabain is mediated by murine α1-Na^+,K^+-ATPase whose affinity for CTS is extremely low (Drummond et al., 2018). It is important to emphasize that both in vitro and in vivo studies demonstrated the suppression of cardiac and renal fibrosis by this microRNA via its intervention with TGF-β/Smad-mediated signaling (Qin et al., 2011; Ramdas, McBride, Denby, & Baker, 2013; van Rooij et al., 2008). Thus, additional investigations should be performed to verify the crosstalk of CTS and TGF-β-independent and -mediated signaling including PI3K/Akt/mTOR pathway in regulation of myofibroblast differentiation.

7. TGF-β affects expression of monovalent ion transporters

The distinct impact of CTS on myofibroblast differentiation demonstrated in the presence and absence of TGF-β might be caused by the action of TGF-β on Na^+,K^+-ATPase expression or/and activity. To examine this hypothesis, we performed microarray analysis of HLF exposed to TGF-β for 24h. We observed that TGF-β changes the expression of 689 genes by >5-fold. Among these differentially expressed genes we found 12 genes encoding monovalent ion transporters including ~100-fold elevation of voltage-sensitive Ca^{2+}-activated Cl^- channel (TMEM16), ~10-fold increase of β1-subunit of the Na^+,K^+-ATPase (ATP1B1) as well as augmented transcription of Na^+-coupled neutral amino acid carrier (SLC38A1) and subunits of intracellular Cl^- channel (CLIC3) and voltage-gated K^+ channels (K_V) (KCTD15, KCNG1). At the same time, we noted by >5-fold down regulation of Kv5.1 (KCNF1), inwardly rectifying K^+ channels (KCNJ2), voltage-gated Na^+ channel (SCN2A) and Na^+/H^+ exchanger type 9 (NHE9) (Table 1).

Does augmented expression of Na^+,K^+-ATPase β1-subunit revealed in TGF-β-treated cells affect monovalent ion handling? To answer this question we examined the action of TGF-β on Na^+,K^+-pump activity estimated by the rate of ^{86}Rb influx and studied Na^+,K^+-pump activity in HLF transfected with siRNA for its β1-subunit. Fig. 5 shows that TGF-β increases Na^+,K^+-ATPase activity by ~40% estimated as the rate of ouabain-sensitive ^{86}Rb influx without any impact on the (ouabain + bumetanide)-resistant ^{86}Rb influx widely employed as a measure of the passive permeability

Table 1 Monovalent ion transporters whose expression in human lung fibroblasts is affected by TGF-β.

Gene name	Fold of change by TGF-β	Gene function
TMEM16A	96.13	Voltage-sensitive calcium-activated chloride channel
KCTD15	10.33	Potassium voltage-gated channel (Kv) tetramerization domain-containing protein 15
KCNU1	9.96	Calcium-activated potassium channel, subfamily M subunit alpha-3 (KCa5)
ATP1B1	9.89	Sodium/potassium-transporting ATPase subunit beta-1
KCNG1	7.70	Potassium voltage-gated channel subfamily G member 1 (Kv6.1)
CLIC3	7.03	Chloride intracellular channel protein 3
SLC38A1	5.22	Sodium-coupled neutral amino acid transporter 1
CLIC6	0.13	Chloride intracellular channel protein 6 (Parchorin)
SCN2A	0.047	Sodium channel protein type 2 subunit alpha (Nav1.2)
KCNF1	0.045	Potassium voltage-gated channel subfamily F member 1 (Kv5.1)
KCNJ2	0.012	Inward rectifier potassium channel 2
SLC9A9	0.010	Sodium/hydrogen exchanger 9 (NHE-9)

Human lung fibroblasts (HLF) were exposed to 1 ng/mL transforming growth factor-β (TGF-β) for 24 h. HLF isolation, their culturing procedure, RNA isolation and microarray procedure were described previously (La, Reed, Koltsova, et al., 2016; La, Reed, Smolyaninova, et al., 2016; Koltsova et al., 2012). The table lists genes whose expression was affected by TGF-β by >5-fold ($p < 0.05$).

for K^+. Importantly, the increment of Na^+,K^+-ATPase activity in TGF-β 1-treated cells was abolished by transfection with β1-siRNA (Fig. 6) thus indicating a key role of over expression of Na^+,K^+-ATPase β1 subunit. It should be noted, however, that side-by-side with regulation of Na^+,K^+-ATPase-mediated ion fluxes (Hasler et al., 1998) a specific amino acid region in the extracellular domain of the β1-subunit is responsible for its trans dimerization and intercellular adhesion (Vagin, Dada, Tokhtaeva, & Sachs, 2012). Keeping these data in mind, additional experiments should be performed for the estimation of the role of TGF-β-dependent overexpression of β1-Na^+,K^+-ATPase in augmented myofibroblast adhesion.

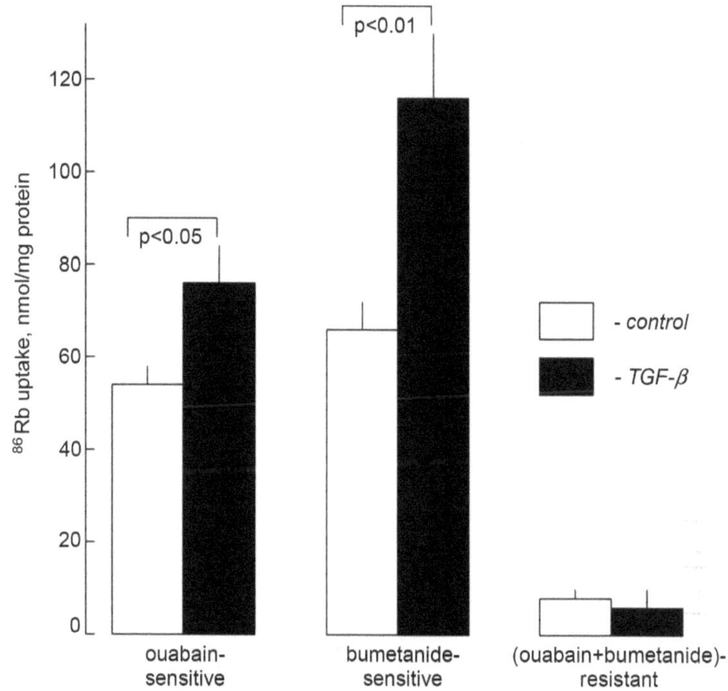

Fig. 5 Effect of TGF-β on the activity of Na^+,K^+-pump (ouabain-sensitive component of the rate of ^{86}Rb influx), Na^+,K^+,Cl^- cotransport (bumetanide-sensitive component of the rate of ^{86}Rb influx) and (ouabain + bumetanide)-resistant ^{86}Rb uptake in human lung fibroblasts (HLF). HLF were cultured in 12-well plates as described elsewhere (La, Reed, Koltsova, et al., 2016; La, Reed, Smolyaninova, et al., 2016) and treated by 1 ng/mL TGF-β for 48 h. Then, this medium was aspirated and 0.25 mL of medium containing 140 mM NaCl, 5 mM KCl, 1 mM $MgCl_2$, 1 mM $CaCl_2$, 5 mM glucose, 20 mM HEPES-tris (pH 7.4), 2 mCi/mL $^{86}Rb \pm 10\,\mu M$ ouabain and 20 μM bumetanide was added for the next 15 min. ^{86}Rb uptake was terminated and quantified as described previously (Orlov, Tremblay, & Hamet, 1996). Means ± S.E. for experiments performed in quadruplicates are shown.

Fig. 5 displays that myofibroblast differentiation triggered by TGF-β was accompanied by ~2-fold activation of the $Na^+,K^+,2Cl^-$ cotransport measured as bumetanide-sensitive component of ^{86}Rb influx. It is known that this electroneutral ion carrier plays a key role in cell volume regulation (Gamba, 2005). Considering this we compared effect of ouabain and bumetanide on the volume of [^{14}C]-urea-available space in control and TGF-β-treated HLF (Koltsova, Akimova, Orlov, & Dulin, 2013). Fig. 6 shows that this parameter in TGF-β-treated HLF was decreased by ~25% compared to control cells. In overwhelming number of cells studied so

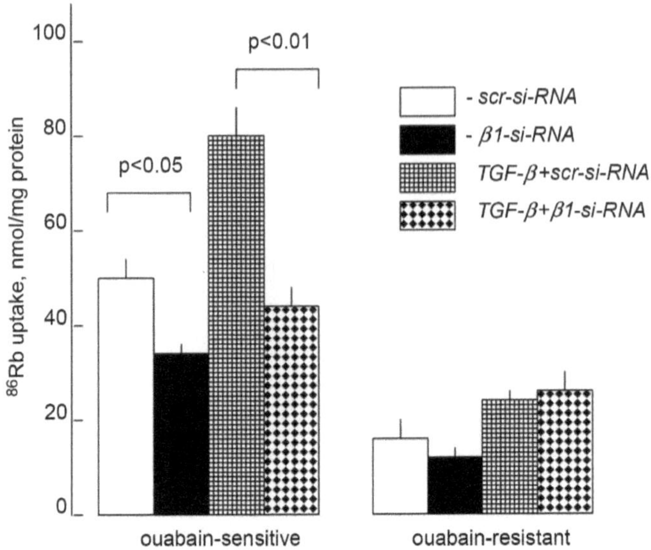

Fig. 6 Effect of TGF-β on si-RNA for Na^+,K^+ ATPase β1 subunit (β1-si-RNA) on the activity of the Na+,K+-pump (ouabain-sensitive component of the rate of ^{86}Rb influx) and ouabain-resistant ^{86}Rb uptake in human lung fibroblasts (HLF). HLF were transfected with β1-si-RNA or scrambled (scr-si-RNA) and the rate of ^{86}Rb influx was estimated as indicated in Fig. 7 legend. Means ± S.E. for experiments performed in triplicates are shown.

far sustained hyperosmotic shrinkage triggers apoptosis (Lang, Lepple-Wienhues, Szabo, Siemen, & Gulbins, 1998; Okada & Maeno, 2001). However, TGF-β does not promote apoptosis in HLF that is consistent with cell-type specific impact of isosmotic shrinkage on cell survival and death (Orlov, Platonova, Hamet, & Grygorczyk, 2013).

We observed that dissipation of transmembrane gradient of monovalent cations triggered by Na^+,K^+-ATPase inhibition in the presence of ouabain was accompanied by elevation of cell volume by ~20% (Fig. 7). Importantly, ouabain-dependent cell swelling was sharply increased in TGF-β-treated HLF compared to control fibroblasts that was consistent with augmented activity of the Na^+,K^+-pump documented in these cells (Fig. 5). We did not reveal any significant impact of bumetanide on cell volume adjustment in control and TGF-β-treated HLF (Fig. 7) thus indicating the lack of net monovalent ion fluxes provided by Na^+,K^+,$2Cl^-$ cotransport. Unlike bumetanide alone, combined addition of ouabain and bumetanide led to 2-fold elevation in cell volume in control as well as in TGF-β-treated HLF. This finding strongly suggests that dissipation of transmembrane

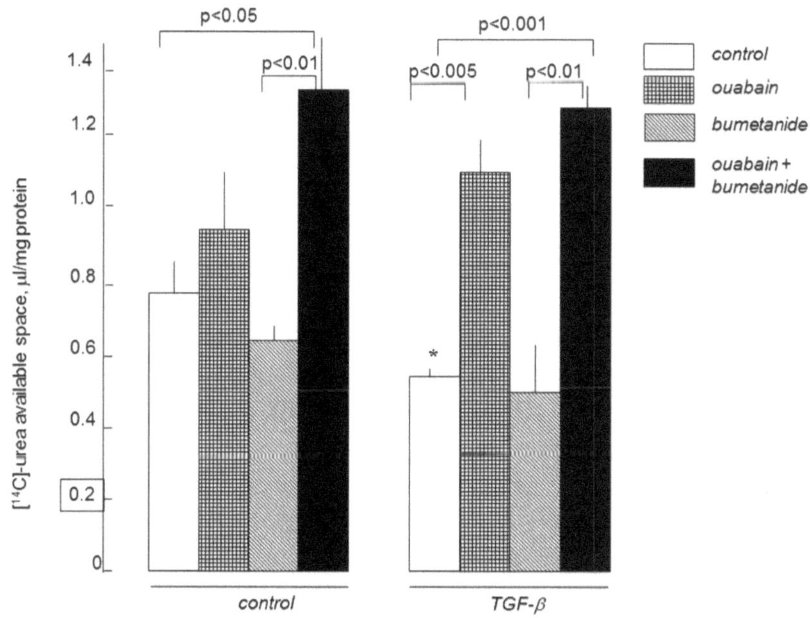

Fig. 7 Effect of TGF-β, ouabain and bumetanide on the volume of human lung fibroblasts estimated as a $[^{14}C]$-urea available space. HLF fibroblasts seeded in 24-well plates were incubated during 24h in DMEM containing 0.2% fetal bovine serum ±1 ng/mL TGF-β, 100 nM ouabain and 20 μM bumetanide; 5 μCi/mL $[^{14}C]$-urea was during the last 2h of incubation. The volume of $[^{14}C]$-urea available space was estimated as described previously (Orlov et al., 1996). Means ± S.E. for experiments performed in quadruplicates are shown.

gradients of monovalent cations triggered by Na+,K+-ATPase inhibition sharply increases the outwardly directed ion fluxes mediated by bumetanide-sensitive $Na^+,K^+,2Cl^-$ cotransport that protects cells from extensive swelling triggered by ouabain.

8. Do monovalent anion transporters affect myofibroblast differentiation?

Table 1 shows that 24-h exposure of HLF to TGF-β resulted in sharp elevation of mRNA encoding TMEM16A or anoctamin-1 (ANO1) identified by three independent research teams as voltage-sensitive Ca^{2+}-activated Cl^- channel (CaCC) (Caputo et al., 2008; Schroeder, Cheng, Jan, & Jan, 2008; Yang et al., 2008). These channels contribute to diverse electrical membrane potential (E_m)-sensitive cellular responses via Ca^{2+}-dependent regulation of intracellular Cl^- concentration ($[Cl^-]_i$)

(Hartzell, Putzier, & Arreola, 2005). Keeping these data in mind we investigated the action of inhibitors of anion transporters on myofibroblast differentiation in TGF-β-treated HLF.

Tannic acid was one of the first identified inhibitors of ANO1 (Namkung, Thiagarajah, Phuan, & Verkman, 2014). Tannic acid is a compound extracted from tara pods, gallnuts and leaves of certain plants and has been used for chemical staining of cellulose fibers (wood, textile), as a fixative-mordant in ultrastructural microscopy studies, as aroma compound in certain drinks, and for treatment of burns in humans (Hupkens, Boxma, & Dokter, 1995). Because the data on toxicity of tannic acid is available (Handler & Baker, 1944), we assessed the effects of tannic acid on myofibroblast differentiation. Fig. 8 displays that addition of 10 μM tannic acid suppresses TGF-β-dependent accumulation in cultured HLF of biochemical markers of myofibroblast differentiation such as smooth muscle actin and fibronectin. Our results are consistent with inhibition by tannic acid of α-smooth muscle actin, N-cadherin and type-1-collagen expression in TGF-β1-treated NIH 3T3 fibroblasts (Pattarayan, Sivanantham, Bethunaickan, Palanichamy, & Rajasekaran, 2018) and in vivo data showing suppression by this compound of carbon tetrachloride-induced liver fibrosis (Chu et al., 2016).

In the overwhelming cell types studied so far $Na^+,K^+,2Cl^-$ and K^+,Cl^- cotransporters mediate inwardly and outwardly directed net ion fluxes thus providing an opposite action on $[Cl^-]_i$ (Markadieu & Delpire, 2014). We did not reveal any action of $Na^+,K^+,2Cl^-$ cotransport inhibition by

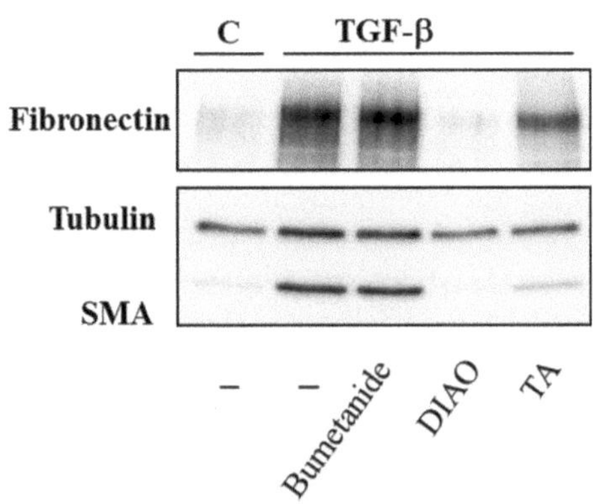

Fig. 8 Effect of anion transport inhibitors on the accumulation of biochemical markers of myofibroblast differentiation (fibronectin and smooth muscle actin, SMA) in TGF-β-treated HLF. HLF were treated with inhibitors of Na^+,K^+,Cl^- cotransport (10 μM bumetanide), K^+,Cl^- cotransport (100 μM DIOA—R(+)-butylindazone, R-(+)-[(2-n-butyl-6,7-dichloro-2-cyclopentyl-2,3-dihydro-1-oxo-1H-inden-5-yl)oxy]acetic acid) and TMEM16A (10 μM tannic acid, TA) with or without 1 ng/mL of TGF-β for 48 h.

10 μM bumetanide whereas inhibition of K^+,Cl^- cotransport by 100 μM DIOA completely abolished an increment of smooth muscle actin and fibronectin expression triggered by TGF-β (Fig. 8). It should be noted, however, that in alveolar type II cells and vascular smooth cells DIOA exhibit diverse side-effects resulting in attenuated cell viability (Anfinogenova et al., 2001; Skriabin, Orlov, Masse, & Berthiaume, 2000). Thus, additional studies should be performed to examine the specificity of antifibrotic action of DIOA and its relation to the suppression of K^+,Cl^- cotransport.

9. Unresolved issues

Data discussed above raise several key questions. (*i*) We demonstrated that suppression of myofibroblast differentiation in TGF-β-treated human lung fibroblast by CTS is mediated by Na^+,K^+-ATPase inhibition and elevation of the $[Na^+]_i/[K^+]_i$ ratio (La, Reed, Koltsova, et al., 2016; La, Reed, Smolyaninova, et al., 2016). What is the relative impact of $[Na^+]_i$,$[K^+]_i$-independent signaling documented in several types of CTS-treated cells (Aperia, 2007; Xie & Askari, 2002) in pro- and antifibrotic actions of these compounds? (*ii*) Transcriptomic changes in CTS-treated myotubes, vascular smooth muscle and endothelial cells are preserved in the presence of Ca^{2+} chelators (Koltsova et al., 2012), inhibitors of Ca^{2+} transporters (Sidorenko et al., 2018) and Ca^{2+}-mediated signaling (Smolyaninova et al., 2017) demonstrating the $[Ca^{2+}]_i$-independent mechanism of transcription regulation. Does the Ca^{2+}-independent mechanism of excitation–transcription coupling contribute to the suppression of TGFBR2 expression in CTS-treated HLF? If so, what is the molecular origin of Ca^{2+}-independent monovalent ion sensor(s) involved in transcription regulation? (*iii*) Does tannic acid inhibit TGF-β-induced differentiation of HLF via suppression of TMEM16A-mediated Cl^- fluxes and $[Cl^-]_i$ changes? Does elevation of $[Na^+]_i/[K^+]_i$ evoked by CTS affect $[Cl^-]_i$? (*iv*) Which ion transporters can be targeted for limiting lung fibrosis and which cannot due to potent systemic effects of their inhibitors? We address these questions to forthcoming studies.

Acknowledgments

This publication was supported by the Russian Scientific Foundation, the Russian Foundation for Fundamental Research, the National Institutes of Health Award R56HL127395, the National Center for Advancing Translational Sciences of the National Institutes of Health Award UL1TR000430, and Dr. Ralph and Marian Falk Medical Research Catalyst Award. The content is solely the responsibility of the authors and does not necessarily represent the official views of the National Institutes of Health.

References

Akimova, O. A., Mongin, A. A., Hamet, P., & Orlov, S. N. (2006). The rapid decline of MTT reduction is not a marker of death signaling in ouabain-treated cells. *Cellular and Molecular Biology*, *52*(8), 71–77.

Akimova, O. A., Tverskoi, A. M., Smolyaninova, L. V., Mongin, A. A., Lopina, O. D., La, J., et al. (2015). Critical role of the a1-Na+,K+-ATPase subunit in insensitivity of rodent cells to cytotoxic action of ouabain. *Apoptosis*, *20*, 1200–1210.

Anfinogenova, Y. J., Rodriguez, X., Grygorczyk, R., Adragna, N., Lauf, P. K., Hamet, P., et al. (2001). Swelling-induced K^+ fluxes in vascular smooth muscle cells are mediated by charybdotoxin-sensitive K^+ channels. *Cellular Physiology and Biochemistry*, *11*, 295–310.

Aperia, A. (2007). New roles for an old Na,K-ATPase emerges as an interesting drug target. *Journal of Internal Medicine*, *261*, 44–52.

Bagrov, A. Y., Fedorova, O. V., Dmitrieva, R. I., Howald, W. N., Hunter, A. P., Kuznetsova, E. A., et al. (1998). Characterization of a urinary bufodielnolide Na,K-ATPase inhibitor in patients after acute myocardial infarction. *Hypertension*, *31*, 1097–1103.

Bagrov, A. Y., Shapiro, J. I., & Fedorova, O. V. (2009). Endogenous cardiotonic steroids: Physiology, pharmacology, and novel therapeutic targets. *Pharmacological Reviews*, *61*, 9–38.

Bauer, N., Muller-Ehmsen, J., Kramer, U., Hambarchian, N., Zobel, C., Schwinger, R. H., et al. (2005). Ouabain-like compound changes rapidly on physical exercise in humans and dogs: Effects of β-blockage and angiotensin-converting enzyme inhibition. *Hypertension*, *45*, 1024–1028.

Bauman, K. A., Wettlaufer, S. H., Okunishi, K., Vannella, K. M., Stoolman, J. S., Huang, S. K., et al. (2010). The antifibrotic effects of plasminogen activation occur via prostaglandin E2 synthesis in humans and mice. *Journal of Clinical Investigation*, *120*, 1950–1960.

Blanco, G. (2015). Subunit heterogeneity as a mechanism for tissue-specific ion regulation. *Seminars in Nephrology*, *25*, 292–303.

Bolivar, J. J., Lazaro, A., Fernandez, S., Stefani, E., Pena-Cruz, V., Lechene, C., et al. (1987). Rescue of a wild-type MDCK cell by a ouabain-resistant mutant. *The American Journal of Physiology*, *253*, C151–C161.

Bonner, J. C., Rice, A. B., Ingram, J. L., Moomaw, C. R., Nyska, A., Bradbury, A., et al. (2002). Susceptibility of cyclooxygense-2-deficient mice to pulmonary fibrosis. *American Journal of Pathology*, *161*, 459–470.

Broekelmann, T. J., Limper, A. H., Colby, T. V., & McDonald, J. A. (1991). Transforming growth factor beta 1 is present at sites of extracellular matrix gene expression in human pulmonary fibrosis. *Proceeding of the National Academy of Science of the USA*, *88*, 6642–6646.

Caputo, A., Caci, E., Ferrera, L., Pedemonte, N., Barsanti, C., Sondo, E., et al. (2008). TMEM16A, a membrane protein associated with calcium-dependent chloride channel activity. *Science*, *322*, 590–594.

Card, J. W., Voltz, J. W., Carey, M. A., Bradbury, A., DeGraff, L. M., Lih, F. B., et al. (2007). Cyclooxygenase-2 deficiency exacerbates bleomycin-induced lung dysfunction but not fibrosis. *American Journal of Respiratory and Critical Care Medicine*, *37*, 300–308.

Chu, X., Wang, H., Jiang, Y.-M., Zhang, Y.-Y., Bao, Y.-F., Zhang, X., et al. (2016). Ameliorative effects of tannic acid on carbon tetrachloride-induced liver fibrosis in vivo and in vitro. *Journal of Pharmacological Sciences*, *130*, 15–23.

Chueh, S.-C., Guh, J.-H., Chen, J., Lai, M.-K., & Teng, C.-M. (2001). Dual effect of ouabain on the regulation of proliferation and apoptosis in human prostatic smooth muscle cells. *Journal of Urology*, *166*, 347–353.

Coleman, D. T., Gray, A. L., Stephens, C. A., Scott, M. L., & Cardelli, J. A. (2016). Repurposed drug screen identifies cardiac glycosides as inhibitors of TGF-beta-induced cancer associated fibroblast differentiation. *Oncotarget*, 7, 32200–32209.

Coward, W. R., Watts, K., Feghali-Bostwick, C. A., Knox, A., & Pang, L. (2009). Defective histone acetylation is responsible for the diminished expression of cyclooxygenase 2 in idiopathic pulmonary fibrosis. *Molecular and Cellular Biology*, *29*, 4325–4339.

Czuwara-Ladykowska, J., Shirasaki, F., Jackers, P., Watson, D. K., & Trojanowska, M. (2001). Fli-1 inhibits collagen type I production in dermal fibroblasts via an Sp1-dependent pathway. *Journal of Biological Chemistry*, *276*, 20839–20848.

Drummond, C. A., Fan, X., Haller, S. T., Kennedy, D. J., Liu, J., & Tian, J. (2018). Na/K-ATPase signaling mediates miR-29b-3p regulation and cardiac fibrosis formation in mice with chronic kidney disease. *PLoS ONE*, *13*, 0197688.

Drummond, C. A., Hill, M. C., Shi, H., Fan, X., Xie, J. X., Haller, S. T., et al. (2016). Na/K-ATPase signaling regulates collagen synthesis through microRNA-29b-3p in cardiac fibroblasts. *Physiological Genomics*, *48*, 220–229.

Elkareh, J., Kennedy, D. J., Yashaswi, B., Vetteth, S., Shidyak, A., Kim, E. G., et al. (2007). Marinobufagenin stimulates fibroblast collagen production and causes fibrosis in experimental uremic cardiomyopathy. *Hypertension*, *49*, 215–224.

Elkareh, J., Periyasamy, S. M., Shidyak, A., Vetteth, S., Schroeder, J., Raju, V., et al. (2009). Marinobufagenin induces increases in procollagen expression in a process involving protein kinase C and Fli-1: Implications for uremic cardiomyopathy. *American Journal of Physiology. Renal Physiology*, *296*, F1219–F1226.

El-Okdi, N., Samili, S., Raju, V., Shidyak, A., Gupta, S., Fedorova, L., et al. (2008). Effects of cardiotonic steroids on dermal collagen synthesis and wound healing. *Journal of Applied Physiology*, *105*, 30–36.

Evans, J. N., Kelley, J., Krill, J., Low, R. B., & Adler, K. B. (1983). The myofibroblast in pulmonary fibrosis. *Chest*, *83*, 97S–98S.

Fedorova, O. V., Emelianov, I. V., Bagrov, K. A., Grigorova, Y. N., Wei, W., Juhasz, O., et al. (2015). Marinobufagenin-induced vascular fibrosis is a likely targets for mineralocorticoid antagonists. *Journal of Hypertension*, *33*, 1602–1610.

Feng, X. H., & Derynck, R. (2005). Specificity and versatility in TGF-beta signaling through Smads. *Annual Review of Cell and Developmental Biology*, *21*, 659–693.

Gabasa, M., Royo, D., Molina-Molina, M., Roca-Ferrer, J., Pujols, L., Picado, C., et al. (2013). Lung myofibroblasts are characterized by down-regulated cyclooxygenase-2 and its main metabolite, prostaglandin E2. *PLoS ONE*, *8*, e65445.

Gamba, G. (2005). Molecular physiology and pathophysiology of electroneutral cation-chloride cotransporters. *Physiological Reviews*, *85*, 423–493.

Garrison, G., Huang, S. K., Okunishi, K., Scott, J. P., Kumar Penke, L. R., Scruggs, A. M., et al. (2013). Reversal of myofibroblast differentiation by prostaglandin E2. *American Journal of Respiratory Cell and Molecular Biology*, *48*, 550–558.

Giri, S. N., Hude, D. M., & Hollinger, M. A. (1993). Effect of antibody to transforming growth factor beta on bleomycin-induced accumulation of lung collagen in mice. *Thorax*, *48*, 959–966.

Giri, S. N., & Hyde, D. M. (1987). Increases in severity of lung damage and mortality by treatment with cyclo and lipoxygenase inhibitors in bleomycin and hyperoxia model of lung injury in humster. *Pathology*, *19*, 150–158.

Grigorova, Y. N., Juhasz, O., Zernetkina, V., Fishbein, K. W., Lakatta, E. G., Fedorova, O. V., et al. (2016). Aortic fibrosis, induced by high salt intake in the absence of hypertensive response, is reduced by monoclonal antibody to marinobufagenin. *American Journal of Hypertension*, *29*, 641–646.

Gundersen, K. (2011). Excitation-transcription coupling in skeletal muscle: The molecular pathways of exercise. *Biological Reviews*, *86*, 564–600.

Haller, S. T., Drummond, C. A., Yan, Y., Liu, J., Tian, J., Malhotra, D., et al. (2014). Passive immunization against marinobufagenin attenuates renal fibrosis and improve renal function in experimental renal disease. *American Journal of Hypertension, 27*, 603–609.

Haloui, M., Taurin, S., Akimova, O. A., Guo, D.-F., Tremblay, J., Dulin, N. O., et al. (2007). Na_i^+-induced c-Fos expression is not mediated by activation of the 5′-promoter containing known transcriptional elements. *The FEBS Journal, 274*, 3257–3267.

Handler, P., & Baker, R. D. (1944). The toxicity of orally administrated tannic acid. *Science, 99*, 393.

Hardie, W. D., Glasser, S. W., & Hagood, J. S. (2009). Emerging concepts in the pathogenesis of lung fibrosis. *American Journal of Pathology, 175*, 3–16.

Hartzell, C., Putzier, I., & Arreola, J. (2005). Calcium-activated chloride channels. *Annual Review of Physiology, 67*, 719–758.

Harvey, A., Montezano, A. C., Lopes, R. A., & Touyz, R. M. (2016). Vascular fibrosis in aging and hypertension: Molecular mechanisms and clinical implications. *Canadian Journal of Cardiology, 32*, 659–668.

Hasler, U., Wang, X., Crambert, G., Bequin, P., Jassier, F., Horisberger, J.-D., et al. (1998). Role of β-suninit domains in the assembly, stable expression, intracellular routing, and functional properties of Na,K-ATPase. *Journal of Biological Chemistry, 273*, 30826–30835.

Hennion, J. P., el-Masri, M. A., Huff, M. O., & el-Mailakh, R. S. (2002). Evaluation of neuroprotection by lithium and valproic acid against ouabain-induced cell damage. *Bipolar Disorders, 4*, 201–206.

Hinz, B. (2006). Masters and servants of the force: The role of matrix adhesions in myofibroblast force perception and transmission. *European Journal of Cell Biology, 85*, 175–181.

Hinz, B., Phan, S. H., Thannickal, V. J., Galli, A., Bochaton-Piallat, M. L., & Gabbiani, G. (2007). The myofibroblast. One function, multiple origin. *American Journal of Pathology, 170*, 1807–1816.

Hodges, R. J., Jenkins, R. G., Wheeler-Jones, C. P., Copeman, D. M., Bottoms, S. E., Bellingan, G. J., et al. (2004). Severity of lung injury in cyclooxygenase-2-deficient mice is dependent on reduced prostaglandin E2 production. *American Journal of Pathology, 165*, 1663–1676.

Huang, F., & Chen, Y.-G. (2012). Regulation of TGF-b receptor activity. *Cell & Bioscience, 2*, 9.

Hupkens, P., Boxma, H., & Dokter, J. (1995). Tannic acid as a topical agent in burns: Historical considerations and implications for new developments. *Burns, 21*, 57–61.

Isaev, N. K., Stelmashook, E. V., Halle, A., Harms, C., Lautenschlager, M., Weih, M., et al. (2000). Inhibition of Na^+,K^+-ATPase activity in cultured cerebellar granule cells prevents the onset of apoptosis induced by low potassium. *Neuroscience Letters, 283*, 41–44.

Kang, H. R., Cho, S. J., Lee, C. G., Homer, R. J., & Elias, J. A. (2007). Transforming growth factor (TGF)-beta 1 stimulates pulmonary fibrosis and inflammation via a Bax-dependent, big-activated pathway that involves matrix metalloprotease-12. *Journal of Biological Chemistry, 282*, 7723–7732.

Keerthisingam, C. B., Jenkins, R. G., Harrison, N. K., Hernandez-Rodriguez, N. A., Booth, H., Laurent, G. J., et al. (2001). Cyclooxygenase-2 deficiency results in a loss of the anti-proliferative response to transforming growth factor-beta in human fibrotic lung fibroblasts and promotes bleomycin-induced pulmonary fibrosis in mice. *American Journal of Pathology, 158*, 1411–1422.

Khalaf, F. K., Dube, P., Mohamed, A., Tian, J., Malhotra, D., Haller, S. T., et al. (2018). Cardiotonic steroids and the sodium trade balance: new insights into trade-off mechanisms mediated by the Na+,K+-ATPase. *International Journal of Molecular Sciences*, (in press).

Klimanova, E. A., Tverskoi, A. M., Koltsova, S. V., Sidorenko, S. V., Lopina, O. D., Tremblay, J., et al. (2017). Time- and dose-dependent actions of cardiotonic steroids on transcriptome and intracellular content of Na+ and K+: A comparative analysis. *Scientific Reports*, 7, 45403. https://doi.org/10.1038/rep45403.

Kolodsick, J. E., Peters-Golden, M., Larios, J., Toews, G. B., Thannickal, V. J., & Moore, B. B. (2003). Prostaglandin E2 inhibits fibroblast to myofibroblast transition via E. prostanoid receptor 2 and cyclic adenosine monophosphate elevation. *American Journal of Respiratory Cell and Molecular Biology*, *29*, 537–544.

Koltsova, S. V., Akimova, O. A., Orlov, S. N., & Dulin, N. O. (2013). Both Na+/K+ pump and Na+,K+,2Cl- cotransport contribute to cell volume control in human lung fibroblasts. *Bulletin of Siberian Medicine*, *12*, 42.

Koltsova, S. V., Tremblay, J., Hamet, P., & Orlov, S. N. (2015). Transcriptomic changes in Ca2+-depleted cells: Role of elevated intracellular [Na+]/[K+] ratio. *Cell Calcium*, *58*, 317–324.

Koltsova, S. V., Trushina, Y., Haloui, M., Akimova, O. A., Tremblay, J., Hamet, P., et al. (2012). Ubiquitous [Na+]i/[K+]i-sensitive transcriptome in mammalian cells: evidence for Ca_i^{2+}-independent excitation-transcription coupling. *PLoS ONE*, 7, e38032.

Kulikov, A., Eva, A., Kirch, U., Boldyrev, A., & Scheiner-Bobis, G. (2007). Ouabain activates signaling pathways associated with cell death in human neuroblastoma. *Biochimica et Biophysica Acta*, *1768*, 1691–1702.

Kurosawa, M., Tani, Y., Nishimura, S., Numazawa, S., & Yoshida, T. (2001). Distinct PKC isozymes regulate buffalin-induced differentiation and apoptosis in human monocyte cells. *American Journal of Physiology-Cell Physiology*, *280*, C459–C464.

La, J., Reed, E. B., Koltsova, S. V., Akimova, O. A., Hamanaka, R. B., Mutlu, R. B., et al. (2016). Regulation of myofibroblast differentiation by cardiac glycosides. *American Journal of Physiology. Lung Cellular and Molecular Physiology*, *310*, L815–L823.

La, J., Reed, E., Smolyaninova, L. V., Akimova, O. A., Orlov, S. N., & Dulin, N. O. (2016). Downregulation of TGF-beta receptor-2 expression and signaling through inhibition of Na/K-ATPase. *PLoS ONE*, *11*, e0168363.

Lang, F., Lepple-Wienhues, A., Szabo, I., Siemen, D., & Gulbins, E. (1998). Cell volume in cell proliferation and apoptotic cell death. In F. Lang (Ed.), *Cell Volume Regulation. Contrib.Nephrol* (pp. 158–168). Basel: Karger.

Leask, A., & Abraham, D. J. (2004). TGF-beta signaling and the fibrotic response. *The FASEB Journal*, *18*, 816–827.

Ledbetter, M. L., Young, G. J., & Wright, E. R. (1986). Cooperation between epithelial cells demonstrated by potassium transfer. *The American Journal of Physiology*, *250*, C306–C313.

Leindecker Stumm, C., Wettlaufer, S. H., Janckar, S., & Peters-Golden, M. (2011). Airway remodeling in murine asthma correlates with a defect in PGE2 synthesis by lung fibroblasts. *American Journal of Physiology. Lung Cellular and Molecular Physiology*, *301*, L636–L644.

Lingrel, J. B. (2010). The physiological significance of the cardiotonic steroid/ouabain-binding site of the Na,K-ATPase. *Annual Review of Physiology*, *72*, 395–412.

Liu, F., Mih, J. D., Shea, B. S., Kho, A. T., Sharif, A. S., Tager, A. M., et al. (2010). Feedback amplification of fibrosis through matrix stiffening and COX-2 suppression. *Journal of Cell Biology*, *190*, 693–706.

Liu, J., & Xie, Z. (2010). The sodium pump and cardiotonic steroids-induced signal transduction protein kinases and calcium-signaling microdomain in regulation of transporter trafficking. *Biochimica et Biophysica Acta*, *1802*, 1237–1245.

Ma, H., Groth, R. D., Wheeler, D. G., Barrett, C. F., & Tsien, R. W. (2011). Excitation-transcription coupling in sympathetic neurons and the molecular mechanism of its initiation. *Neuroscience Research*, *70*, 2–8.

Markadieu, N., & Delpire, E. (2014). Physiology and pathophysiology of SLC12A1/2 transporters. *Pflügers Archiv: European Journal of Physiology*, *466*, 91–105.

McConkey, D. J., Lin, Y., Nutt, L. K., Ozel, H. Z., & Newman, R. A. (2000). Cardiac glycosides stimulate Ca^{2+} increases and apoptosis in androgen-independent, metastatic human prostate adenocarcinoma cells. *Canser Research*, *60*, 3807–3812.

Nakao, S., Ogata, Y., Modeer, T., Segawa, M., Furuyama, S., & Sugiya, H. (2001). Bradykinin induces a rapid cyclooxygenase-2 mRNA expression via Ca2+ mobilization in human gingival fibroblasts primed with interleukin-1 beta. *Cell Calcium*, *29*, 446–452.

Namkung, W., Thiagarajah, J. R., Phuan, P. W., & Verkman, A. S. (2014). Inhibition of Ca^{2+}-activated Cl^- channels by gallotannins as a possible molecular basis for health benefits of red wine and green tea. *The FASEB Journal*, *24*, 4178–4186.

Noble, P. W., & Homer, R. J. (2005). Back to the future: Historical perspective on the pathogenesis of idiopathic fibrosis. *American Journal of Respiratory Cell and Molecular Biology*, *33*, 113–120.

Okada, Y., & Maeno, E. (2001). Apoptosis, cell volume regulation and volume-regulatory chloride channels. *Molecular and Integrative Physiology CBP. Part A*, *130*, 377–383.

Olson, A. L., Swigris, J. J., Lezotte, D. C., Norris, J. M., Wilson, C. G., & Brown, K. K. (2003). Mortality from pulmonary fibrosis increased in United States from 1992 to 2003. *American Journal of Respiratory and Critical Care Medicine*, *176*, 277–284.

Orlov, S. N., Akimova, O. A., & Hamet, P. (2005). Cardiotonic steroids: Novel mechanisms of Na_i^+-mediated and -independent signaling involved in the regulation of gene expression, proliferation and cell death. *Current Hypertension Reviews*, *1*, 243–257.

Orlov, S. N., & Hamet, P. (2015). Salt and gene expression: evidence for Na_i^+,K_i^+-mediated signaling pathways. *Pflügers Archiv: European Journal of Physiology*, *467*, 489–498.

Orlov, S. N., Klimanova, E. A., Tverskoi, A. M., Vladychenskaya, E. A., Smolyaninova, L. V., & Lopina, O. D. (2017). Na_i^+,K_i^+-dependent and -independent signaling triggered by cardiotonic steroids: Facts and artifacts. *Molecules*, *22*, 635. https://doi.org/10.3390/molecules22040635.

Orlov, S. N., La, J., Smolyaninova, L. V., & Dulin, N. O. (2017). Na+,K+-ATPase as a target for treatment of tissue fibrosis. *Current Medicinal Chemistry*, *24*. https://doi.org/10.2174/0929867324666170619105407.

Orlov, S. N., Platonova, A. A., Hamet, P., & Grygorczyk, R. (2013). Cell volume and monovalent ion transporters: Their role in the triggering and progression of the cell death machinery. *American Journal of Physiology-Cell Physiology*, *305*, C361–C372.

Orlov, S. N., Thorin-Trescases, N., Kotelevtsev, S. V., Tremblay, J., & Hamet, P. (1999). Inversion of the intracellular Na^+/K^+ ratio blocks apoptosis in vascular smooth muscle at a site upstream of caspase-3. *Journal of Biological Chemistry*, *274*, 16545–16552.

Orlov, S. N., Thorin-Trescases, N., Pchejetski, D., Taurin, S., Farhat, N., Tremblay, J., et al. (2004). Na^+/K^+ pump and endothelial cell survival: $[Na^+]_i/[K^+]_i$-independent necrosis triggered by ouabain, and protection against apoptosis mediated by elevation of $[Na^+]_i$. *Pflügers Archiv: European Journal of Physiology*, *448*, 335–345.

Orlov, S. N., Tremblay, J., & Hamet, P. (1996). Cell volume in vascular smooth muscle is regulated by bumetanide-sensitive ion transport. *The American Journal of Physiology*, *270*, C1388–C1397.

Pattarayan, D., Sivanantham, A., Bethunaickan, R., Palanichamy, R., & Rajasekaran, S. (2018). Tannic acid modulates fibroblast proliferation and differentiation in response to pro-fibrotic stimuli. *Journal of Cellular Biochemistry*, *119*(8), 6732–6742. https://doi.org/10.1002/jcb.26866.

Perne, A., Muellner, M. K., Steinrueck, M., Craig-Mueller, N., Mayerhofer, J., Schwarzinger, I., et al. (2009). Cardiotonic glycosides induce cell death in human cells by inhibiting general protein synthesis. *PLoS ONE*, *4*, e8292.

Petkova, D. K., Clelland, C. A., Ronan, J. E., Lewis, S., & Knox, A. J. (2003). Reduced expression of cyclooxygenase (COX) in idiopathic pulmonary fibrosis and sarcoidosis. *Histopathology*, *43*, 381–386.

Phan, S. H. (1996). Role of the myofibroblast in pulmonary fibrosis. *Kidney International. Supplement*, *96*, S46–S48.

Phan, S. H. (2014). Fibroblast phenotypes in pulmonary fibrosis. *American Journal of Respiratory Cell and Molecular Biology*, *29*, S87–S92.

Phan, S. H., & Kunkel, S. L. (1992). Lung cytokine production in bleomycin-induced pulmonary fibrosis. *Experimental Lung Research*, *18*, 29–43.

Polanska, U. M., & Orimo, A. (2013). Carcinoma-associated fibroblasts: Non-neoplastic tumour-promoting mesenchymal cells. *Journal of Cellular Physiology*, *228*, 1651–1657.

Qin, W., Chung, A. C., Huang, X. R., Meng, X. M., Hui, D. S., Yu, C. M., et al. (2011). TGF-beta/Smad3 signaling promotes renal fibrosis by inhibiting miR-29. *Journal of the American Society of Nephrology*, *22*, 1462–1467.

Ramdas, V., McBride, M., Denby, L., & Baker, A. H. (2013). Canonical transforming growth factor-beta signaling regulates disintegrin metalloprotease expression in experimental renal fibrosis via miR-29. *American Journal of Pathology*, *183*, 1885–1896.

Riganti, C., Campia, I., Kopecka, J., Gazzano, E., Doublier, S., Aldieri, E., et al. (2011). Pleiotropic effects of cardioactive glycosides. *Current Medicinal Chemistry*, *18*, 872–885.

Ronnov-Jensen, L., Petersen, O. W., & Bissell, M. J. (1996). Cellular changes involved in conversion of normal to malignant beast: Importance of the stromal reaction. *Physiology Reversion*, *76*, 69–125.

Rosen, H., Glukmann, V., Feldman, T., Fridman, E., & Lichtstein, D. (2004). Cardiac steroids induce changes in recycling of the plasma membrane in human NT2 cells. *Molecular Biology of the Cell*, *15*, 1044–1054.

Sandbo, N., Kregel, S., Taurin, S., Bhorade, S., & Dulin, N. O. (2009). Critical role of serum response factor in pulmonary myofibroblast differentiation induced by TGF-β. *American Journal of Respiratory Cell and Molecular Biology*, *41*, 332–338.

Sandbo, N., Lau, A., Kach, J., Ngam, C., Yau, D., & Dulin, N. O. (2011). Delayed stress fiber formation mediates pulmonary myofibroblast differentiation in response to TGF-β. *American Journal of Physiology. Lung Cellular and Molecular Physiology*, *301*, L656–L666.

Santana, A., Saxena, B., Noble, N. A., Gold, L. I., & Marshall, B. C. (1995). Increased expression of transforming growth factor beta isoforms (beta 1, beta 2, beta 3) in bleomycin-induced pulmonary fibrosis. *American Journal of Respiratory Cell and Molecular Biology*, *13*, 34–44.

Schatzmann, H. J. (1953). Herzglykoside als Hennstoffe fьr den aktiven Kalium- und Natriumtransport durch die erythrocytenmembran. *Helvetica Physiologica et Pharmacologica Acta*, *11*, 346–354.

Scheiner-Bobis, G. (2002). The sodium pump. Its molecular properties and mechanics of ion transport. *European Journal of Biochemistry*, *269*, 2424–2433.

Schoner, W., & Scheiner-Bobis, G. (2007). Endogenous and exogenous cardiac glycosides: Their role in hypertension, salt metabolism, and cell growth. *American Journal of Physiology-Cell Physiology*, *293*, C509–C536.

Schroeder, B. C., Cheng, T., Jan, Y. N., & Jan, L. Y. (2008). Expression cloning of TMEM16A as a calcium-activated chloride channel subunit. *Cell*, *134*, 1019–1029.

Segel, G. B., & Lichtman, M. A. (1980). The apparent discrepancy of ouabain inhibition of cation transport and lymphocyte proliferation is explained by time-dependency of ouabain binding. *Journal of Cellular Physiology*, *104*, 21–26.

Sidorenko, S. V., Klimanova, E. A., Milovanova, K., Lopina, O. D., Kapilevich, L. V., Chibalin, A. V., et al. (2018). Transciptomic changes in C2C12 myotubes triggered by electrical stimulation: Role of Ca_i^{2+}-mediated and Ca_i^{2+}-independent signaling and elevated [Na+]i/[K+]i ratio. *Cell Calcium, 76*, 72–86.

Sime, P. J., Xing, Z., Graham, F. L., Csaky, K. G., & Gauldie, J. (1997). Adenovector-mediated gene transfer to activate transforming growth factor beta1 induces prolonged severe fibrosis in lung. *Journal of Clinical Investigation, 100*, 768–776.

Skou, J. C. (1957). The influence of some cations on an adenosine triphosphatase from peripheral nerves. *Biochimica et Biophysica Acta, 23*, 394–401.

Skou, J. C. (1960). Further investigation on a Mg^{2+} + Na^+-activated adenosine triphosphatase possibly related to the active transport of Na^+ and K^+ across the nerve cell membrane. *Biochimica et Biophysica Acta, 42*, 6–23.

Skriabin, G., Orlov, S. N., Masse, C., & Berthiaume, Y. (2000). Phloretin inhibits Na^+ and K^+ uptake in cultured alveolar type II cells by reduction of cellular ATP content. *Experimental Lung Research, 26*, 319–333.

Smolyaninova, L. V., Koltsova, S. V., Sidorenko, S. V., & Orlov, S. N. (2017). Augemented gene expression triggered by Na+,K+-ATPase inhibition: role of Ca2+-mediated and -independent excitation-transcription coupling. *Cell Calcium, 68*, 5–13.

Song, J., & Shi, W. (2018). The concomitant apoptosis and EMT underlie the fundamental functions of TGF-β. *Acta Biochimica et Biophysica Sinica, 50*, 91–97.

Taurin, S., Dulin, N. O., Pchejetski, D., Grygorczyk, R., Tremblay, J., Hamet, P., et al. (2002). C-Fos expression in ouabain-treated vascular smooth muscle cells from rat aorta: Evidence for an intracellular-sodium-mediated, calcium-independent mechanism. *Journal of Physiology (London), 543*, 835–847.

Thannickai, V. J., & Horowitz, J. C. (2006). Evolving concepts of apoptosis in idiopathic pulmonary fibrosis. *Proceedings of the American Thoracic Society, 3*, 350–356.

Tomasek, J. J., Gabbiani, G., Hinz, B., Chaponnier, C., & Brown, R. A. (2002). Myofibroblasts and mechano-regulation of connective tissue remodelling. *Nature Reviews Molecular Cell Biology, 3*, 349–363.

Trevisi, L., Visentin, B., Cusinato, F., Pighin, I., & Luciani, S. (2004). Antiapoptotic effect of ouabain on human umbilical endothelial cells. *Biochemical and Biophysical Research Communications, 321*, 716–721.

Tverskoi, A. M., Sidorenko, S. V., Klimanova, E. A., Akimova, O. A., Smolyaninova, L. V., Lopina, O. D., et al. (2016). Effects of ouabain on proliferation of human endothelial cells correlate with Na+,K+-ATPase activity and intracellular ratio of Na+ and K+. *Biochemistry (Moscow), 81*, 876–883.

Vagin, O., Dada, L. A., Tokhtaeva, E., & Sachs, G. (2012). The Na-K-ATPase a1b1 heterodimer as cell adhesion molecule in epithelia. *American Journal of Physiology-Cell Physiology, 302*, C1271–C1281.

van Rooij, E., Sutherland, L. B., Thatcher, J. E., DiMaio, J. M., Naseem, R. H., Marshall, W. S., et al. (2008). Dysregulation of microRNAs after myocardial infarction reveals a role of miR-29 in cardiac fibrosis. *Proceeding of the National Academy of Science of the USA, 105*, 13027–13032.

Vancheri, C., Sortino, M. A., Tomaselli, V., Condorelli, F., Mastruzzo, C., Bellistri, G., et al. (2000). Different expression of TNF-alpha receptors and prostagalndin E2 production in normal and fibrotic lung fibroblasts: Potential implications for the evolution of inflammatory process. *American Journal of Respiratory Cell and Molecular Biology, 22*, 628–634.

Wang, Q., Wang, Y., Hyde, D. M., Gotwals, P. J., Koteliansky, V. E., Ryan, S. T., et al. (1999). Reduction of bleomycin-induced lung fibrosis by transforming growth factor beta soluble receptor in hamsters. *Thorax, 54*, 805–812.

Wettlaufer, S. H., Scott, J. B., McEachin, R. C., Peters-Golden, M., & Huang, S. K. (2016). Reversal of the transciptome by prostaglandin E2 during myofibroblast dedifferentiation. *American Journal of Respiratory Cell and Molecular Biology, 54*, 114–127.

Wilborn, J., Crofford, L. J., Burdick, M. D., Kunkel, S. L., Strieter, R. M., & Peters-Golden, M. (1995). Cultured lung fibroblasts isolated from patients with idiopathic pulmonary fibrosis have a diminished capacity to synthesize prostaglandin E2 and to express cyclooxygenase-2. *Journal of Clinical Investigation*, *95*, 1861–1868.

Wood, E. H., & Moe, G. K. (1938). Studies on the effect of digitalis glycosides on potassium ion loss from the heart. *The American Journal of Physiology*, *123*, 219–220.

Wynn, T. A. (2008). Cellular and molecular mechanisms of fibrosis. *The Journal of Pathology*, *214*, 199–210.

Xie, Z., & Askari, A. (2002). Na^+/K^+-ATPase as a signal transducer. *European Journal of Biochemistry*, *269*, 2434–2439.

Xie, J. K., Shapiro, A. P., & Shapiro, J. I. (2014). The trade-off between dietary salt and cardiovascular disease: A role for Na.K-ATPase signaling. *Frontiers in Endocrinology*, *5*, 97.

Yang, Y. D., Cho, H., Koo, J. Y., Tak, M. H., Cho, Y., Shim, W. S., et al. (2008). TMEM16A confers receptor-activated calcium-dependent chloride conductance. *Nature*, *455*, 1210–1215.

Zhang, H. Y., & Phan, S. H. (1999). Inhibition of myofibroblast apoptosis by transforming growth factor(1). *American Journal of Respiratory Cell and Molecular Biology*, *21*, 658–665.

Zhou, X., Jiang, G., Zhao, A., Bondeva, T., Hirzel, P., & Balla, T. (2001). Inhibition of Na, K-ATPase activates PI3 kinase and inhibits apoptosis in LLC-PK1 cells. *Biochemical and Biophysical Research Communications*, *285*, 46–51.

Zhu, Y., Hua, P., Rafig, S., Waffner, E. J., Duffey, M. E., & Lance, P. (2002). Ca2+ and PKC-dependent stimulation of PGE2 synthesis by deoxycholic acid in human colonic fibroblasts. *American Journal of Physiology. Gastrointestinal and Liver Physiology*, *283*, G503–G510.

CHAPTER THREE

Mechanosensitive ATP release in the lungs: New insights from real-time luminescence imaging studies

Ryszard Grygorczyk[a,b,*], Francis Boudreault[b], Ju Jing Tan[a,b], Olga Ponomarchuk[b,c], Masahiro Sokabe[d], Kishio Furuya[d]

[a]Department of Medicine, Université de Montréal, Montréal, QC, Canada
[b]Centre de recherche, Centre hospitalier de l'Université de Montréal (CRCHUM), Montréal, QC, Canada
[c]Faculty of Biology, M.V. Lomonosov Moscow State University, Moscow, Russia
[d]Mechanobiology Laboratory, Nagoya University, Graduate School of Medicine, Nagoya, Japan
*Corresponding author: e-mail address: ryszard.grygorczyk@umontreal.ca

Contents

Abstract

Extracellular ATP and other nucleotides are important autocrine/paracrine mediators that stimulate purinergic receptors and regulate diverse processes in the normal lungs. They are also associated with pathogenesis of a number of respiratory diseases and

Current Topics in Membranes, Volume 83
ISSN 1063-5823
https://doi.org/10.1016/bs.ctm.2019.02.001

clinical complications including acute respiratory distress syndrome and ventilator induced lung injury. Mechanical forces are major stimuli for cellular ATP release but precise mechanisms responsible for this release are still debated. The present review intends to provide the current state of knowledge of the mechanisms of ATP release in the lung. Putative pathways of the release, including the contribution of cell membrane injury and cell lysis are discussed addressing their strength, weaknesses and missing evidence that requires future study. We also provide an overview of the recent technical advances in studying cellular ATP release in vitro and ex vivo. Special attention is given to new insights into lung ATP release obtained with the real-time luminescence ATP imaging. This includes recent data on stretch-induced mechanosensitive ATP release in a model and primary cells of lung alveoli in vitro as well as inflation-induced ATP release in airspaces and pulmonary blood vessels of lungs, ex vivo.

1. Introduction

Intercellular communication initiated by the release of purines as signaling molecules appears to be the most primitive and widespread autocrine and paracrine mechanism in most, if not all, organs and tissues (Burnstock, 2006). The downstream purinergic signaling system consists of a large family of G protein-coupled adenosine (Ado) receptors, P2Y receptors, as well as ionotropic P2X receptors, Fig. 1. Several families of extracellular enzymes contribute to signal diversification by converting/metabolizing one nucleotide species into another, while at the same time, providing a mechanism for the termination of purinergic signals (Abbracchio, Burnstock, Verkhratsky, & Zimmermann, 2009; Lazarowski & Boucher, 2009). With over 50 membrane components or processes involved, it is considered as one of the most versatile signaling system (Novak, 2011). Range of action of the extracellular nucleotides can considerably expand by stimulating the release of other extracellular messengers, which in turn activate additional cellular mechanisms through their own receptors. These include neurotransmitters, hormones, growth factors and other proteins including enzymes, cytokines, lipid mediators, nitric oxide and reactive oxygen species (Zimmermann, 2016). Despite the fundamental importance of purinergic signaling in a variety of physiological and pathophysiological processes in all organs and tissues, the mechanism of the initiating step of this cascade—nucleotide release remains poorly understood (Burnstock, 2008, 2009; Lazarowski, Sesma, Seminario-Vidal, & Kreda, 2011; Praetorius & Leipziger, 2009). This contrasts with enormous progress in molecular and pharmacological characterization of purinergic receptors and purine metabolizing ecto-enzymes (Jacobson & Muller, 2016; Zimmermann, 2000; Zimmermann, Zebisch, & Strater, 2012).

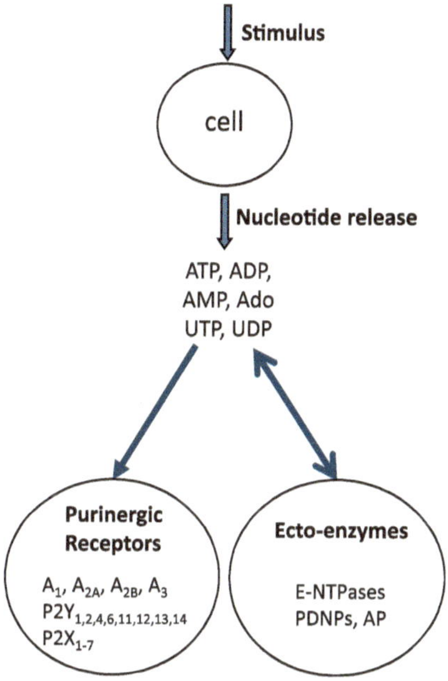

Fig. 1 Purinergic signaling system. Purinergic signaling cascade is initiated by the release of adenosine and uridine nucleotides and their metabolites to the extracellular space. Release pathway may include non-lytic mechanisms (exocytosis, ionic channels) but also plasma membrane injury and cell lysis. Purinergic receptor families include: metabotropic A_1, A_{2A}, A_{2B}, A_3—adenosine receptors, $P2Y_{1,\ 2,\ 4,\ 6,\ 11,\ 12,\ 13,\ 14}$—receptors, and ionotropic $P2X_{1-7}$—receptors. Extracellular nucleotides are subjected to hydrolysis and interconversion by secreted and cell surface-attached enzymes that include: ecto-nucleoside triphosphate diphosphohydrolases (E-NTPase), nucleotide pyrophosphatase/phosphodiesterase (PDNP) and alkaline phosphatase (AP).

Purinergic signaling plays important role in vast range of processes in normal and diseased lungs (Burnstock, Brouns, Adriaensen, & Timmermans, 2012; Picher & Boucher, 2011). Extracellular such as ATP, UTP, UDP and Ado modulate multiple components of innate lung defense by controlling ion transport and airway surface liquid homeostasis (Button, Okada, Frederick, Thelin, & Boucher, 2013; Lazarowski et al., 2004; Tarran, 2004), mucus secretion, and ciliary beat (Bucheimer & Linden, 2004; Kreda et al., 2007; Paradiso, Ribeiro, & Boucher, 2001; Ramminger et al., 1999; Stutts, Fitz, Paradiso, & Boucher, 1994). Nucleotides also regulate surfactant secretion by alveolar type 2 (AT2) pneumocytes (Andreeva, Kutuzov, & Voyno-Yasenetskaya, 2007; Patel et al., 2005). Extracellular nucleotide signaling is modified in and contributes to several respiratory pathologies. In clinical settings, significantly elevated levels of extracellular ATP and other purines were found in mechanically ventilated lungs, as detected in vivo in broncho-alveolar lavage fluids (Verbrugge, de Jong, Keijzer, Vazquez de Anda, & Lachmann, 1999), contributing to altered luminal nucleotide composition and purinoreceptor expression profile as well as promoting IL-6 release and lung edema (Douillet, Robinson, Milano, Boucher, & Rich, 2006; Rich, Douillet, Mahler, Husain, & Boucher, 2003). Accumulating evidence supports major role of extracellular ATP and purinergic signaling in pathophysiology of asthma and chronic obstructive pulmonary diseases (COPD) (Adriaensen & Timmermans, 2004; Aliagas et al., 2018; Idzko et al., 2007; Lommatzsch et al., 2010; Mortaz, Folkerts, Nijkamp, & Henricks, 2010; Pelleg, Schulman, & Barnes, 2016), lung emphysema (Mortaz et al., 2009), and pulmonary fibrosis (Della, Cabiati, Rocchiccioli, Del, & Morales, 2013; Muller et al., 2017). In the airways of cystic fibrosis patients, purinergic control of mucus and pathogen clearance are dysregulated, in part, due to extensive remodeling of purine enzymatic network. The changes involve altered activity, expression, and tissue distribution of NTPDase1 and NTPDase3, the two ectonucleotidases critical for the regulation of ATP on airway surfaces (Fausther, Pelletier, Ribeiro, Sevigny, & Picher, 2010).

Extracellular ATP in the lungs may originate not only from cell-regulated release but also from cell membrane injury and cell lysis. Clinical trials have shown that improperly delivered mechanical ventilation (MV) may worsen or cause lung injury (Dreyfuss, Soler, Basset, & Saumon, 1988; Sugiura, McCulloch, Wren, Dawson, & Froese, 1994). With large tidal volumes, MV produces alveolar overdistension that can be injurious by direct physical disruption of anatomical structures, cell-cell contacts, cell membranes, and stimulate inflammation (Cereda et al., 2013; Uhlig & Uhlig, 2011). The resulting ventilator induced lung injury is characterized by an increased alveolar permeability, pulmonary edema, infiltration of neutrophils, and the release of inflammatory mediators, including ATP (Jaecklin et al., 2011; Nickles et al., 2014; Rich et al., 2003; Tschumperlin, Oswari, & Margulies, 2000). Massive ATP release due to disruption or permeabilization of cell plasma membrane saturates the ATP-hydrolyzing enzyme system (CD39/CD73) (Hasan et al., 2018), leading to its sustained high levels acting as a "danger signal" a component of DAMPs (damage-associated molecular patterns), activating proinflammatory responses via immune cell's purinergic receptors (Hasan, Blankman, & Nieman, 2017; Mishra, 2013).

To identify ATP release pathways and to map time-dependent distribution of secreted ATP, accurate measurement techniques are required that could provide quantitative spatial and temporal information on extracellular

ATP. Only recently advanced ATP imaging approaches have been developed enabling precise investigation of cellular ATP release in real time and at the single cell level in vitro and ex vivo, capable of providing new insights into the release pathways.

2. Approaches to bioluminescence imaging of cellular ATP release

Several techniques were developed to detect and measure extracellular ATP, including non-imaging approaches such as amperometric ATP electrodes, use of P2X receptor-mediated ion currents as biosensor, as well as fluorescence-based or luciferin-luciferase (LL) bioluminescence methods that may permit ATP imaging, reviewed by Praetorius and Leipziger (2009), Rajendran, Dane, Conley, and Tantama (2016)). The standard approach to study ATP efflux from cells mainly involves collecting aliquots of bulk extracellular solution and subsequent off-line ATP analysis by highly sensitive LL bioluminescence assay. It detects light generated by ATP-dependent luciferase-mediated substrate oxidation, e.g., of D-luciferin, by the commonly used firefly luciferase. The assay could be used for real-time measurements of bulk ATP, e.g., by placing cell cultures covered with LL-containing medium directly in the luminometer measurement chamber. Development of cell surface-attached luciferase allowed monitoring local ATP concentration at the extracellular cell surface in real time (Beigi, Kobatake, Aizawa, & Dubyak, 1999). However, none of these approaches provides spatial information that would allow not only identification of specific sites or cells releasing ATP but also its quantification and time-dependent spreading toward other cells. Such information is relevant to understand the mechanisms of the release and autocrine/paracrine effects of secreted ATP. In the past decade much effort has been devoted toward direct visualization of extracellular ATP in living systems in real time.

2.1 Real-time luminescence imaging of ATP release

One of the experimental approaches that is compatible with live specimens and provides spatio-temporal insights into the ATP-release process involves real-time visualization of ATP-dependent LL bioluminescence. In this approach, a specimen such as substrate-attached cells is covered with DMEM culture medium containing the LL mixture. After stimulation, e.g., by brief stretching of the flexible substrate, the released ATP reacts with LL, producing photons, Fig. 2A.

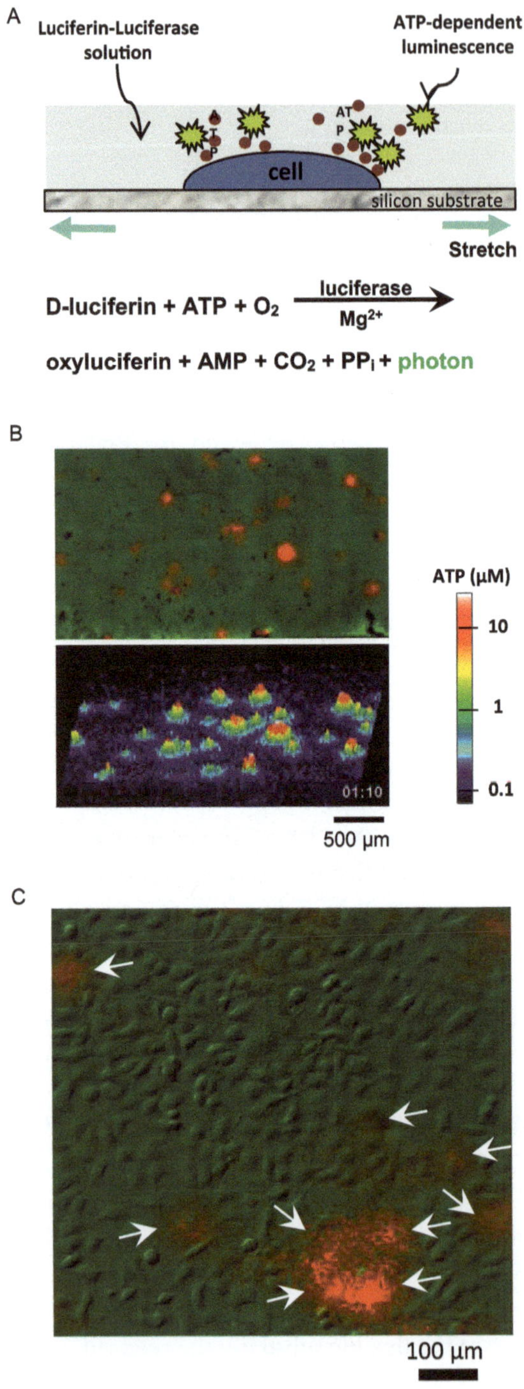

Fig. 2 See figure legend on opposite page.

The major challenge to imaging rapid release of ATP is low-light signal generated by bioluminescence reaction, which imposes substantial constraints on spatial and temporal resolution of ATP detection by this approach. However, continuing improvements in low-light imaging technology allowed high-sensitivity light detection, sufficient for real-time imaging of ATP-dependent luminescence, first reported by Furuya, Harada, and Sokabe (2008) and then used in several subsequent studies (Grygorczyk, Furuya, & Sokabe, 2013; Murata et al., 2014; Takahara et al., 2014; Yamamoto et al., 2011). Their luminescence imaging microscopy system combines a cooled electron-multiplying charge-coupled device (EMCCD) camera coupled with a water-cooled image intensifier and low-magnification, high numerical aperture (NA) objective. It provides a signal-to-noise ratio sufficient to study ATP release from single cells with ~10 nM detection sensitivity and 100 ms temporal resolution (Furuya, Sokabe, & Grygorczyk, 2014). An example image of stretch-induced ATP release from A549 cells is shown in Fig. 2B and C. It provides spatio-temporal information on absolute ATP concentration and its diffusion from the release points, that is not possible to obtain with standard luminometry. The system is also equipped with a unique infra-red (IR) differential interference contrast (DIC) imaging, allowing simultaneous identification of ATP-releasing cells and assessing the extent of cell distension during stretch-stimulation. Further instrumental details of the approach could be found in Furuya et al. (2014). The imaging system described here enabled us to investigate ATP release in several cell types and tissues including ex vivo lungs (Furuya et al., 2016), providing new insights into the release mechanisms and purinergic signaling in different biological systems.

Fig. 2 Principles of bioluminescence ATP release imaging. (A) A concept of ATP release imaging via ATP-dependent luciferin-luciferase (LL) luminescence in vitro. Cells are covered with DMEM medium containing LL mixture and stimulated by brief (1 s) stretching. Released ATP reacts with LL producing photons. (B) Example image of stretch-induced ATP release from A549 cells. The image is an overlay of DIC IR image of cells grown on silicon substrate (green), and ATP-dependent LL luminescence (red). Image below shows local ATP concentration as pseudo-3D intensity profile where the pseudo-colored luminescence intensity was converted to height. (C) Stretch-induced ATP release from A549 cells recorded at higher magnification with 20× objective reveals that ATP release sites are single cells. The image overlay (green—DIC IR image, red—ATP-dependent luminescence) reveals that ATP is released from single cells (indicated by arrows). *Panel C: The image was taken 3 s after 25% stretch, from Grygorczyk, R., Furuya, K., & Sokabe, M. (2013). Imaging and characterization of stretch-induced ATP release from alveolar A549 cells.* The Journal of Physiology, 591, *1195–1215.*

Importantly, the LL is non-toxic for cells and they remain viable for many hours of experimentation. However, similar to the standard LL luminometry, when performing live cell luminescence imaging, attention should be given to potential interference with LL reaction when pharmacological or other interventions (e.g., pH or ion concentration changes) are applied. Bioluminescence signal calibration and verification of its stability over time should be performed for each experimental condition to ensure proper performance of the imaging system.

2.2 Simplified systems for bioluminescence ATP imaging

In the luminescence microscopy system described above, the high sensitivity of light detection was achieved by combining EMCCD camera and specialized low-noise liquid-cooled image intensifier, that results in an overall complexity, precluding its wider use by many laboratories. However, recent improvements in the sensitivity of EMCCD cameras allow observation of stronger ATP release events without image intensifier, albeit with reduced ATP detection sensitivity ($\sim$10 μM). We have used such simplified system successfully in part of our study of stretch-induced ATP release in primary rat alveolar AT2 cells, described in Section 4.2. Other alternative approaches involve enhancing the light signal captured by the camera. Typical modern microscopes are designed with infinity optics allowing their versatile use with different optical components. However, this comes at the cost of reduced light gathering power, a critical factor for low-light imaging. Therefore, in an alternative approach to avoid these limitations, we used a simple optical system directly mounted on the EMCCD camera, instead of microscope. The system consists of two lenses with a shorter optical path and bigger NA-to-magnification ratio, providing wide field of view (FOV) imaging with significantly boosted light-gathering power, largely exceeding (>50-fold) those of standard microscopes. This straightforward and cost-affordable approach enabled us to image the whole stretch chamber (approx. 20×20 mm) with cultured rat primary AT2 cells to easily and accurately track the kinetics of stretch-induced secretion of ATP and to map and quantify its time-dependent distribution. Quantitative ATP imaging is required to gain new insights into the mechanisms of cellular ATP release, which are otherwise difficult to assess with a standard microscope that visualizes secreted ATP in spatially restricted regions of interest and lacks the ability of quantifying released ATP in the entire chamber/specimen.

3. Mechanisms of ATP release

3.1 ATP release stimuli

Release of ATP into the extracellular space is a common cellular response to various physical stimuli, which may involve mechanical cell stretch and deformation caused by physical tissue distension (Furuya et al., 2016; Grygorczyk et al., 2013; Sadananda, Kao, Liu, Mansfield, & Burcher, 2012), fluid- or air-flow shear stress (Tarran, Button, & Boucher, 2006; Yamamoto et al., 2011), cell distortion by tension forces at the air-liquid interface (Ramsingh et al., 2011), mechanical stress between motile cilia and mucus layer on airway surfaces (Button et al., 2013), or in vitro hypotonic cell swelling or tilting cell culture plate (Grygorczyk & Hanrahan, 1997a). While mechanical cell stimulation appears by far the most widespread inducer of ATP secretion in all cell types (Mikolajewicz, Mohammed, Morris, & Komarova, 2018), other known stimuli include agonist-dependent receptor-mediated release, where agonists may include thrombin via Ca^{2+}, and Rho-mediated signaling (Seminario-Vidal et al., 2009), and purines themselves (via P2-receptors), resulting in ATP-induced ATP release (Anderson, Bergher, & Swanson, 2004). A spectrum of other nucleotides is co-released with ATP, including adenosine and uridine nucleotides (Kreda et al., 2010; Tatur, Groulx, Orlov, & Grygorczyk, 2007; Tatur, Kreda, Lazarowski, & Grygorczyk, 2008), that can also amplify ATP release by synergistic autocrine/paracrine action on purinergic receptors (Tatur et al., 2007). Interestingly, light-induced ATP release mediated by photoreceptor melanopsin has been recently reported in lens epithelial cells (Pintor, 2018).

3.2 Non-lytic ATP release pathways

ATP release pathways were discussed in previous comprehensive reviews (Lazarowski, Boucher, & Harden, 2003; Praetorius & Leipziger, 2009). There are two major pathways of cellular ATP release: non-lytic (or cell-regulated) and lytic. The non-lytic mode of ATP release is generally divided into conductive (ionic channels) and non-conductive pathways such as exocytosis, with the latter further divided into regulated (Ca^{2+}-dependent) and non-regulated exocytosis (Lazarowski et al., 2011). Cell lysis resulting from plasma membrane rupture allowing the free discharge of cytoplasmic ATP can be induced by physical or chemical stressors; the damage can be transient or can lead to cell death (Skotak, Wang, & Chandra, 2012). Presently, most research has focused on regulated non-lytic release mechanisms, while release involving plasma membrane injury and/or cell lysis has so far attracted less attention. This mode of release, however, may be taking place not only in pathological conditions, but might be an important contributing factor in certain physiological situations as exemplified by shear stress and hypoxia-induced ATP release from RBCs, as discussed below.

3.2.1 ATP-conducting channels

Historically several ion channels have been implicated in conductive release of cellular ATP, including cystic fibrosis transmembrane conductance regulator (CFTR), P-glycoprotein, connexins hemichannels and pannexins, calcium homeostasis modulator 1 (CALHM1), volume regulated anion channels (VRAC), voltage dependent anion channel (VDAC), and VDAC-like maxi anion channels, reviewed in Taruno (2018). In the case of the CFTR, studies by several independent groups with the patch clamp, lipid bilayer and luminometry techniques, have not revealed any detectable CFTR-mediated or CFTR-regulated ATP release in several epithelial and non-epithelial cells (Grygorczyk & Hanrahan, 1997a, 1997b; Grygorczyk, Tabcharani, & Hanrahan, 1996; Hazama et al., 1999; Li, Ramjeesingh, & Bear, 1996; Reddy et al., 1996; Watt, Lazarowski, & Boucher, 1998).

VRAC and Maxi-Cl channels were implicated in cell volume regulation and ATP release from swollen cells (Okada, Okada, Islam, & Sabirov, 2018; Osei-Owusu, Yang, Vitery, & Qiu, 2018; Sabirov, Merzlyak, Islam, Okada, & Okada, 2016). Both have been recently identified at the molecular level with LRRC8 protein as a main component of VRAC (Qiu et al., 2014; Voss et al., 2014) and organic anion transporter CLCO2A1 (prostaglandin transporter) as a key component of Maxi-Cl channel (Sabirov et al., 2017). Both channels possess a pore, wide enough to pass variety of organic compounds like ATP, glutamate and taurine. It is still debatable whether VRAC is a conduit of ATP release (Okada et al., 2018), but it was suggested that variation of heteromeric subunit structures in LRRC8 hexamer may alter the properties of VRAC allowing for ATP permeation (Osei-Owusu et al., 2018). Activity of both channels is associated with cell proliferation, migration, angiogenesis and apoptosis, and thus the physiological and pathophysiological relevance of these ATP release pathways requires further research.

Pannexins are probably most extensively studied putative ATP channels in the past decade. In particular, Pannex 1-mediated ATP release was reported in a wide range of cell types, including excitable and non-excitable cells, such as airway epithelial cells and RBCs. The release can be activated by diverse mechanisms including mechanical stimuli, hypotonicity, intracellular Ca^{2+} and receptor-mediated (cAMP) signaling, reviewed in Chiu, Schappe, Desai and Bayliss (2018). However, the role of cAMP signaling pathway in stimulating ATP release (via Pannexin-1 or CFTR) in RBCs was contradicted by recent studies (Keller et al., 2017; Sikora, Orlov, Furuya, & Grygorczyk, 2014). The unresolved issues concerning the role of Pannex 1 in ATP release include its regulatory mechanisms and basic electrophysiological properties such as ATP permeability and unitary channel conductance, which may depend on the mode of channel activation (Taruno, 2018).

Calcium homeostasis modulator 1 (CAL HM1) is another example of non-selective channel with large pore similar to that of connexins and pannexins, and implicated in ATP release from type 2 taste receptor cells (Taruno et al., 2013) and airway epithelial cells (Workman et al., 2017).

P2X7R is an ATP-gated cation channel found predominantly (but not exclusively) on immune cells but also on AT1 cells (Chen et al., 2004; Mishra, 2013). P2X7R activation results in a number of downstream events, including the release of pro-inflammatory mediators and cell death. Sustained activation of P2X7R causes permeabilization of plasma membrane due to the formation of cytolytic non-selective pore, which allows fluxes of large molecules (up to 900 Da), leakage of metabolites, ultimately leading to cell death. The nature of such permeability pore is still debated. It may involve a hemichannel protein pannexin-1; but recent reports suggest that it is an intrinsic property of the P2X7R itself (Di Virgilio, Schmalzing, & Markwardt, 2018). Importantly, it was shown that activation of P2X7R induces release of large amounts of ATP in a model of human embryonic kidney cells via ATP-induced ATP release mechanism (Pellegatti, Falzoni, Pinton, Rizzuto, & Di Virgilio, 2005).

Regardless of the molecular nature of ATP-conducting channels, they all require pore of large dimensions (0.6–1.1 nm, (Sabirov & Okada, 2005)), resulting in poor selectivity and large conductance (hundreds of pS) for small ions such as K^+, Na^+, Cl^- and Ca^{2+}, as exemplified by the well characterized VDAC channel (Colombini, 2012). Opening of such large ATP permeable pores unavoidably results in massive influx of Ca^{2+}, Na^+ and efflux of K^+ ions downstream their electrochemical gradients, greatly perturbing cell homeostasis. None of the currently known putative plasma membrane ATP channels can allow selective permeation of ATP over small ions, preserving cell homeostasis. Thus, conductive ATP secretion appears as a crude homeostasis-perturbing approach to "cell-regulated" purinergic signaling. On the other hand, such non-selective conductive release of cytoplasmic ATP can occur without specialized endogenous membrane proteins as a result of injurious mechanical insult or membrane permeabilization by pathogen-derived pore-forming toxins.

3.2.2 Ca^{2+}-dependent exocytosis of ATP-containing vesicles

Vesicular release of ATP is well established mechanism in neuronal and endocrine cells (Praetorius & Leipziger, 2009). Recently accumulated evidence supports such a release pathway also in several epithelial cells of different origin (Feranchak et al., 2010; Kreda et al., 2010; Lazarowski et al., 2011). In particular in lung epithelial cells, our earlier study using flow-through chamber and collecting perfusate aliquots for ATP determination with luminometry demonstrated that hypotonic shock induced transient ATP release from human A549 alveolar and 16HBE14o^{-} bronchial epithelial cells. The release correlated tightly with intracellular Ca^{2+} elevations was significantly diminished in cells loaded with Ca^{2+} chelator BAPTA-AM and was not affected by different inhibitors of putative ATP-conducting channels (NPPB, Gd). Consistent with exocytosis, the release was abolished by temperature reduction to 10 °C (Boudreault & Grygorczyk, 2004). It was subsequently demonstrated that several adenosine and uridine nucleotides are co-released with ATP in a Ca^{2+}-dependent manner (Tatur et al., 2008), and that stress-induced ATP release from alveolar cells is amplified by the synergistic autocrine/paracrine action of co-released uridine (via $P2Y_6$ receptor) and adenosine nucleotides (Tatur et al., 2007). Similar mechanism of purinergic signal propagation may operate in other cell types as a more general paradigm. Involvement of exocytosis process in hypotonic shock-induced ATP release from A549 cells was most directly demonstrated by observing single exocytotic events of quinacrine-stained, ATP-containing vesicles with total internal reflection fluorescence (TIRF) microscopy in the study by Akopova et al. (2012).

In the lung epithelial cells, similar to neuronal and endocrine cells, ATP often is not the only vesicular cargo but is co-stored, e.g., with surfactant proteins in the lamellar bodies of alveolar type 2 cells (Fois et al., 2018), and co-released upon stimulation. Similarly, ATP and other nucleotides (ADP, AMP) were found as a co-cargo that was released from mucin granules in airway epithelial cells (Kreda et al., 2010). Our recent ATP imaging data with A549 and Calu-3 cells provide further support for such a co-release mechanism.

3.3 ATP release by cell membrane injury and cell lysis

ATP release associated with plasma membrane permeabilization/injury and cell lysis has not been fully appreciated and remains under-investigated. Such a release mechanism is particularly relevant in pathological conditions (e.g., infections) and clinical complications such as ventilator induced lung injury (VILI), where cell overstretch and extensive tissue damage occurs. The plasma membrane integrity is essential for cell survival; therefore, cells possess several evolutionarily conserved mechanisms to rapidly repair the injured membranes and ensure survival. Eukaryotic cells when challenged by mechanical stress or pore forming proteins (native or from invading pathogens), reseal in few seconds and can repair breaches in their plasma membrane of surprisingly large area (Bischofberger, Gonzalez, & van der Goot, 2009). The membrane "holes" may be of different nature, lipidic pores not delimited by protein boundaries typically arise as a result of mechanical stress (Shigematsu, Koshiyama, & Wada, 2015), while membrane damage caused by pore-forming proteins give rise to proteinaceous or proteolipidic pores with well-defined boundaries (Etxaniz, Gonzalez-Bullon, Martin, & Ostolaza, 2018). Depending on the size and nature of the pore the repair mechanisms vary. Patching and clogging processes are involved in the repair of mechanical/physical damage, while shedding (ectocytosis) and endocytosis are implicated in the elimination of pore-forming toxins (Etxaniz et al., 2018; Jimenez & Perez, 2017). Membrane patching is triggered by extracellular Ca^{2+} influx into wounded cells activating lysosomal exocytosis. Clogging involves an accumulation of proteins around the lesion, forming a barrier that prevents the loss of cytoplasmic contents to the extracellular medium and isolates the area of membrane damage. Shedding allows to isolate and expel the toxin pore-damaged membrane area in the form of vesicles called "toxosomes" or "ectosomes." Endocytosis is an alternative mechanism by which small pores formed by toxins are removed by internalizing the damaged area, including the pore.

Considering the above processes, it is clear that in many situations membrane wounding is transient and could be reversed rapidly by repair mechanisms preventing cell lysis and death. Transient membrane permeabilization during wounding followed by the repair processes will be accompanied by short-lived release of cytoplasmic ATP that may display kinetics and extent similar to that observed with regulated processes (exocytosis, channels). Thus, distinction between the two modes of release (non-lytic and lytic) may be difficult in some situations. Moreover, cells may show transient uptake of lysed/dead cell markers (e.g., propidium iodide), complicating interpretation of the results. Since short-lived membrane injury can often be encountered during the cell's life, the associated ATP release should be considered as normal and important mechanism contributing to intercellular purinergic signaling in eukaryotic cells. Clearly, such a release is often confronted in ATP release studies and in some cases may represent a legitimate physiologically relevant mechanism. In red blood cells hemolysis was shown to be a primary release mechanism, where it may contribute to local blood flow regulation (Grygorczyk & Orlov, 2017; Sikora et al., 2014; Thomas, 2014). In bone cells regulation of mechanically stimulated ATP release was attributed to a balance of injury and repair (Mikolajewicz, Zimmermann, Willie, & Komarova, 2018).

4. Mechanosensitive ATP release in lung epithelial cells

In Sections 4 and 5, we provide an overview of our recent studies on ATP release performed with real-time bioluminescence ATP imaging. Lung epithelial cells are subjected to several forms of mechanical stresses during the breathing cycle, sneezing and coughing, that includes airflow shear stress, cell stretch, and cell distortion by tension forces at the air-liquid interface, all of which can evoke ATP release. In majority of the following in vitro studies, we used physiologically relevant stretch stimuli or inflation with ex vivo lungs.

4.1 Stretch-induced ATP release in A549 alveolar cells

Stretch is a particularly relevant physiological stimulus in the lungs (Edwards, 2001), where expansion of alveoli during lung inflation leads to significant thinning of the alveolar walls that involves stretch and deformation of the alveolar epithelium. Stretch is also recognized as important physiological stimulus of surfactant secretion by alveolar type 2 cells (Ashino, Ying, Dobbs, & Bhattacharya, 2000; Hildebran, Goerke, & Clements, 1981; Mason & Voelker, 1998; Patel et al., 2005), which may involve synergistic effects of co-released ATP, a potent surfactant secretagogue (Dietl & Haller, 2005; Dietl, Haller, & Frick, 2012). With model human alveolar A549 cells grown on a flexible substrate, we found that 10–30% stretch pulse of 1 s duration induced ATP release in a limited number of discrete sites. Fig. 3A shows an example with cells stimulated by 21% stretch. Each release site was identified on IR images as a single cell. Responses were rapid and some could be noted within 200–300 ms from the start of stretch stimulus. The sequential images clearly show temporal and spatial changes of

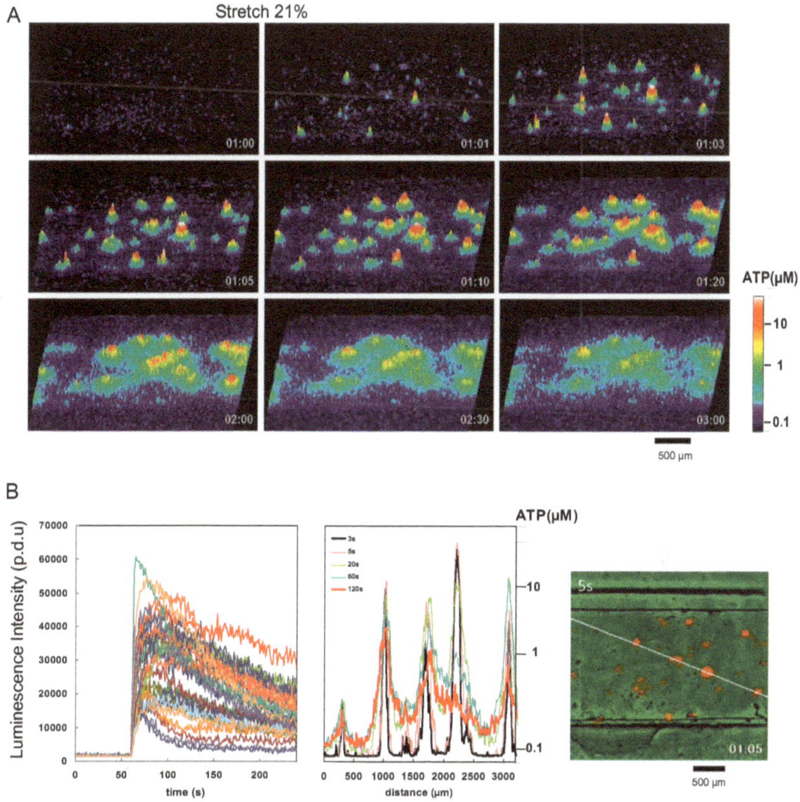

Fig. 3 ATP release from A549 cells after single stretch. (A) Sequence of images showing ATP-dependent luminescence around active cells at different time points after 21% stretch of 1-s duration applied at 1 min. The experiment elapsed time (in min: sec) is indicated in the lower right corner. The release is shown as a pseudo-3D intensity profile where the pseudo-colored luminescence intensity was converted to height. Note the multiple ATP-release sites and high (micromolar) ATP concentration in their vicinity sufficient for autocrine/paracrine stimulation of P2Y receptors on neighboring cells. The images disclose the cell culture area of approximately 3 × 2 mm. (B) Left graph, time-course of ATP release for all 35 responding cells shown in (A). Middle graph, spreading of released ATP measured along a white line drawn on the image on the right. Color lines represent corresponding ATP concentration profiles observed at 3–120 s after the release. Peak ATP concentration reached over 10 μM for some cells. At 2 min high sub-micromolar [ATP] is observed at large >250 μm distances from the release sites, sufficient for paracrine purinergic receptors activation. Image on the right, depicts an overlay of luminescence (red) and IR (green) images taken at 5 s after stretching. *From Furuya, K., Sokabe, M., & Grygorczyk, R. (2014). Real-time luminescence imaging of cellular ATP release.* Methods, 66, *330–344.*

luminescence intensity, which is calibrated to absolute ATP concentration. Under static conditions (no perfusion), extracellular ATP concentration in the proximity (~50 μm radius) of active cells often reached 10 μM, sufficient to activate P2 receptors on neighboring cells. The time-course of ATP release from individual cells in this experiment is illustrated on the left graph in Fig. 3B. Kinetic analysis demonstrated that the duration of ATP release was in the range from 10 to 82 s (Grygorczyk et al., 2013). The middle graph in Fig. 3B shows time-dependent changes in released ATP concentration along the line crossing five active release sites, shown in the right image. It is evident from the graph that after the release, extracellular ATP spreads from the release sites up to 1 mm, and its concentration stays at a high submicromolar level for several minutes. These spreads are important for ATP actions as local paracrine mediator. For example, our results suggest that ATP release from single alveolar cell may be sufficient to initiate purinergic signaling in the entire human lung alveolus of average ~200 μm radius. ATP is a potent surfactant secretagogue and P2 receptor stimulation by high extracellular ATP might be relevant for stimulating exocytosis of surfactant-containing lamellar bodies, which require prolonged elevation of cytoplasmic Ca^{2+} to fully open a fusion pore and proper dispersal of their contents (Dietl et al., 2012). In addition, ATP stored in lamellar bodies themselves may contribute to autocrine stimulation of the surfactant secretion process (Fois et al., 2018).

4.2 Stretch induces coordinated ATP release in rat primary alveolar type 2 cells

Stretch-induced ATP release in rat primary AT2 cells reveals some features not observed in A549 cells. Fig. 4A demonstrates that 1 s stretch of 17% induces release of ATP from individual small clusters of cells, rather than single cells, in a 3-day-old AT2 cell culture. The release is seen as a localized ATP-dependent luminescence signal that peaks at ~10- to 20-s following the stretch, image (b) in Fig. 4A. However, stronger stretch (44%) induces stronger release from higher number of sites leading to formation of large clouds of released ATP with initially well-defined contour that coincided with substrate areas covered by cells, Fig. 4B, image (b). Defined contour of the released ATP cloud, suggests coordinated ATP release from cells within the active region, which may involve intracellular Ca^{2+} signaling between neighboring cells through gap junctions and by extracellular ATP (Isakson, Evans, & Boitano, 2001). This contrasts with A549 cells, where even at the strongest stretches the release occurred from single cells

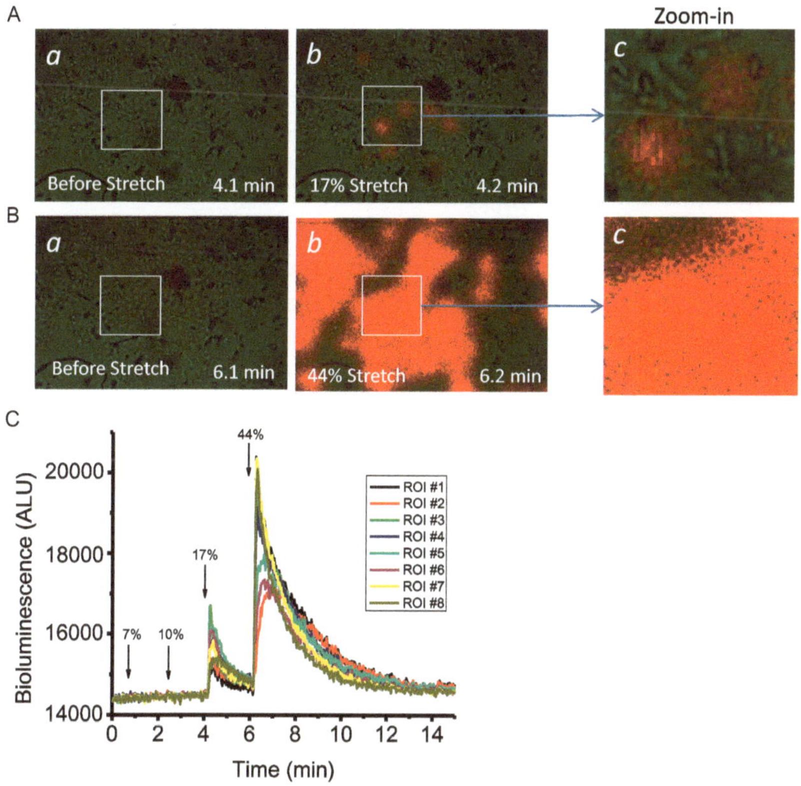

Fig. 4 Stretch-induced ATP release from rat primary AT2 cells. Examples of ATP release-dependent LL luminescence in 3-day-old AT2 cell culture induced by 1 s stretch of 17% and 44%, (A) and (B), respectively. The images show overlay of ATP-dependent luminescence (red), and phase-contrast transmitted light image (green) of AT2 cells grown in the 2 × 18 mm grove of the silicon stretch chamber. Image (a) was taken before stretch while (b) after the stretch. Images (c) show ATP-release responses in a zoom-in area marked by a rectangle in (b). ATP release was viewed in real-time using simplified imaging system consisting of Zeiss Axio Observer microscope (5 × lens) and Photometrics® Evolve™ 512 EMCCD camera without image intensifier. Localized small areas of ATP release for 17% stretch (image A, b), correspond to clusters of cells. Clouds of massive ATP release and larger area of active release were observed for stronger 44% stretch, image (B, b) (at 6.2 min, acquired 6 s after the stretch) and zoom-in image (B, c). (C) Time-course of ATP-release responses to a sequence of four stretches of increasing amplitude (from 7% to 44%, arrows) given at 2-min interval. The graph reports bioluminescence changes measured in eight different regions of interest. Data are from the same experiment as in (A) and (B). Only stretches above 10% induced detectable ATP release.

that were scattered randomly amongst inactive cells, which may indicate the absence or deficiency of gap junctional communication in these cells. Following the release, ATP cloud spreads due to diffusion and covers entire FOV of AT2 cell culture within <1 min after stretch (Fig. 4B, c). Full time-course of ATP-dependent luminescence for a sequence of stretches of 7–44% in this experiment is shown on Fig. 4C. While minimal responses were seen at 10% stretch, only stretches of ≥17% produced clearly detectable ATP release in this experiment. The release is transient and local concentration of secreted ATP at the release site decreases over time due to consumption by LL reaction and diffusion out of the FOV. Kinetically all release responses were similar displaying a rapid rise followed by exponential decay suggesting a release mechanism that may involve rapid exocytosis of a relatively uniform population of ATP-containing vesicles. This was not the case for ATP release from A549 cells that was stimulated by hypotonic swelling or in submucosal Calu-3 cells stimulated by stretch, where kinetically different components of the release can be observed.

4.3 Two components of stretch-induced ATP release in Calu-3 cells

Calu-3 are mucus secreting cells derived from human airway epithelial submucosal glands. In polarized cell cultures MUC5AC mucin-containing granules could be found subapically in large fraction of these cells. Study by Kreda et al. revealed coordinated release of mucins and nucleotides (ATP, ADP, AMP) in Calu-3 cells involving Ca^{2+}-regulated exocytosis of mucin granules (Kreda et al., 2007, 2010). Direct nucleotide release from mucin-secreting goblet cells provides paracrine purinergic signals for mucin hydration and regulation of airway mucociliary clearance mechanism.

Using real-time luminescence ATP imaging we found that stretch-stimulation of Calu-3 cells induced two types of ATP-release responses, a fast response peaking at 10–15 s, similar to that observed in stretch-stimulated A549 cells and a slowly rising response that peaked at 60–90 s after the stretch, Fig. 5A (Supplementary Video 1 in the online version at https://doi.org/10.1016/bs.ctm.2019.02.001). Fast component may represent rapid vesicular ATP release, while the slow long-lasting component may be consistent with ATP co-release from mucin granules, where delayed discharge of granular ATP is expected due to its slow diffusion through the secreted mucus, Fig. 5B. However, further detailed studies will be required to prove such a mechanism of prolonged ATP secretion.

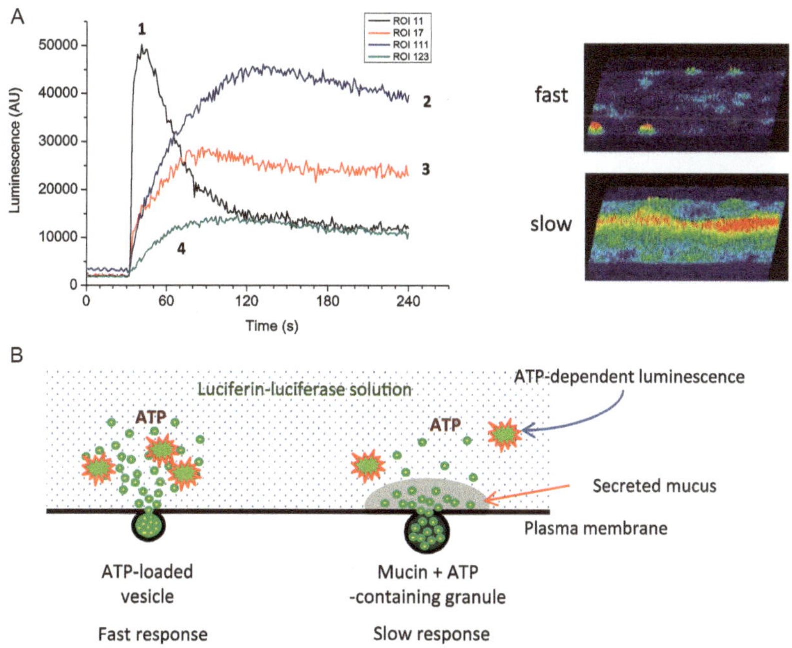

Fig. 5 Stretch-induced ATP release from Calu-3 cells. (A) Example time-course of ATP-release responses in Calu-3 cells stimulated by 1 s stretch of 28%. ATP-dependent LL luminescence is shown in arbitrary light units (ALU). Note the presence of fast (trace 1) and slow responses (traces 2–4). The rising phase of the slow release responses was 16–28-fold slower compared to the fast response. The images on the right show a pseudo-3D intensity profile of ATP-dependent luminescence observed with Calu-3 cell culture in the stretch chamber. Upper image was taken at 10 s and lower image at ~90 s after the stretch, corresponding to peak response of the fast and slow responses, respectively. See Supplementary Video 1 in the online version at https://doi.org/10.1016/bs.ctm.2019.02.001 for the full time-course of ATP release in this experiment. (B) A cartoon illustrating the concept of delayed ATP release from mucin granules.

4.4 Hypotonic stress-induced ATP release in A549 cells—Contribution of cell lysis

Hypotonic shock is known to evoke ATP release from many cell types and was commonly used as a surrogate of mechanical stimulation in earlier studies by several investigators. In our previous studies we used a flow-through chamber and standard luminometry to investigate kinetics of hypotonic stress-induced ATP release (Boudreault & Grygorczyk, 2004; Tatur et al., 2007). However, no spatial information could be obtained with such approach. Bioluminescence ATP imaging with its superior spatial and temporal resolution can provide new insights into the release process and the underlying mechanisms. In experiment illustrated in Fig. 6, A549 cells were

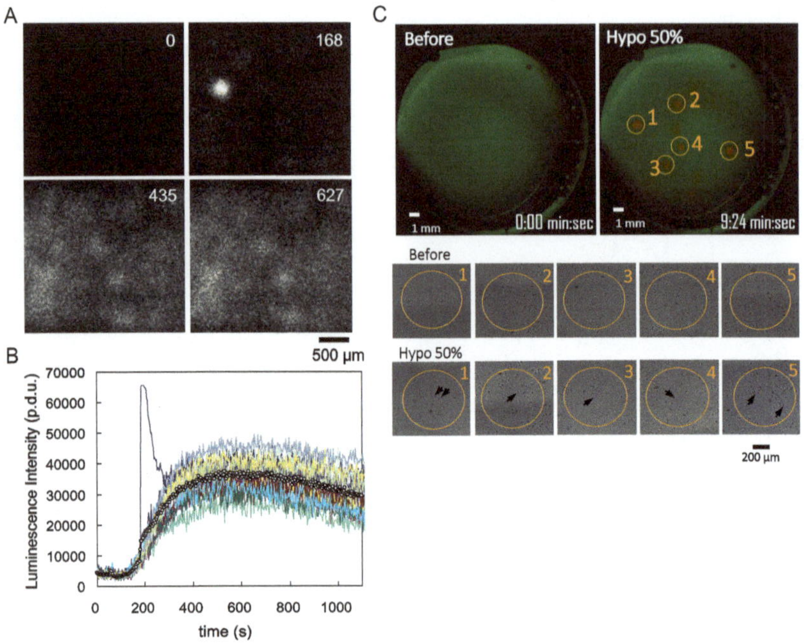

Fig. 6 Hypotonic stress-induced ATP release in A549 cells. (A) 50% Hypotonic shock-induced ATP release in lung alveolar A549 cells. The responses were slow-rising and of diffused appearance, so that ATP-releasing cells could not be clearly identified except for one isolated release spike at 168 s. (B) Time-course of ATP-dependent luminescence observed in several different regions of A549 cell culture in the experiment shown in (A). The slow responses reached maximum at about 600 s. (C) Co-localization of ATP release regions and hypo-induced PI^+ staining of A549 cells. ATP secretion from A549 cells grown in 0.1% FBS for 48 h (fragile cells) on glass coverslip was imaged with LL bioluminescence and a simplified imaging system consisting of large field of view optics and Photometrics® Evolve™ 512 EMCCD camera. Secretion was stimulated by cell swelling in extracellular solution of reduced tonicity (hypo 50%) under static condition (no perfusion). Regions of prominent extracellular ATP level are identified by circles (1–5) on the Hypo 50% image. Plasma membrane permeability to large molecules was assessed prior and post-treatment with propidium iodide (PI) uptake evaluated with an epifluorescence-equipped Zeiss Axio Observer microscope (10× objective), lower panel. Images of the entire coverslip were stitched with open-source ImageJ software. A549 cells that were PI^+ before hypo are shown in red, while those that become PI^+ after hypo 50% are shown in blue. All five ATP release regions on image Hypo 50% co-localize with hypo-induced PI^+ cells (arrows) identified on the lower panel images. *Panel B: Images were obtained with 4× lens and 500-ms exposure time, from Furuya, K., Sokabe, M., & Grygorczyk, R. (2014). Real-time luminescence imaging of cellular ATP release.* Methods, 66, *330–344.*

exposed to hypotonic shock (50%) and bioluminescence imaging was subsequently performed under static condition (no perfusion). We found two types of release events, a slowly rising ATP release response of diffuse appearance in addition to several random but highly localized ATP-release spikes. An example is shown in Fig. 6A and B (Furuya et al., 2014). The extracellular ATP resulting from the slow release reached maximum in 500–700 s, but spikes appeared mostly at the beginning of hypotonic stimulation. To further examine the nature of the ATP release spikes, in a subsequent study we used wide FOV ATP imaging system and propidium iodide (PI) fluorescence to detect lysed cells, Fig. 6C. We compared ATP release in cells that were rendered fragile (by culturing them for >24 h in low serum medium) and thus prone to cell lysis, and normal control cells cultured in serum-containing medium. While we observed similar slow ATP secretion component in both cell cultures, number of punctuate ATP secretion events was significantly increased in fragile cells (Ponomarchuk, Boudreault, Orlov, & Grygorczyk, 2017). Importantly, all punctuate ATP secretion sites co-localized with PI-positive cells, consistent with cell lysis. Total cell lysis per coverslip caused by acute cell swelling constituted <2% and the amount of normalized ATP release per cell did not differ in control and fragile A549 cells, indicating that contribution of ATP release due to cell lysis is negligible in these experiments. The nature of the slow ATP release remains not fully resolved in this study but likely results from exocytosis of ATP-loaded vesicles (Akopova et al., 2012; Boudreault & Grygorczyk, 2004) and co-release of ATP co-stored with surfactants in lamellar bodies (Fois et al., 2018).

5. Inflation-induced ATP release in ex vivo lungs

While ATP release has been investigated extensively in vitro using isolated lung epithelial cells (Boudreault & Grygorczyk, 2004; Grygorczyk et al., 2013; Kreda et al., 2007; Lazarowski et al., 2011; Okada, Nicholas, Kreda, Lazarowski, & Boucher, 2006), cultured lung cells may inadequately replicate native cell responses because of phenotypic transformations. In the following study to investigate inflation-induced purinergic cell signaling in the native setting in real time, we have applied bioluminescence ATP imaging to lungs of rat, ex vivo.

5.1 ATP release in airspaces

Detection of alveolar ATP release requires instillation into lung airspaces a liquid containing LL reagent, thus limiting study to liquid-filled "edematous" lungs (Furuya et al., 2016). With this approach, we found that brief inflation (1 s, ~20 cm H_2O positive pressure via trachea) induced transient (<30 s) ATP release in a limited number of air-inflated alveolar sacs during their recruitment/opening, Fig. 7A. Released ATP reached

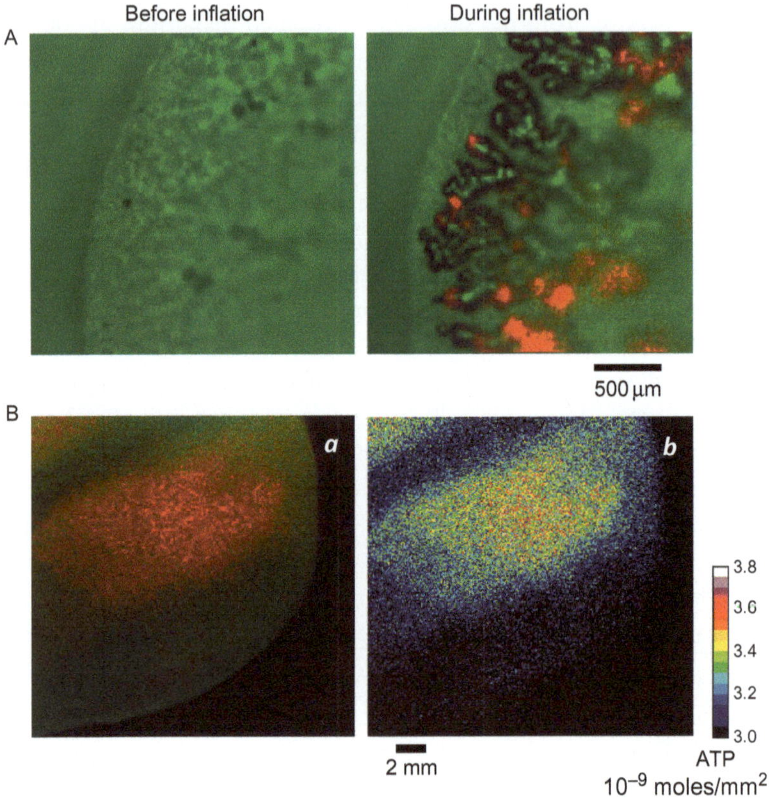

Fig. 7 Inflation-induced ATP release in airspaces of ex vivo rat lung. (A) The overlay images show ATP-dependent luminescence (red) and infra-red (IR) tissue images (green), before and during pressure pulse of ~20 cm H_2O. Air-inflated strings of alveolar clusters appear as dark areas on the image. (B) Bioluminescence ATP imaging of "dry" rat lung with LL reagent introduced by nebulization. (a) Overlay image of ATP-LL luminescence (red) and transmitted light image of the lung tissue (green). (b) Tissue distribution of ATP (in femto-moles/mm^2) is shown on pseudocolor scale. DMEM containing isotonicity-adjusted LL with 10 mM ATP was introduced into rat lungs via a compressor nebulizer (MedPro AMG 705-470) during five consecutive lung inflations (−20 cm H_2O, 4 s) in artificial thoracic chamber. Bioluminescence was recorded with a simplified wide FOW imaging system, exposure time 10 s. *Adapted from Furuya, K., Tan, J. J., Boudreault, F., Sokabe, M., Berthiaume, Y., & Grygorczyk, R. (2016). Real-time imaging of inflation-induced ATP release in the ex vivo rat lung.* American Journal of Physiology. Lung Cellular and Molecular Physiology, 311, *L956–L969.*

concentrations in the order of $\sim 10^{-6}$ M, relevant for autocrine/paracrine signaling. Closer analysis revealed that released ATP remained spatially restricted to irregular polygonal shapes the size of alveolar sacs (equivalent diameter of 40–160 μm), suggesting that the elementary unit of ATP responses consist of a single alveolar sacs or their clusters. Only limited ATP diffusion to neighboring alveolar sacs was occasionally observed. ATP release was stimulus-dependent, prolonged (100-s) inflation evoked long-lasting ATP release, which terminated upon alveoli deflation/de-recruitment while cyclic inflation/suction produced cyclic ATP release (Furuya et al., 2016). Several physical stimuli associated with edematous lung inflation such as cell stretch, cell deformation by tension forces at the air-liquid interface and shear stress caused by liquid movement, may contribute to ATP release in lung airspaces observed in this study. The exact role of these different stimuli to ATP release requires future investigations.

In future studies limitations of the edematous lung model could be avoided by delivering LL reagent into lung airspaces with nebulizer. Our preliminary experiments indicate feasibility of such approach, where extracellular ATP can be detected in "dry lungs" after nebulization of LL/ATP mixture into airspaces, Fig. 7B.

5.2 ATP release in pulmonary blood vessels

In the lumen of blood vessels ATP contributes to local control of blood flow through interaction with purinergic receptors of vascular endothelial cells and stimulation of nitric oxide synthesis, a powerful relaxant of vascular smooth muscle. Plasma membrane of endothelial cells sense changes in shear stress and stretching and transduce these changes into intracellular biochemical signals (Yamamoto & Ando, 2018), activating multiple signaling pathways including purinergic signaling via the release of ATP.

To visualize ATP release in alveolar blood capillaries of rat lung, LL was introduced into their lumen via pulmonary artery. Inflation induced transient ATP release that displayed a more diffuse pattern, different from that observed in airspaces. Despite diffuse appearance, the responses clearly consisted of several patch-like areas (70–260 μm) corresponding to the size of alveolar sacs or their clusters, Fig. 8. The responses subsided quickly after lung deflation. We have verified that with LL reagent introduced into blood capillaries, ATP-dependent luminescence signal originates exclusively from the lumen of blood vessels without cross-leakage into lung airspaces (Furuya et al., 2016). In the absence of red blood cells in the lung perfusate, released

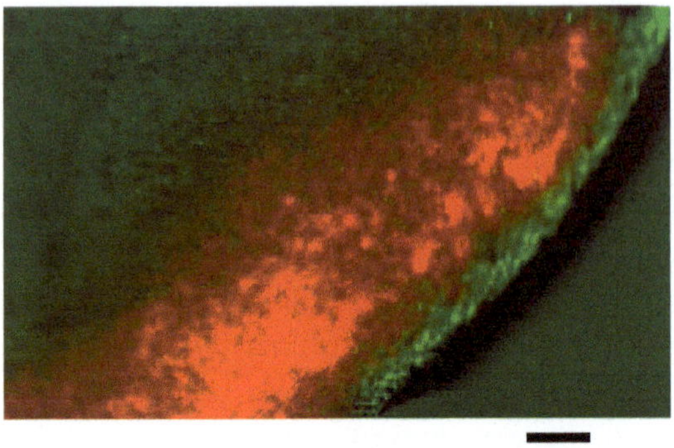

Fig. 8 Inflation-induced ATP release in pulmonary blood capillaries of ex vivo rat. The overlay image shows ATP-dependent luminescence (red) and infra-red (IR) tissue images (green). Patch-like pattern of ATP-dependent luminescence is seen after brief inflation (~30 s duration, ~10 cm H_2O). Images were acquired with Olympus MVX10 macro-view microscope with 2× objective at 0.63× zoom. *From Furuya, K., Tan, J. J., Boudreault, F., Sokabe, M., Berthiaume, Y., & Grygorczyk, R. (2016). Real-time imaging of inflation-induced ATP release in the ex vivo rat lung.* American Journal of Physiology. Lung Cellular and Molecular Physiology, 311, *L956–L969.*

ATP originates exclusively from endothelial cells. Fluid shear stress is recognized as important mechanical stimuli acting on vascular endothelial cells augmenting mitochondrial ATP generation and triggering its release (Yamamoto et al., 2011; Yamamoto, Imamura, & Ando, 2018). However, in pulmonary capillaries, significant distension of endothelial cells during lung inflation might be expected, which may also contribute to mechanosensitive release of ATP. Stretch-induced ATP release from human pulmonary microvascular endothelial cells was indeed confirmed in vitro by Ito, Furuya, Sokabe, and Hasegawa (2016). The study demonstrates that inflation induces ATP release in both the compartments—alveoli and the surrounding blood capillary network. The functional units of ATP release consist of alveolar sacs or their clusters. The findings directly confirm that lung inflation initiates purinergic signaling in the intact lung.

6. Conclusions and future directions

Mechanosensitive ATP release appears as a prevailing mode of triggering purinergic signaling cascade in the lungs. Diverse mechanical stimuli and release pathways exist. Exocytosis of ATP-loaded vesicles and several types of intracellular ATP-containing granules is a well-accepted route of cell-regulated ATP release, supported by wealth of experimental evidence. The physiological and pathophysiological roles and often basic biophysical properties of several putative ATP channels still remain unsettled and require further studies. Recent molecular identification of VRAC and Maxi-Cl channels should allow study of their structure-function relationships, better characterization of their regulation, biophysical/electrophysiological properties and potential role as ATP conduit.

Cell lysis and transient membrane injury emerge as important archaic mechanism of ATP release contributing to intercellular purinergic signaling in the lungs and other organs. Such a release may represent for some cells (e.g., RBCs) a genuine physiologically relevant mechanism that should be considered in clinical and translational studies in the field. Importantly, in some situations it may represent a dominating mode of the release surpassing by far, the contribution of cell-regulated release pathways. Therefore, better understanding of the mechanisms responsible for plasma membrane injury and repair will facilitate development of specific therapeutic approaches promoting resealing of damaged cell membranes (Andrews, Almeida, & Corrotte, 2014), or minimizing injury by increasing cell resilience to injurious stimuli, by, e.g., mechanical priming (Mikolajewicz, Mohammed, et al., 2018; Mikolajewicz, Zimmermann, et al., 2018). This is particularly important in the context of lung injury induced by mechanical ventilation and emphasizes critical role and benefits of using lung protective ventilation protocols to prevent/attenuate injury. It is also worth to consider interventions that will minimize spreading and downstream effects of released ATP in adjacent tissues. Massive release from damaged cells/tissue will result in high sustained levels of extracellular ATP that is required for P2X7 channel activation and formation of lytic pores on neighboring intact cells, consequently inducing secondary P2X7-mediated ATP release. As a protective measure, use of ATP scavengers such as apyrase has been shown to significantly improve survival of ventilated mice (Eckle et al., 2007), while antagonists of P2X7R attenuated the severity of lung injury by inhibiting NLRP3 inflammasome. These are highly promising approaches that should be further explored in future studies.

Recent bioluminescence ATP imaging studies revealed that physiologically relevant stretch induces ATP release from alveolar and endothelial cells, which could spread and exert paracrine effects on other cells at significant distances. In the intact ex vivo lung, real-time ATP visualization confirmed directly that inflation induces ATP release in airspaces and in the lumen of pulmonary blood vessels. Future technical improvements of

ATP imaging in lung airspaces should consider use of nebulized LL reagent to avoid the limitations of the LL liquid-instilled edematous lung model. Ultimately, while technically challenging, direct visualization of lung ATP release in live animals using bioluminescence imaging and intravital microscopy should be considered (Rodriguez-Tirado et al., 2016).

Reference

Abbracchio, M. P., Burnstock, G., Verkhratsky, A., & Zimmermann, H. (2009). Purinergic signalling in the nervous system: An overview. *Trends in Neurosciences*, *32*, 19–29.

Adriaensen, D., & Timmermans, J. P. (2004). Purinergic signalling in the lung: Important in asthma and COPD? *Current Opinion in Pharmacology*, *4*, 207–214.

Akopova, I., Tatur, S., Grygorczyk, M., Luchowski, R., Gryczynski, I., Gryczynski, Z., et al. (2012). Imaging exocytosis of ATP-containing vesicles with TIRF microscopy in lung epithelial A549 cells. *Purinergic Signal*, *8*, 59–70.

Aliagas, E., Munoz-Esquerre, M., Cuevas, E., Careta, O., Huertas, D., Lopez-Sanchez, M., et al. (2018). Is the purinergic pathway involved in the pathology of COPD? Decreased lung CD39 expression at initial stages of COPD. *Respiratory Research*, *19*, 103.

Anderson, C. M., Bergher, J. P., & Swanson, R. A. (2004). ATP-induced ATP release from astrocytes. *Journal of Neurochemistry*, *88*, 246–256.

Andreeva, A. V., Kutuzov, M. A., & Voyno-Yasenetskaya, T. A. (2007). Regulation of surfactant secretion in alveolar type II cells. *American Journal of Physiology. Lung Cellular and Molecular Physiology*, *293*, L259–L271.

Andrews, N. W., Almeida, P. E., & Corrotte, M. (2014). Damage control: Cellular mechanisms of plasma membrane repair. *Trends in Cell Biology*, *24*, 734–742.

Ashino, Y., Ying, X., Dobbs, L. G., & Bhattacharya, J. (2000). [Ca(2+)](i) oscillations regulate type II cell exocytosis in the pulmonary alveolus. *American Journal of Physiology. Lung Cellular and Molecular Physiology*, *279*, L5–13.

Beigi, R., Kobatake, E., Aizawa, M., & Dubyak, G. R. (1999). Detection of local ATP release from activated platelets using cell surface-attached firefly luciferase. *The American Journal of Physiology*, *276*, C267–C278.

Bischofberger, M., Gonzalez, M. R., & van der Goot, F. G. (2009). Membrane injury by pore-forming proteins. *Current Opinion in Cell Biology*, *21*, 589–595.

Boudreault, F., & Grygorczyk, R. (2004). Cell swelling-induced ATP release is tightly dependent on intracellular calcium elevations. *The Journal of Physiology*, *561*, 499–513.

Bucheimer, R. E., & Linden, J. (2004). Purinergic regulation of epithelial transport. *The Journal of Physiology*, *555*, 311–321.

Burnstock, G. (2006). Purinergic signalling. *British Journal of Pharmacology*, *147*(Suppl. 1), S172–S181.

Burnstock, G. (2008). Unresolved issues and controversies in purinergic signalling. *The Journal of Physiology*, *586*, 3307–3312.

Burnstock, G. (2009). Purinergic signalling: Past, present and future. *Brazilian Journal of Medical and Biological Research*, *42*, 3–8.

Burnstock, G., Brouns, I., Adriaensen, D., & Timmermans, J. P. (2012). Purinergic signaling in the airways. *Pharmacological Reviews*, *64*, 834–868.

Button, B., Okada, S. F., Frederick, C. B., Thelin, W. R., & Boucher, R. C. (2013). Mechanosensitive ATP release maintains proper mucus hydration of airways. *Science Signaling*, *6*, ra46.

Cereda, M., Emami, K., Xin, Y., Kadlecek, S., Kuzma, N. N., Mongkolwisetwara, P., et al. (2013). Imaging the interaction of atelectasis and overdistension in surfactant-depleted lungs. *Critical Care Medicine*, *41*, 527–535.

Chen, Z., Jin, N., Narasaraju, T., Chen, J., McFarland, L. R., Scott, M., et al. (2004). Identification of two novel markers for alveolar epithelial type I and II cells. *Biochemical and Biophysical Research Communications, 319*, 774–780.

Chiu, Y. H., Schappe, M. S., Desai, B. N., & Bayliss, D. A. (2018). Revisiting multimodal activation and channel properties of Pannexin 1. *The Journal of General Physiology, 150*, 19–39.

Colombini, M. (2012). VDAC structure, selectivity, and dynamics. *Biochimica et Biophysica Acta, 1818*, 1457–1465.

Della, L. V., Cabiati, M., Rocchiccioli, S., Del, R. S., & Morales, M. A. (2013). The role of the adenosinergic system in lung fibrosis. *Pharmacological Research, 76*, 182–189.

Di Virgilio, F., Schmalzing, G., & Markwardt, F. (2018). The elusive P2X7 macropore. *Trends in Cell Biology, 28*, 392–404.

Dietl, P., & Haller, T. (2005). Exocytosis of lung surfactant: From the secretory vesicle to the air-liquid interface. *Annual Review of Physiology, 67*, 595–621.

Dietl, P., Haller, T., & Frick, M. (2012). Spatio-temporal aspects, pathways and actions of Ca(2+) in surfactant secreting pulmonary alveolar type II pneumocytes. *Cell Calcium, 52*, 296–302.

Douillet, C. D., Robinson, W. P., III, Milano, P. M., Boucher, R. C., & Rich, P. B. (2006). Nucleotides induce IL-6 release from human airway epithelia via P2Y2 and p38 MAPK-dependent pathways. *American Journal of Physiology. Lung Cellular and Molecular Physiology, 291*, L734–L746.

Dreyfuss, D., Soler, P., Basset, G., & Saumon, G. (1988). High inflation pressure pulmonary edema. Respective effects of high airway pressure, high tidal volume, and positive end-expiratory pressure. *The American Review of Respiratory Disease, 137*, 1159–1164.

Eckle, T., Fullbier, L., Wehrmann, M., Khoury, J., Mittelbronn, M., Ibla, J., et al. (2007). Identification of ectonucleotidases CD39 and CD73 in innate protection during acute lung injury. *Journal of Immunology, 178*, 8127–8137.

Edwards, Y. S. (2001). Stretch stimulation: Its effects on alveolar type II cell function in the lung. *Comparative Biochemistry and Physiology. Part A, Molecular & Integrative Physiology, 129*, 245–260.

Etxaniz, A., Gonzalez-Bullon, D., Martin, C., & Ostolaza, H. (2018). Membrane repair mechanisms against permeabilization by pore-forming toxins. *Toxins (Basel), 10*(6), 234. https://doi.org/10.3390/toxins10060234.

Fausther, M., Pelletier, J., Ribeiro, C. M., Sevigny, J., & Picher, M. (2010). Cystic fibrosis remodels the regulation of purinergic signaling by NTPDase1 (CD39) and NTPDase3. *American Journal of Physiology. Lung Cellular and Molecular Physiology, 298*, L804–L818.

Feranchak, A. P., Lewis, M. A., Kresge, C., Sathe, M., Bugde, A., Luby-Phelps, K., et al. (2010). Initiation of purinergic signaling by exocytosis of ATP-containing vesicles in liver epithelium. *The Journal of Biological Chemistry, 285*, 8138–8147.

Fois, G., Winkelmann, V. E., Bareis, L., Staudenmaier, L., Hecht, E., Ziller, C., et al. (2018). ATP is stored in lamellar bodies to activate vesicular P2X4 in an autocrine fashion upon exocytosis. *The Journal of General Physiology, 150*, 277–291.

Furuya, K., Harada, K., & Sokabe, M. (2008). Three types of ATP-release in mammary epithelial cells revealed by ATP imaging. *Purinergic Signalling, 4*(Suppl. 5), S91.

Furuya, K., Sokabe, M., & Grygorczyk, R. (2014). Real-time luminescence imaging of cellular ATP release. *Methods, 66*, 330–344.

Furuya, K., Tan, J. J., Boudreault, F., Sokabe, M., Berthiaume, Y., & Grygorczyk, R. (2016). Real-time imaging of inflation-induced ATP release in the ex vivo rat lung. *American Journal of Physiology. Lung Cellular and Molecular Physiology, 311*, L956–L969.

Grygorczyk, R., Furuya, K., & Sokabe, M. (2013). Imaging and characterization of stretch-induced ATP release from alveolar A549 cells. *The Journal of Physiology, 591*, 1195–1215.

Grygorczyk, R., & Hanrahan, J. W. (1997a). CFTR-independent ATP release from epithelial cells triggered by mechanical stimuli. *The American Journal of Physiology, 272*, C1058–C1066.

Grygorczyk, R., & Hanrahan, J. W. (1997b). Cystic fibrosis transmembrane conductance regulator and adenosine triphosphate [response]. *Science, 275*, 1325–1326.

Grygorczyk, R., & Orlov, S. N. (2017). Effects of hypoxia on erythrocyte membrane properties-implications for intravascular hemolysis and purinergic control of blood flow. *Frontiers in Physiology, 8*, 1110.

Grygorczyk, R., Tabcharani, J. A., & Hanrahan, J. W. (1996). CFTR channels expressed in CHO cells do not have detectable ATP conductance. *The Journal of Membrane Biology, 151*, 139–148.

Hasan, D., Blankman, P., & Nieman, G. F. (2017). Purinergic signalling links mechanical breath profile and alveolar mechanics with the pro-inflammatory innate immune response causing ventilation-induced lung injury. *Purinergic Signal, 13*, 363–386.

Hasan, D., Satalin, J., van der Zee, P., Kollisch-Singule, M., Blankman, P., Shono, A., et al. (2018). Excessive extracellular ATP desensitizes P2Y2 and P2X4 ATP receptors provoking surfactant impairment ending in ventilation-induced lung injury. *International Journal of Molecular Sciences, 19*.

Hazama, A., Shimizu, T., Ando-Akatsuka, Y., Hayashi, S., Tanaka, S., Maeno, E., et al. (1999). Swelling-induced, CFTR-independent ATP release from a human epithelial cell line: Lack of correlation with volume-sensitive cl(−) channels. *The Journal of General Physiology, 114*, 525–533.

Hildebran, J. N., Goerke, J., & Clements, J. A. (1981). Surfactant release in excised rat lung is stimulated by air inflation. *Journal of Applied Physiology: Respiratory, Environmental and Exercise Physiology, 51*, 905–910.

Idzko, M., Hammad, H., van Nimwegen, M., Kool, M., Willart, M. A., Muskens, F., et al. (2007). Extracellular ATP triggers and maintains asthmatic airway inflammation by activating dendritic cells. *Nature Medicine, 13*, 913–919.

Isakson, B. E., Evans, W. H., & Boitano, S. (2001). Intercellular Ca^{2+} signaling in alveolar epithelial cells through gap junctions and by extracellular ATP. *American Journal of Physiology. Lung Cellular and Molecular Physiology, 280*, L221–L228.

Ito, S., Furuya, K., Sokabe, M., & Hasegawa, Y. (2016). Cellular ATP release in the lung and airway. *AIMS Biophysics, 3*, 571–584.

Jacobson, K. A., & Muller, C. E. (2016). Medicinal chemistry of adenosine, P2Y and P2X receptors. *Neuropharmacology, 104*, 31–49.

Jaecklin, T., Engelberts, D., Otulakowski, G., O'Brodovich, H., Post, M., & Kavanagh, B. P. (2011). Lung-derived soluble mediators are pathogenic in ventilator-induced lung injury. *American Journal of Physiology Lung Cellular and Molecular Physiology, 300*, L648–L658.

Jimenez, A. J., & Perez, F. (2017). Plasma membrane repair: The adaptable cell life-insurance. *Current Opinion in Cell Biology, 47*, 99–107.

Keller, A. S., Diederich, L., Panknin, C., DeLalio, L. J., Drake, J. C., Sherman, R., et al. (2017). Possible roles for ATP release from RBCs exclude the cAMP-mediated Panx1 pathway. *American Journal of Physiology. Cell Physiology, 313*, C593–C603.

Kreda, S. M., Okada, S. F., Van Heusden, C. A., O'neal, W., Gabriel, S., Abdullah, L., et al. (2007). Coordinated release of nucleotides and mucin from human airway epithelial Calu-3 cells. *Journal of Physiology (London), 584*, 245–259.

Kreda, S. M., Seminario-Vidal, L., Van Heusden, C. A., O'neal, W., Jones, L., Boucher, R. C., et al. (2010). Receptor-promoted exocytosis of airway epithelial mucin granules containing a spectrum of adenine nucleotides. *The Journal of Physiology, 588*, 2255–2267.

Lazarowski, E. R., & Boucher, R. C. (2009). Purinergic receptors in airway epithelia. *Current Opinion in Pharmacology*, *9*, 262–267.
Lazarowski, E. R., Boucher, R. C., & Harden, T. K. (2003). Mechanisms of release of nucleotides and integration of their action as P2X- and P2Y-receptor activating molecules. *Molecular Pharmacology*, *64*, 785–795.
Lazarowski, E. R., Sesma, J. I., Seminario-Vidal, L., & Kreda, S. M. (2011). Molecular mechanisms of purine and pyrimidine nucleotide release. *Advances in Pharmacology*, *61*, 221–261.
Lazarowski, E. R., Tarran, R., Grubb, B. R., Van Heusden, C. A., Okada, S., & Boucher, R. C. (2004). Nucleotide release provides a mechanism for airway surface liquid homeostasis. *The Journal of Biological Chemistry*, *279*, 36855–36864.
Li, C., Ramjeesingh, M., & Bear, C. E. (1996). Purified cystic fibrosis transmembrane conductance regulator (CFTR) does not function as an ATP channel. *The Journal of Biological Chemistry*, *271*, 11623–11626.
Lommatzsch, M., Cicko, S., Muller, T., Lucattelli, M., Bratke, K., Stoll, P., et al. (2010). Extracellular adenosine triphosphate and chronic obstructive pulmonary disease. *American Journal of Respiratory and Critical Care Medicine*, *181*, 928–934.
Mason, R. J., & Voelker, D. R. (1998). Regulatory mechanisms of surfactant secretion. *Biochimica et Biophysica Acta*, *1408*, 226–240.
Mikolajewicz, N., Mohammed, A., Morris, M., & Komarova, S. V. (2018). Mechanically stimulated ATP release from mammalian cells: Systematic review and meta-analysis. *Journal of Cell Science*, *131*.
Mikolajewicz, N., Zimmermann, E. A., Willie, B. M., & Komarova, S. V. (2018). Mechanically stimulated ATP release from murine bone cells is regulated by a balance of injury and repair. *eLife*, 7, 1–23.
Mishra, A. (2013). New insights of P2X7 receptor signaling pathway in alveolar functions. *Journal of Biomedical Science*, *20*, 26.
Mortaz, E., Braber, S., Nazary, M., Givi, M. E., Nijkamp, F. P., & Folkerts, G. (2009). ATP in the pathogenesis of lung emphysema. *European Journal of Pharmacology*, *619*, 92–96.
Mortaz, E., Folkerts, G., Nijkamp, F. P., & Henricks, P. A. (2010). ATP and the pathogenesis of COPD. *European Journal of Pharmacology*, *638*, 1–4.
Muller, T., Fay, S., Vieira, R. P., Karmouty-Quintana, H., Cicko, S., Ayata, C. K., et al. (2017). P2Y6 receptor activation promotes inflammation and tissue remodeling in pulmonary fibrosis. *Frontiers in Immunology*, *8*, 1028.
Murata, N., Ito, S., Furuya, K., Takahara, N., Naruse, K., Aso, H., et al. (2014). Ca^{2+} influx and ATP release mediated by mechanical stretch in human lung fibroblasts. *Biochemical and Biophysical Research Communications*, *453*, 101–105.
Nickles, H. T., Sumkauskaite, M., Wang, X., Wegner, I., Puderbach, M., & Kuebler, W. M. (2014). Mechanical ventilation causes airway distension with proinflammatory sequelae in mice. *American Journal of Physiology. Lung Cellular and Molecular Physiology*, *307*, L27–L37.
Novak, I. (2011). Purinergic signalling in epithelial ion transport: Regulation of secretion and absorption. *Acta Physiologica (Oxford, England)*, *202*, 501–522.
Okada, S. F., Nicholas, R. A., Kreda, S. M., Lazarowski, E. R., & Boucher, R. C. (2006). Physiological regulation of ATP release at the apical surface of human airway epithelia. *The Journal of Biological Chemistry*, *281*, 22992–23002.
Okada, Y., Okada, T., Islam, M. R., & Sabirov, R. Z. (2018). Molecular identities and ATP release activities of two types of volume-regulatory anion channels, VSOR and Maxi-Cl. *Current Topics in Membranes*, *81*, 125–176.
Osei-Owusu, J., Yang, J., Vitery, M. D. C., & Qiu, Z. (2018). Molecular biology and physiology of volume-regulated anion channel (VRAC). *Current Topics in Membranes*, *81*, 177–203.

Paradiso, A. M., Ribeiro, C. M., & Boucher, R. C. (2001). Polarized signaling via purinoceptors in normal and cystic fibrosis airway epithelia. *The Journal of General Physiology*, *117*, 53–67.

Patel, A. S., Reigada, D., Mitchell, C. H., Bates, S. R., Margulies, S. S., & Koval, M. (2005). Paracrine stimulation of surfactant secretion by extracellular ATP in response to mechanical deformation. *American Journal of Physiology. Lung Cellular and Molecular Physiology*, *289*, L489–L496.

Pelleg, A., Schulman, E. S., & Barnes, P. J. (2016). Extracellular adenosine 5′-triphosphate in obstructive airway diseases. *Chest*, *150*, 908–915.

Pellegatti, P., Falzoni, S., Pinton, P., Rizzuto, R., & Di Virgilio, F. (2005). A novel recombinant plasma membrane-targeted luciferase reveals a new pathway for ATP secretion. *Molecular Biology of the Cell*, *16*, 3659–3665.

Picher, M., & Boucher, R. C. (2011). *Purinergic regulation of respiratory diseases*. Springer Netherlands.

Pintor, J. (2018). Light-induced ATP release from the lens. *Purinergic Signal*, *14*, 499–504.

Ponomarchuk, O., Boudreault, F., Orlov, S. N., & Grygorczyk, R. (2017). Mechanosensitive ATP release involves a non-conductive pathway: Evidence from large field of view real-time imaging. *Biophysical Journal*, *112*(Suppl. 1), 311A.

Praetorius, H. A., & Leipziger, J. (2009). ATP release from non-excitable cells. *Purinergic Signal*, *5*, 433–446.

Qiu, Z., Dubin, A. E., Mathur, J., Tu, B., Reddy, K., Miraglia, L. J., et al. (2014). SWELL1, a plasma membrane protein, is an essential component of volume-regulated anion channel. *Cell*, *157*, 447–458.

Rajendran, M., Dane, E., Conley, J., & Tantama, M. (2016). Imaging adenosine triphosphate (ATP). *The Biological Bulletin*, *231*, 73–84.

Ramminger, S. J., Collett, A., Baines, D. L., Murphie, H., McAlroy, H. L., Olver, R. E., et al. (1999). P2Y2 receptor-mediated inhibition of ion transport in distal lung epithelial cells. *British Journal of Pharmacology*, *128*, 293–300.

Ramsingh, R., Grygorczyk, A., Solecki, A., Cherkaoui, L. S., Berthiaume, Y., & Grygorczyk, R. (2011). Cell deformation at the air-liquid interface induces Ca^{2+}-dependent ATP release from lung epithelial cells. *American Journal of Physiology. Lung Cellular and Molecular Physiology*, *300*, L587–L595.

Reddy, M. M., Quinton, P. M., Haws, C., Wine, J. J., Grygorczyk, R., Tabcharani, J. A., et al. (1996). Failure of the cystic fibrosis transmembrane conductance regulator to conduct ATP. *Science*, *271*, 1876–1879.

Rich, P. B., Douillet, C. D., Mahler, S. A., Husain, S. A., & Boucher, R. C. (2003). Adenosine triphosphate is released during injurious mechanical ventilation and contributes to lung edema. *The Journal of Trauma*, *55*, 290–297.

Rodriguez-Tirado, C., Kitamura, T., Kato, Y., Pollard, J. W., Condeelis, J. S., & Entenberg, D. (2016). Long-term high-resolution intravital microscopy in the lung with a vacuum stabilized imaging window. *Journal of Visualized Experiments*, 1–13.

Sabirov, R. Z., Merzlyak, P. G., Islam, M. R., Okada, T., & Okada, Y. (2016). The properties, functions, and pathophysiology of maxi-anion channels. *Pflügers Archiv*, *468*, 405–420.

Sabirov, R. Z., Merzlyak, P. G., Okada, T., Islam, M. R., Uramoto, H., Mori, T., et al. (2017). The organic anion transporter SLCO2A1 constitutes the core component of the Maxi-Cl channel. *The EMBO Journal*, *36*, 3309–3324.

Sabirov, R. Z., & Okada, Y. (2005). ATP release via anion channels. *Purinergic Signal*, *1*, 311–328.

Sadananda, P., Kao, F. C., Liu, L., Mansfield, K. J., & Burcher, E. (2012). Acid and stretch, but not capsaicin, are effective stimuli for ATP release in the porcine bladder mucosa: Are ASIC and TRPV1 receptors involved? *European Journal of Pharmacology*, *683*, 252–259.

Seminario-Vidal, L., Kreda, S., Jones, L., O'neal, W., Trejo, J., Boucher, R. C., et al. (2009). Thrombin promotes release of ATP from lung epithelial cells through coordinated activation of rho- and Ca^{2+}-dependent signaling pathways. *The Journal of Biological Chemistry, 284*, 20638–20648.

Shigematsu, T., Koshiyama, K., & Wada, S. (2015). Effects of stretching speed on mechanical rupture of phospholipid/cholesterol bilayers: Molecular dynamics simulation. *Scientific Reports, 5*, 15369.

Sikora, J., Orlov, S. N., Furuya, K., & Grygorczyk, R. (2014). Hemolysis is a primary ATP-release mechanism in human erythrocytes. *Blood, 124*, 2150–2157.

Skotak, M., Wang, F., & Chandra, N. (2012). An in vitro injury model for SH-SY5Y neuroblastoma cells: Effect of strain and strain rate. *Journal of Neuroscience Methods, 205*, 159–168.

Stutts, M. J., Fitz, J. G., Paradiso, A. M., & Boucher, R. C. (1994). Multiple modes of regulation of airway epithelial chloride secretion by extracellular ATP. *The American Journal of Physiology, 267*, C1442–C1451.

Sugiura, M., McCulloch, P. R., Wren, S., Dawson, R. H., & Froese, A. B. (1994). Ventilator pattern influences neutrophil influx and activation in atelectasis-prone rabbit lung. *Journal of Applied Physiology (Bethesda, MD: 1985)*, 77, 1355–1365.

Takahara, N., Ito, S., Furuya, K., Naruse, K., Aso, H., Kondo, M., et al. (2014). Real-time imaging of ATP release induced by mechanical stretch in human airway smooth muscle cells. *American Journal of Respiratory Cell and Molecular Biology, 51*, 772–782.

Tarran, R. (2004). Regulation of airway surface liquid volume and mucus transport by active ion transport. *Proceedings of the American Thoracic Society, 1*, 42–46.

Tarran, R., Button, B., & Boucher, R. C. (2006). Regulation of normal and cystic fibrosis airway surface liquid volume by phasic shear stress. *Annual Review of Physiology, 68*, 543–561.

Taruno, A. (2018). ATP release channels. *International Journal of Molecular Sciences, 19*.

Taruno, A., Vingtdeux, V., Ohmoto, M., Ma, Z., Dvoryanchikov, G., Li, A., et al. (2013). CALHM1 ion channel mediates purinergic neurotransmission of sweet, bitter and umami tastes. *Nature, 495*, 223–226.

Tatur, S., Groulx, N., Orlov, S. N., & Grygorczyk, R. (2007). Ca^{2+}-dependent ATP release from A549 cells involves synergistic autocrine stimulation by coreleased uridine nucleotides. *The Journal of Physiology, 584*, 419–435.

Tatur, S., Kreda, S., Lazarowski, E., & Grygorczyk, R. (2008). Calcium-dependent release of adenosine and uridine nucleotides from A549 cells. *Purinergic Signal, 4*, 139–146.

Thomas, S. L. (2014). Intravascular hemolysis: The sacrifice of few.... *Blood, 124*, 2011–2012.

Tschumperlin, D. J., Oswari, J., & Margulies, A. S. (2000). Deformation-induced injury of alveolar epithelial cells. Effect of frequency, duration, and amplitude. *American Journal of Respiratory and Critical Care Medicine, 162*, 357–362.

Uhlig, U., & Uhlig, S. (2011). Ventilation-induced lung injury. *Comprehensive Physiology, 1*, 635–661.

Verbrugge, S. J., de Jong, J. W., Keijzer, E., Vazquez de Anda, G., & Lachmann, B. (1999). Purine in bronchoalveolar lavage fluid as a marker of ventilation-induced lung injury. *Critical Care Medicine, 27*, 779–783.

Voss, F. K., Ullrich, F., Munch, J., Lazarow, K., Lutter, D., Mah, N., et al. (2014). Identification of LRRC8 heteromers as an essential component of the volume-regulated anion channel VRAC. *Science, 344*, 634–638.

Watt, W. C., Lazarowski, E. R., & Boucher, R. C. (1998). Cystic fibrosis transmembrane regulator-independent release of ATP. Its implications for the regulation of P2Y2 receptors in airway epithelia. *The Journal of Biological Chemistry, 273*, 14053–14058.

Workman, A. D., Carey, R. M., Chen, B., Saunders, C. J., Marambaud, P., Mitchell, C. H., et al. (2017). CALHM1-mediated ATP release and ciliary beat frequency modulation in nasal epithelial cells. *Scientific Reports*, *7*, 6687.

Yamamoto, K., & Ando, J. (2018). Emerging role of plasma membranes in vascular endothelial mechanosensing. *Circulation Journal*, *82*, 2691–2698.

Yamamoto, K., Furuya, K., Nakamura, M., Kobatake, E., Sokabe, M., & Ando, J. (2011). Visualization of flow-induced ATP release and triggering of Ca^{2+} waves at caveolae in vascular endothelial cells. *Journal of Cell Science*, *124*, 3477–3483.

Yamamoto, K., Imamura, H., & Ando, J. (2018). Shear stress augments mitochondrial ATP generation that triggers ATP release and Ca(2+) signaling in vascular endothelial cells. *American Journal of Physiology. Heart and Circulatory Physiology*, *315*, H1477–H1485.

Zimmermann, H. (2000). Extracellular metabolism of ATP and other nucleotides. *Naunyn-Schmiedeberg's Archives of Pharmacology*, *362*, 299–309.

Zimmermann, H. (2016). Extracellular ATP and other nucleotides-ubiquitous triggers of intercellular messenger release. *Purinergic Signal*, *12*, 25–57.

Zimmermann, H., Zebisch, M., & Strater, N. (2012). Cellular function and molecular structure of ecto-nucleotidases. *Purinergic Signal*, *8*, 437–502.

CHAPTER FOUR

Structural model of a2-subunit N-terminus and its binding interface for Arf-GEF CTH2: Implication for regulation of V-ATPase, CTH2 function and rational drug design

Vladimir Marshansky[a,*], Hiroyuki Hosokawa[b], Maria Merkulova[b], Anastasia Bakulina[c], Phat Vinh Dip[d,†], Youg Raj Thaker[d,‡], Ajay Bjargava[e], James R. Tonra[f], Dennis A. Ausiello[g], Gerhard Grüber[d]
[a]Neuro-Horizon Pharma, Sharon, MA, United States
[b]Center for Systems Biology, Program in Membrane Biology, MGH, Boston, MA, United States
[c]Novosibirsk State University, Novosibirsk, Russia
[d]Nanyang Technological University, School of Biological Sciences, Singapore, Republic of Singapore
[e]The Francis Crick Institute, London, United Kingdom
[f]BeyondSpring Pharmaceuticals, New York, NY, United States
[g]Center for Assessment for Technology and Continuous Health, MGH, Boston, MA, United States
*Corresponding author: e-mail addresses: vlad@neuro-horizonpharma.com; vlad@neurohorizonpharma.com

Contents

† Current address: Department of Biology, Massachusetts Institute of Technology, Cambridge, MA, United States.
‡ Current address: School of Biological Science, Protein Structure and Disease Mechanisms, University of Essex, Colchester, United Kingdom.

Current Topics in Membranes, Volume 83
ISSN 1063-5823
https://doi.org/10.1016/bs.ctm.2019.01.008

Abstract

We have previously identified the interaction between mammalian V-ATPase a2-subunit isoform and cytohesin-2 (CTH2) and studied molecular details of binding between these proteins. In particular, we found that six peptides derived from the N-terminal cytosolic domain of a2 subunit ($a2N_{1-402}$) are involved in interaction with CTH2 (Merkulova, Bakulina, Thaker, Grüber, & Marshansky, 2010). However, the actual 3D binding interface was not determined in that study due to the lack of high-resolution structural information about a-subunits of V-ATPase. Here, using a combination of homology modeling and NMR analysis, we generated the structural model of complete $a2N_{1-402}$ and uncovered the CTH2-binding interface. First, using the crystal-structure of the bacterial *M. rubber* I_{cyt}-subunit of A-ATPase as a template (Srinivasan, Vyas, Baker, & Quiocho, 2011), we built a homology model of mammalian $a2N_{1-352}$ fragment. Next, we combined it with the determined NMR structures of peptides $a2N_{368-395}$ and $a2N_{386-402}$ of the C-terminal section of $a2N_{1-402}$. The complete molecular model of $a2N_{1-402}$ revealed that six CTH2 interacting peptides are clustered in the distal and proximal lobe sub-domains of $a2N_{1-402}$. Our data indicate that the proximal lobe sub-domain is the major interacting site with the Sec7 domain of first CTH2 protein, while the distal lobe sub-domain of $a2N_{1-402}$ interacts with the PH-domain of second CTH2. Indeed, using Sec7/Arf-GEF activity assay we experimentally confirmed our model. The interface formed by peptides $a2N_{1-17}$ and $a2N_{35-49}$ is involved in specific interaction with Sec7 domain and regulation of GEF activity. These data are critical for understanding of the cross-talk between V-ATPase and CTH2 as well as for the rational drug design to regulate their function.

1. Introduction

The V-ATPases are ubiquitous proton pumps, that use the energy of ATP to translocate protons from cytosol to intracellular compartments or extracellular space. V-ATPases maintain pH homeostasis at the cellular and the whole organism level, and also play a critical role in cellular function via direct interaction with a variety of proteins, whose functions are unrelated to pH homeostasis. The direct and indirect roles of V-ATPase were previously reviewed by us (Marshansky & Futai, 2008; Marshansky, Rubinstein, & Grüber, 2014) and others (Forgac, 2007; Hinton, Bond, & Forgac, 2009).

V-ATPases are very complex multi-subunit enzymes that function as proton-pumping rotary nano-motors (Marshansky et al., 2014; Marshansky & Futai, 2008). Functional V-ATPases are composed of two parts: a cytoplasmic V_1- and a transmembrane V_O-sector, which may dissociate from each other in response to some stimuli. This dissociation results in reduced ATPase activity and a shutting down of proton translocation of V-ATPase, and is one of the main mechanisms of down-regulation of V-ATPase function (Marshansky et al., 2014; Marshansky & Futai, 2008). Each of the two sectors are composed of multiple different subunits. In mammalian cells eight different proteins are combined in the following stoichiometry $A_3B_3C_1D_1E_3F_1G_3H_1$ to form a V_1-sector; while at least six different proteins $a_1c_5c''_1d_1e_1Ac45_1$ form the transmembrane V_O-sector. Proton pumping across the membrane occurs via coupling of ATP-hydrolysis with rotary-mechanism of protons translocation in the interface between the rotating c-ring and the stationary a-subunit of V-ATPase. The rotation of the c-ring is driven by ATP hydrolysis catalyzed by the stationary A_3B_3 headpiece of the V_1-sector. In order to accomplish an efficient transfer of ATP hydrolysis energy to c-ring rotation, the rotation of A_3B_3 headpiece is prevented by three peripheral stalks (Zhang et al., 2008). Each of these stalks is composed by a G/E heterodimer, which anchors the A_3B_3 headpiece to the membrane through either: (i) direct binding to the N-terminal tail of a-subunit or (ii) indirectly through subunit C. In particular, while two of three G/*E*-stalks are directly bound to a-subunit, the third G/E-stalk is bound to subunit C, which itself interacts simultaneously with a-subunit and the second G/E-peripheral stalk (Oot, Couoh-Cardel, Sharma, & Wilkens, 2017; Oot & Wilkens, 2012; Zhang et al., 2008). It is noteworthy that during the disassembly of V_1-sector from V_O-sector, the protein-protein interactions between a-subunit and G/E-stalks as well as C-subunit and G/E-stalk are destabilized through a yet unknown molecular mechanism (Oot et al., 2017).

In mammals there are four a-subunit isoforms, (a1, a2, a3 and a4) that contain two major domains: a cytosolic N-terminal domain (aN ~ 400 aa) and a membrane-integrated C-terminal domain (aC ~ 400 aa), containing eight transmembrane spanning helices (Marshansky, 2007; Toei, Toei, & Forgac, 2011). While a-subunit isoforms are highly homologous to one another, nevertheless they perform non-redundant functions. Initially we found, that the cytosolic N-terminal domain of a2-subunit ($a2N_{1-402}$) directly interacts with cytohesin-2 (CTH2), which acts as Arf-GEF (GDP/GTP exchange factor) and activator of Arf-family small GTPases. Since CTH2 and Arf small GTPases are the key regulators of receptors signaling, vesicular trafficking and actin cytoskeleton rearrangement (Hurtado-Lorenzo et al., 2006; Merkulova, Bakulina, Thaker, Grüber, & Marshansky, et al., 2010), these findings provided new insights into the functional link between V-ATPase dependent organellar acidification and vesicular trafficking. Importantly, our recent study revealed that the other three a-subunit isoforms (a1, a3 and a4) also bind to CTH2, suggesting its ubiquitous nature and cell biological significance of signaling between V-ATPase and CTH2 (Merkulova et al., 2011).

Cytohesin-2 (CTH2, also known as ARNO) together with cytohesins-1, -3 and -4, belongs to cytohesin subfamily of Arf-GEFs, activators of Arf small GTPases (Casanova, 2007). Cytohesins are highly homologous proteins composed of four distinct structural domains: (i) an N-terminal coiled-coil domain; (ii) a central Sec7-domain; (iii) a pleckstrin homology (PH) domain; and (iv) a C-terminal polybasic domain. In our previous study we addressed the molecular details of binding between CTH2 and $a2N_{1-402}$ (Merkulova et al., 2010) and uncovered very complex interactions between these proteins. We found multiple binding sites within the Sec7 domain of CTH2 as the strongest interaction sites with a-subunit of V-ATPase (Merkulova et al., 2010). Moreover, homology modeling of CTH2 performed in our study also suggested the multisite binding and complex character of interactions between these two proteins (Merkulova et al., 2010). Indeed, pull-down experiments using synthetic peptides demonstrated that six $a2N_{1-402}$ specific peptides $a2N_{1-17}$, $a2N_{35-49}$, $a2N_{198-214}$, $a2N_{215-230}$, $a2N_{313-331}$ and $a2N_{386-402}$ are involved in the interaction with full-length CTH2 (Merkulova et al., 2010). Intriguingly, these peptide motifs were not clustered but instead randomly distributed throughout the sequence of $a2N_{1-402}$ of V-ATPase (Merkulova et al., 2010). Additional pull-down and surface plasmon resonance experiments with purified domains of CTH2 revealed that peptide $a2N_{1-17}$ interacted specifically and strongly with Sec7 domain, while peptide $a2N_{198-214}$ interacted with PH domain (Merkulova et al., 2010). This data suggested that the epitope formed by $a2N_{1-17}$ is most likely involved in interaction with Sec7 domain, while the epitope formed by $a2N_{198-214}$ is probably part of a different binding site, involving the PH-domain of CTH2. However, this hypothesis could not be verified due to the lack of structural features of the $a2N_{1-402}$ subunit of V-ATPase.

Previously, a high-resolution crystal structure of cytosolic N-terminus of I-subunit (I_{cyt}) of *Meiothermus ruber* A-ATP synthase was solved (Srinivasan, Vyas, Baker, & Quiocho, 2011). This protein is homologous to yeast Vph1 and mammalian a-subunit isoforms of V-ATPase. Overall, I_{cyt} is composed of a curved long central α-helical bundle capped on both ends by two lobes with similar α/β architecture that shows remarkable structural similarity with yeast C-subunit of V-ATPase (Srinivasan et al., 2011). The crystal structure of I_{cyt} also fits well into the corresponding place in an electron microscopy density map of intact A-ATP synthase from *Thermus thermophilus* bacteria (Lau & Rubinstein, 2012; Srinivasan et al., 2011). Importantly, extensive analysis of the multiple alignments of all currently known a-subunits of V-ATPase demonstrated the evolutionarily conserved character of these proteins and their structures. This in turn opens the possibility to generate homology models of N-terminal cytosolic domains of a-subunit isoforms from other species including mammals.

Here, we used the crystal structure of I_{cyt} from *Meiothermus ruber* to build a homology model of mouse a2-subunit ($a2N_{1–352}$) of V-ATPase. Moreover, we also solved the NMR structures of two synthetic overlapping peptides derived from $a2N_{339–402}$ in order to determine the structural features of this unknown part of the protein and to build a structural model of the entire $a2N_{1–402}$ cytosolic tail. Using this model, we determined the distribution of CTH2 interacting epitopes clustered in two spatially separate binding regions on 3D structural model of $a2N_{1–402}$. We also outlined and analyzed a *G*/*E*-stalks binding interfaces on $a2N_{1–402}$ based on previously published data from the Forgac laboratory (Qi & Forgac, 2008). Our data suggest the complex interaction between a2-subunit V-ATPase and CTH2 that may modulate binding between a2-subunit and G/E-subunits of the peripheral stalks. We propose this interplay between these interactions as a molecular mechanism of cross-talk between trans-membrane V-ATPase complex, cytosolic CTH2 and Arf small GTPases.

2. Results

2.1 The structure of *Meiothermus ruber* I_{cyt} subunit as a template for mouse $a2N_{1–352}$ homology model

The first high-resolution structure of A-ATP synthase *M. ruber* I_{cyt} has been solved previously (Srinivasan et al., 2011), which opened the possibility to generate homology models of cytosolic N-terminal domains of a-subunits from other species. We performed multiple amino acid sequence alignment of a-subunit homologues from various species in order to assess a general overview of a-subunit evolution. This analysis revealed that high level conservation of amino acid sequences of cytosolic N-terminal tail of a-subunits from different species is not required for preservation of the overall secondary and tertiary structures. Thus, we concluded that I_{cyt} could be used as a template to build a reliable model of 3D structure of the mouse a2N protein.

2.2 The homology model of $a2N_{1-352}$ structure

The homology modeling of $a2N_{1-352}$ was performed using the crystal structure of I_{cyt} from *M. ruber* (PDB ID: 2RRK) as a template as described in Methods. The crystal structure of *M. rubber* I_{cyt} represents the amino acids 1–301 (I_{cyt_1-301}), whereby the C-terminal 50 amino acids of I_{cyt} including the residues 302–351 ($I_{cyt\_302-351}$) remained unresolved (Srinivasan et al., 2011). Since I_{cyt_1-301} is an independently folded structural unit, it most likely represents the separate domain of *M. ruber* I subunit. In the mouse a2N sequence this separate domain corresponds to $a2N_{1-352}$ region (Fig. 1). The homology-modeled structure of mouse $a2N_{1-352}$ adopts the same overall shape of a barbell as its I_{cyt} template and contains the same three major structural elements (Fig. 2A–C, Supplementary Movie S1 in the online version at https://doi.org/10.1016/bs.ctm.2019.01.008). The domain composition of $a2N_{1-402}$ is shown in Fig. 2A and the color and amino acids composition of the structural elements also correspond to the ribbon structures shown in Fig. 2B and C, respectively. The proximal lobe (*PL*) with the mixed α/β architecture is formed by two distinct amino acid regions 1–42 aa (*PL*, red) and 322–352 aa (*PL*, red). The distal lobe (*DL*) with the similar mixed architecture is formed by a single continuous amino acid stretch 173–244 aa (*DL*, blue). The central bar domain (*BD*) is also formed by two distinct amino acid regions 43–172 aa (*BD*, green) and 245–321 aa (*BD*, green) which contains two anti-parallel rows of long α-helices, that link two lobes together (Figs. 1 and 2B and C). The proximal lobe consists of four anti-parallel β-strands (Fig. 1, strands 1, 2, 6 and 7) and two α-helices (Fig. 1, helices I and XI) that fold into two-layer α/β sandwich. The strands-2 and 7 are not visible in the model since these regions assigned by PyMOL as loops, Swiss-PDB Viewer (Guex, Peitsch, & Schwede, 2009) assigns β-strands for these regions, while PsiPred predicts β-strand for strand-2 (Fig. 1, in blue). Thus we suggest that this is not a significant difference between the model and the template. The distal lobe is composed

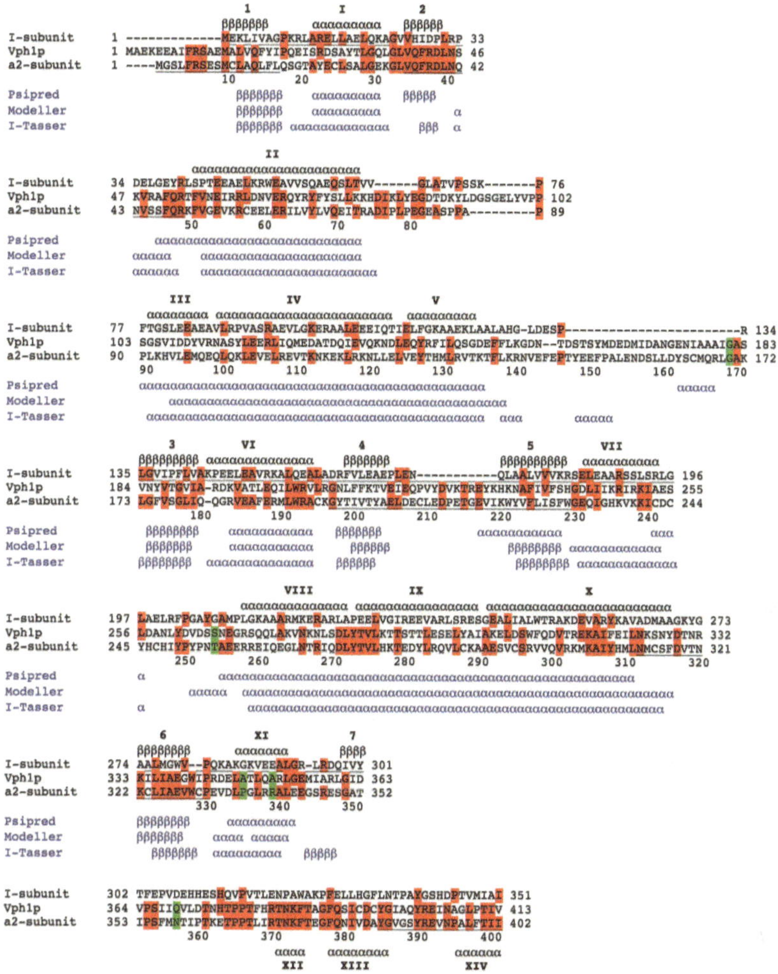

Fig. 1 Amino acid sequence alignment of the cytosolic N-terminal regions from bacterial *M. ruber* subunit I, fungal *S. cerevisiae* Vph1p and mammalian *M. musculus* a2-subunit isoform. The alignment is performed as described in Methods. Identical residues are highlighted in red. The secondary structures of α-helices (α) and β-strands (β) on top in black are from bacterial I_{cyt} crystal structure. The secondary structures of mouse $a2N_{1–402}$, predicted by PsiPred, or assigned by PyMol for Modeler and I-Tasser models are on the bottom in blue. These secondary structures look similar to each other and to the secondary structure of I_{cyt}, however while two long continuous α-helices are present in a2N model, the corresponding two regions in I_{cyt} each contains three α-helices (III, IV, V and VIII, IX, X) disrupted by kinks. The α-helices in the very bottom row are from NMR solution structures of peptides $a2N_{368–395}$ and $a2N_{386–402}$. Five residues of Vph1 including G_{181}, S_{266}, A_{347}, A_{351}, and Q_{369} (Qi & Forgac, 2008) and the corresponding mouse residues G_{170}, T_{255}, P_{336}, R_{340} and N_{358} are highlighted in green. Cytohesin-2 interacting peptides $a2N_{1–17}$, $a2N_{35–49}$, $a2N_{198–214}$, $a2N_{215–230}$, $a2N_{313–331}$ and $a2N_{386–402}$ are underlined on $a2N_{1–402}$ sequence, while discontinuous proximal $I_{cyt\_1–33}$ and $I_{cyt\_274–301}$ and continuous $I_{cyt\_135–196}$ distal lobe regions are underlined on I_{cyt} sequence. They also are shown as separate stack of amino acid sequences. Numbers below a2-subunit lines are from a2N amino acid sequence. The very bottom stack of amino acid sequences ($a2N_{353–402}$ in a2N) is an appendage region of these proteins with previously unknown structure.

of three antiparallel β-strands (Fig. 1, strands 3, 4 and 5) and two α-helices (Fig. 1, helices VI and VII) that fold into similar two-layer α/β sandwich. The "extra" 51 residues of a2N are predominantly located in very N-terminus (additional 9 aa) and contribute to the formation of the longer loops between α-helices II and III, α-helix V and β-strand 3, and between β-strands 4 and 5 (Figs. 1 and 2). Thus the overall structure of $a2N_{1-352}$ model is very similar to the crystal structure of I_{cyt}.

2.3 The NMR structure of $a2N_{353-402}$ region

The cytosolic N-terminal tail of mouse a2-subunit ($a2N_{1-402}$) has been predicted to span 1–402 aa (Fig. 1) (Merkulova et al., 2010). Thus the homology model of $a2N_{1-352}$ is not a complete part of the cytosolic N-terminal tail, since the last 50 aa ($a2N_{353-402}$) could not been modeled using I_{cyt_1-301} as a template. Since, this part of $a2N_{1-402}$ is connecting the rest of the tail to its predicted first transmembrane domain, we called this part of the protein the stem domain (*SD*) (Figs. 2A and 3A, 353–402 aa in white). In addition, the region 386–402 of mouse V-ATPase subunit a2-isoform has been shown to be essential for binding to CTH2 Arf GEF small GTPase (Merkulova et al., 2010). Here were synthesized the overlapping peptides $a2N_{339-395}$ and $a2N_{386-402}$, which were derived from the $a2N_{339-402}$ region (Fig. 3A, in cyan). The solution structure of entire $a2N_{386-402}$ (in cyan) and part of $a2N_{339-395}$ (in cyan), shown as $a2N_{368-395}$ (in gray), were solved by NMR spectroscopy. Amino acids of the peptides were sequentially assigned using both NOESY and TOCSY data. 2D TOCSY and 2D NOESY raw data was processed using in-built Topspin software (Bruker). Fig. 4A–B shows the assigned NH region of the 2D NOESY spectrum of $a2N_{339-395}$ and $a2N_{386-402}$ (Fig. 3A, in cyan). Primary sequence amino acid marking was followed by the assignment of cross peaks by overlaying 2D TOCSY and 2D NOESY spectra (Fig. 4A–B). The secondary structure elements of $a2N_{339-395}$ were analyzed based on $^1H\alpha$ chemical shifts with respect to the random coil values. In case of $a2N_{339-395}$ the data indicate an α-helical formation between the amino acids N_{373}–Y_{386} (Fig. 4C). Identified cross peaks in the HN-HN region are shown in Fig. 4E, indicating α-helical features of $a2N_{339-395}$. HN–HN, Hα–HN(i, i+3), Hα–HN(i, i+4), and Hα–Hβ (i, i+3) connectivities were plotted from the assigned NOESY spectrum (Fig. 4E). Hα–HN(i, i+3), indicating α-helical formation between the residues G_{378} and Y_{387}, whereby

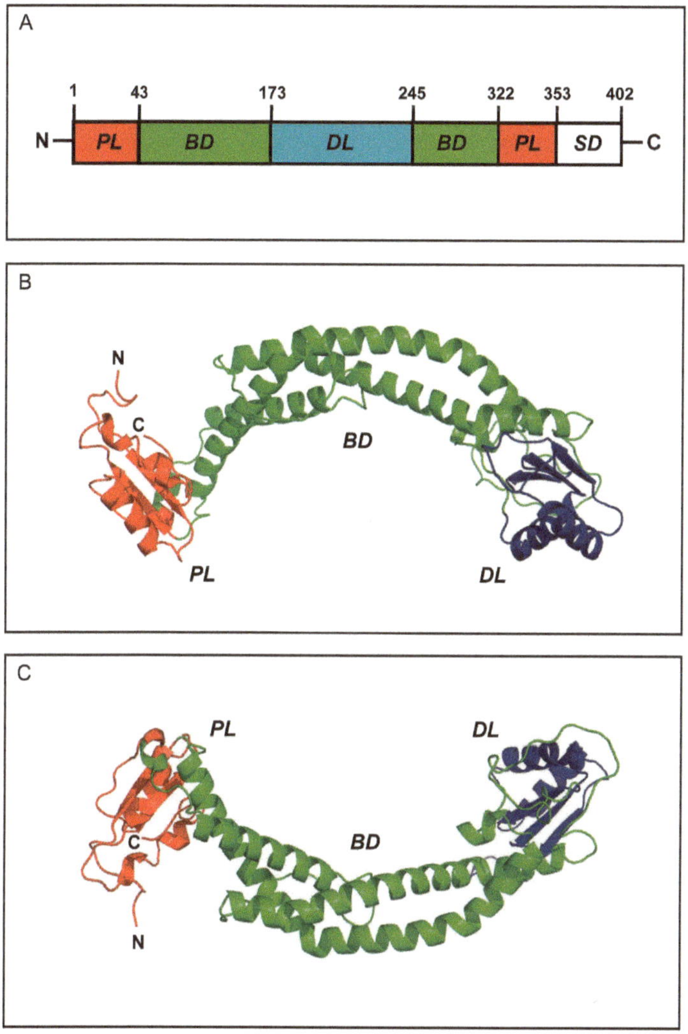

Fig. 2 Structural homology model of a2N$_{1-352}$. (A) Schematic representation of structural domains of a2N$_{1-402}$. These structures are indicated as follows: (i) proximal lobe (*PL*) formed by 1–42 aa and by 322–352 aa in red; (ii) bar domain (*BD*) formed by 43–172 aa and 245–321 aa in green; (iii) distal lobe (*DL*) formed by 173–244 aa in blue; and (v) stem domain (*SD*) formed by 353–402 aa in white. (B, C) Ribbon representations of a homology model of a2N$_{1-352}$. Structural domains are colored as above. It should be noted that stem domain is not present in this model. (B) Bottom view from the membrane toward V$_1$-sector. (C) Top view from the V$_1$-sector toward the membrane. This view is rotated by 180° vertically relative to the view in B.

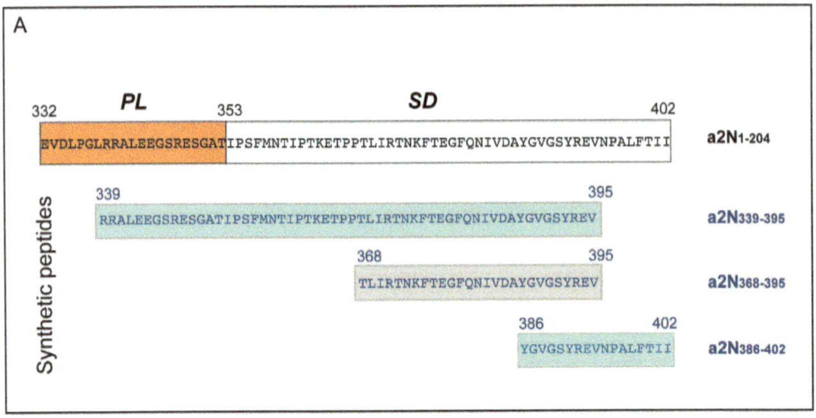

B

	$a2N_{368\text{-}395}$	$a2N_{386\text{-}402}$
Distance restraints		
Total	235	175
Intraresidue (i – j = 0)	66	45
Sequential (|i-j| = 1)	77	79
Medium-range (2 ≤ |i–j| ≤ 4)	92	51
Long-range (|*i*–*j*| ≥ 5)	0	0
Average number of violations		
Distance violations > 5 Å	0	0
Ramachandran plot[2] (%)		
Residues in most favored regions	92.2	59.7
Residues in additionally allowed regions	7.8	31.9
Residues in generously allowed regions	0.0	8.3
Residues in disallowed regions	0.0	0.0
Average RMSD to Mean (Å)		
Residues 378-387, Backbone (C^{α}, C′, and N)	0.297	0.68

Fig. 3 Synthesis and NMR analysis of peptides overlapping and corresponding to stem domain of $a2N_{1–402}$. (A) The C-terminal amino acid sequence of $a2N_{1–402}$. The end of proximal lobe (PL) domain (332–353 aa) is shown in peach and stem domain (SD) is in white. Overlapping synthetic peptides $a2N_{339–395}$ and $a2N_{386–402}$ are shown in cyan. The C-terminal part of $a2N_{339–395}$ peptide (called $a2N_{368–395}$), for which structure was solved by NMR spectroscopy is shown in gray. (B) Table of structural statistics for the amino acids of $a2N_{368–395}$ and $a2N_{386–402}$ peptides determined by NMR analysis in solution.

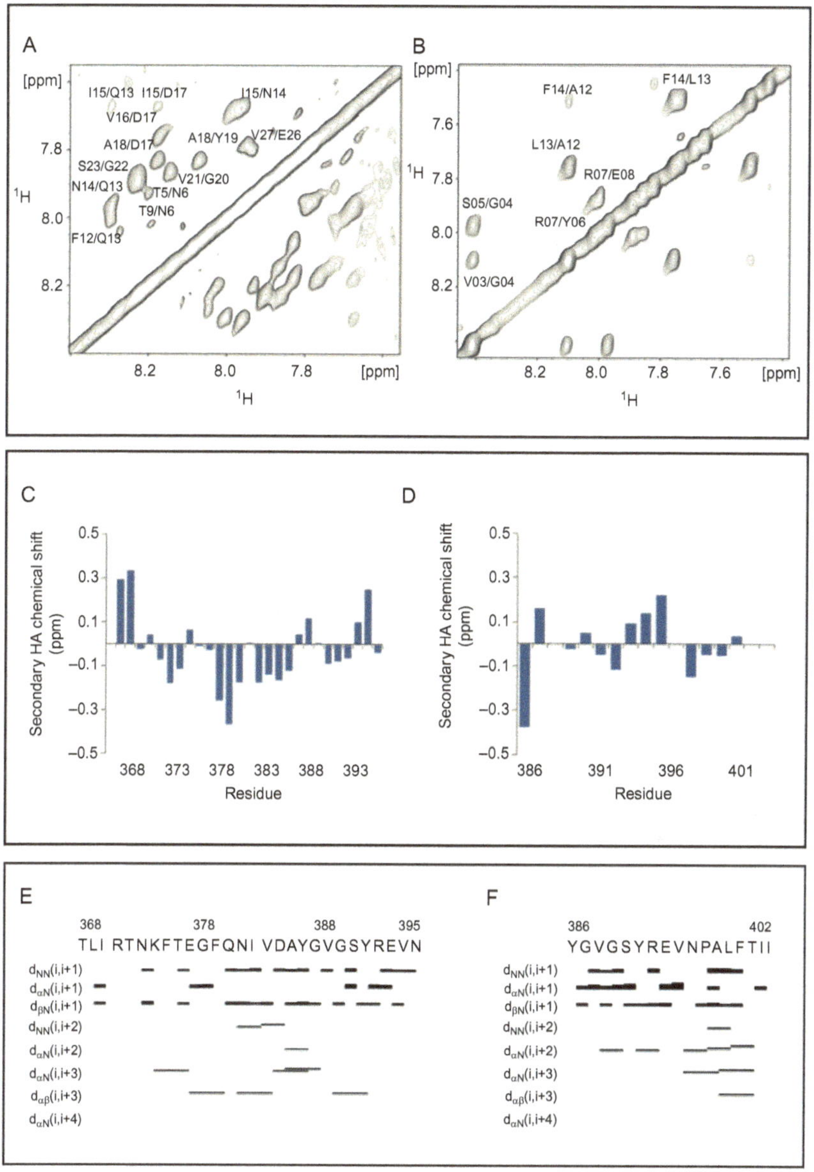

Fig. 4 NMR data and analysis of $a2N_{339–395}$ and $a2N_{386–402}$ peptides. Assignment of cross-peaks in the NOESY spectrum of (A) $a2N_{339–395}$ and (B) $a2N_{386–402}$ Peak picking was done in Sparky 3.1 software and peaks were identified based on TOCSY spectrum. Since the 2D NOESY assignment of $a2N_{339–395}$ reveals an unstructured N-terminal part, the calculation was performed only for the C-terminal region with the residues 368–395, called $a2N_{368–395}$. The amino acid sequence of (C) $a2N_{368–395}$ and (D) $a2N_{386–402}$ peptides and their secondary structure elements based on 1Hα chemical shifts with respect to the random coil values. NOESY connectivity plot of peptides (E) $a2N_{368–395}$ and (F) $a2N_{386–402}$ are indicative of the residues connected in space revealing the presence of a helical structure.

Hα–HN(i, i+2) between T_{372} and F_{375} indicates a potential α-helical turn. Data from assigned 2D NOESY spectra, torsion angle calculated from HA values by TALOS software and primary amino acid sequence were used as input for the automated structure calculation by Cyana 3.0 package (Herrmann, Güntert, & Wüthrich, 2002). Since the 2D NOESY assignment of $a2N_{339–395}$ reveals an unstructured N-terminal part, the calculation was performed only for the C-terminal region with the residues 368–395, called $a2N_{368–395}$ (Fig. 3A, in gray). Out of 100 generated structures the 20 lowest energy structures were taken for further analysis. In total an ensemble of 20 calculated structures resulted in an overall mean root square deviation (RMSD) of 0.297 Å for the residues 378–387 (Fig. 3B). All the structures of $a2N_{368–395}$ have energies lower than $-100\,kcal\,mol^{-1}$, no NOE violations greater than 0.3 Å and no dihedral violations greater than 5°. The statistics for 20 structures are shown in Fig. 3B. The structure of $a2N_{368–395}$ shows a total length of 43.05 Å and forms a flexible N-terminal region from residues 368–371 with a helix extending from amino acids 372–375, followed by a short loop from 376 to 378 and a second helix between the residues 379–387 (Fig. 5A and B), while the remaining C-terminal region is flexible. Molecular surface electrostatic potential of $a2N_{368–395}$ reveals an amphiphilic surface (Fig. 5C and D). At one side of the peptide, residues E_{377}, D_{384} and E_{393} are forming a negative charged surface (Fig. 5C). The opposite side reflects an amphiphilic surface potential, formed by the charged residues at the N- and C-termini as well as the hydrophobic residues F_{375}, F_{379}, I_{382}, V_{383}, A_{385}, Y_{386} V_{388}, and Y_{391} (Fig. 5D).

In parallel, the secondary structure elements of $a2N_{386–402}$ showed α-helical formation between the amino acids P_{396} and I_{402} (Fig. 6A and B). HN–HN, Hα–HN(i, i+3), Hα–HN(i, i+4), and Hα–Hβ (i, i+3) connectivities were plotted from the assigned NOESY spectrum (Fig. 4F) reflecting α-helical formation in the C-terminus. In total an ensemble of 20 calculated structures resulted in an overall mean root square deviation (RMSD) of 0.68 Å (Fig. 3B). All these structures have energies lower than $-100\,kcal\,mol^{-1}$, no NOE violations greater than 0.3 Å and no dihedral violations greater than 5° (Fig. 3B). The $a2N_{386–402}$ peptide contains an unstructured N-terminal region formed by the amino acid residues 386–395 and a C-terminal α-helical region from residues 396–402 (Fig. 6B). Since $a2N_{386–402}$ in the intact protein is predicted to be located on the border of the cytoplasmic and membrane-embedded parts of a2-subunit (Merkulova et al., 2010), we propose that this C-terminal helix of $a2N_{386–402}$ corresponds in fact to the beginning of the first transmembrane helix of the a2-subunit of V-ATPase.

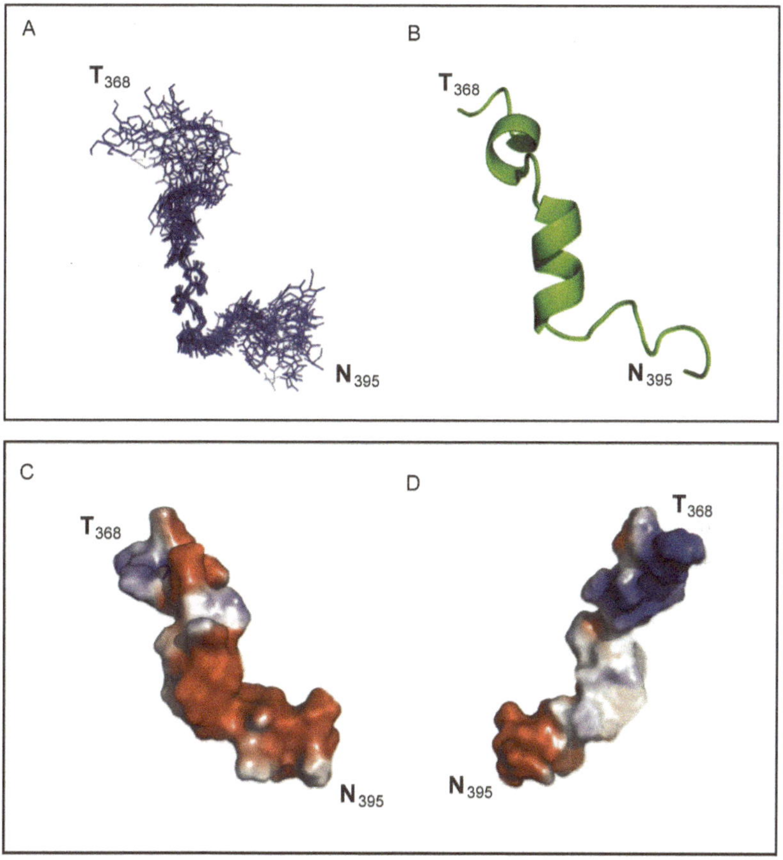

Fig. 5 NMR structure of $a2N_{368–395}$ peptide. (A) NMR structure of the peptide $a2N_{368–395}$ showing superimposition of 20 structures calculated by Cyana 3.1 package and revealing a partially α-helical formation. (B) Cartoon representation of a single NMR structure of $a2N_{368–395}$. (C, D) The molecular surface electrostatic potential of peptide (C) $a2N_{368–395}$ and (D) its 180 view respectively generated by Pymol (DeLano, 2002). The positive potentials are drawn in blue, negative in red and hydrophobic in light gray.

2.4 The complete model of $a2N_{1–402}$

In order to build a 3D model of the entire $a2N_{1–402}$ region, the homology model of $a2N_{1–352}$ was combined with NMR structures of the two peptides $a2N_{368–395}$ (Fig. 5) and $a2N_{386–402}$ (Fig. 6) as described in Methods (Fig. 7, Supplementary Movie S2 in the online version at https://doi.org/10.1016/bs.ctm.2019.01.008). In comparison with the $a2N_{1–352}$ structural model, three additional α-helices can be assigned in the $a2N_{1–402}$ model (Fig. 1, helices XII, XIII, and XIV), which are connected by unstructured linkers (Fig. 7A, B, and D). According to $a2N_{1–402}$ model only α-helix XIII lays

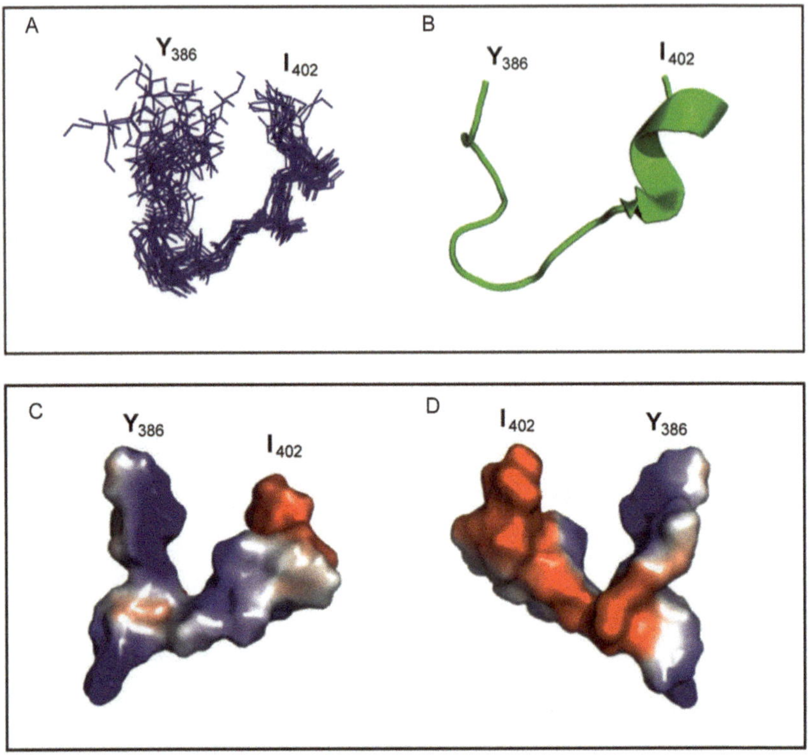

Fig. 6 NMR structure of $a2N_{386–402}$ peptide. (A) NMR structure of the peptide $a2N_{386–402}$ showing superimposition of 20 structures calculated by Cyana 3.1 package and revealing a partially α-helical formation. (B) Cartoon representation of a single NMR structure of $a2N_{386–402}$. (C, D) The molecular surface electrostatic potential of peptide (C) $a2N_{386–402}$ and (D) its 180 view respectively generated by Pymol (DeLano, 2002). The positive potentials are drawn in blue, negative in red and hydrophobic in light gray.

close to α-helix I of $a2N_{1–352}$ region. Overall, $a2N_{353–402}$ region represents an appendage joined to the $a2N_{1–352}$ structural core through a flexible linker formed by amino acids 353–371 between β-strand 7 and α-helix XII (Fig. 1). This appendage will most likely contributes to the formation of the short stem domain (SD), which connects the cytosolic and membrane-embedded parts in intact full-length a-subunit of V-ATPase as observed on electron microscopy images (Lau & Rubinstein, 2010, 2012).

The calculated $a2N_{1–402}$ molecular model can be fitted well into the density of the barbell-shaped collar of the 23 Å resolution EM map of the related *P. furiosus* A-ATP synthase (EMD-1542) (Vonck, Pisa, Morgner, Brutschy, & Müller, 2009) with a correlation coefficient being 0.8702, as calculated using the Chimera program and as shown Fig. 8A and B. The fit takes into consideration that the regions $a2N_{140–173}$ and $a2N_{206–222}$,

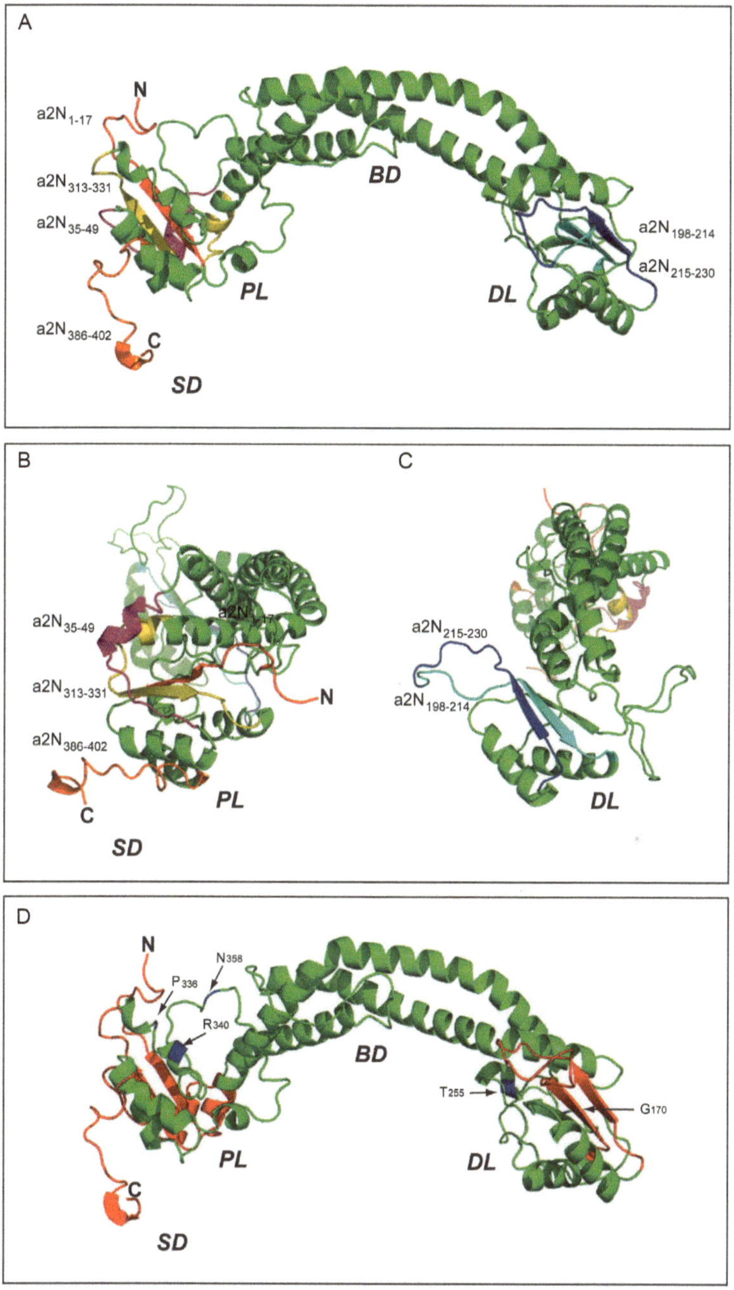

Fig. 7 Structural model of complete a2N$_{1–402}$ and identification of CTH2 and G/*E*-peripheral stalks binding sites. (A–D) Structural molecular model of a2N$_{1–402}$ and identification of two distinct CTH2 binding sites. (A) Ribbon representation of a top view of a

which are segments of the eukaryotic V-ATPase subunit and absent in *M. ruber* A-ATP synthase I_{cyt} template, are presented as unstructured regions in the final $a2N_{1-402}$ model (Fig. 7A, C, and D). In addition, the structures of the subunits E and H of the *T. thermophilus* and *Pyrococcus horikoshii* OT3 A-ATP synthase are well accommodated by the density of the straight and bent peripheral stalk with a correlation of 0.7827 and 0.7704, respectively. The structure of the H^+-translocating *c*-ring (Murata, Yamato, Kakinuma, Leslie, & Walker, 2005; Vonck et al., 2009) was fitted with a correlation coefficient being 0.8601.

2.5 The structural model and features of CTH2 binding interface on a2-subunit of V-ATPase

While six CTH2 interacting peptides are evenly distributed in amino acid sequence of $a2N_{1-402}$ (Merkulova et al., 2010), our complete $a2N_{1-402}$ model revealed that epitopes formed by these peptides are forming just two spatially distinct binding sites for two molecules of CTH2 (Fig. 7A–D). The CTH2 interacting peptides $a2N_{1-17}$, $a2N_{35-49}$, $a2N_{313-331}$ and $a2N_{386-402}$ are clustered together in the proximal lobe (*PL*) into the first binding site (Fig. 7A and D left and B). Three of these peptides $a2N_{1-17}$, $a2N_{35-49}$ and $a2N_{313-331}$ form a continuous surface from the β-strands 1, 2 and 6, and the adjacent loops (Figs. 1 and 7A, B, and D), while $a2N_{386-402}$ peptide is separate and forms an additional binding surface (Figs. 1, 5A and B, 7A and D left and B). The second CTH2 binding site is composed by $a2N_{198-214}$

molecular model of $a2N_{1-402}$ showing six cytohesin-2 binding peptides as follows: (i) $a2N_{1-17}$ (in red); (ii) $a2N_{35-49}$ (in purple); (iii) $a2N_{198-214}$ (in blue); (iv) $a2N_{215-230}$ (in cyan); (v) $a2N_{313-331}$ (in yellow) and (vi) $a2N_{386-402}$ (in orange). Note, that α-helix XIII of appendage region $a2N_{353-402}$ (lower left part with peptide $a2N_{386-402}$) lays close to α-helix I of core region $a2N_{1-352}$. (B) View of proximal lobe (*PL*) and details of the first cytohesin-2 binding site formed by $a2N_{1-17}$, $a2N_{35-49}$, $a2N_{313-331}$ and $a2N_{386-402}$ peptides. Note that three peptides $a2N_{1-17}$, $a2N_{35-49}$ and $a2N_{313-331}$ form a continuous surface while $a2N_{386-402}$ peptide is separate. This view is rotated by 90° vertically and then counter-clockwise by 90° horizontally relatively to the view in A. (C) View of distal lobe (*DL*) and details of the second CTH2 binding site formed by $a2N_{198-214}$ and $a2N_{215-230}$ peptides. This view is rotated clockwise 90° horizontally relatively to the view in A. (D) Ribbon representation of a top view of a molecular model of $a2N_{1-402}$. A ll six CYH2 binding peptides are colored in red, while five amino acids G_{170}, T_{255}, P_{336}, R_{340} and N_{358} identified in G/E-stalks binding vicinity are colored in blue (Qi & Forgac, 2008). Amino acids P_{336}, R_{340} and N_{358} forming first G/E-stalk binding site in *PL* are on the left, while amino acids G_{170} and T_{255} forming second G/E-stalk binding site in *DL* are on the right.

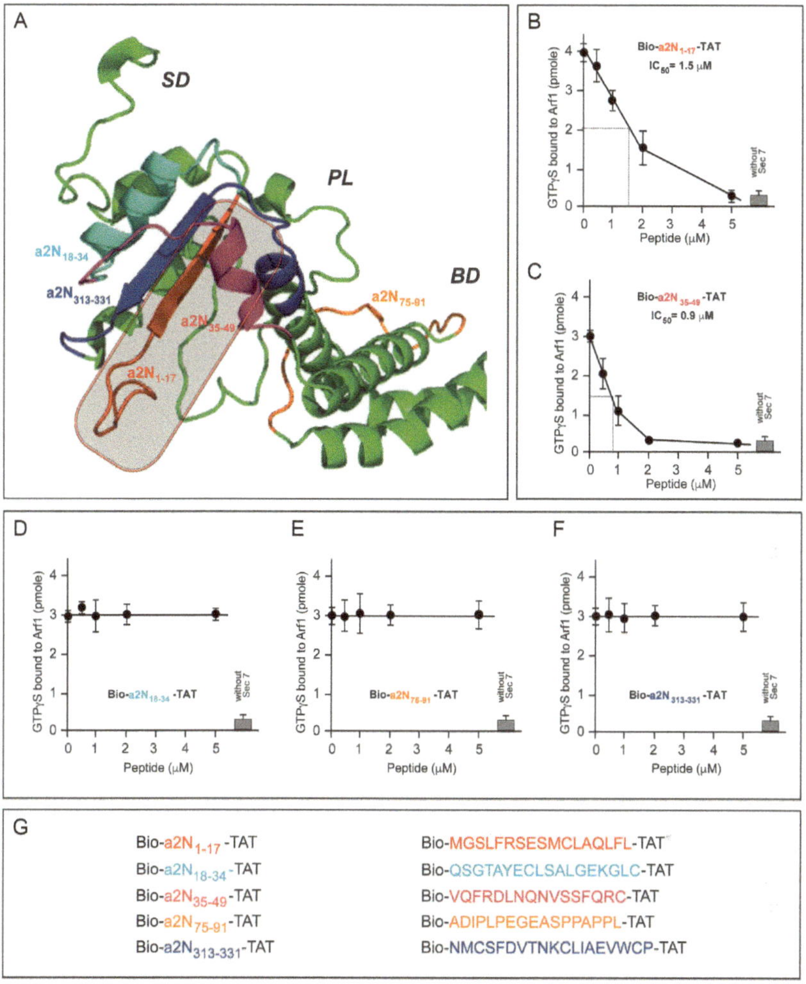

Fig. 8 Identification of V-ATPase interface involved in binding and signaling with CTH2. (A) Two V-ATPase derived peptides a2N$_{1-17}$ and a2N$_{35-49}$ are forming a specific interaction interface on proximal lobe sub-domain of a2N (shown by gray rectangle) that is involved in interaction with Sec7 domain of CTH2. (B, C) Experimental testing of our model demonstrated a potent inhibition of Sec7/Arf1-GEF activity by both (B) Bio-a2N$_{1-17}$-TAT (IC_{50} = 1.5 μM) and (C) Bio-a2N$_{35-49}$-TAT (IC_{50} = 0.9 μM) interface forming peptides. (D,E,F) The V-ATPase-derived peptides: (D) Bio-a2N$_{18-34}$-TAT; (E) Bio-a2N$_{75-91}$-TAT and (F) Bio-a2N$_{313-331}$-TAT that are not involved in the formation of binding-competent interface, are unable inhibit enzymatic GEF activity of Sec7 domain. (G) Sequences of V-ATPase-derived synthetic peptides tested in enzymatic GEF-activity assay of CTH2 Sec7 domain.

and $a2N_{215-230}$ peptides which form a continuous surface from the β-strands 4 and 5, and the loop between them in the distal lobe (*DL*) of $a2N_{1-402}$ (Figs. 1 and 7A, D right, and C). Thus, both binding sites have a very similar predominantly β-sheet architecture, except for the presence of an additional unique α-helical region XIV of $a2N_{386-402}$ peptide in the proximal lobe binding interface. Importantly, according to $a2N_{1-402}$ model, both sites are not buried inside the protein molecule but are exposed and readily accessible to binding to CTH2.

2.6 Mapping of the G/E peripheral stalk interaction sites on structural model of $a2N_{1-402}$

Previously, Forgac and co-workers performed extensive site-specific cross-linking experiments with the N-terminal domain of yeast Vph1p, a homolog of mammalian V-ATPase a-subunit (Qi & Forgac, 2008). Following five amino acid residues in Vph1 G_{181}, S_{266}, A_{347}, A_{351} and Q_{369} that are located in proximity or interacting with G/E peripheral stalks of V-ATPase were identified (Qi & Forgac, 2008). We applied this data to $a2N_{1-402}$ structural model, in order to reconstitute the G/*E*-binding interface on the mouse a2-subunit of V-ATPase. According to multiple sequence alignment, G_{181}, S_{266}, A_{347}, A_{351}, and Q_{369} amino acids of Vph1p correspond to G_{170}, T_{255}, P_{336}, R_{340} and N_{358} in mouse a2N sequence (Fig. 1). In the structural model of $a2N_{1-402}$ the G_{170} residue is located in the unstructured loop, connecting α-helix V of the central region and first β-strand 3 of the distal lobe (Figs. 1 and 7D, right). The residue of T_{255} is in close proximity and a forming part of the short predicted α-helix connecting unstructured region behind the last α-helix VII of the distal lobe and α-helix VIII of the central bar region (Figs. 1 and 7D, right). Thus, both amino acids G_{170} and T_{255} come close together in the structural model and most likely are forming one G/E-binding interface in the distal lobe of $a2N_{1-402}$ (Fig. 7D, right). The next two of five residues P_{336} and R_{340} are located in the proximal lobe of $a2N_{1-402}$ forming a single continuous interface in the same α-helix XI (Figs. 1 and 7D, left). In close proximity to this surface is residue N_{358}, which is located in the middle of the unstructured linker that connects core and appendage domains of $a2N_{1-402}$. Thus, all three amino acids P_{336}, R_{340} and N_{358} come close together in the structural model and most likely are forming another G/E-binding interface on the proximal lobe of $a2N_{1-402}$ (Fig. 7D, left).

2.7 Identification of V-ATPase interface involved in binding with CTH2 and modulation of its GEF activity

According to our studies, two V-ATPase derived peptides $a2N_{1-17}$ and $a2N_{35-49}$ are forming a specific interaction interface on proximal lobe (*PL*) sub-domain of a2N that is involved in interaction with Sec7 domain of CTH2 (Fig. 8A, shown by gray rectangle). To confirm our model, we therefore performed an additional GEF activity experiments using recombinant Sec7 CTH2 domain and a2N V-ATPase derived synthetic peptides (Bio-$a2N_{1-17}$-TAT; Bio-$a2N_{18-34}$-TAT; Bio-$a2N_{35-49}$-TAT; Bio-$a2N_{75-91}$-TAT and Bio-$a2N_{313-331}$-TAT) (Fig. 8G). Experimental testing of our model demonstrated a potent inhibition of Sec7/Arf1-GEF activity by both Bio-$a2N_{1-17}$-TAT ($IC_{50}=1.5\,\mu M$) (Fig. 8B) and Bio-$a2N_{35-49}$-TAT ($IC_{50}=0.9\,\mu M$) (Fig. 8C), peptides forming an interface highlighted in gray rectangle in Fig. 8A. However, peptides of a2N V-ATPase that is not involved in formation of a binding-competent interface: Bio-$a2N_{18-34}$-TAT (Fig. 8D), Bio-$a2N_{75-91}$-TAT (Fig. 8E), Bio-$a2N_{313-331}$-TAT (Fig. 8F), are not able to modulate the function of CTH2 Sec7 domain in a Sec7/Arf1-GEF activity assay. These data indicate that the interface formed by $a2N_{1-17}$ and $a2N_{35-49}$ of V-ATPase (Fig. 8A) is specifically involved in the interaction with Sec7 domain and regulation of enzymatic GEF activity of CTH2.

3. Discussion

In this study we generated the structural model of $a2N_{1-402}$ and uncovered two binding interfaces for Arf GEF CTH2. The complete molecular model of $a2N_{1-402}$ revealed that all CTH2 interacting peptides are located in distinct regions of the proximal lobe (*PL*) and distal lobe (*DL*) of $a2N_{1-402}$ and there are no interacting epitopes in the central bar domain (*BD*) of the protein (Fig. 7). The average distance between two CTH2 binding sites on $a2N_{1-402}$ was estimated as ~73 Å. The first CTH2 binding interface located in *PL* is formed in part by the externally exposed $a2N_{1-17}$, and $a2N_{35-49}$ peptides (Figs. 7A,B, and 9). According to our previous studies these two peptides are specifically interacting with Sec7 domain of cytohesin-2 (Merkulova et al., 2010). Therefore, here we conclude that this first *PL* located binding interface is the major CTH2 binding site, which is involved in interaction with catalytic Sec7 domain of first CTH2. Two other

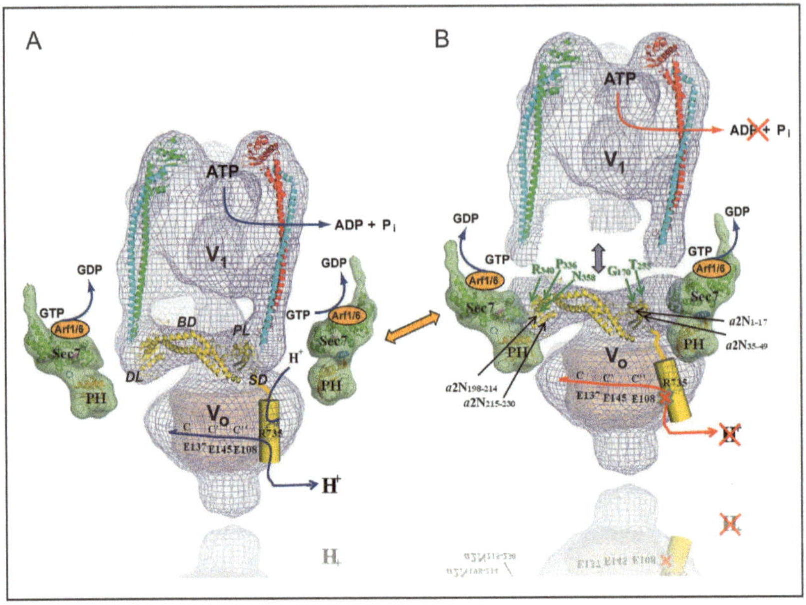

Fig. 9 Model of interactions between CTH2 and V-ATPase: Implication for cross-talk and regulation of CTH2 and V-ATPase functions. (A–B) The 3D reconstruction EM map of the related A-ATP synthase of the archaeon *P. furiosus* enzyme. The presented molecular model of $a2N_{1-402}$ (yellow), the *c*-ring (wheat) and G/E-peripheral stalks (G, cyan and E, green/red) were used for the fitting. Cytohesin-2 (CTH2) (green) has been accommodated based on the interactions described in the Figs. 7A–C. (A) This model proposes that binding of CTH2 to the V-ATPase may destabilize the binding of both G/E peripheral stalks to a-subunit, which could promote the disassembly of the V_1- and V_O-sectors and inactivation of H^+-pumping of the V-ATPase. (B) This model proposes that ion-translocation in the interface of the *c*-ring and the C-terminal membrane-embedded domain of a-subunit (yellow cylinder), could be affected by the binding of the CTH2 to the proximal lobe of N-terminal domain of 2-subunit of V-ATPase.

peptides $a2N_{313-331}$ and $a2N_{386-402}$ are also located in this region. However, these peptides do not interact with Sec7 domain, while they are interacting with full-length CTH2 in pull-down experiments (Merkulova et al., 2010). We conclude, that epitopes formed by these peptides are most likely involved in the interaction with regulatory elements of CTH2 such as Sec7/PH-linker, PH- and/or PB-domains, which are located in the close proximity to the Sec7 domain (Merkulova et al., 2010). Finally, based on our previous data showing low affinity interaction of peptide $a2N_{198-214}$ with purified PH-domain, we hypothesized that the second *DL* located epitope should be involved in interaction with only regulatory PH-domain of second CTH2 (Figs. 7A and C, and 9).

What would be the molecular role and cell biological functional significance of such a complex and differential interaction between a2-subunit of V-ATPase and two molecules of CTH2? Previously, we reported that interaction between a2-subunit and CTH2 is depending upon V-ATPase driven acidification of the endosomal lumen which suggests, that V-ATPase itself could function as pH-sensing receptor (Hurtado-Lorenzo et al., 2006). Thus, we indeed identified V-ATPase as a novel signaling receptor that on one hand, is sensing the acidification status of endosomal lumen, and on the other hand, could transmit this information across the membrane and modulate activity of the Arf-GEF CTH2 and cognate Arf small GTPases. The Arf family of small GTPases (Arf1—Arf6) belongs to the Ras-superfamily small GTPases that function as "molecular switches" and regulate multiple cellular functions. CTH2 as an Arf-GEF is a regulatory protein, whose main function is the activation of Arf small GTPases and thus regulating multiple down-stream effectors. Previously, we suggest that cytohesin/Arf's might function as "molecular on/off switches" for self-regulation of the V-ATPase function (Marshansky & Futai, 2008). In particular we proposed that cross-talk and signaling between a2-subunit and CTH2 might be involved in self-regulated acidification-dependent disassembly of V-ATPase and shutting down of its function as a proton-pumping nano-motor (Marshansky & Futai, 2008). Here using structural insights from $a2N_{1-402}$ model we test this hypothesis.

Insights from structural model of $a2N_{1-402}$*: Regulatory role of CTH2 in modulation of V-ATPase function.* Importantly, our structural model of $a2N_{1-402}$ and identification of its interacting interfaces with both CTH2 and G/E-stalks provide basis for analysis of the potential molecular mechanism of self-regulation of V-ATPase function. Cryo-electron microscopy study of the intact holo-complex of V-ATPase demonstrates that a-subunit and C-subunit of V-ATPase are oriented parallel to the membrane surface forming so called "colar" separating V_1- and V_O-sectors of V-ATPase (Lau & Rubinstein, 2010, 2012). Thus, one of the major structural roles of the a-subunit as part of V-ATPase complex is to hold together V_1- and V_O-sectors during its rotational catalysis, in particular via its interaction with G/E peripheral stalks (Forgac, 2007). This model is supported by various cross-linking, two-hybrid, co-immunoprecipitation and electron microscopy experiments showing that cytosolic N-terminal tail of a-subunit directly interacts with A-, G-, E-, H– and C-subunits of V_1- sector of V-ATPase (Forgac, 2007). If the a-subunit is involved in such multiple interaction with other subunits of V_1-sector of V-ATPase, the question arises of how and when subunit a interacts with CTH2? One possibility is that CTH2 interacts with V_O-sector only after disassembly of the V_1/V_O-complex complex when a-subunit becomes more exposed for its binding. In this scenario CTH2 may share the same binding interfaces with other subunits of V_1-sector, such as G/E, competing for their binding with a-subunit, and thus, preventing the assembly of yet unassembled V-ATPase or the re-assembly of previously disassembled V-ATPase. In another scenario the interaction between CTH2 and a-subunit may take place in the intact V_1V_O-ATPase. In this case the G/E peripheral stalk subunits of V_1-sector and CTH2 would have different and not overlapping binding interfaces on $a2N_{1–402}$, which however, might trigger structural rearrangements upon CTH2 binding resulting in the disassembly of V_1/V_O-complex.

The structural model of $a2N_{1–402}$ and identification of its interacting interfaces with both CTH2 and G/E peripheral stalks helps to evaluate these possibilities. Our data demonstrated that there are two non-identical and non-overlapping sites on $a2N_{1–402}$ for both G/E stalks and CTH2. In case of G/E stalks, each of these two sites directly bind two out of three G/E peripheral stalks in intact V-ATPase complex (Zhang et al., 2008). The first G/E-binding interface is located in proximal lobe and is formed by α-helix XI and unstructured linker that connects core and appendage domains of $a2N_{1–402}$ (Fig. 7D, left, *PL*). The second G/E-binding interface is located in distal lobe and is formed by the unstructured loop and the short predicted α-helix, preceding α-helix VII (Fig. 7D, right, *DL*). Similarly, CTH2 has also got two distinct and non-overlapping binding sites. While the CTH2 binding interface is in close proximity to the G/E interaction sites, they are not, however, identical since the G/E- and CTH2 binding motifs form separate non-overlapping interaction surfaces (Fig. 7D). This non-overlapping character of CTH2 and G/E-binding surfaces identified in $a2N_{1–402}$ model suggests that CTH2 might be recruited and interacts with intact V_1/V_O-complex. This possibility was recently confirmed in our experiments with intact *S. cerevisiae* V-ATPase and a truncated form of CTH2 (Hosokawa et al., 2013). Thus our data indicate that CTH2 does not compete with G/E subunits for binding to a2-subunit of V-ATPase. However, its binding may compromise the stability of interaction between the a2-subunit and the G/E peripheral stalks. Thus, according to this model we conclude, that the acidification-dependent recruitment of CTH2 to V-ATPase previously uncovered in our studies (Hurtado-Lorenzo et al., 2006) is most likely involved in regulating disassembly of intact V_1/V_O-complex rather than being involved in prevention of assembly of V-ATPase (Fig. 9A). We have also recently proposed an alternative mechanism of V-ATPase regulation by CTH2 (Dip, Saw, Roessle, Marshansky, & Grüber, 2012). According to this model, binding of the CTH2 to the proximal lobe of the N-terminal domain of Vph1 (yeast homolog of a-subunit) of V-ATPase could affect ion-translocation in the interface of the *c*-ring and the C-terminal membrane-embedded domain of a-subunit (Fig. 9B).

In summary, combining of homology modeling and NMR structural analysis give rise to the structural model of $a2N_{1-402}$, which allow to uncover the molecular features of CTH2-binding interfaces on V-ATPase. Our model reveals the existence of two CTH2 binding sites located on the distal and proximal lobes of $a2N_{1-402}$. We also determined that while Sec7-domain is a major interaction site on the proximal lobe of $a2N_{1-402}$, its distal lobe most likely interacts with only the PH-domain of CTH2. The further analysis revealed, that these two CTH2 binding sites are in a close proximity to two G/E-binding sites on subunit a. Thus, we hypothesize, that binding of CTH2 to the V-ATPase complex may destabilize binding of both G/E peripheral stalks to the a-subunit, triggering disassembly of V-ATPase into a V_1 and V_O-complex (Fig. 9A) and/or shutting down proton pumping activity of the V-ATPase nano-motor (Fig. 9B). However, this hypothesis is only the first step to uncover the molecular details of the interaction, interplay and signaling between V-ATPase, CTH2 and Arf family small GTPases. While recently performed multi-sequence alignment analysis of interaction sites between V-ATPase and CTH2 have uncovered the evolutionary conservation of their binding-mechanism (unpublished data), these models are awaiting experimental testing and confirmation using alternative molecular and cell biological approaches. Our final goal in uncovering these interactions is a structure-based design of small molecules that would specifically interfere with interactions between V-ATPase and CTH2. These small molecules can be used as drugs to modulate: (i) either function of V-ATPase or (ii) signaling and function of CTH2 and Arf family small GTPases in variety of diseases, since both V-ATPase and CTH2/Arf6 small GTPases are emerging as important drug targets (Marshansky et al., 2014).

It is noteworthy, that an accessory ATP6AP2 subunit of eukaryotic V-ATPase was identified as (pro)renin receptor (PRR) which is playing a central role in the activation of the local renin-angiotensin system (RAS) (Danser, 2009; Ichihara & Kinouchi, 2011; Marshansky et al., 2014). This transmembrane ATP6AP2/PRR subunit together with a2-subunit form part of V_O-sector and both are crucial for the biogenesis of active V-ATPase. In particular, the N-terminal extracellular domain (ECD) of

ATP6AP2/PRR, which is also implicated in binding with prorenin or renin, is required for V-ATPase biogenesis and function (Kinouchi et al., 2013). Previously, it was uncovered that over-activation of PRR is strongly associated with cardiovascular hypertensive and kidney diseases (Danser & Deinum, 2005; Ichihara, Kaneshiro, Takemitsu, Sakoda, & Itoh, 2007). Thus, future studies in this area could lead to the development of novel therapeutic approaches for the treatment of hypertension and its complications (Li, Zhang, & Zhuo, 2017; Sun, Danser, & Lu, 2017). Moreover, the V-ATPase is also emerging as an important drug target, able to control signaling and trafficking of another receptors such as EGFR/ErbB, Fz/LRP6 and Notch, and thus useful for treatment of lung cancers among others (Bill et al., 2012; Lu et al., 2013; Marshansky et al., 2014; McGuire, Cotter, Stransky, & Forgac, 2016; Stransky, Cotter, & Forgac, 2016).

On the other hand, cytohesin2 (CTH2) and Arf6 small GTPase have emerged as critical regulators of Slit2–Robo4–paxillin–GIT1 network which controls vascular stability and leak (Jones et al., 2009; London & Li, 2011). Inhibition of CTH2/Arf6 function can significantly reduce vascular permeability, which identified them as a new therapeutic target for ameliorating diseases involving the vascular system. In particular, cardiovascular hypertension and acute lung injury (ALI)/adult respiratory distress syndrome (ARDS), diseases in which inflammation, cytokine storm and vascular leak are present (Jones et al., 2009; London & Li, 2011), may be effectively treated by drugs targeting these proteins.

4. Methods

4.1 Bioinformatics analysis and a2N$_{1-402}$ homology modeling

The prediction of the secondary structure of a2N$_{1-402}$ was performed using PsiPred on-line server (http://bioinf.cs.ucl.ac.uk/PSIPRED/) (Buchan et al., 2010). The model of 1–352 part of a2 was built by homology modeling with MODELLER software (Eswar et al., 2006). The crystal structure of recently solved cytoplasmic N-terminal domain of subunit I, homolog of subunit a, of V-ATPase (PDB ID 2RRK) was used as a template for homology modeling (Srinivasan et al., 2011). The alignment for homology modeling is based on profile-profile alignment of a2 homologs alignment and subunit I homologs alignment. Also subunit I secondary structure and a2 PSIPRED predicted secondary structure were taken into account (Buchan et al., 2010). For joining the homology a2N$_{1-352}$ model and the

NMR peptide structures, Autodock Vina docking program was used to build several complexes of a2N$_{1-352}$ model and a2N$_{378-387}$ fragment from NMR structure of a2N$_{368-395}$ peptide (Trott & Olson, 2010).

Then the NMR structure of the entire a2N$_{368-395}$ peptide was fitted to these complexes and the complex was selected where the peptide did not intersect with the protein. The final model was built by Modeler based on two templates: (1) the aforementioned complex of a2N$_{1-352}$ and a2N$_{368-395}$ peptide, and (2) NMR structure of a2N$_{386-402}$ peptide. Lastly, in the final model dihedral angles of Glu$_{393}$ were manually changed for better fit into the cryo-EM map segment corresponding to I subunit of the *T. thermophilus* V-ATPase (Lau & Rubinstein, 2012).

Also I-TASSER software was used for building a2N$_{1-402}$ model (Roy, Kucukural, & Zhang, 2010). We made some modification to I-TASSER, which allow to set several preferred templates, and used 2RRK and NMR peptides structures as templates. a2N$_{1-352}$ parts of I-TASSER models were very similar to corresponded parts of the Modeler homology model. But all I-TASSER models were not in good agreement with Cryo-EM map and our knowledge about a2N structure, with 353–402 part interacting with BD domain (data not shown). The possible reason is following: I-TASSER is optimized for globular proteins and cannot take into account proximity of membrane which influences a2N$_{1-402}$ structure.

4.2 NMR data collection, processing and peptide structure determination

NMR spectra of a2N$_{339-395}$ and a2N$_{386-402}$, respectively, were collected at temperature ranging from 288 to 298 K on Avance cryo-probed 600 or 700 MHz instruments (Bruker, Billerica, MA). All NMR data were processed using Bruker Avance spectrometer in-built software Topspin program. To elucidate the structure of a2N$_{339-395}$ and a2N$_{386-402}$, data were collected on Bruker Avance at 600 MHz or 700 MHz. Pulse calibrations and other parameter including temperature, buffer and peptide concentration were optimized before making final measurements. 10% D_2O was used to lock the NMR signal in each experiment. The two dimensional (2D) ^{1}H NMR spectra including total correlation spectroscopy (TOCSY) and nuclear overhauser enhancement spectroscopy (NOESY) were obtained at the temperature of 298 K. 2D experiments with TOCSY were performed with mixing time of 60–80 ms, whereby NOESY used 200–300 ms mixing time. Baseline corrections were applied wherever necessary. The proton chemical shift was referenced to the methyl signal ofDSS (2,2-dimethyl-2-silapentane-5-sufonate

[Cambridge Isotope Laboratories]), an external standard reference to 0 ppm. Peak-assignment and data analysis of the Fourier transformed spectra were performed with SPARKY 3.1 program (Kneller & Goddard, 1997). Assignments were carried out according to the classical procedure including spin system identification and sequential assignment (Wüthrich, 1986). Inter proton distance were obtained from the NOESY spectra. NOESY peaks were categorized as strong, medium and weak based on the signal intensity and were translated into distance constraints as 3.0, 4.0 and 5.0 Å, respectively. Dihedral angle restraints were calculated from the spectra assigned chemical shift using torsion angle likelihood obtained from shift and sequence similarity (TALOS) (Cornilescu, Delaglio, & Bax, 1999). The secondary structures were predicted using the chemical shift index (CSI) and NOE pattern. Structure calculations were performed by using Cyana program which uses simulated annealing with molecular dynamics in torsion angle space (Güntert, Mumenthaler, & Wüthrich, 1997). Final Cyana ensemble structures were visualized by Pymol (DeLano, 2002).

4.3 Reagents, peptides synthesis, labeling, and purification

If not otherwise specified, all reagents were purchased from Sigma. Peptide synthesis resins and Fmoc-protected amino acids were purchased from EMD Chemicals. All peptides were synthesized, purified by HPLC, and analyzed by mass spectrometry as previously described (Hosokawa et al., 2013). The following a2N-derived peptides were synthesized:

a2N$_{339–395}$
(RRALEEGSRESGATIPSFMNIIPTKETPPTRIRTNKFTEGFQNIVDAYGVGSYREV)

a2N$_{386–402}$
(YGVGSYREVNPALFTII)

Bio- a2N$_{1–17}$-TAT
(Bio-MGSLFRSESMCLAQLFL-YGRKKRRQRRR)

Bio-a2N$_{18–34}$-TAT
(Bio-QSGTAYECLSALGEKGLC-YGRKKRRQRRR)

Bio-a2N$_{35–49}$-TAT
(Bio-VQFRDLNQNVSSFQRC-YGRKKRRQRRR)

Bio-a2N$_{75–91}$-TAT
(Bio-ADIPLPEGEASPPAPPL-YGRKKRRQRRR)

Bio-a2N$_{313–331}$-TAT
(Bio-NMCSFDVTNKCLIAEVWCP–YGRKKRRQRRR)

4.4 Recombinant proteins and GEF activity assay

In these experiments, we used radiolabel-based assay to determine the GDP/GTP exchange activity of Sec7 domain and cytohesin-2 (CTH2). This assay allows the steady-state enzymatic GEF activity analysis of Sec7 domain and CTH2 with delta17-Arf1 in the presence of PIP2-containing liposomes. Preparation of recombinant proteins (CTH2, Sec7, delta17-Arf1) and experimental procedures of GEF activity assay were previously described (Hosokawa et al., 2013).

Acknowledgments

This study was supported by A*STAR BMRC (09/1/22/19/609) (to GG). PD is grateful to receive the Singapore International Graduate Award (SINGA).

Competing financial interests

The authors declare that they have no competing financial interests.

References

Bill, A., Schmitz, A., Konig, K., Heukamp, L. C., Hannam, J. S., & Famulok, M. (2012). Anti-proliferative effect of cytohesin inhibition in gefitinib-resistant lung cancer cells. *PLoS One*, 7(7), e41179. https://doi.org/10.1371/journal.pone.0041179.

Buchan, D. W., Ward, S. M., Lobley, A. E., Nugent, T. C., Bryson, K., & Jones, D. T. (2010). Protein annotation and modelling servers at University College London. *Nucleic Acids Research*, *38*, W563–W568 (Web Server issue). (doi:gkq427 [pii] https://doi.org/10.1093/nar/gkq427).

Casanova, J. E. (2007). Regulation of Arf activation: The Sec7 family of guanine nucleotide exchange factors. *Traffic*, *8*(11), 1476–1485 (doi:TRA634 [pii]10.1111/j.1600-0854.2007.00634.x).

Cornilescu, G., Delaglio, F., & Bax, A. (1999). Protein backbone angle restraints from searching a database for chemical shift and sequence homology. *Journal of Biomolecular NMR*, *13*(3), 289–302.

Danser, A. H. (2009). (Pro)renin receptor and vacuolar H+-ATPase. *Hypertension*, *54*(2), 219–221. https://doi.org/10.1161/HYPERTENSIONAHA.109.135236.

Danser, A. H., & Deinum, J. (2005). Renin, prorenin and the putative (pro)renin receptor. *Hypertension*, *46*(5), 1069–1076. https://doi.org/10.1161/01.HYP.0000186329.92187.2e.

DeLano, W. (2002). *The PyMOL molecular graphics system*. DeLano Scientific. CA: Palo Alto http://www.pymol.org.

Dip, P. V., Saw, W. G., Roessle, M., Marshansky, V., & Grüber, G. (2012). Solution structure of subunit a, a (104-363), of the Saccharomyces cerevisiae V-ATPase and the importance of its C-terminus in structure formation. *Journal of Bioenergetics and Biomembranes*, *44*(3), 341–350. https://doi.org/10.1007/s10863-012-9442-3.

Eswar, N., Webb, B., Marti-Renom, M. A., Madhusudhan, M. S., Eramian, D., Shen, M. Y., et al. (2006). Comparative protein structure modeling using Modeller.. *Current Protocols in Bioinformatics*, *15*(1), 5.6.1–5.6.30. https://doi.org/10.1002/0471250953.bi0506s15.

Forgac, M. (2007). Vacuolar ATPases: Rotary proton pumps in physiology and pathophysiology. *Nature Reviews Molecular Cell Biology*, *8*(11), 917–929 (doi:nrm2272 [pii]10.1038/nrm2272).

Guex, N., Peitsch, M. C., & Schwede, T. (2009). Automated comparative protein structure modeling with SWISS-MODEL and Swiss-PdbViewer: A historical perspective. *Electrophoresis*, *30*(Suppl. 1), S162–S173. https://doi.org/10.1002/elps.200900140.

Güntert, P., Mumenthaler, C., & Wüthrich, K. (1997). Torsion angle dynamics for NMR structure calculation with the new program DYANA. *Journal of Molecular Biology*, *273*(1), 283–298 (doi:S0022-2836(97)91284-5 [pii]10.1006/jmbi.1997.1284).

Herrmann, T., Güntert, P., & Wüthrich, K. (2002). Protein NMR structure determination with automated NOE assignment using the new software CANDID and the torsion angle dynamics algorithm DYANA. *Journal of Molecular Biology*, *319*(1), 209–227 (doi:S0022283602002413).

Hinton, A., Bond, S., & Forgac, M. (2009). V-ATPase functions in normal and disease processes. *Pflügers Archiv—European Journal of Physiology*, *457*(3), 589–598. https://doi.org/10.1007/s00424-007-0382-4.

Hosokawa, H., Dip, P. V., Merkulova, M., Bakulina, A., Zhuang, Z., Khatri, A., et al. (2013). The N termini of a-subunit isoforms are involved in signaling between vacuolar H^+-ATPase (V-ATPase) and cytohesin-2. *The Journal of Biological Chemistry*, *288*(8), 5896–5913. https://doi.org/10.1074/jbc.M112.409169.

Hurtado-Lorenzo, A., Skinner, M., El Annan, J., Futai, M., Sun-Wada, G. H., Bourgoin, S., et al. (2006). V-ATPase interacts with ARNO and Arf6 in early endosomes and regulates the protein degradative pathway. *Nature Cell Biology*, *8*(2), 124–136 (doi:ncb1348 [pii] 10.1038/ncb1348).

Ichihara, A., Kaneshiro, Y., Takemitsu, T., Sakoda, M., & Itoh, H. (2007). The (pro)renin receptor and the kidney. *Seminars in Nephrology*, *27*(5), 524–528. https://doi.org/10.1016/j.semnephrol.2007.07.005.

Ichihara, A., & Kinouchi, K. (2011). Current knowledge of (pro)renin receptor as an accessory protein of vacuolar H^+-ATPase. *Journal of the Renin-Angiotensin-Aldosterone System*, *12*(4), 638–640. https://doi.org/10.1177/1470320311429227.

Jones, C. A., Nishiya, N., London, N. R., Zhu, W., Sorensen, L. K., Chan, A. C., et al. (2009). Slit2-Robo4 signalling promotes vascular stability by blocking Arf6 activity. *Nature Cell Biology*, *11*(11), 1325–1331. https://doi.org/10.1038/ncb1976.

Kinouchi, K., Ichihara, A., Sano, M., Sun-Wada, G. H., Wada, Y., Ochi, H., et al. (2013). The role of individual domains and the significance of shedding of ATP6AP2/(pro)renin receptor in vacuolar H^+-ATPase biogenesis. *PLoS One*, *8*(11), e78603. https://doi.org/10.1371/journal. pone.0078603.

Kneller, D., & Goddard, T. (1997). *SPARKY 3.105*. San Francisco, CA: University of California.

Lau, W. C., & Rubinstein, J. L. (2010). Structure of intact *Thermus thermophilus* V-ATPase by cryo-EM reveals organization of the membrane-bound V(O) motor. *Proceedings of the National Academy of Sciences of the United States of America*, *107*(4), 1367–1372 (doi:0911085107 [pii] 10.1073/pnas.0911085107).

Lau, W. C., & Rubinstein, J. L. (2012). Subnanometre-resolution structure of the intact *Thermus thermophilus* H^+-driven ATP synthase. *Nature*, *481*(7380), 214–218 (doi:nature10699 [pii] 10.1038/nature10699).

Li, X. C., Zhang, J., & Zhuo, J. L. (2017). The vasoprotective axes of the renin-angiotensin system: Physiological relevance and therapeutic implications in cardiovascular, hypertensive and kidney diseases. *Pharmacological Research*, *125*(Pt. A), 21–38. https://doi.org/10.1016/j.phrs.2017.06.005.

London, N. R., & Li, D. Y. (2011). Robo4-dependent slit signaling stabilizes the vasculature during pathologic angiogenesis and cytokine storm. *Current Opinion in Hematology*, *18*(3), 186–190. https://doi.org/10.1097/MOH.0b013e328345a4b9.

Lu, Q., Lu, S., Huang, L., Wang, T., Wan, Y., Zhou, C. X., et al. (2013). The expression of V-ATPase is associated with drug resistance and pathology of non-small-cell lung cancer. *Diagnostic Pathology*, *8*, 145. https://doi.org/10.1186/1746-1596-8-145.

Marshansky, V. (2007). The V-ATPase a2-subunit as a putative endosomal pH-sensor. *Biochemical Society Transactions*, *35*(Pt. 5), 1092–1099. (doi:BST0351092 [pii] 10.1042/BST0351092).

Marshansky, V., & Futai, M. (2008). The V-type H^{+}-ATPase in vesicular trafficking: Targeting, regulation and function. *Current Opinion in Cell Biology*, *20*(4), 415–426. (doi:S0955-0674(08)00056-2 [pii] https://doi.org/10.1016/j.ceb.2008.03.015).

Marshansky, V., Rubinstein, J. L., & Grüber, G. (2014). Eukaryotic V-ATPase: Novel structural findings and functional insights. *Biochimica et Biophysica Acta*, *1837*(6), 857–879. https://doi.org/10.1016/j.bbabio.2014.01.018.

McGuire, C., Cotter, K., Stransky, L., & Forgac, M. (2016). Regulation of V-ATPase assembly and function of V-ATPases in tumor cell invasiveness. *Biochimica et Biophysica Acta*, *1857*(8), 1213–1218. https://doi.org/10.1016/j.bbabio.2016.02.010.

Merkulova, M., Bakulina, A., Thaker, Y. R., Grüber, G., & Marshansky, V. (2010). Specific motifs of the V-ATPase a2-subunit isoform interact with catalytic and regulatory domains of ARNO. *Biochimica et Biophysica Acta*, *1797*(8), 1398–1409. (doi:S0005-2728(10)00055-1 [pii] https://doi.org/10.1016/j.bbabio.2010.02.009).

Merkulova, M., Hurtado-Lorenzo, A., Hosokawa, H., Zhuang, Z., Brown, D., Ausiello, D. A., et al. (2011). Aldolase directly interacts with ARNO and modulates cell morphology and acidic vesicle distribution. *American Journal of Physiology. Cell Physiology*, *300*(6), C1442–C1455. (doi:ajpcell.00076.2010 [pii] https://doi.org/10.1152/ajpcell.00076.2010).

Murata, T., Yamato, I., Kakinuma, Y., Leslie, A. G., & Walker, J. E. (2005). Structure of the rotor of the V-type Na^{+}-ATPase from *Enterococcus hirae*. *Science*, *308*(5722), 654–659 (doi:1110064 [pii] 10.1126/science.1110064).

Oot, R. A., Couoh-Cardel, S., Sharma, S., & Wilkens, S. (2017). Breaking up and making up: The secret life of the vacuolar H^{+} -ATPase. *Protein Science*, *26*, 896–909. https://doi.org/10.1002/pro.3147.

Oot, R. A., & Wilkens, S. (2012). Subunit interactions at the V1-Vo interface in the yeast vacuolar ATPase. *The Journal of Biological Chemistry*, (doi:M112.343962 [pii] 10.1074/jbc.M112.343962).

Qi, J., & Forgac, M. (2008). Function and subunit interactions of the N-terminal domain of subunit a (Vph1p) of the yeast V-ATPase. *The Journal of Biological Chemistry*, *283*(28), 19274–19282 (doi:M802442200 [pii] 10.1074/jbc.M802442200).

Roy, A., Kucukural, A., & Zhang, Y. (2010). I-TASSER: A unified platform for automated protein structure and function prediction. *Nature Protocols*, *5*(4), 725–738. (doi:nprot.2010.5 [pii] https://doi.org/10.1038/nprot.2010.5).

Srinivasan, S., Vyas, N. K., Baker, M. L., & Quiocho, F. A. (2011). Crystal structure of the cytoplasmic N-terminal domain of subunit I, a homolog of subunit a, of V-ATPase. *Journal of Molecular Biology*, *412*(1), 14–21. (doi:S0022-2836(11)00768-6 [pii] https://doi.org/10.1016/j.jmb.2011.07.014).

Stransky, L., Cotter, K., & Forgac, M. (2016). The function of V-ATPases in cancer. *Physiological Reviews*, *96*(3), 1071–1091. https://doi.org/10.1152/physrev.00035.2015.

Sun, Y., Danser, A. H. J., & Lu, X. (2017). (Pro)renin receptor as a therapeutic target for the treatment of cardiovascular diseases? *Pharmacological Research*, *125*(Pt. A), 48–56. https://doi.org/10.1016/j.phrs.2017.05.016.

Toei, M., Toei, S., & Forgac, M. (2011). Definition of membrane topology and identification of residues important for transport in subunit a of the vacuolar ATPase. *The Journal of Biological Chemistry*, *286*(40), 35176–35186. (doi:M111.273409 [pii] https://doi.org/10.1074/jbc.M111.273409).

Trott, O., & Olson, A. J. (2010). AutoDock Vina: Improving the speed and accuracy of docking with a new scoring function, efficient optimization, and multithreading. *Journal of Computational Chemistry*, *31*(2), 455–461. https://doi.org/10.1002/jcc.21334.

Vonck, J., Pisa, K. Y., Morgner, N., Brutschy, B., & Müller, V. (2009). Three-dimensional structure of A1A0 ATP synthase from the hyperthermophilic archaeon *Pyrococcus furiosus* by electron microscopy. *The Journal of Biological Chemistry*, *284*(15), 10110–10119 (doi:M808498200 [pii] 10.1074/jbc.M808498200).

Wüthrich, K. (1986). *NMR of proteins and nucleic acids*. New York: WileyXV, (292p).

Zhang, Z., Zheng, Y., Mazon, H., Milgrom, E., Kitagawa, N., Kish-Trier, E., et al. (2008). Structure of the yeast vacuolar ATPase. *The Journal of Biological Chemistry*, *283*(51), 35983–35995(doi:M805345200 [pii] 10.1074/jbc.M805345200).

CHAPTER FIVE

Ubiquitous and cell type-specific transcriptomic changes triggered by dissipation of monovalent cation gradients in rodent cells: Physiological and pathophysiological implications

Elizaveta A. Klimanova[a,b,*], Svetlana V. Sidorenko[a,b], Larisa V. Smolyaninova[a,b], Leonid V. Kapilevich[b], Svetlana V. Gusakova[c], Olga D. Lopina[a], Sergei N. Orlov[a,b,c]
[a]Faculty of Biology, Lomonosov Moscow State University, Moscow, Russia
[b]National Research Tomsk State University, Tomsk, Russia
[c]Siberian State Medical University, Tomsk, Russia
*Corresponding author: e-mail address: klimanova.ea@yandex.ru

Contents

Current Topics in Membranes, Volume 83
ISSN 1063-5823
https://doi.org/10.1016/bs.ctm.2019.01.006

Abstract

Elevation of $[Na^+]_i/[K^+]_i$-ratio is considered as one of the major signals triggering transcriptomic changes in various cells types. In this study, we identified ubiquitous and cell type-specific Na^+_i,K^+_i-sensitive genes by comparative analysis of transcriptomic changes in ouabain-treated rat aorta smooth muscle cells and rat aorta endothelial cells (RASMC and RAEC, respectively), rat cerebellar granule cells (RCGC), and mouse C2C12 myoblasts. Exposure of the cells to ouabain increased intracellular Na^+ content by ~14, 8, 7, and 6-fold and resulted in appearance of 7577, 2698, 2120, and 1146 differentially expressed transcripts in RAEC, RASMC, C2C12, and RCGC, respectively. Eighty-three genes were found as the intersection of the four sets of identified transcripts corresponding to each cell type and are classified as ubiquitous. Among the 10 top upregulated ubiquitous transcripts are the following: *Dusp6*, *Plk3*, *Trib1*, *Ccl7*, *Mafk*, *Atf3*, *Ptgs2*, *Cxcl1*, *Spry4*, and *Coq10b*. Unique transcripts whose expression is cell-specific include 4897, 1523, 789, and 494 transcripts for RAEC, RASMC, C2C12, and RCGC, respectively. The role of gene expression and signal pathways induced by dissipation of transmembrane gradient of monovalent cations in the development of various diseases is discussed with special attention to cardiovascular and pulmonary illnesses.

Abbreviations

Atf3	activating transcription factor 3
C2C12	murine skeletal muscle
Ccl7	chemokine (C-C motif) ligand 7
Cxcl1	chemokine (C-X-C motif) ligand 1
CYP	cytochrome P450
Dusp6	dual specificity phosphatase 6
Egr1	gene that encodes early growth response protein 1
Fos	gene that encodes a leucine zipper DNA-binding FOS-type nuclear phosphoprotein
GO	gene ontology
KEGG	kyoto encyclopedia of genes and genomes
Mafk	V-maf musculoaponeurotic fibrosarcoma oncogene family; protein K
NMDA	*N*-methyl-D-aspartate
Plk3	polo-like kinase 3
Ptgs2	prostaglandin-endoperoxide synthase 2, also known as cyclooxygenase 2
RAEC	rat aorta endothelial cells
RASMC	rat aorta smooth muscle cells
RCGC	rat cerebellum granule cells
Slco4a1	solute carrier organic anion transporter family member 4A1
Spry4	sprouty RTK signaling antagonist 4, a member of developmental proteins involved in cell signaling that works by inhibiting the MAPK/ERK pathway
Trib1	tribbles pseudokinase 1

1. Introduction

Electrochemical gradients of monovalent cations across the plasma membrane (high intracellular potassium, $[K^+]_i$ *versus* low intracellular sodium, $[Na^+]_i$) are created by the Na^+,K^+-pump (Na^+,K^+-ATPase) and determine a large variety of physiologically important processes. These processes include maintenance of electrical membrane potentials (E_m), cell volume regulation, secondary transport of monovalent and divalent ions (such as chloride, calcium, and phosphate), and accumulation of nutrients (glucose, amino acids, nucleotides) and other relevant molecules (Scheiner-Bobis, 2002). In excitable cells, E_m affects the triggering and spreading of action potential spikes and the formation of membrane lipid microdomains (Malinsky, Tanner, & Opekarova, 2016).

The Na^+,K^+-pump is an integral plasma membrane enzyme consisting of catalytic α and regulatory β and γ subunits that transports three sodium ions out of the cell and two potassium ions in opposite direction driven by the energy of one ATP molecule. Cardiotonic steroids are potent and specific inhibitors of the Na^+,K^+-ATPase (Akera, 1977). Among them the water soluble ouabain (g-strophanthin) is widely used for *in vitro* studies. It was shown that along with the above-listed "classic" Na^+_i,K^+_i-dependent cellular processes, sustained elevation of the $[Na^+]_i/[K^+]_i$-ratio in ouabain-treated vascular smooth muscle cells, cardiomyocytes, hepatocytes, renal epithelial, and neuronal cells causes differential expression of *c-Fos* and other immediate response genes (IRGs), as well as cell type-specific late response genes such as tumor growth factor-β, the α1- and β1-subunits of Na^+,K^+-ATPase, myosin light chain, skeletal muscle actin, atrial natriuretic factor, and mortalin (for review, see Orlov & Hamet, 2006; Orlov & Hamet, 2015; Taurin, Hamet, & Orlov, 2003). Employing Affymetrix-based technology, we compared the action of elevated $[Na^+]_i/[K^+]_i$-ratio on the set of differentially expressed genes in vascular smooth muscle cells from rat aorta, the human adenocarcinoma cell line HeLa, and human umbilical vein endothelial cells. To augment $[Na^+]_i$ and reduce $[K^+]_i$, cells were treated with ouabain or placed in K^+-free medium. We found that in these cell types the changes in expression of hundreds of transcripts were highly correlated between the two treatments, thus suggesting a key role of elevation of the $[Na^+]_i/[K^+]_i$-ratio rather than Na^+_i,K^+_i-independent signaling (Koltsova et al., 2012). Later, this conclusion was confirmed in comparative analysis of the action of ouabain and another cardiotonic steroid, marinobufagenin, on the kinetics of transcriptomic changes and gain of Na^+_i and loss of K^+_i in human endothelial cells (Klimanova et al., 2017).

It should be noted that comparative analysis of transcriptomic changes triggered by ouabain in human and rodent cells is complicated for several reasons. *First*, the apparent affinity of the housekeeping α1-Na^+,K^+-ATPase isoform for ouabain and other cardiotonic steroids in rodents is ~1000-fold lower than in other mammalian species (Lingrel, Argüello, Van Huysse, & Kuntzweiler, 1997). *Second*, we revealed that kinetics of elevation of $[Na^+]_i$ in ouabain-treated human cells is retarded compared to ouabain-treated rodent counterparts that affect transcriptomic changes triggered by ouabain (Koltsova et al., 2012). This conclusion is consistent with the slow kinetics of ouabain interaction with the human α1-Na^+,K^+-ATPase isoform demonstrated in early investigations (Akashi, Loussararian, Adelman, Saito, & Koeffler, 1990). *Third*, long-term exposure to ouabain resulted in the death of human astrocytes as well as endothelial and vascular smooth muscle cells but did not affect the survival of the rat and mouse cells of the same origin (Akimova et al., 2015). Keeping these observations in mind, in the present study we identified ubiquitous and cell type-specific Na^+_i,K^+_i-sensitive genes by comparative analysis of transcriptomic changes in ouabain-treated rat aorta smooth muscle cells and rat aorta endothelial cells (RASMC and RAEC, respectively), rat cerebellar granule cells (RCGC), and mouse C2C12 myoblasts.

2. Materials and methods

2.1 Cell cultures

Murine C2C12 cells (RRID: CVCL_0188) subjected to fewer than 10 passages were obtained from the Mammalian Cultures Collection of the Institute of Cytology of the Russian Academy of Sciences (Saint Petersburg, Russia). The cells were seeded at a density of 3×10^4 cells/well in 6-well plates containing Dulbecco's Modified Eagle's Medium (DMEM) supplemented with 5 mM glucose, 10% heat-inactivated fetal bovine serum (FBS), 100 U/mL penicillin, and 100 μg/mL streptomycin and kept at 37 °C in a humidified 5% CO_2 atmosphere. 3 to 5 days after plating, the cells reached 70–80% confluence and were subjected to differentiation in DMEM containing 5 mM glucose, antibiotics, 2% calf serum, and 1 nM insulin. The differentiation medium was changed daily. Cell morphology was evaluated by phase-contrast microscopy at ×100 magnification without preliminary fixation. Consistent with previous publications (Burch et al., 2010; Fujita, Nedachi, & Kanzaki, 2007; Nedachi, Fujita, & Kanzaki, 2008; Park et al., 2008; Whitham et al., 2012), we found that 7-day exposure to medium depleted of serum-derived growth factors and supplied with insulin resulted in myogenic differentiation of C2C12 cells indicated by cell fusion, formation of prolonged multinucleated myotubes, and appearance of troponin T (Danilov et al., 2017).

Endothelial cells from rat aorta (RAEC) were kindly provided by Dr. Thorin Trescases (Institute of Cardiology, University of Montreal, Canada). These cells were isolated and passaged three to four times as described elsewhere (Akimova, Hamet, & Orlov, 2006; Voghel et al., 2007).

Rat aorta smooth muscle cells (RASMC) were isolated from rat aorta according to the procedures outlined in the Guide for the Care and Use of Experimental Animals endorsed by the Canadian Institutes of Health Research and accepted by the Institutional Animal Protection Committee of CRCHUM. These cells maintain a number of characteristics of primary cultured rat vascular smooth muscle cells (RVSMC) including high expression of smooth muscle-specific α-actin, SM22 protein, and myosin light chain kinase (Davis et al., 2003). The RASMC were maintained in Dulbecco's Modified Eagle Medium (DMEM, Invitrogen, Carlsbad, CA) supplemented with 10% fetal bovine serum (FBS) and 100 U/mL penicillin and 100 mg/mL streptomycin. RASMC cell culture was maintained in a humidified atmosphere with 5% CO_2, balance air at 37 °C. To establish quiescence, cells were incubated for 24 h in medium in which concentration of FBS was reduced to 0.2%.

Rat cerebellum granule cells (RCGC) were isolated from 7-day-old rat pups under sterile conditions. The cells were dissociated with 0.25% trypsin-EDTA (Paneco, Russia) for 15 min at 37 °C. The trypsin was inactivated by incubation with 10% fetal bovine serum (Paneco, Russia). Then the cells were cultivated in the following medium: Neurobasal Medium (Gibco, USA), 2% Supplement B-27 (Gibco, USA), 0.5 mM GlutaMAX (Gibco, USA), 100 U/mL penicillin, and 100 mg/mL streptomycin (Paneco, Russia), 20 mM KCl. The cells were plated at a density of 1.2×10^5 cells/cm^2 onto 6-well plates pretreated with poly-L-ornithine (Sigma, USA). The cells were maintained at 37 °C, 90% humidity, 5% CO_2 for 7 days. Before the treatment, the medium was replaced with fresh medium containing 5 mM KCl to increase the ability of ouabain (Sigma, USA) to bind to Na,K-ATPase.

Cell viability was quantitatively assessed by a lactate dehydrogenase (LDH) release assay and the Alamar Blue assay following the manufacturer's protocols. Proapoptotic changes were quantified by measuring activity of caspase-3 and chromatin cleavage. For more details, see Akimova et al. (2015) and Danilov et al. (2017).

2.2 Measurement of intracellular Na^+ and K^+

C2C12 cells in 6-well plates were transferred onto ice, the experimental medium was quickly removed, and the cells were washed three times with 3 mL of ice-cold 0.1 M $MgCl_2$ solution in doubly deionized (DDI) water. Then, 1.5 mL of 5% trichloroacetic acid (TCA) in DDI water was added to each well, followed by incubation at 4 °C overnight for complete extraction of ions from myotubes. Cell precipitates were suspended and centrifuged for 5 min at 18,000 relative centrifugal force. The supernatants were transferred to test tubes and stored at −20°C. The cell pellets were resuspended in 0.75 mL of 0.1 M NaOH and incubated at 65 °C for 1 h for complete protein dissolution. The resulting protein solutions were used for protein quantification by the Lowry protein assay (Lowry, Rosebrough, Farr, & Randall, 1951). The Na^+ and K^+ contents in TCA extracts were measured by flame atomic absorption spectrometry using a Kvant-2m1 spectrometer (Cortec, Russia) with propane-air mixture in accordance with the manual. Solutions of KCl (0.5–4 mg/L K^+) and NaCl (0.05–2 mg/L Na^+) in 5% TCA in DDI water were used for calibration. The Na^+ and K^+ contents in each well were normalized to protein amount in the same well.

Cellular contents of exchangeable K^+ and Na^+ in RAEC and RASMC were measured as accumulation of ^{86}Rb and ^{22}Na, respectively. To establish isotope equilibrium, cells growing in 12-well plates were preincubated for 3 h in DMEM containing 0.5 μCi/mL $^{86}RbCl$ or 3 μCi/mL $^{22}NaCl$, and then ouabain was added and incubated for the next 3 h. Then the cells were transferred onto ice and washed four times with 2 mL of ice-cold medium W containing 100 mM $MgCl_2$ and 10 mM HEPES-Tris buffer (pH 7.4). The washing medium was aspirated, and the cells were lysed in a solution of 1% SDS and 4 mM EDTA. Radioactivity of the incubation media and cell lysates was quantified, and the intracellular cation content was calculated as A/a^*m, where A was the radioactivity of the samples (cpm), a was the specific radioactivity of ^{86}Rb (K^+) and ^{22}Na in the medium (cpm/nmol), and m was the protein content (mg). For more details, see Akimova et al. (2005).

Intracellular contents of Na^+ and K^+ in RCGC were detected by inductively coupled plasma mass-spectrometry (ICP-MS). Cells were seeded onto a six-well plate at a density of 1.2×10^5 cells/cm^2. After 3h of ouabain treatment, the medium was discarded, and the cells were washed twice with cold 0.1 M $MgCl_2$ and lysed in 200 μL trichloroacetic acid. The lysates were transferred to acid-washed and metal-free 0.5 mL clear microtubes and incubated for 12h at 4°C. After the incubation, the samples were brought to 500 μL volume with doubly deionized water. The samples were analyzed on a Varian 820-MS instrument (CA, USA). Concentrations of the selected elements (ng/g) were measured by the external calibration method. Data analysis was performed using ICP-MS Expert version 2.1 b-107 (USA). Na^+ and K^+ contents are expressed as part per million (ppm) per sample.

2.3 Microarray analysis

RNA isolation and microarray assay. Total RNA was isolated using miRNeasy Micro Kits (Qiagen) according to the manufacturer's recommendations, including DNase I step, and eluted in 40 μL of nuclease-free water. Purity and concentration of RNA were measured using a NanoDrop ND-1000 instrument. RNA Integrity Number (RIN) was evaluated using an Agilent 2100 Bioanalyzer (the RIN of intact RNA is 10, and the RIN of fully degraded RNA is 1; RIN score >7 is generally considered sufficient for microarray analysis). Microarray analysis was performed on GeneChip® MoGene 1.0 ST, GeneChip® Rat Gene 1.0 ST, and GeneChip® Rat Gene 2.0 ST Arrays (Affymetrix, ThermoFisher Scientific) according to the manufacturer's recommendations using 500 ng of input RNA. Scans of microarrays were converted into CEL files by the scanner software and then jointly preprocessed in Affymetrix Expression Console (build 1.4.1.46) using the RMA-Sketch method. One-way ANOVA for each pair of condition groups was performed using the Affymetrix Transcriptome Analysis Console (version 3.1.0.5).

2.4 Transcriptome and statistical analysis

Transcriptome analysis. The data were filtered to identify transcripts with statistically significant variation of expression among the groups that are modulated >20%, with multiple testing correction by the false discovery rate (FDR). These data were analyzed using WebGestalt (Wang, Duncan, Shi, & Zhang, 2013; Zhang, Kirov, & Snoddy, 2005). In the case of Gene Set Enrichment Analysis, we used the following parameters: minimum and maximum number of Entrez Gene IDs in category—5 and 2000, respectively; significant level—FDR < 0.05; number of permutations—1000; number of categories with the outputted leading-edge genes—20.

Statistical analysis. All data are expressed as means ± standard error and analyzed by Dunnett's multiple comparison one-way ANOVA test. The threshold for significance was $P < 0.05$.

3. Results

3.1 Change of intracellular Na^+ and K^+ content in ouabain-treated cells

Keeping in mind the low affinity of rodent α1-Na^+,K^+-ATPase for cardiotonic steroids, we exposed cells to 1 (RCGC) or 3 mM (RASMC, RAEC, C2C12 myotubes) for two (C2C12) or three (other cells) hours. In preliminary experiments, we revealed that this procedure does not affect cell survival (data not shown) and increased intracellular Na^+ content in RAEC, RASMC, C2C12, and RCGC by ~14, 8, 7, and 6-fold, respectively (Fig. 1A). As expected, in all the cell types the gain of Na^+ was accompanied by sharp decline of intracellular K^+ content (Fig. 1B). The highest elevation of $[Na^+]_i/[K^+]_i$-ratio was observed in RAEC (by 14-fold) and the lowest in RCGC (by sixfold). It is important to note that cell type-specific differences

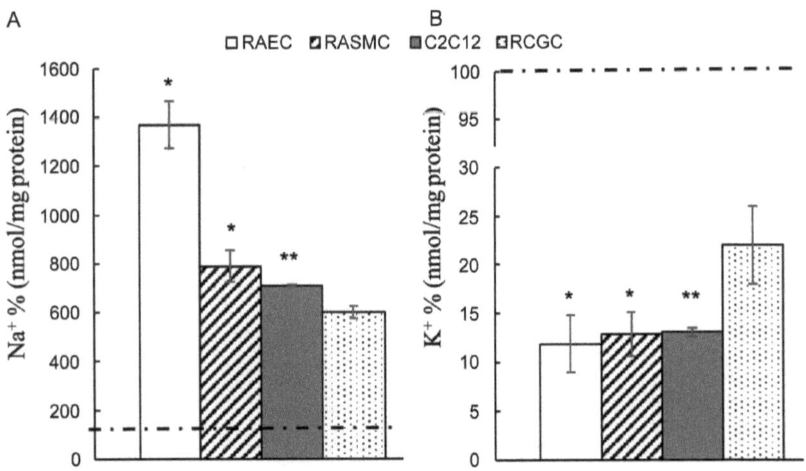

Fig. 1 Effect of ouabain on intracellular Na^+ (A) and K^+ (B) content in different cell types. The cells were treated by 1 (RCGC) or 3 mM ouabain (other cells) during two (C2C12) or three hours (other cells). Baseline values obtained in the absence of ouabain (56 nmol Na^+/mg protein and 722 nmol K^+/mg protein in RAEC; 111 nmol Na^+/mg protein and 1211 nmol K^+/mg protein in RASMC; 132 nmol Na^+/mg protein and 943 nmol K^+/mg protein in C2C12; 1.15 ppm Na^+ per sample and 11.5 ppm K^+ per sample in RCGC) were taken as 100%. Means ± S.E. obtained in three independent experiments are shown. * and **—$P < 0.001$ and 0.0001 compared to control, respectively.

in increments of $Na^+{}_i$ elevation were mainly due to the differences in baseline $Na^+{}_i$ content (see Fig. 1), whereas absolute values of $[Na^+]_i$ in ouabain-treated cells were about the same (766, 876, and 936 nmol/mg protein in RAEC, RASMC, and C2C12, respectively).

3.2 Transcriptomic changes triggered by ouabain

Microarray analysis performed in three independent experiments show that exposure to ouabain resulted in appearance of 2121, 2698, 7577, and 1147 differentially expressed transcripts in C2C12, RASMC, RAEC, and RCGC, respectively (Table 1, for complete list of differentially expressed genes see Supplementary Table S1 in the online version at https://doi.org/10.1016/bs.ctm.2019.01.006). We noted that a number of differentially expressed transcripts under the described conditions correlate with changes of intracellular $[Na^+]_i/[K^+]_i$-ratio: the greater elevation of the $[Na^+]_i/[K^+]_i$-ratio detected in RAEC correlates with highest number of differentially expressed transcripts. These results are consistent with comparative analysis of the action of ouabain and marinobufagenin in human umbilical vein endothelial cells (HUVEC). In those cells, long-term exposure to 3 nM ouabain and 30 nM marinobufagenin resulted in elevation of the $[Na^+]_i/[K^+]_i$-ratio by 14- and 3-fold, respectively; the number of differentially expressed transcripts was two time higher in ouabain-treated cells compared to cells treated with marinobufagenin (Klimanova et al., 2017). With the exception of RAEC, the increments of the content of upregulated transcripts were much higher compared to decrements detected in downregulated ones (Table 1 and Supplementary Table S1 in the online version at https://doi.org/10.1016/bs.ctm.2019.01.006).

3.2.1 *Ubiquitous ouabain-sensitive transcripts*

Eighty-three transcripts were found as the intersection of the four sets of identified transcripts corresponding to each cell type and classified as ubiquitous (Fig. 2). The complete list of ubiquitous transcripts is represented in Table 2. Among the top upregulated ubiquitous transcripts are the following. *Ptgs2* (cyclooxygenase 2) exhibits sharp elevation of its content in all the investigated cell types (fold of change (FC) 10.85, 9.46, 5.76, and 16.59 for RAEC, RASMC, C2C12, and RCGC, respectively). In RAEC, higher than *Ptgs2* increments were demonstrated by *Dusp6* (dual specificity phosphatase 6; FC 32.56), *Plk3* (polo-like kinase 3; FC 26.60), *Trib1* (tribbles pseudokinase 1; FC 15.59), *Ccl7* (chemokine (C-C motif) ligand 7; FC 14.23),

Table 1 Transcriptomic changes in ouabain-treated cells.

	Number of differentially expressed transcripts	Upregulated transcripts		Downregulated transcripts	
		Number	Maximal changes	Number	Maximal changes
RAEC	7577	3234	35.94	4343	27.03
RASMC	2698	1117	9.46	1581	4.15
C2C12	2120	1013	10.99	1107	4.35
RCGC	1146	445	18.82	701	2.68

Transcripts content was changed by more than 1.2-fold ($P < 0.05$).

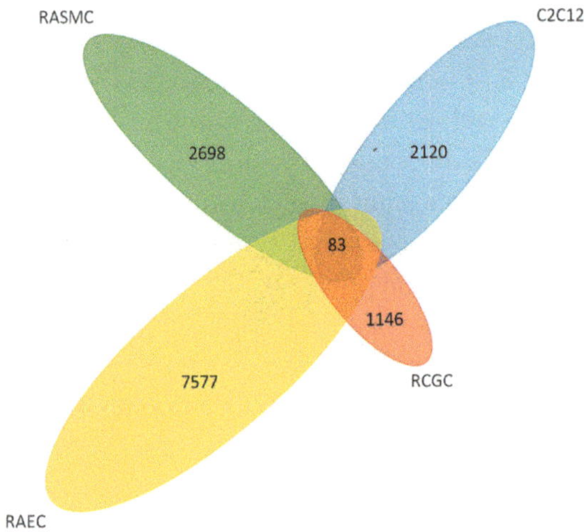

Fig. 2 Venn diagram representing distribution of ubiquitous differentially expressed transcripts identified in RAEC, RASMC, C2C12, and RCGC treated by ouabain as indicated in the legend to Fig. 1. The total numbers of upregulated and downregulated transcripts identified in ouabain-treated cells are presented in Table 1.

Mafk (V-maf musculoaponeurotic fibrosarcoma oncogene family, protein K; FC 12.79), and *Atf3* (activating transcription factor 3; FC 12.33).

Annotation of ubiquitous unigenes based on Gene Ontology (GO) and Kyoto Encyclopedia of Genes and Genomes (KEGG) databases. According to the WEB-based GEne SeT AnaLysis Toolkit, the list of ubiquitous gene (Table 2) contains 83 user IDs in which 82 user IDs are unambiguously mapped to the unique Entrez Gene IDs and one user ID is mapped to multiple Entrez Gene IDs or could not be mapped to any Entrez Gene ID. The 82 transcripts are annotated in three main GO categories: biological process, cellular

Table 2 List of ubiquitous differentially regulated transcripts identified in ouabain-treated RAEC, RASMC, C2C12, and RCGC.

Gene symbol	RAEC fold of change	RASMC fold of change	C2C12 fold of change	RCGC fold of change
Dusp6	32.56	3.36	3.35	3.55
Plk3	26.60	3.29	3.02	2.63
Trib1	15.59	2.67	1.73	4.33
Ccl7	14.23	1.85	1.36	6.88
Mafk	12.79	2.94	1.59	1.78
Atf3	12.33	4.72	1.95	3.02
Ptgs2	10.85	9.46	5.76	16.59
Cxcl1	9.05	2.97	1.35	5.62
Spry4	6.97	1.84	1.36	3.48
Coq10b	6.05	1.71	1.4	4.32
Jun	6.03	1.28	1.9	2.78
Abl2	5.89	1.96	2.09	1.28
Slc25a25	4.72	2.18	2.79	2.92
Zbtb2	4.51	2.26	1.72	1.33
Cyr61	4.45	1.71	2.06	8.09
Per1	4.44	2.72	1.75	3.9
Zswim4	3.75	2.91	2.42	1.49
Sgms2	3.71	1.59	1.27	1.76
Dusp16	3.70	2.33	1.64	1.43
Junb	3.55	4.03	1.65	4.22
Inhba	3.25	1.68	1.3	4.94
Flrt2	3.06	1.99	1.25	1.68
Gmeb2	2.88	1.85	1.35	1.65
Pnrc1	2.72	2.24	1.47	2.87
Klf5	2.69	1.76	1.33	1.62
Rbm7	2.52	1.54	−1.29	1.75

Continued

Table 2 List of ubiquitous differentially regulated transcripts identified in ouabain-treated RAEC, RASMC, C2C12, and RCGC.—cont'd

Gene symbol	RAEC fold of change	RASMC fold of change	C2C12 fold of change	RCGC fold of change
Nfkbia	2.50	2.46	2.14	3.14
Chd1	2.50	1.35	1.25	1.51
Gem	2.48	2.32	2	8.45
Rasa2	2.44	1.39	1.22	1.46
Zfp36	2.38	3.52	3.48	7.56
Topors	2.38	1.54	1.37	1.36
Ccnl1	2.29	2.10	2.04	3.56
Chka	2.22	2.79	2.46	1.92
Il1rap	2.20	1.32	1.25	2.15
Ccnt1	2.14	1.53	1.34	1.6
Arl5b	2.12	3.24	1.51	2.12
Pip5k1a	2.09	1.71	1.23	1.53
Rictor	2.03	1.54	1.4	1.35
Btg1	1.99	1.75	1.92	1.51
Insig1	1.84	1.99	2.02	1.87
Rassf1	1.84	1.59	1.62	1.55
Atp7a	1.79	1.52	1.37	1.34
Yod1	1.78	1.71	1.53	1.59
Jmjd1c	1.75	1.49	1.51	1.27
Klf6	1.71	1.61	2.11	5.52
Epc1	1.69	1.34	1.49	1.36
Clk4	1.68	1.55	1.44	−1.7
Ldlr	1.57	1.50	1.46	2.31
Slc39a14	1.57	1.24	1.39	2.14
Idi1	1.56	1.93	1.3	1.85
Fgf2	1.55	−1.26	−1.23	1.54

Table 2 List of ubiquitous differentially regulated transcripts identified in ouabain-treated RAEC, RASMC, C2C12, and RCGC.—cont'd

Gene symbol	RAEC fold of change	RASMC fold of change	C2C12 fold of change	RCGC fold of change
Mxd1	1.55	1.83	1.58	2.01
Htr2a	1.48	2.01	1.39	1.55
Jag1	1.46	1.63	1.42	1.72
Taf7	1.43	1.32	1.46	1.42
Bach1	1.35	1.26	1.32	1.45
Otud1	1.33	−1.41	1.69	1.26
Arid5b	1.29	1.40	1.68	2.35
B4galt5	1.22	1.32	1.7	1.41
Wdr53	−1.27	−1.35	−1.31	−1.4
Errfi1	−1.30	3.14	1.87	1.54
Mrps31	−1.32	−1.23	−1.21	−1.28
Dnajc6	−1.38	1.24	−1.23	1.26
Mcat	−1.50	−1.42	−1.26	−1.77
Asf1a	−1.50	−1.28	−1.33	−1.5
Apaf1	−1.54	1.21	1.39	−1.31
Palb2	−1.55	−1.34	−1.3	−1.78
Pex26	−1.68	−1.28	−1.29	−1.34
Poldip2	−1.80	−1.25	−1.37	−1.33
Asb8	−1.82	−1.84	−1.28	−1.85
Cep68	−1.83	−1.77	−1.3	−1.63
Invs	−2.04	−1.22	−1.21	1.27
Spata6	−2.17	−1.28	−1.25	−1.79
Aaas	−2.56	−1.21	−1.26	−1.22
Ssbp1	−2.92	−1.81	−1.66	−1.3
Rgma	−2.95	−1.66	−1.38	−1.91
Krcc1	−3.10	−1.37	−1.24	−2.57

Continued

Table 2 List of ubiquitous differentially regulated transcripts identified in ouabain-treated RAEC, RASMC, C2C12, and RCGC.—cont'd

Gene symbol	RAEC fold of change	RASMC fold of change	C2C12 fold of change	RCGC fold of change
Dnajc28	−3.30	−1.67	−1.24	−1.7
Ubac1	−3.55	−1.22	−1.3	−1.47
Golim4	−3.95	−1.40	−1.27	−1.84
Zfand1	−4.92	−1.31	−1.26	−1.41
Ptcd2	−5.61	−2.90	−1.45	−1.41

Table lists ubiquitous differentially expressed transcripts whose content was changed by more than 1.2-fold ($P<0.05$).

component, molecular function (Fig. 3). The category "biological process" is divided into 12 types. Transcript expression changes detected in our experiments are linked to metabolic processes (52.4%), biological regulation (50.0%), response to stimulus (41.5%), multicellular organismal processes (35.4%), developmental processes (34.2%), localization (30.5%), cell communication (29.3%), cellular component organization (26.8%), cell proliferation (18.3%), reproduction (9.8%), multi-organism processes (9.8%), and growth (8.5%). The category "cellular components" is divided into 19 types, which are distributed as follows: nucleus (34.2%), membrane (23.2%), macromolecular complex (22.0%), membrane-enclosed lumen (19.5%), endomembrane system (13.4%), cytosol (12.2%), vesicle (9.8%), extracellular space (7.3%), and Golgi apparatus (7.3%). In the category "molecular functions" ubiquitous transcripts are distributed among 13 types, the main being protein binding (42.7%), nucleic acid binding (17.0%), ion binding (13.4%), nucleotide binding (8.5%), transferase activity (8.5%), hydrolase activity (8.5%), and enzyme regulator activity (7.3%).

Among the 82 unique Entrez Gene IDs, 41 IDs are annotated to the selected functional categories, which are used for enrichment analysis. Based on chosen analysis parameters, no enriched category is identified.

3.2.2 Cell type-specific ouabain-sensitive transcripts

Along with the set of ubiquitous genes ouabain treatment results in appearance of unique transcripts whose differential expression is cell type-specific. Table 3 shows that this category includes 4897, 1523, 789, and 494 transcripts for RAEC, C2C12, RASMC, and RCGC, respectively. Complete

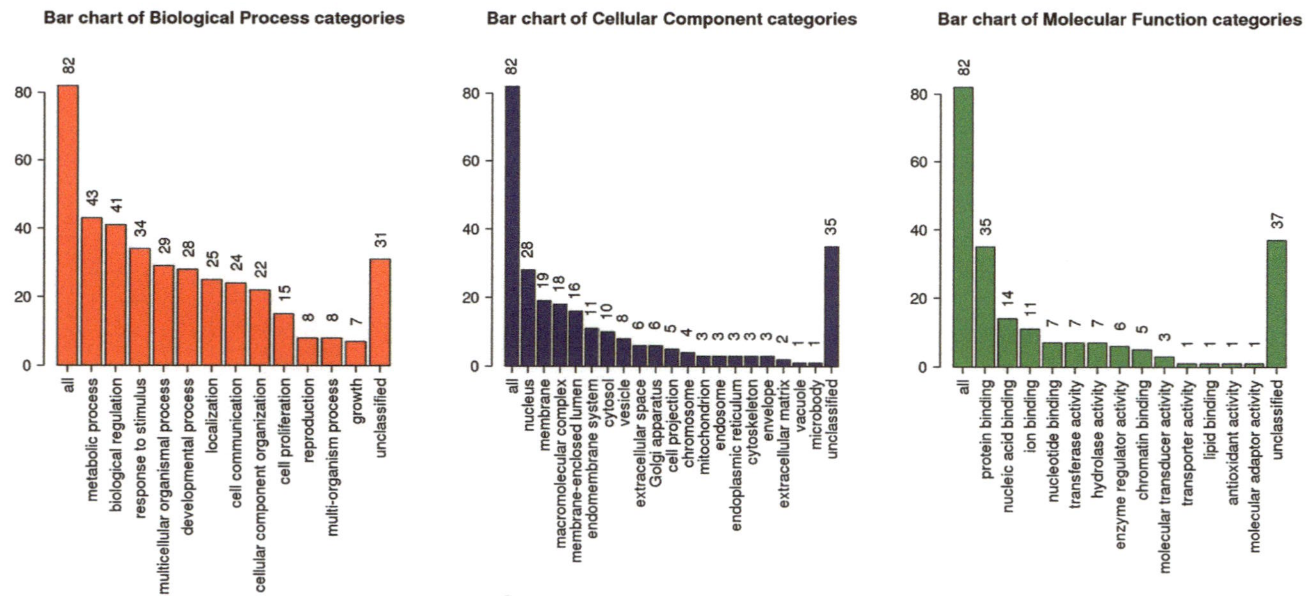

Fig. 3 Significantly enriched GO terms in each of the biological processes, cellular components, and molecular functions for the ubiquitous genes identified in ouabain-treated RAEC, RASMC, C2C12, and RCGC.

lists of cell-specific transcripts are presented in Supplementary Table S2 in the online version at https://doi.org/10.1016/bs.ctm.2019.01.006.

Annotation of cell-specific unigenes based on Gene Ontology (GO) and Kyoto Encyclopedia of Genes and Genomes (KEGG) databases. The GO database allowed us to classify cell type-specific genes into three major categories: biological processes (Supplementary Table S3 in the online version at https://doi.org/10.1016/bs.ctm.2019.01.006), cellular components (Supplementary Table S4 in the online version at https://doi.org/10.1016/bs.ctm.2019.01.006), and molecular functions (Supplementary Table S5 in the online version at https://doi.org/10.1016/bs.ctm.2019.01.006). We did not observe any drastic cell type-specific differences in the distribution of differentially expressed genes according to these three categories. We also noted that the distribution of these cell-specific transcripts occurs in about the same way as we revealed in the analysis of the ubiquitous ouabain-sensitive transcripts (Fig. 3).

Analysis of the total number of transcripts whose expression is changed after cell treatment by ouabain demonstrated existence of 63 signal pathways in all cell types, and only two of them were negative regulated—*DNA replication* (RAEC) and *olfactory transduction* (C2C12). Common signal pathways for all four cell types are *HTLV-I infection, leishmaniasis*, and *transcriptional misregulation in cancer* (Table 4).

Applying only cell-specific transcripts to KEGG enrichment is not correct, since ubiquitous transcripts (not considered in this case) can be participants of several signaling pathways simultaneously. Therefore, total differentially expressed transcripts for each ouabain-treated cell type were

Table 3 Cell-specific differentially expressed transcripts detected in ouabain-treated RAEC, RASMC, C2C12, and RCGC.

	Number of cell type-specific differentially expressed transcripts	Upregulated transcripts		Downregulated transcripts	
		Number	Maximal changes	Number	Maximal changes
RAEC	4522	1792	21.45	2730	14.29
RASMC	758	254	3.38	504	3.49
C2C12	1140	414	5.64	726	4.35
RCGC	465	181	12.97	284	2.51

Table lists cell-specific differentially expressed transcripts whose content was changed by more than 1.2-fold ($P < 0.05$).

Table 4 63 significant KEGG pathways enriched of the total differentially expressed genes identified in ouabain-treated RAEC, RASMC, C2C12, and RCGC. Upregulated and downregulated pathways are shown in regular and *italics*, respectively.

	RAEC		RASMC		C2C12		RNC	
Pathways names	**#**	**FDR**	**#**	**FDR**	**#**	**FDR**	**#**	**FDR**
Adipocytokine signaling pathway	–	–	14	2.23E–02	–	–	–	–
AGE-RAGE signaling pathway in diabetic complications	47	3.34E–03	19	8.56E–03	–	–	10	1.87E–02
Alcoholism	–	–	–	–	30	0.00E+00	–	–
Aldosterone synthesis and secretion	–	–	8	3.86E–02	–	–	–	–
Amoebiasis	–	–	–	–	–	–	6	4.61E–02
Amphetamine addiction	–	–	–	–	9	4.91E–02	–	–
Apoptosis	–	–	28	1.27E–02	–	–	–	–
Bladder cancer	–	–	10	2.67E–02	–	–	–	–
Chagas disease (American trypanosomiasis)	45	5.98E–03	21	1.12E–02	–	–	5	2.23E–02
Chemokine signaling pathway	72	3.62E–03	20	2.48E–02	–	–	10	1.89E–02
Chronic myeloid leukemia	–	–	15	2.17E–02	–	–	–	–
Circadian rhythm	19	3.78E–02	9	1.14E–02	–	–	–	–
Complement and coagulation cascades	17	2.94E–02	–	–	–	–	–	–
Cytokine-cytokine receptor interaction	66	0.00E+00	32	9.03E–04	–	–	11	8.05E–03
Cytosolic DNA-sensing pathway	–	–	–	–	–	–	6	3.55E–02

Continued

Table 4 63 significant KEGG pathways enriched of the total differentially expressed genes identified in ouabain-treated RAEC, RASMC, C2C12, and RCGC. Upregulated and downregulated pathways are shown in regular and *italics*, respectively.—cont'd

	RAEC		RASMC		C2C12		RNC	
Pathways names	**#**	**FDR**	**#**	**FDR**	**#**	**FDR**	**#**	**FDR**
DNA replication	25	1.72E–02	–	–	–	–	–	–
EGFR tyrosine kinase inhibitor resistance	–	–	–	–	–	–	8	2.91E–02
Endocrine resistance	–	–	13	2.96E–02	–	–	–	–
Endocytosis	–	–	29	2.23E–02	–	–	–	–
ErbB signaling pathway	52	2.37E–04	18	1.10E–02	13	8.61E–03	–	–
FoxO signaling pathway	–	–	26	1.08E–02	–	–	10	4.66E–02
Hematopoietic cell lineage	24	2.76E–04	8	1.42E–02	–	–	–	–
Hepatitis B	–	–	32	1.20E–02	23	3.56E–02	10	2.97E–02
Herpes simplex infection	–	–	34	1.42E–02	–	–	11	1.25E–02
HIF-1 signaling pathway	48	2.03E–02	–	–	–	–	9	4.45E–02
Hippo signaling pathway	–	–	–	–	16	2.61E–02	–	–
HTLV-I infection	113	2.24E–02	50	0.00E+00	28	0.00E+00	18	1.42E–03
Inflammatory bowel disease (IBD)	15	1.15E–02	–	–	–	–	5	4.43E–02
Influenza A	–	–	26	2.73E–02	–	–	10	1.46E–02
Insulin resistance	–	–	17	1.86E–02	–	–	–	–

Intestinal immune network for IgA production	8	1.21E–02	–	–	–	–	–	–
Jak-STAT signaling pathway	59	6.04E–04	23	0.00E + 00	–	–	7	4.86 E–02
Legionellosis	–	–	12	1.41E–02	–	–	5	1.78E–02
Leishmaniasis	21	3.88E–02	12	1.83E–02	6	1.63E–02	5	2.06E–02
Malaria	12	2.14E–02	8	2.16E–02	–	–	5	2.05E–02
MAPK signaling pathway	–	–	45	3.15E–02	36	3.50E–04	–	–
Measles	–	–	–	–	–	–	10	4.58E–02
Metabolic pathways	–	–	132	4.71E–02	–	–	–	–
MicroRNAs in cancer	–	–	26	5.06E–03	–	–	13	4.55E–02
Neurotrophin signaling pathway	–	–	17	3.16E–02	–	–	–	–
NF-kappa B signaling pathway	37	2.24E–02	16	1.97E–03	–	–	9	1.20E–02
NOD-like receptor signaling pathway	–	–	25	2.63E–03	–	–	13	1.44E–02
Non-alcoholic fatty liver disease (NAFLD)	–	–	16	1.54E–03	–	–	–	–
Olfactory transduction	–	–	–	–	52	0.00E + 00	–	–
Osteoclast differentiation	–	–	22	5.08E–03	17	3.10E–02	8	1.17E–02
Oxytocin signaling pathway	–	–	14	1.01E–02	–	–	11	1.53E–02
Pathways in cancer	–	–	60	1.07E–02	–	–	24	8.05E–03
Pertussis	24	4.78E–02	8	1.33E–02	–	–	6	1.64E–02
PI3K-Akt signaling pathway	–	–	35	1.51E–03	–	–	17	3.07E–02

Continued

Table 4 63 significant KEGG pathways enriched of the total differentially expressed genes identified in ouabain-treated RAEC, RASMC, C2C12, and RCGC. Upregulated and downregulated pathways are shown in regular and *italics*, respectively.—cont'd

	RAEC		RASMC		C2C12		RNC	
Pathways names	**#**	**FDR**	**#**	**FDR**	**#**	**FDR**	**#**	**FDR**
Prion diseases	15	2.66E–02	5	4.66E–02	–	–	–	–
Prolactin signaling pathway	–	–	8	2.15E–02	–	–	–	–
Rheumatoid arthritis	27	0.00E+00	18	4.90E–03	–	–	–	–
Salmonella infection	33	0.00E+00	14	1.00E–02	–	–	6	1.74E–02
Signaling pathways regulating pluripotency of stem cells	–	–	20	1.50E–02	–	–	10	3.13E–02
Small cell lung cancer	41	5.00E–03	17	4.92E–03	–	–	–	–
Systemic lupus erythematosus	–	–	–	–	22	0.00E+00	–	–
TGF-beta signaling pathway	–	–	17	1.04E–02	–	–	–	–
Th17 cell differentiation	–	–	19	1.24E–02	–	–	12	1.02E–02
TNF signaling pathway	62	0.00E+00	33	0.00E+00	–	–	10	0.00E+00
Toll-like receptor signaling pathway	41	2.92E–02	20	4.99E–03	–	–	7	1.55E–02
Transcriptional misregulation in cancer	72	0.00E+00	33	0.00E+00	28	1.40E–03	14	1.63E–02
Tuberculosis	62	2.15E–02	24	4.71E–03	–	–	10	3.37E–02
Viral carcinogenesis	–	–	–	–	31	1.91E–03	–	–

KEGG analysis was performed using following parameters: minimum number of Entrez Gene IDs in category—5; maximum number of Entrez Gene IDs in category—2000; significant level—FDR <0.05; number of permutation—1000; number of categories with the outputted leading-edge genes—20. FDR = 0.00E+00 means high significant level.

annotated by KEGG. In this case the cell-specific pathway means the pathway is regulated strictly in a certain cell type. As shown in Table 4, the largest number of such pathways (13) is in RASMC: *adipocytokine signaling pathway, aldosterone synthesis and secretion, apoptosis, bladder cancer, chronic myeloid leukemia, endocrine resistance, endocytosis, insulin resistance, metabolic pathways, neurotrophin signaling pathway, non-alcoholic fatty liver disease (NAFLD), prolactin signaling pathway, TGF-beta signaling pathway.* And the least number of cell-specific pathways (4) is in RAEC: *amoebiasis, cytosolic DNA-sensing pathway, EGFR tyrosine kinase inhibitor resistance, measles.*

4. Physiological implications of the elevation of $[Na^+]_i/[K^+]_i$-ratio

Along with the inhibition of Na^+,K^+-ATPase by augmented production of endogenous cardiotonic steroids, dissipation of transmembrane gradients in electrically excitable cells might be triggered by sustained activation of voltage-gated and Ca^{2+}-sensitive Na^+- and K^+-channels. Here, we discuss two tissue-specific examples.

4.1 Excitation of neuronal cells

Almost 30 years ago, several research teams reported that expression of *Egr1*, *Fos*, and other ouabain-sensitive immediate response genes listed in Supplementary Table S1 in the online version at https://doi.org/10.1016/bs.ctm.2019.01.006 is strongly increased in neuronal cells subjected to excitation by diverse stimuli including neurotransmitters, depolarization, and light (Cole, Saffen, Baraban, & Worley, 1989; Hunt, Pini, & Evan, 1987; Morgan & Curran, 1989). The mechanisms of this phenomenon, widely employed for the identification of excited brain areas, remain largely unknown. It has been shown that activation of *N*-methyl-D-aspartate (NMDA) receptors, functioning as glutamate-gated ion channels, contributes to elevation of *Egr1* mRNA content in excited neuronal cells. This link is so prominent that monitoring *Egr1* expression has been proposed as a biological assay for NMDA receptor activity. Given the key role of NMDA receptor activation in regulating synaptic strength, *Egr1* accumulation has also been connected with learning and memory (for review, see Thiel, Mayer, Müller, Stefano, & Rössler, 2010). Indeed, experiments performed on *Egr1*-deficient mice showed their inability to form long-term memory in a variety of behavioral tasks (Jones et al., 2001). It is generally accepted that *Egr1* expression in neuronal cells is triggered by NMDA receptor-mediated elevation of $[Ca^{2+}]_i$ and activation of Ca^{2+}-mediated signaling (Alberini, 2009; Kobrinsky, 2015). However, ion currents through NMDA channels are mainly mediated by monovalent cations ($P_{Na} \sim P_K \gg P_{Ca}$). In addition, activation of glutamate-gated Na^+-selective α-amino-3-hydroxy-5-methyl-4-isoxazolepropionic acid (AMPA)-sensitive receptor precedes activation of NMDA receptors (Bredt & Nicoll, 2003). Viewed collectively, these data suggest that along with transient increment of $[Ca^{2+}]_i$, long-term synaptic activity is accompanied by elevation of the $[Na^+]_i/[K^+]_i$-ratio. This hypothesis is concordant with numerous reports showing that neuronal activity is accompanied by elevation of $[K^+]$ and attenuation of $[Na^+]$ in cerebrospinal fluid (for review, see Deitmer & Rose, 2010; Orlov & Mongin, 2007; Somjen, 2002; Verkhratsky, Noda, Parpura, & Kirischuk, 2013). In primary cultures of rat cerebellar granule cells loaded with the Na^+-sensitive fluorescent indicator SBFI, activation of NMDA receptors by glutamate resulted in reversible elevation of $[Na^+]_i$ from 5 to 60 mM (Kiedrowski, Wroblewski, & Costa, 1994). Using tissue slices of mouse cerebellum loaded with SBFI, it was shown that in dendrites of Purkinje neurons glutamate transiently increases $[Na^+]_i$ by 5–10 mM *via* activation of AMPA receptors (Bennay, Langer, Meier, Kafitz, & Rose, 2008). In Purkinje neurons, transient increments of $[Na^+]_i$ were also evoked by brief (50 Hz/100 ms) electrical pulse stimulation (Bennay et al., 2008). In dopamine cells of the *substantia nigra pars compacta*, electrical hyperpolarization induced a rise of $[Na^+]_i$ from 10 to 25 mM (Knöpfel, Guatteo, Bernardi, & Mercuri, 1998). The relative impact of Na^+_i/K^+_i-mediated, Ca^{2+}-independent excitation-transcription coupling in learning and memory formation should be examined further.

4.2 Intensive exercise

Calculating femoral arterial-venous Na^+ and K^+ concentration differences, it was found that in humans intense knee extensor exercise rapidly increased intracellular Na^+ concentration ($[Na^+]_i$) in skeletal muscle from 13 to 23 mM and reduced $[K^+]_i$ from 162 to 129 mM (Sjøgaard, Adams, & Saltin, 1985). Using flame photometry, Murphy and coworkers found that in rat soleus muscle 60 min of intermittent running resulted in about twofold elevation of $[Na^+]_i$ and attenuation of $[K^+]_i$ by 10–20% (Murphy, Nielsen, & Clausen, 2008). Recently, we reported that electrical pulse stimulation (EPS) of C2C12 myotubes resulted in about fivefold elevation of $[Na^+]_i/[K^+]_i$-ratio (Danilov et al., 2017). Considering this, it might be proposed that along with the above-listed signaling pathways, elevation of $[Na^+]_i/[K^+]_i$-ratio detected in exercising skeletal muscle and EPS-treated myotubes contributes to myokine production (Kapilevich et al., 2015). This hypothesis is supported by several observations. *First*, sustained inhibition of Na^+,K^+-ATPase affects expression of hundreds of genes in the presence of extracellular and intracellular Ca^{2+}-chelators (Koltsova et al., 2012) and inhibitors of $Ca^{2+}{}_i$-mediated signaling pathways (Smolyaninova, Koltsova, Sidorenko, & Orlov, 2017). *Second*, in cultured vascular smooth muscle cells subjected to hypoxia and glucose-deprivation, transcriptomic changes are partially caused by activation of $[Na^+]_i/[K^+]_i$-mediated, HIF-1α-independent signaling (Koltsova et al., 2014). *Third*, we recently reported that in C2C12 cells short-term electrical pulse stimulation triggers expression of potent transcription regulators including immediate response genes such as *Fos, Egr1, Myc, Egr2, Atf3, Fosl1, Fosb*, 12 histone clusters, and 29 microRNAs (Sidorenko et al., 2018). These genes might be considered as upstream intermediates of differential expression of late response genes including myokines whose expressions were slightly increased by 2-h electrical pulse stimulation (Sidorenko et al., 2018) and further potentiated after prolongation of EPS to 48 h (Nedachi et al., 2008; Nikolić et al., 2012; Whitham et al., 2012). This hypothesis is consistent with profound accumulation of IRGs in skeletal muscle biopsies obtained from humans after intensive resistance exercises (Broholm et al., 2011). Importantly, the same genes were also found in ouabain-treated C2C12 cells (Supplementary Table S1 in the online version at https://doi.org/10.1016/bs.ctm.2019.01.006), thus suggesting an impact of $Na^+{}_i/K^+{}_i$-mediated, Ca^{2+}-independent excitation-transcription coupling in myokine production.

5. Functional characterization of ubiquitous and cell type-specific Na_i^+/K_i^+-sensitive genes and their pathophysiological implications

5.1 Ubiquitous genes

Among the top 10 upregulated *ubiquitously expressed ouabain-sensitive* genes we found *Dusp6, Plk3, Trib1, Atf3, Ccl7, Cxcl1, Ptgs2, Mafk, Spry4*, and *Coq10b*. The product of *Dusp6* is inducible MAPK phosphatase, which displays a high level of substrate specificity for Erk1/2 (Keyse, 2008). It is known that the *Dusp6* gene is upregulated in the case of obesity and insulin resistance and plays an important role in the development of obesity-related hyperglycemia (Wu et al., 2010); patients with multiple sclerosis also demonstrated overexpression of *Dusp6* in optic nerves (Domercq et al., 2011).

A rise of *Dusp6* expression was observed also at ischemia/reperfusion in human endothelial cells, which resulted in their apoptosis (Yang, Xie, & Liu, 2012). It was noted that *Dusp6* may be associated with cancer development, and its effect depends upon the cell type. *Dusp6* was shown to provide cell growth-promoting effect in myeloid leukemia (Arora et al., 2012), glioblastoma (Messina et al., 2011), and thyroid carcinoma cells (Lee et al., 2012), and in contrast it displayed suppressive effect in pancreatic (Furukawa, Sunamura, Motoi, Matsuno, & Horii, 2003) and lung cancers (Zhang et al., 2010).

Plk3 encodes Polo-like kinase 3, which is considered as an important regulator of cell cycle progression. It is commonly accepted that *Plk3* is upregulated in response to various stress stimuli like DNA damage (Xie, Wang, et al., 2001, Xie, Wu, et al., 2001) and ionizing radiation (Keam et al., 2018; Kis et al., 2006; Zhou et al., 2006). Plk3 can also display chaperon activity, providing degradation of misfolded prion protein aggregates (Wang et al., 2015). It is difficult to establish clear significance of Plk3 in tumorigenesis (Helmke, Becker, & Strebhardt, 2016) because its effect (like the effect of Dusp6) depends upon cell and tumor types.

Trib1 encodes an adapter protein that is involved in protein degradation; it also regulates MAP-kinases, despite the absence of its own kinase activity (Eyers, Keeshan, & Kannan, 2017). The product of this gene was shown to increase cholesterol efflux from low-density lipoproteins and to suppress macrophage foam cell formation (Fu, Zhao, & Huang, 2017); it also positively regulates hepatocyte NF4-α, which appears to support normal liver physiology (Soubeyrand, Martinuk, & McPherson, 2017). Analysis of gene expression in patients with quiescent systemic lupus erythematosus revealed overexpression of *Trib1* in B-cells. This was later explained by data showing that the product of *Trib1* is negative regulator of B-cells, displaying immunoregulatory action on them (Simoni et al., 2018). Protein encoding by *Trib1* provides migration and invasion of colorectal cancer cells, and its overexpression correlates with poor prognosis in these patients (Wang et al., 2017). There are other data demonstrating increased level of *Trib1* expression for some other types of cancer (Eyers et al., 2017); in glioma it is considered as factor that is responsible for the development of cell radioresistance (Tang et al., 2015).

The protein encoded by *Atf3* is activating transcription factor 3, a member of the cyclic adenosine monophosphate response element binding family that acts as either a repressor or an activator of transcription *via* the formation of homodimers or heterodimers with other ATF/CREB members. Its expression in cells is usually increased in response to stress. Its physiological function is connected with DNA damage repair and cell-cycle regulation (Rohini, Haritha Menon, & Selvamurugan, 2018). Being involved in activation of p53 that is a suppressor of tumor development, it participates in carcinogenesis. It was also shown to exert protective effects against myocardial ischemia-reperfusion injury (Yang, Yang, Fan, & Yang, 2016), and it might participate in the development of cardiac hypertrophy (Zhou et al., 2014).

Products of the *Ccl7* and *Cxcl1* genes are small proteins of the CC and CXC chemokine families, respectively. Protein CCL7, also known as monocyte-chemotactic protein 3 (MCP3), is widely expressed in multiple cell types (Liu, Cai, Liu, Wu, & Xiong, 2018) including certain tumor cell lines and macrophages (Opdenakker, Froyen, Fiten, Proost, & Vandamme, 1993). It can participate in anti-inflammatory responses through binding to its receptors to mediate the recruitment of immune cells (Liu et al., 2018). *Cxcl1* encodes C-X-C motif ligand 1 proteins previously called GRO1 oncogene or melanoma growth stimulating activity alpha (MGSA-α). CXCL1 is secreted by human melanoma cells, has mitogenic properties, and is implicated in the pathogenesis of melanoma (Anisowicz, Bardwell, & Sager, 1987; Richmond & Thomas, 1988). CXCL1 is expressed by macrophages, neutrophils, and epithelial cells (Becker, Quay, Koren, & Haskill, 1994; Iida & Grotendorst, 1990) and has neutrophil chemoattractant activity; it is involved in the processes of angiogenesis, arteriogenesis, inflammation, wound healing, and tumorigenesis (Devalaraja et al., 2000; Haghnegahdar et al., 2000; Owen et al., 1997). Like many other chemokines, CXCL1 plays an important role in the development of atherosclerosis, and expression of CXCL1 can be downregulated by miR-181b, which inhibits nuclear factor-κB activation in atherosclerotic endothelium, thus attenuating the adhesive properties of endothelial cells and exerting early atheroprotective effects (Hartmann, Schober, & Weber, 2015).

Ptg2 encodes an inducible isoform of cyclooxygenase (COX-2). Induction of COX-2 by any stimulus quickly results in the biosynthesis mainly of prostaglandins E2 (PGE2) and prostacyclin. These prostanoids, in turn, are involved in diverse physiological and pathophysiological processes including renal failure (Rios, Vargas-Robles, Gámez-Méndez, & Escalante, 2012), control of blood pressure, endothelial thromboresistance, inflammation, pulmonary diseases, (Dubois et al., 1998) and tumorigenesis (Hashemi Goradel, Najafi, Salehi, Farhood, & Mortezaee, 2018). We recently observed that protection against fibrosis of cultured human lung fibroblasts by ouabain and K^+-free medium is accompanied by sharp elevation of COX-2 mRNA and protein content (La et al., 2016). Previously, it was shown that the content of PGE2 is sharply decreased in bronchoalveolar lavage from patients with idiopathic pulmonary fibrosis (Borok et al., 1991; Ozaki et al., 1990).

The product of *Mafk* belongs to the super-enhancers-mediated networks (Yuan et al., 2017). Super-enhancers are associated with genes that are involved in cell identity and can play an important role in transcriptional misregulation in cancer (Hnisz et al., 2013). For example, Mafk induces epithelial-mesenchymal transition and promotes tumorigenic growth of triple-negative breast cancer cells (Okita et al., 2017).

The protein encoded by *Spry4* is known as a negative regulator of the MAPK/ERK signaling cascade. It can play the role of an oncogene in testicular germ cell tumors by activating the PI3K/Akt signaling pathway (Das, Furu, Evensen, Haugen, & Haugen, 2018). However, there are data demonstrating that Spry4 possesses suppressive activity in relation to breast cancer (Jing et al., 2016), as downregulation of this gene was associated with breast cancer progression (Tian et al., 2018). Besides this, in glioma cells *Spry4* was significantly downregulated (Chai et al., 2017). By inhibiting ERK signaling, proteins Spry1 and Spry4 have opposing effects on Akt activation and in this way regulate proliferation and differentiation in many cell types including endothelial (Impagnatiello et al., 2001) and C2C12 (de Alvaro, Martinez, Rojas, & Lorenzo, 2005) cells. Spry4 was shown to inhibit Akt activation and proliferation of vascular smooth cells, their migration, and marker gene expression (Yang et al., 2013).

The protein encoded by *Coq10b* (coenzyme Q-binding protein; COQ10 homolog B) participates in coenzyme Q (ubiquinone, UQ) transport to mitochondria, which can affect the function of the respiratory chain (Barros et al., 2005), and it serves also as an important antioxidant; biochemical alterations that have been observed in UQ-deficient mutants include respiratory chain defects and altered ROS metabolism (Ying Wang & Hekimi, 2013). It is known that coenzyme Q10 decreases the injury of endothelium that is produced by oxidized low-density lipoprotein and the formation of atherosclerotic plaques (Tsai et al., 2012).

Thus, analysis of the 10 top upregulated ubiquitous genes led us to the conclusion that dissipation of the gradient of monovalent cations affected mainly genes participating in regulation of gene transcription, cell proliferation, and tumorigenesis (*Dusp6*, *Plk3*, *Trib1*, *Mafk*, *Atf3*, *Spry4*).

5.2 Cell type-specific genes

It is known that cardiovascular and pulmonary diseases are interrelated. The risk of cardiovascular diseases rises in patients with chronic obstructive pulmonary disease. Numerous factors, in particular hypoxia systemic inflammation and oxidative stress promote development of these diseases (Maclay, McAllister, & MacNee, 2007). The endothelium appears to play an important role in the development of both diseases. The primary function of blood vessel endothelium is to coordinate delivery of blood flow to all other organs and tissues in accord with metabolic demand. During its functioning, it interacts with vascular smooth muscle cells, whose dysfunction affects their coordinative interaction and contributes in the development of cardiac and pulmonary illnesses. It is known also that connective tissue remodeling mediated by smooth muscle cells can result in the development of hypertension (Mecham et al., 1987). Taking this into account, we limit this section to the top 10 upregulated cell-specific transcripts detected in endothelial and smooth muscle cells (Table 5).

Olr1 (or *Lox1*) encodes a low-density lipoprotein receptor that belongs to the C-type lectin superfamily. The encoded protein binds, internalizes, and degrades oxidized low-density lipoprotein. It may also be involved in the regulation of Fas-induced apoptosis. Mutations of this gene have been associated with atherosclerosis, risk of myocardial infarction, and may modify the risk of Alzheimer's disease. Olr1 appears to play a significant role in atherogenesis (Sawamura et al., 1997). *In vivo* overexpression of *Olr1* occurs in hypertension, hyperlipidemia, and diabetes. Besides its main ligands, Olr1 is able to recognize apoptotic or aged cells, activated thrombocytes, and bacteria. All these properties led to considering *Olr1* as an important participant of vascular disorders (Chen, Masaki, & Sawamura, 2002).

Slc9a3 encodes Na^+/H^+ exchanger 3 (or solute carrier family 9 member 3). This antiporter catalyzes the electroneutral exchange of extracellular sodium for intracellular proton across cellular membranes and plays a key role in the regulation of intracellular pH, cell volume, and proliferative properties of many cell types, in particular endothelial cells (Bussolino et al., 1989; Roos & Boron, 1981). Protective effects of tenapanor (previously referred to as RDX5791 or AZD1722), an inhibitor of Na^+/H^+ exchanger 3 against Na^+-driven cardiac and renal damage in rodents, are well-documented (Spencer et al., 2014). The Na^+/H^+ exchanger is often associated with hypertension and organ ischemia, but its role in the development of these pathologies is not completely clear (Bobulescu, Di Sole, & Moe, 2005).

Table 5 List of top 10 upregulated cell-specific differentially expressed genes identified in ouabain-treated RAEC, RASMC, C2C12 and RCGC.

RAEC (fold of change)		**RASMC (fold of change)**		**C2C12 (fold of change)**		**RCGC (fold of change)**	
LOC100364099	21.45	LOC500007	3.38	Hist1h3b	5.64	Npas4	12.97
Olr1	16.65	Mobkl2c	2.98	Tmem95	5.07	Procr	4.12
RGD1563441	15.96	Jmjd3	2.84	Mir1938	3.89	Rd3l	3.94
Slc9a3	9.68	RGD1562174	2.82	Hist1h3c	3.85	Cxcl6	3.86
Slco4a1	8.75	LOC691431	2.59	Hist1h4n	3.82	Vgf	3.32
LOC100364783	8.44	Tbx3	2.36	Hist1h2bj	3.8	Rnf122	2.94
Epgn	7.86	Socs3	2.34	Hist1h2ag	3.44	Mstn	2.87
Dcx	7.65	Rsad2	2.25	Hist1h2bk	3.16	LOC102552920	2.45
Prl6a1	7.58	RGD1566124	2.18	Mir697	2.88	Xirp1	2.35
Cyp1a1	7.44	RGD1309969	2.14	Hist1h3e	2.76	Fgf9	2.33

Table lists ubiquitous differentially expressed transcripts whose content was changed by >1.2-fold ($P<0.05$).

Increased level of *Slco4a1* (solute carrier organic anion transporter family member 4A1) was found in four subtypes of human lung cancer cell lines (Watanabe et al., 2010). Idiopathic interstitial pneumonias that are group of lung diseases associated with inflammation and fibrosis are associated with overexpression of some genes including *Slco4a1* in lung tissue (Chen et al., 2014). Some data indicate that *Slco4a1* is a hypoxia response gene (Applebaum et al., 2016).

In ouabain-treated RAEC, we observed sevenfold increment of *Cyp1a1* mRNA encoding cytochrome P450A1. This isoform of P-450 is highly induced in vascular endothelium (Garrick, Woodin, & Stegeman, 2005), and its overexpression is associated with reactive oxygen species (ROS) production (Kopf & Walker, 2010). Puntarulo and Cederbaum, using microsomes enriched in various CYP isoforms (CYP1A1, CYP1A2, CYP2B6, CYP3A4), demonstrated that CYP1A1 plays an important role in the process of NADPH oxidation and formation ROS (Puntarulo & Cederbaum, 1998). Dioxin-like halogenated aromatic hydrocarbons (HAHs), which are toxic environmental pollutants, are now considered among factors inducing hypertension. Chronic exposition of mice to the prototypical HAH 2,3,7,8,-tetrachlorodibenzo-p-dioxin resulted in ROS formation in vessels that cause endothelial dysfunction and hypertension. Overexpression of *CYP1A1* mRNA under these conditions was observed in aorta and mesenteric arteries of the endothelium (Phillip G. Kopf et al., 2010). Beside this, it was demonstrated that *CYP1A1* mRNA level is higher in the lungs of smoker and in lung tumors (McLemore et al., 1990). These data suggest that overexpression of *CYP1A1* is a risk factor for development of vascular and pulmonary diseases.

Using exome-wide association studies (EWASs), it was shown that expression of *Mobkl2c* (or *Mob3c*) encoding polyvalent metal ion binding protein associates with hypertension among a Japanese population (Yamada et al., 2018). It is known also that *Mobkl2c* was induced by Salmonella infection into chicken embryo fibroblasts (Szmolka, Wiener, Matulova, Varmuzova, & Rychlik, 2015). A human homolog of *Mobkl2c* codes for a kinase activator acting as a tumor suppressor (Chow, Hao, & Yang, 2009).

Jmjd3 encoding histone H3K27 demethylase is known as an epigenetic modifier (Xiang et al., 2007). De Santa and coworkers showed that JMJD3 is expressed in macrophages in response to secretion of inflammatory cytokines. It binds to Polycomb group (PcG) target genes and regulates their transcriptional activity (De Santa et al., 2007). PcG protein provides gene silencing in an H3 lysine 27 trimethylation (H3K27me3)-dependent manner (Boyer et al., 2006). Therefore, JMJD3 may be considered as a linker between chronic inflammation and reprogramming of the genome, *i.e.*, altered expression of genetic information occurring without changes in DNA sequence. It is known that cardiac diseases are often coupled with epigenetic modifications (Lorenzen, Martino, & Thum, 2012).

The product of *Tbx3* (T-box 3) is a member of the T-box transcription factor family, and this protein is involved in maintenance and induction of cell pluripotency (Klingenstein et al., 2016; Russell et al., 2015). *Tbx3* is known to play a crucial role in the development of the cardiac conduction system that initiates and propagates the electric impulse that is necessary for well-coordinated work of the heart (van Weerd et al., 2014). Enhanced expression of *Tbx3* was sufficient to generate cardiac conduction system cells both *in vitro* (Rentschler et al., 2012) and *in vivo* (Kapoor, Liang, Marbán, & Cho, 2013). This is a very interesting fact because one of the current problems of biology and medicine is creation of innovative strategies for heart regeneration and repair (Nam, Song, & Olson, 2013).

Socs3 (suppressor of cytokine signaling 3) is a cytokine-inducible inhibitor of the JAK2/STAT3 pathway. It is involved in the development of numerous pathologies of the cardiac-vascular system: it decreases myocyte resistance to acute cardiac injury caused by coxsackievirus B3 (Yasukawa et al., 2003), participates in the development of left ventricular remodeling after myocardial infarction (Oba et al., 2012), plays an important role in the development of myocardial ischemia/reperfusion injury (Nagata et al., 2015), and is connected with development of atherosclerosis (Ortiz-Muñoz et al., 2009; Tang et al., 2005). Pulmonary artery remodeling is one of the events promoting development of pulmonary arterial hypertension (Ascites syndrome, AS) progression in broiler chickens. Transcriptome analysis of AS broiler arteries has demonstrated that *Socs3* is among upregulated genes within pathways regulating pulmonary artery remodeling (Yang, Cao, et al., 2016). Besides this, age-related myocardial hypertrophy, fibrosis, and inflammation are associated with age-related increasing SOCS3 activity (Aboulhoda, 2017). *Socs3* is a hypoxia-inducible gene: in rat pulmonary vascular smooth muscle cells, *Socs3* mRNA was significantly enhanced by hypoxic stimulation (Bai et al., 2006), and hypoxia-induced expression of *Socs3* was observed also in rat embryonic cardiomyocytes (Gu, Kong, Yu, Bai, & Xiao, 2011).

Rsad2 (radical SAM domain-containing 2) is also known as viperin (virus inhibitory protein, endoplasmic reticulum-associated, interferon-inducible). Expression of this gene was increased during the development of human and murine atherosclerosis. It was upregulated in human coronary artery smooth muscle cells by proinflammatory stimuli (Olofsson et al., 2005). Inflammatory processes play an important role in pathophysiology of cardiovascular diseases. Change in expression of STAT1-dependent genes (including *Rsad2*) reflect proinflammatory and proatherogenic profile (Chmielewski et al., 2014; Chmielewski, Piaszyk-Borychowska, Wesoly, & Bluyssen, 2016).

6. Conclusion

Our data indicate that elevation of the $[Na^+]_i/[K^+]_i$-ratio triggered by ouabain results in transcriptomic changes in four different types of rodent cells: vascular smooth muscle cells and endothelial cells of rat aorta (RASMC and RAEC, respectively), rat cerebellum granule cells (RCGC), and mouse C2C12 myoblasts. We noted that these transcriptomic changes correlate with changes of intracellular $[Na^+]_i/[K^+]_i$-ratio: greater elevation of this ratio is associated with the highest number of differentially expressed transcripts.

Acknowledgments

This work was supported by the Russian Foundation for Basic Research (project no. 18-34-00344) and by the Russian Scientific Foundation (project no. 16-15-10026).

References

Aboulhoda, B. E. (2017). Age-related remodeling of the JAK/STAT/SOCS signaling pathway and associated myocardial changes: From histological to molecular level. *Annals of Anatomy - Anatomischer Anzeiger, 214*, 21–30. Retrieved November 11, 2018, from http://www.ncbi.nlm.nih.gov/pubmed/28782583.

Akashi, M., Loussararian, A. H., Adelman, D. C., Saito, M., & Koeffler, H. P. (1990). Role of lymphotoxin in expression of interleukin 6 in human fibroblasts. Stimulation and regulation. *The Journal of Clinical Investigation, 85*(1), 121–129. American Society for Clinical Investigation. Retrieved November 19, 2018, from https://www.jci.org/articles/view/114401.

Akera, T. (1977). Membrane adenosinetriphosphatase: A digitalis receptor? *Science (New York, N.Y.), 198*(4317), 569–574. American Association for the Advancement of Science. Retrieved October 29, 2018, from http://www.ncbi.nlm.nih.gov/pubmed/144320.

Akimova, O. A., Bagrov, A. Y., Lopina, O. D., Kamernitsky, A. V., Tremblay, J., Hamet, P., et al. (2005). Cardiotonic steroids differentially affect intracellular Na+ and [Na+]i/[K+]i-independent signaling in C7-MDCK cells. *The Journal of Biological Chemistry, 280*(1), 832–839. American Society for Biochemistry and Molecular Biology. Retrieved September 9, 2018, from http://www.ncbi.nlm.nih.gov/pubmed/15494417.

Akimova, O., Hamet, P., & Orlov, S. N. (2006). The rapid decline of MTT reduction is not a marker of death signaling in ouabain-treated cells. *Cellular and Molecular Biology, 52*(8), 68–74. Retrieved September 9, 2018, from https://www.researchgate.net/publication/6299822.

Akimova, O. A., Tverskoi, A. M., Smolyaninova, L. V., Mongin, A. A., Lopina, O. D., La, J., et al. (2015). Critical role of the α1-Na+, K+-ATPase subunit in insensitivity of rodent cells to cytotoxic action of ouabain. *Apoptosis*, *20*(9), 1200–1210. Retrieved October 8, 2018, from http://www.ncbi.nlm.nih.gov/pubmed/26067145.

Alberini, C. M. (2009). Transcription factors in long-term memory and synaptic plasticity. *Physiological Reviews*, *89*(1), 121–145. American Physiological Society. Retrieved November 20, 2018, from http://www.physiology.org/doi/10.1152/physrev.00017.2008.

Anisowicz, A., Bardwell, L., & Sager, R. (1987). Constitutive overexpression of a growth-regulated gene in transformed Chinese hamster and human cells. *Proceedings of the National Academy of Sciences of the United States of America*, *84*(20), 7188–7192. National Academy of Sciences. Retrieved November 12, 2018, from http://www.ncbi.nlm.nih.gov/pubmed/2890161.

Applebaum, M. A., Jha, A. R., Kao, C., Hernandez, K. M., DeWane, G., Salwen, H. R., et al. (2016). Integrative genomics reveals hypoxia inducible genes that are associated with a poor prognosis in neuroblastoma patients. *Oncotarget*, 7(47), 76816–76826. Retrieved November 8, 2018, from http://www.oncotarget.com/fulltext/12713.

Arora, D., Köthe, S., van den Eijnden, M., Hooft van Huijsduijnen, R., Heidel, F., Fischer, T., et al. (2012). Expression of protein-tyrosine phosphatases in Acute Myeloid Leukemia cells: FLT3 ITD sustains high levels of DUSP6 expression. *Cell Communication and Signaling: CCS*, *10*(1), 19. Retrieved October 26, 2018, from http://biosignaling.biomedcentral.com/articles/10.1186/1478-811X-10-19.

Bai, L., Yu, Z., Qian, G., Qian, P., Jiang, J., Wang, G., et al. (2006). SOCS3 was induced by hypoxia and suppressed STAT3 phosphorylation in pulmonary arterial smooth muscle cells. *Respiratory Physiology & Neurobiology*, *152*(1), 83–91. Retrieved November 11, 2018, from http://www.ncbi.nlm.nih.gov/pubmed/16125475.

Barros, M. H., Johnson, A., Gin, P., Marbois, B. N., Clarke, C. F., & Tzagoloff, A. (2005). The Saccharomyces cerevisiae COQ10 gene encodes a START domain protein required for function of coenzyme Q in respiration. *The Journal of Biological Chemistry*, *280*(52), 42627–42635. Retrieved October 31, 2018, from http://www.jbc.org/lookup/doi/10.1074/jbc.M510768200.

Becker, S., Quay, J., Koren, H. S., & Haskill, J. S. (1994). Constitutive and stimulated MCP-1, GRO alpha, beta, and gamma expression in human airway epithelium and bronchoalveolar macrophages. *American Journal of Physiology—Lung Cellular and Molecular Physiology*, *266*(3), L278–L286. Retrieved November 12, 2018, from http://www.ncbi.nlm.nih.gov/pubmed/8166297.

Bennay, M., Langer, J., Meier, S. D., Kafitz, K. W., & Rose, C. R. (2008). Sodium signals in cerebellar Purkinje neurons and Bergmann glial cells evoked by glutamatergic synaptic transmission. *GLIA*, *56*(10), 1138–1149. John Wiley & Sons, Ltd. Retrieved November 20, 2018, from http://doi.wiley.com/10.1002/glia.20685.

Bobulescu, I. A., Di Sole, F., & Moe, O. W. (2005). Na+/H+ exchangers: Physiology and link to hypertension and organ ischemia. *Current Opinion in Nephrology and Hypertension*, *14*(5), 485. NIH Public Access. Retrieved November 8, 2018, from https://www.ncbi.nlm.nih.gov/pmc/articles/PMC2861558/.

Borok, Z., Gillissen, A., Buhl, R., Hoyt, R. F., Hubbard, R. C., Ozaki, T., et al. (1991). Augmentation of functional prostaglandin E levels on the respiratory epithelial surface by aerosol administration of prostaglandin E. *The American Review of Respiratory Disease*, *144*(5), 1080–1084. American Lung Association. Retrieved November 20, 2018, from http://www.atsjournals.org/doi/abs/10.1164/ajrccm/144.5.1080.

Boyer, L. A., Plath, K., Zeitlinger, J., Brambrink, T., Medeiros, L. A., Lee, T. I., et al. (2006). Polycomb complexes repress developmental regulators in murine embryonic stem cells. *Nature*, *441*(7091), 349–353. Nature Publishing Group. Retrieved November 9, 2018, from http://www.nature.com/articles/nature04733.

Bredt, D. S., & Nicoll, R. A. (2003). AMPA receptor trafficking at excitatory synapses. *Neuron*, *40*(2), 361–379. Cell Press. Retrieved November 20, 2018, from https://www.sciencedirect.com/science/article/pii/S0896627303006408.

Broholm, C., Laye, M. J., Brandt, C., Vadalasetty, R., Pilegaard, H., Pedersen, B. K., et al. (2011). LIF is a contraction-induced myokine stimulating human myocyte proliferation. *Journal of Applied Physiology*, *111*(1), 251–259. American Physiological Society Bethesda, MD. Retrieved November 20, 2018, from http://www.physiology.org/doi/10.1152/japplphysiol.01399.2010.

Burch, N., Arnold, A.-S., Item, F., Summermatter, S., Brochmann Santana Santos, G., Christe, M., et al. (2010). Electric pulse stimulation of cultured murine muscle cells reproduces gene expression changes of trained mouse muscle. (G. Nogales-Gadea, Ed.)*PLoS One*, *5*(6), e10970 Public Library of Science. Retrieved September 9, 2018, from http://dx.plos.org/10.1371/journal.pone.0010970.

Bussolino, F., Wang, J. M., Turrini, F., Alessi, D., Ghigo, D., Costamagna, C., et al. (1989). Stimulation of the Na+/H+ exchanger in human endothelial cells activated by granulocyte- and granulocyte-macrophage-colony-stimulating factor. Evidence for a role in proliferation and migration. *Journal of Biological Chemistry*, *264*(31), 18284–18287. American Society for Biochemistry and Molecular Biology Inc. Retrieved November 8, 2018, from https://moh-it.pure.elsevier.com/en/publications/stimulation-of-the-nah-exchanger-in-human-endothelial-cells-activ.

Chai, Z., Fan, H., Li, Y., Song, L., Jin, X., Yu, J., et al. (2017). miR-1908 as a novel prognosis marker of glioma via promoting malignant phenotype and modulating SPRY4/RAF1 axis. *Oncology Reports*, *38*(5), 2717–2726. Retrieved October 30, 2018, from http://www.ncbi.nlm.nih.gov/pubmed/29048686.

Chen, H., Fang, X., Zhu, H., Li, S., He, J., Gu, P., et al. (2014). Gene expression profile analysis for different idiopathic interstitial pneumonias subtypes. *Experimental Lung Research*, *40*(8), 367–379. Taylor & Francis. Retrieved November 8, 2018, from http://www.tandfonline.com/doi/full/10.3109/01902148.2014.933985.

Chen, M., Masaki, T., & Sawamura, T. (2002). LOX-1, the receptor for oxidized low-density lipoprotein identified from endothelial cells: Implications in endothelial dysfunction and atherosclerosis. *Pharmacology & Therapeutics*, *95*(1), 89–100. Pergamon. Retrieved November 6, 2018, from https://www.sciencedirect.com/science/article/pii/S016372580200236X.

Chmielewski, S., Olejnik, A., Sikorski, K., Pelisek, J., Błaszczyk, K., Aoqui, C., et al. (2014). STAT1-dependent signal integration between IFNγ and TLR4 in vascular cells reflect pro-atherogenic responses in human atherosclerosis [N. K. Dhillon, Ed.]. *PLoS One*, *9*(12), e113318. Public Library of Science. Retrieved November 12, 2018, from https://dx.plos.org/10.1371/journal.pone.0113318.

Chmielewski, S., Piaszyk-Borychowska, A., Wesoly, J., & Bluyssen, H. A. R. (2016). STAT1 and IRF8 in vascular inflammation and cardiovascular disease: Diagnostic and therapeutic potential. *International Reviews of Immunology*, *35*(5), 434–454. Taylor & Francis. Retrieved November 12, 2018, from https://www.tandfonline.com/doi/full/10.3109/08830185.2015.1087519.

Chow, A., Hao, Y., & Yang, X. (2009). Molecular characterization of human homologs of yeast MOB1. *International Journal of Cancer*, *126*(9). NA-NA. Retrieved November 9, 2018, from http://www.ncbi.nlm.nih.gov/pubmed/19739119.

Cole, A. J., Saffen, D. W., Baraban, J. M., & Worley, P. F. (1989). Rapid increase of an immediate early gene messenger RNA in hippocampal neurons by synaptic NMDA receptor activation. *Nature*, *340*(6233), 474–476. Nature Publishing Group. Retrieved November 20, 2018, from http://www.nature.com/doifinder/10.1038/340474a0.

Danilov, K., Sidorenko, S., Milovanova, K., Klimanova, E., Kapilevich, L. V., & Orlov, S. N. (2017). Electrical pulse stimulation decreases electrochemical Na^{+} and

K^+ gradients in C2C12 myotubes. *Biochemical and Biophysical Research Communications, 493*(2), 875–878.

Das, M. K., Furu, K., Evensen, H. F., Haugen, Ø. P., & Haugen, T. B. (2018). Knockdown of SPRY4 and SPRY4-IT1 inhibits cell growth and phosphorylation of Akt in human testicular germ cell tumours. *Scientific Reports, 8*(1), 2462. Retrieved October 30, 2018, from http://www.ncbi.nlm.nih.gov/pubmed/29410498.

Davis, A., Hogarth, K., Fernandes, D., Solway, J., Niu, J., Kolenko, V., et al. (2003). Functional significance of protein kinase A activation by endothelin-1 and ATP: Negative regulation of SRF-dependent gene expression by PKA. *Cellular Signalling, 15*(6), 597–604. Pergamon. Retrieved September 9, 2018, from https://www.sciencedirect.com/science/article/pii/S0898656802001481.

de Alvaro, C., Martinez, N., Rojas, J. M., & Lorenzo, M. (2005). Sprouty-2 overexpression in C2C12 cells confers myogenic differentiation properties in the presence of FGF2. *Molecular Biology of the Cell, 16*(9), 4454–4461. Retrieved November 15, 2018, from http://www.molbiolcell.org/doi/10.1091/mbc.e05-05-0419.

De Santa, F., Totaro, M. G., Prosperini, E., Notarbartolo, S., Testa, G., & Natoli, G. (2007). The histone H3 lysine-27 demethylase Jmjd3 links inflammation to inhibition of polycomb-mediated gene silencing. *Cell, 130*(6), 1083–1094. Cell Press. Retrieved November 9, 2018, from https://www.sciencedirect.com/science/article/pii/S0092867407010823.

Deitmer, J. W., & Rose, C. R. (2010). Ion changes and signalling in perisynaptic glia. *Brain Research Reviews, 63*(1–2), 113–129. Elsevier. Retrieved November 20, 2018, from https://www.sciencedirect.com/science/article/pii/S0165017309001088.

Devalaraja, R. M., Nanney, L. B., Du, J., Qian, Q., Yu, Y., Devalaraja, M. N., et al. (2000). Delayed wound healing in CXCR2 knockout mice. *The Journal of Investigative Dermatology, 115*(2), 234–244. Elsevier. Retrieved November 12, 2018, from http://www.ncbi.nlm.nih.gov/pubmed/10951241.

Domercq, M., Alberdi, E., Sánchez-Gómez, M. V., Ariz, U., Pérez-Samartín, A., & Matute, C. (2011). Dual-specific phosphatase-6 (Dusp6) and ERK mediate AMPA receptor-induced oligodendrocyte death. *Journal of Biological Chemistry, 286*(13), 11825–11836. Retrieved October 26, 2018, from http://www.jbc.org/lookup/doi/10.1074/jbc.M110.153049.

Dubois, R. N., Abramson, S. B., Crofford, L., Gupta, R. A., Simon, L. S., Van De Putte, A., et al. (1998). Cyclooxygenase in biology and disease. *The FASEB Journal, 12*(12), 1063–1073. Retrieved November 12, 2018, from http://www.fasebj.org/doi/10.1096/fasebj.12.12.1063.

Eyers, P. A., Keeshan, K., & Kannan, N. (2017). Tribbles in the 21st century: The evolving roles of tribbles pseudokinases in biology and disease. *Trends in Cell Biology, 27*(4), 284–298. Elsevier. Retrieved October 29, 2018, from http://www.ncbi.nlm.nih.gov/pubmed/27908682.

Fu, Y., Zhao, Y., & Huang, B. (2017). Tribbles homolog 1 enhances cholesterol efflux from oxidized low-density lipoprotein-loaded THP-1 macrophages. *Experimental and Therapeutic Medicine, 14*(1), 862–866. Retrieved October 29, 2018, from http://www.ncbi.nlm.nih.gov/pubmed/28673011.

Fujita, H., Nedachi, T., & Kanzaki, M. (2007). Accelerated de novo sarcomere assembly by electric pulse stimulation in C2C12 myotubes. *Experimental Cell Research, 313*(9), 1853–1865. Academic Press. Retrieved September 9, 2018, from https://www.sciencedirect.com/science/article/pii/S0014482707001061.

Furukawa, T., Sunamura, M., Motoi, F., Matsuno, S., & Horii, A. (2003). Potential tumor suppressive pathway involving DUSP6/MKP-3 in pancreatic cancer. *The American Journal of Pathology, 162*(6), 1807–1815. Retrieved October 26, 2018, from http://linkinghub.elsevier.com/retrieve/pii/S0002944010643155.

Garrick, R. A., Woodin, B. R., & Stegeman, J. J. (2005). Cytochrome p4501a induced differentially in endothelial cells cultured from different organs of Anguilla rostrata. *In Vitro Cellular & Developmental Biology. Animal*, *41*, 57–63. (1071–2690 (Print)). Retrieved November 9, 2018, from http://www.ncbi.nlm.nih.gov/pubmed/15926861.

Gu, Q., Kong, Y., Yu, Z.-B., Bai, L., & Xiao, Y.-B. (2011). Hypoxia-induced SOCS3 is limiting STAT3 phosphorylation and NF-κB activation in congenital heart disease. *Biochimie*, *93*(5), 909–920. Retrieved November 12, 2018, from http://linkinghub.elsevier.com/retrieve/pii/S0300908411000654.

Haghnegahdar, H., Du, J., Wang, D., Strieter, R. M., Burdick, M. D., Nanney, L. B., et al. (2000). The tumorigenic and angiogenic effects of MGSA/GRO proteins in melanoma. *Journal of Leukocyte Biology*, *67*(1), 53–62. Retrieved November 12, 2018, from http://www.ncbi.nlm.nih.gov/pubmed/10647998.

Hartmann, P., Schober, A., & Weber, C. (2015). Chemokines and microRNAs in atherosclerosis. *Cellular and Molecular Life Sciences: CMLS*, *72*(17), 3253–3266. Springer Basel. Retrieved November 20, 2018, from http://link.springer.com/10.1007/s00018-015-1925-z.

Hashemi Goradel, N., Najafi, M., Salehi, E., Farhood, B., & Mortezaee, K. (2018). Cyclooxygenase-2 in cancer: A review. *Journal of Cellular Physiology*, *234*(5), 5683–5699. Wiley-Blackwell. Retrieved November 12, 2018, from http://doi.wiley.com/10.1002/jcp.27411.

Helmke, C., Becker, S., & Strebhardt, K. (2016). The role of Plk3 in oncogenesis. *Oncogene*, *35*(2), 135–147. Nature Publishing Group. Retrieved October 26, 2018, from http://www.nature.com/articles/onc2015105.

Hnisz, D., Abraham, B. J., Lee, T. I., Lau, A., Saint-André, V., Sigova, A. A., et al. (2013). Super-enhancers in the control of cell identity and disease. *Cell*, *155*(4), 934–947. NIH Public Access. Retrieved October 30, 2018, from http://www.ncbi.nlm.nih.gov/pubmed/24119843.

Hunt, S. P., Pini, A., & Evan, G. (1987). Induction of c-fos-like protein in spinal cord neurons following sensory stimulation. *Nature*, *328*(6131), 632–634. Nature Publishing Group. Retrieved November 20, 2018, from http://www.nature.com/doifinder/10.1038/328632a0.

Iida, N., & Grotendorst, G. R. (1990). Cloning and sequencing of a new gro transcript from activated human monocytes: Expression in leukocytes and wound tissue. *Molecular and Cellular Biology*, *10*(10), 5596–5599. Retrieved November 12, 2018, fromhttp://www.ncbi.nlm.nih.gov/pubmed/2078213.

Impagnatiello, M. A., Weitzer, S., Gannon, G., Compagni, A., Cotten, M., & Christofori, G. (2001). Mammalian sprouty-1 and -2 are membrane-anchored phosphoprotein inhibitors of growth factor signaling in endothelial cells. *The Journal of Cell Biology*, *152*(5), 1087–1098. Rockefeller University Press. Retrieved November 15, 2018, from http://www.ncbi.nlm.nih.gov/pubmed/11238463.

Jing, H., Liaw, L., Friesel, R., Vary, C., Hua, S., & Yang, X. (2016). Suppression of Spry4 enhances cancer stem cell properties of human MDA-MB-231 breast carcinoma cells. *Cancer Cell International*, *16*(1), 19. Retrieved October 30, 2018, from http://www.ncbi.nlm.nih.gov/pubmed/26973433.

Jones, M. W., Errington, M. L., French, P. J., Fine, A., Bliss, T. V. P., Garel, S., et al. (2001). A requirement for the immediate early gene Zif268 in the expression of late LTP and long-term memories. *Nature Neuroscience*, *4*(3), 289–296. Nature Publishing Group. Retrieved November 20, 2018, from http://www.nature.com/articles/nn0301_289.

Kapilevich, L. V., Kironenko, T. A., Zaharova, A. N., Kotelevtsev, Y. V., Dulin, N. O., & Orlov, S. N. (2015). Skeletal muscle as an endocrine organ: Role of [Na+]i/[K+]i-mediated excitation-transcription coupling. *Genes & Diseases*, *2*(4), 328–336. Elsevier.

Retrieved November 20, 2018, from https://www.sciencedirect.com/science/article/pii/S2352304215000665.

Kapoor, N., Liang, W., Marbán, E., & Cho, H. C. (2013). Direct conversion of quiescent cardiomyocytes to pacemaker cells by expression of Tbx18. *Nature Biotechnology*, *31*(1), 54–62. Nature Publishing Group. Retrieved November 11, 2018, from http://www.nature.com/articles/nbt.2465.

Keam, S. P., Gulati, T., Gamell, C., Caramia, F., Arnau, G. M., Huang, C., et al. (2018). Biodosimetric transcriptional and proteomic changes are conserved in irradiated human tissue. *Radiation and Environmental Biophysics*, *57*(3), 241–249. Springer Berlin Heidelberg. Retrieved October 26, 2018, from http://link.springer.com/10.1007/s00411-018-0746-5.

Keyse, S. M. (2008). Dual-specificity MAP kinase phosphatases (MKPs) and cancer. *Cancer and Metastasis Reviews*, *27*(2), 253–261. Springer US. Retrieved October 26, 2018, from http://link.springer.com/10.1007/s10555-008-9123-1.

Kiedrowski, L., Wroblewski, J. T., & Costa, E. (1994). Intracellular sodium concentration in cultured cerebellar granule cells challenged with glutamate. *Molecular Pharmacology*, *45*(5), 1050–1054. Retrieved November 20, 2018, from http://molpharm.aspetjournals.org/content/45/5/1050.short.

Kis, E., Szatmári, T., Keszei, M., Farkas, R., Ésik, O., Lumniczky, K., et al. (2006). Microarray analysis of radiation response genes in primary human fibroblasts. *International Journal of Radiation Oncology, Biology, Physics*, *66*(5), 1506–1514. Elsevier. Retrieved October 26, 2018, from https://www.sciencedirect.com/science/article/pii/S0360301606027490.

Klimanova, E. A., Tverskoi, A. M., Koltsova, S. V., Sidorenko, S. V., Lopina, O. D., Tremblay, J., et al. (2017). Time- and dose dependent actions of cardiotonic steroids on transcriptome and intracellular content of Na + and K +: A comparative analysis. *Scientific Reports*, 7, 45403.

Klingenstein, M., Raab, S., Achberger, K., Kleger, A., Liebau, S., & Linta, L. (2016). TBX3 knockdown decreases reprogramming efficiency of human cells. *Stem Cells International*, *2016*, 1–7. Retrieved November 9, 2018, from http://www.ncbi.nlm.nih.gov/pubmed/26697078.

Knöpfel, T., Guatteo, E., Bernardi, G., & Mercuri, N. B. (1998). Hyperpolarization induces a rise in intracellular sodium concentration in dopamine cells of the substantia nigra pars compacta. *European Journal of Neuroscience*, *10*(5), 1926–1929. John Wiley & Sons, Ltd. Retrieved November 20, 2018, from http://doi.wiley.com/10.1046/j.1460-9568.1998.00195.x.

Kobrinsky, E. (2015). Heterogeneity of calcium channel/cAMP-dependent transcriptional activation. *Current Molecular Pharmacology*, *8*(1), 54–60. Bentham Science Publishers. Retrieved November 20, 2018, from https://www.ingentaconnect.com/content/ben/cmp/2015/00000008/00000001/art00010.

Koltsova, S. V., Shilov, B., Birulina, J. G., Akimova, O. A., Haloui, M., Kapilevich, L. V., et al. (2014). Transcriptomic changes triggered by hypoxia: Evidence for HIF-1α-independent, [Na+]i/[K+]i-mediated, excitation-transcription coupling. *PLoS One*, *9*(11), e110597. Public Library of Science. Retrieved November 20, 2018, from http://go.galegroup.com/ps/anonymous?id=GALE%7CA418706963&sid=googleScholar&v=2.1&it=r&linkaccess=abs&issn=19326203&p=AONE&sw=w.

Koltsova, S. V., Trushina, Y., Haloui, M., Akimova, O. A., Tremblay, J., Hamet, P., et al. (2012). Ubiquitous [Na+]i/[K+]i-sensitive transcriptome in mammalian cells: Evidence for Ca2+i-Independent excitation-transcription coupling [Y. Tsuji, Ed.]. *PLoS One*, 7(5), e38032. Retrieved October 8, 2018, from http://www.ncbi.nlm.nih.gov/pubmed/22666440.

Kopf, P. G., Scott, J. A., Agbor, L. N., Boberg, J. R., Elased, K. M., Huwe, J. K., et al. (2010). Cytochrome P4501A1 is required for vascular dysfunction and hypertension

induced by 2,3,7,8-tetrachlorodibenzo-p-dioxin. *Toxicological Sciences*, *117*(2), 537–546. Retrieved November 8, 2018, from http://www.ncbi.nlm.nih.gov/pubmed/20634294.
Kopf, P. G., & Walker, M. K. (2010). 2,3,7,8-Tetrachlorodibenzo-p-dioxin increases reactive oxygen species production in human endothelial cells via induction of cytochrome P4501A1. *Toxicology and Applied Pharmacology*, *245*(1), 91–99. Retrieved November 9, 2018, from http://www.ncbi.nlm.nih.gov/pubmed/20171976.
La, J., Reed, E. B., Koltsova, S., Akimova, O., Hamanaka, R. B., Mutlu, G. M., et al. (2016). Regulation of myofibroblast differentiation by cardiac glycosides. *American Journal of Physiology—Lung Cellular and Molecular Physiology*, *310*(9), L815–L823. American Physiological Society Bethesda, MD. Retrieved November 20, 2018, from http://www.physiology.org/doi/10.1152/ajplung.00322.2015.
Lee, J. U., Huang, S., Lee, M. H., Lee, S. E., Ryu, M. J., Kim, S. J., et al. (2012). Dual specificity phosphatase 6 as a predictor of invasiveness in papillary thyroid cancer. *European Journal of Endocrinology*, *167*(1), 93–101. Retrieved October 26, 2018, from https://eje.bioscientifica.com/view/journals/eje/167/1/93.xml.
Lingrel, J. B., Argüello, J. M., Van Huysse, J., & Kuntzweiler, T. A. (1997). Cation and cardiac glycoside binding sites of the Na,K-ATPase. *Annals of the New York Academy of Sciences*, *834*, 194–206. John Wiley & Sons, Ltd. Retrieved November 15, 2018, from http://doi.wiley.com/10.1111/j.1749-6632.1997.tb52251.x.
Liu, Y., Cai, Y., Liu, L., Wu, Y., & Xiong, X. (2018). Crucial biological functions of CCL7 in cancer. *PeerJ*, *6*, e4928. PeerJ Inc. Retrieved November 15, 2018, from https://peerj.com/articles/4928.
Lorenzen, J. M., Martino, F., & Thum, T. (2012). Epigenetic modifications in cardiovascular disease. *Basic Research in Cardiology*, *107*(2), 245. Springer-Verlag. Retrieved November 9, 2018, from http://link.springer.com/10.1007/s00395-012-0245-9.
Lowry, O. H., Rosebrough, N. J., Farr, L., & Randall, R. J. (1951). Protein measurement with the Folin phenol reagent. *Journal of Biological Chemistry*, *193*, 265–275.
Maclay, J. D., McAllister, D. A., & MacNee, W. (2007). Cardiovascular risk in chronic obstructive pulmonary disease. *Respirology*, *12*(5), 634–641. Wiley/Blackwell (10.1111). Retrieved November 8, 2018, from http://doi.wiley.com/10.1111/j.1440-1843.2007.01136.x.
Malinsky, J., Tanner, W., & Opekarova, M. (2016). Transmembrane voltage: Potential to induce lateral microdomains. *Biochimica et Biophysica Acta (BBA) - Molecular and Cell Biology of Lipids*, *1861*(8), 806–811. Elsevier. Retrieved October 29, 2018, from https://www.sciencedirect.com/science/article/pii/S1388198116300336.
McLemore, T. L., Adelberg, S., Liu, M. C., McMahon, N. A., Yu, S. J., Hubbard, W. C., et al. (1990). Expression of CYP1A1 gene in patients with lung cancer: Evidence for cigarette smoke-induced gene expression in normal lung tissue and for altered gene regulation in primary pulmonary carcinomas. *Journal of the National Cancer Institute*, *82*(16), 1333–1339. Oxford University Press. Retrieved November 9, 2018, from https://academic.oup.com/jnci/article-lookup/doi/10.1093/jnci/82.16.1333.
Mecham, R. P., Whitehouse, L. A., Wrenn, D. S., Parks, W. C., Griffin, G. L., Senior, R. M., et al. (1987). Smooth muscle-mediated connective tissue remodeling in pulmonary hypertension. *Science (New York, N.Y.)*, *237*(4813), 423–426. American Association for the Advancement of Science. Retrieved November 8, 2018, from http://www.ncbi.nlm.nih.gov/pubmed/3603030.
Messina, S., Frati, L., Leonetti, C., Zuchegna, C., Di Zazzo, E., Calogero, A., et al. (2011). Dual-specificity phosphatase DUSP6 has tumor-promoting properties in human glioblastomas. *Oncogene*, *30*(35), 3813–3820. Retrieved October 26, 2018, from http://www.nature.com/articles/onc201199.
Morgan, J. I., & Curran, T. (1989). Stimulus-transcription coupling in neurons: Role of cellular immediate-early genes. *Trends in Neurosciences*, *12*(11), 459–462. Elsevier Current

Trends. Retrieved November 20, 2018, from https://www.sciencedirect.com/science/article/pii/0166223689900969.

Murphy, K. T., Nielsen, O. B., & Clausen, T. (2008). Analysis of exercise-induced Na+-K+exchange in rat skeletal muscle in vivo. Experimental physiology. *93*(12), 1249–1262. Wiley/Blackwell (10.1111). Retrieved November 20, 2018, from http://doi.wiley.com/10.1113/expphysiol.2008.042457.

Nagata, T., Yasukawa, H., Kyogoku, S., Oba, T., Takahashi, J., Nohara, S., et al. (2015). Cardiac-specific SOCS3 deletion prevents in vivo myocardial ischemia reperfusion injury through sustained activation of cardioprotective signaling molecules [D. Tang, Ed.]. *PLoS One*, *10*(5), e0127942. Retrieved November 11, 2018, from http://www.ncbi.nlm.nih.gov/pubmed/26010537.

Nam, Y.-J., Song, K., & Olson, E. N. (2013). Heart repair by cardiac reprogramming. *Nature Medicine*, *19*(4), 413–415. Retrieved November 9, 2018, from http://www.nature.com/articles/nm.3147.

Nedachi, T., Fujita, H., & Kanzaki, M. (2008). Contractile C 2 C 12 myotube model for studying exercise-inducible responses in skeletal muscle. *American Journal of Physiology—Endocrinology and Metabolism*, *295*(5), E1191–E1204. American Physiological Society. Retrieved September 9, 2018, from http://www.physiology.org/doi/10.1152/ajpendo.90280.2008.

Nikolić, N., Skaret Bakke, S., Tranheim Kase, E., Rudberg, I., Flo Halle, I., Rustan, A. C., et al. (2012). Electrical pulse stimulation of cultured human skeletal muscle cells as an in vitro model of exercise [M. Moran, Ed.]. *PLoS One*, 7(3), e33203. Public Library of Science. Retrieved March 19, 2018, from http://dx.plos.org/10.1371/journal.pone.0033203.

Oba, T., Yasukawa, H., Hoshijima, M., Sasaki, K., Futamata, N., Fukui, D., et al. (2012). Cardiac-specific deletion of SOCS-3 prevents development of left ventricular remodeling after acute myocardial infarction. *Journal of the American College of Cardiology*, *59*(9), 838–852. Retrieved November 11, 2018, from http://www.ncbi.nlm.nih.gov/pubmed/22361405.

Okita, Y., Kimura, M., Xie, R., Chen, C., Shen, L. T.-W., Kojima, Y., et al. (2017). The transcription factor MAFK induces EMT and malignant progression of triple-negative breast cancer cells through its target GPNMB. *Science Signaling*, *10*(474), eaak9397. Retrieved October 30, 2018, from http://www.ncbi.nlm.nih.gov/pubmed/28400538.

Olofsson, P. S., Jatta, K., Wågsäter, D., Gredmark, S., Hedin, U., Paulsson-Berne, G., et al. (2005). The antiviral cytomegalovirus inducible gene 5/viperin is expressed in atherosclerosis and regulated by proinflammatory agents. *Arteriosclerosis, Thrombosis, and Vascular Biology*, *25*(7), e113–e116. Retrieved November 12, 2018, from https://www.ahajournals.org/doi/10.1161/01.ATV.0000170130.85334.38.

Opdenakker, G., Froyen, G., Fiten, P., Proost, P., & Vandamme, J. (1993). Human monocyte chemotactic protein-3 (MCP-3): Molecular cloning of the cDNA and comparison with other chemokines. *Biochemical and Biophysical Research Communications*, *191*(2), 535–542. Academic Press. Retrieved November 12, 2018, from https://www.sciencedirect.com/science/article/pii/S0006291X83712519?via%3Dihub.

Orlov, S., & Hamet, P. (2006). Intracellular monovalent ions as second messengers. *Journal of Membrane Biology*, *210*(3), 161–172. Springer-Verlag. Retrieved November 19, 2018, from http://link.springer.com/10.1007/s00232-006-0857-9.

Orlov, S., & Hamet, P. (2015). Salt and gene expression: Evidence for [Na+]i/[K+]i-mediated signaling pathways. *Pflügers Archiv—European Journal of Physiology*, *467*(3), 489–498. Springer Berlin Heidelberg. Retrieved November 19, 2018, from http://link.springer.com/10.1007/s00424-014-1650-8.

Orlov, S. N., & Mongin, A. A. (2007). Salt-sensing mechanisms in blood pressure regulation and hypertension. *American Journal of Physiology—Heart and Circulatory Physiology*, *293*(4),

H2039–H2053. American Physiological Society. Retrieved November 20, 2018, from http://www.physiology.org/doi/10.1152/ajpheart.00325.2007.

Ortiz-Muñoz, G., Martin-Ventura, J. L., Hernandez-Vargas, P., Mallavia, B., Lopez-Parra,- V., Lopez-Franco, O., et al. (2009). Suppressors of cytokine signaling modulate JAK/ STAT-mediated cell responses during atherosclerosis. *Arteriosclerosis, Thrombosis, and Vascular Biology, 29*(4), 525–531. Retrieved November 11, 2018, from https://www.ahajournals.org/doi/10.1161/ATVBAHA.108.173781.

Owen, J. D., Strieter, R., Burdick, M., Haghnegahdar, H., Nanney, L., Shattuck-Brandt, R., et al. (1997). Enhanced tumor-forming capacity for immortalized melanocytes expressing melanoma growth stimulatory activity/growth-regulated cytokine beta and gamma proteins. *International Journal of Cancer, 73*(1), 94–103. Retrieved November 12, 2018, from http://www.ncbi.nlm.nih.gov/pubmed/9334815.

Ozaki, T., Moriguchi, H., Nakamura, Y., Kamei, T., Yasuoka, S., & Ogura, T. (1990). Regulatory effect of prostaglandin E2 on fibronectin release from human alveolar macrophages. *The American Review of Respiratory Disease, 141*(4 Pt. 1), 965–969. American Lung Association. Retrieved November 20, 2018, from http://www.atsjournals.org/doi/abs/10.1164/ajrccm/141.4_Pt_1.965.

Park, H., Bhalla, R., Saigal, R., Radisic, M., Watson, N., Langer, R., et al. (2008). Effects of electrical stimulation in C2C12 muscle constructs. *Journal of Tissue Engineering and Regenerative Medicine, 2*(5), 279–287. Wiley-Blackwell. Retrieved September 9, 2018, from http://doi.wiley.com/10.1002/term.93.

Puntarulo, S., & Cederbaum, A. I. (1998). Production of reactive oxygen species by microsomes enriched in specific human cytochrome P450 enzymes. *Free Radical Biology and Medicine, 24*(7–8), 1324–1330. Pergamon. Retrieved November 8, 2018, from https://www.sciencedirect.com/science/article/pii/S0891584997004632.

Rentschler, S., Yen, A. H., Lu, J., Petrenko, N. B., Lu, M. M., Manderfield, L. J., et al. (2012). Myocardial notch signaling reprograms cardiomyocytes to a conduction-like phenotype. *Circulation, 126*(9), 1058–1066. NIH Public Access. Retrieved November 11, 2018, from http://www.ncbi.nlm.nih.gov/pubmed/22837163.

Richmond, A., & Thomas, H. G. (1988). Melanoma growth stimulatory activity: Isolation from human melanoma tumors and characterization of tissue distribution. *Journal of Cellular Biochemistry, 36*(2), 185–198. Wiley-Blackwell. Retrieved November 12, 2018, from http://doi.wiley.com/10.1002/jcb.240360209.

Rios, A., Vargas-Robles, H., Gámez-Méndez, A. M., & Escalante, B. (2012). Cyclooxygenase-2 and kidney failure. *Prostaglandins & Other Lipid Mediators, 98*(3–4), 86–90. Elsevier. Retrieved November 12, 2018, from https://www.sciencedirect.com/science/article/pii/S1098882311001183.

Rohini, M., Haritha Menon, A., & Selvamurugan, N. (2018). Role of activating transcription factor 3 and its interacting proteins under physiological and pathological conditions. *International Journal of Biological Macromolecules, 120*, 310–317. Elsevier. Retrieved October 30, 2018, from https://www.sciencedirect.com/science/article/pii/S0141813018333610?via%3Dihub.

Roos, A., & Boron, W. F. (1981). Intracellular pH. *Physiological Reviews, 61*(2), 296–434. Retrieved November 8, 2018, from http://www.physiology.org/doi/10.1152/physrev.1981.61.2.296.

Russell, R., Ilg, M., Lin, Q., Wu, G., Lechel, A., Bergmann, W., et al. (2015). A dynamic role of TBX3 in the pluripotency circuitry. *Stem Cell Reports, 5*(6), 1155–1170. Retrieved November 9, 2018, from http://www.ncbi.nlm.nih.gov/pubmed/26651606.

Sawamura, T., Kume, N., Aoyama, T., Moriwaki, H., Hoshikawa, H., Aiba, Y., et al. (1997). An endothelial receptor for oxidized low-density lipoprotein. *Nature, 386*(6620), 73–77. Nature Publishing Group. Retrieved November 6, 2018, from http://www.nature.com/doifinder/10.1038/386073a0.

Scheiner-Bobis, G. (2002). The sodium pump. *European Journal of Biochemistry*, *269*(10), 2424–2433. John Wiley & Sons, Ltd. Retrieved November 19, 2018, from http://doi.wiley.com/10.1046/j.1432-1033.2002.02909.x.

Sidorenko, S., Klimanova, E., Milovanova, K., Lopina, O. D., Kapilevich, L. V., Chibalin, A. V., et al. (2018). Transcriptomic changes in C2C12 myotubes triggered by electrical stimulation: Role of Ca2+i-mediated and Ca2+i-independent signaling and elevated [Na+]i/[K+]i ratio. *Cell Calcium*, *76*, 72–86. Retrieved October 25, 2018, from http://www.ncbi.nlm.nih.gov/pubmed/30300758.

Simoni, L., Delgado, V., Ruer-Laventie, J., Bouis, D., Soley, A., Heyer, V., et al. (2018). Trib1 is overexpressed in systemic lupus erythematosus, while it regulates immunoglobulin production in murine B cells. *Frontiers in Immunology*, *9*, 373. Retrieved October 29, 2018, from http://journal.frontiersin.org/article/10.3389/fimmu.2018.00373/full.

Sjøgaard, G., Adams, R. P., & Saltin, B. (1985). Water and ion shifts in skeletal muscle of humans with intense dynamic knee extension. *American Journal of Physiology*, *248*(2 Pt. 2), R190–R196. Retrieved November 20, 2018, from http://www.ncbi.nlm.nih.gov/pubmed/3970234.

Smolyaninova, L. V., Koltsova, S. V., Sidorenko, S. V., & Orlov, S. N. (2017). Augmented gene expression triggered by Na+,K+-ATPase inhibition: Role of Ca2+i-mediated and −independent excitation-transcription coupling. *Cell Calcium*, *68*, 5–13. Churchill Livingstone. Retrieved November 20, 2018, from https://www.sciencedirect.com/science/article/pii/S0143416017301513.

Somjen, G. G. (2002). Ion regulation in the brain: Implications for pathophysiology. *The Neuroscientist*, *8*(3), 254–267. Sage PublicationsSage CA: Thousand Oaks, CA. Retrieved November 20, 2018, from http://journals.sagepub.com/doi/10.1177/1073858402008003011.

Soubeyrand, S., Martinuk, A., & McPherson, R. (2017). TRIB1 is a positive regulator of hepatocyte nuclear factor 4-alpha. *Scientific Reports*, 7(1), 5574. Nature Publishing Group. Retrieved October 29, 2018, from http://www.nature.com/articles/s41598-017-05768-1.

Spencer, A. G., Labonte, E. D., Rosenbaum, D. P., Plato, C. F., Carreras, C. W., Leadbetter, M. R., et al. (2014). Intestinal inhibition of the Na+/H+ exchanger 3 prevents cardiorenal damage in rats and inhibits Na+ uptake in humans. *Science Translational Medicine*, *6*(227), 227ra36. American Association for the Advancement of Science. Retrieved November 12, 2018, from http://stm.sciencemag.org/content/6/227/227ra36.short.

Szmolka, A., Wiener, Z., Matulova, M. E., Varmuzova, K., & Rychlik, I. (2015). Gene Expression profiles of chicken embryo fibroblasts in response to salmonella enteritidis infection. (D. Chakravortty, Ed.)*PLoS One*, *10*(6), e0127708 Public Library of Science. Retrieved November 9, 2018, from https://dx.plos.org/10.1371/journal.pone.0127708.

Tang, J., Kozaki, K., Farr, A. G., Martin, P. J., Lindahl, P., Betsholtz, C., et al. (2005). The absence of platelet-derived growth factor-B in circulating cells promotes immune and inflammatory responses in atherosclerosis-prone ApoE −/− mice. *The American Journal of Pathology*, *167*(3), 901–912. Elsevier. Retrieved November 11, 2018, from https://www.sciencedirect.com/science/article/pii/S0002944010620615.

Tang, B., Wu, W., Zhang, Q., Sun, Y., Cui, Y., Wu, F., et al. (2015). Inhibition of tribbles protein-1 attenuates radioresistance in human glioma cells. *Scientific Reports*, *5*(1), 15961. Retrieved October 29, 2018, from http://www.ncbi.nlm.nih.gov/pubmed/26521947.

Taurin, S., Hamet, P., & Orlov, S. N. (2003). Na/K pump and intracellular monovalent cations: Novel mechanism of excitation-transcription coupling involved in inhibition of apoptosis. *Molekuliarnaia Biologiia*, *37*(3), 371–381. Retrieved November 19, 2018, from http://www.ncbi.nlm.nih.gov/pubmed/12815944.

Thiel, G., Mayer, S. I., Müller, I., Stefano, L., & Rössler, O. G. (2010). Egr-1—A Ca2 +-regulated transcription factor. *Cell Calcium*, *47*(5), 397–403. Churchill Livingstone. Retrieved November 20, 2018, from https://www.sciencedirect.com/science/article/pii/S0143416010000333.

Tian, Y., Fu, X., Li, Q., Wang, Y., Fan, D., Zhou, Q., et al. (2018). MicroRNA-181 serves an oncogenic role in breast cancer via the inhibition of SPRY4. *Molecular Medicine Reports*, *18*, 5603–5613. Spandidos Publications. Retrieved October 30, 2018, from http://www.spandidos-publications.com/10.3892/mmr.2018.9572.

Tsai, K.-L., Huang, Y.-H., Kao, C.-L., Yang, D.-M., Lee, H.-C., Chou, H.-Y., et al. (2012). A novel mechanism of coenzyme Q10 protects against human endothelial cells from oxidative stress-induced injury by modulating NO-related pathways. *The Journal of Nutritional Biochemistry*, *23*(5), 458–468. Elsevier. Retrieved November 12, 2018, from https://www.sciencedirect.com/science/article/pii/S0955286311000659.

van Weerd, J. H., Badi, I., van den Boogaard, M., Stefanovic, S., van de Werken, H. J. G., Gomez-Velazquez, M., et al. (2014). A large permissive regulatory domain exclusively controls *Tbx3* expression in the cardiac conduction system. *Circulation Research*, *115*(4), 432–441. Retrieved November 11, 2018, from https://www.ahajournals.org/doi/10.1161/CIRCRESAHA.115.303591.

Verkhratsky, A., Noda, M., Parpura, V., & Kirischuk, S. (2013). Sodium fluxes and astroglial function. *Advances in Experimental Medicine and Biology*, *961*, 295–305. Springer, Boston, MA. Retrieved November 20, 2018, from http://link.springer.com/10.1007/978-1-4614-4756-6_25.

Voghel, G., Thorin-Trescases, N., Farhat, N., Nguyen, A., Villeneuve, L., Mamarbachi, A. M., et al. (2007). Cellular senescence in endothelial cells from atherosclerotic patients is accelerated by oxidative stress associated with cardiovascular risk factors. *Mechanisms of Ageing and Development*, *128*(11–12), 662–671. Elsevier. Retrieved September 9, 2018, from https://www.sciencedirect.com/science/article/pii/S0047637407001595.

Wang, J., Duncan, D., Shi, Z., & Zhang, B. (2013). WEB-based GEne SeT AnaLysis toolkit (WebGestalt): Update 2013. *Nucleic Acids Research*, *41*(W1), W77–W83. Retrieved October 23, 2018, from http://academic.oup.com/nar/article/41/W1/W77/1105552/WEBbased-GEne-SeT-AnaLysis-Toolkit-WebGestalt.

Wang, Y., & Hekimi, S. (2013). Molecular genetics of ubiquinone biosynthesis in animals. *Critical Reviews in Biochemistry and Molecular Biology*, *48*(1), 69–88. Taylor & Francis. Retrieved November 15, 2018, from http://www.tandfonline.com/doi/full/10.3109/10409238.2012.741564.

Wang, H., Tian, C., Fan, X.-Y., Chen, L.-N., Lv, Y., Sun, J., et al. (2015). Polo-like kinase 3 (PLK3) mediates the clearance of the accumulated PrP mutants transiently expressed in cultured cells and pathogenic PrP(Sc) in prion infected cell line via protein interaction. *The International Journal of Biochemistry & Cell Biology*, *62*, 24–35. Retrieved October 26, 2018, from https://linkinghub.elsevier.com/retrieve/pii/S1357272515000540.

Wang, Y., Wu, N., Pang, B., Tong, D., Sun, D., Sun, H., et al. (2017). TRIB1 promotes colorectal cancer cell migration and invasion through activation MMP-2 via FAK/Src and ERK pathways. *Oncotarget*, *8*(29), 47931–47942. Impact Journals. Retrieved October 29, 2018, from http://www.oncotarget.com/fulltext/18201.

Watanabe, T., Miura, T., Degawa, Y., Fujita, Y., Inoue, M., Kawaguchi, M., et al. (2010). Comparison of lung cancer cell lines representing four histopathological subtypes with gene expression profiling using quantitative real-time PCR. *Cancer Cell International*, *10*(1), 2. BioMed Central. Retrieved November 8, 2018, from http://cancerci.biomedcentral.com/articles/10.1186/1475-2867-10-2.

Whitham, M., Chan, M. H. S., Pal, M., Matthews, V. B., Prelovsek, O., Lunke, S., et al. (2012). Contraction-induced interleukin-6 gene transcription in skeletal muscle is regulated by c-Jun terminal kinase/activator protein-1. *Journal of Biological Chemistry*, *287*(14), 10771–10779. American Society for Biochemistry and Molecular Biology. Retrieved September 9, 2018, from http://www.ncbi.nlm.nih.gov/pubmed/22351769.

Wu, Z., Jiao, P., Huang, X., Feng, B., Feng, Y., Yang, S., et al. (2010). MAPK phosphatase-3 promotes hepatic gluconeogenesis through dephosphorylation of forkhead box O1 in mice. *The Journal of Clinical Investigation*, *120*(11), 3901–3911. Retrieved October 26, 2018, from http://www.jci.org/articles/view/43250.

Xiang, Y., Zhu, Z., Han, G., Lin, H., Xu, L., & Chen, C. D. (2007). JMJD3 is a histone H3K27 demethylase. *Cell Research*, *17*(10), 850–857. Retrieved November 9, 2018, from http://www.ncbi.nlm.nih.gov/pubmed/17923864.

Xie, S., Wang, Q., Wu, H., Cogswell, J., Lu, L., Jhanwar-Uniyal, M., et al. (2001). Reactive oxygen species-induced phosphorylation of p53 on serine 20 is mediated in part by polo-like kinase-3. *The Journal of Biological Chemistry*, *276*(39), 36194–36199. Retrieved October 26, 2018, from http://www.jbc.org/lookup/doi/10.1074/jbc.M104157200.

Xie, S., Wu, H., Wang, Q., Cogswell, J. P., Husain, I., Conn, C., et al. (2001). Plk3 functionally links DNA damage to cell cycle arrest and apoptosis at least in part via the p53 pathway. *The Journal of Biological Chemistry*, *276*(46), 43305–43312. Retrieved October 26, 2018, from http://www.jbc.org/lookup/doi/10.1074/jbc.M106050200.

Yamada, Y., Kato, K., Oguri, M., Horibe, H., Fujimaki, T., Yasukochi, Y., et al. (2018). Identification of 13 novel susceptibility loci for early-onset myocardial infarction, hypertension, or chronic kidney disease. *International Journal of Molecular Medicine*, *42*(5), 2415–2436. Spandidos Publications. Retrieved November 9, 2018, from http://www.spandidos-publications.com/10.3892/ijmm.2018.3852.

Yang, F., Cao, H., Xiao, Q., Guo, X., Zhuang, Y., Zhang, C., et al. (2016). Transcriptome analysis and gene identification in the pulmonary artery of broilers with ascites syndrome [J. West, Ed.]. *PLoS One*, *11*(6), e0156045. Retrieved November 11, 2018, from http://www.ncbi.nlm.nih.gov/pubmed/27275925.

Yang, X., Gong, Y., Tang, Y., Li, H., He, Q., Gower, L., et al. (2013). Spry1 and Spry4 differentially regulate human aortic smooth muscle cell phenotype via Akt/FoxO/myocardin signaling [G. Pintus, Ed.]. *PLoS One*, *8*(3), e58746. Public Library of Science. Retrieved November 15, 2018, from https://dx.plos.org/10.1371/journal.pone.0058746.

Yang, D., Xie, P., & Liu, Z. (2012). Ischemia/reperfusion-induced MKP-3 impairs endothelial NO formation via inactivation of ERK1/2 pathway [C. Gaetano, Ed.]. *PLoS One*, 7(7), e42076. Retrieved October 26, 2018, from http://dx.plos.org/10.1371/journal.pone.0042076.

Yang, C.-J., Yang, J., Fan, Z.-X., & Yang, J. (2016). Activating transcription factor 3—An endogenous inhibitor of myocardial ischemia-reperfusion injury (Review). *Molecular Medicine Reports*, *13*(1), 9–12. Spandidos Publications. Retrieved November 20, 2018, from https://www.spandidos-publications.com/10.3892/mmr.2015.4529.

Yasukawa, H., Yajima, T., Duplain, H., Iwatate, M., Kido, M., Hoshijima, M., et al. (2003). The suppressor of cytokine signaling–1 (SOCS1) is a novel therapeutic target for enterovirus-induced cardiac injury. *Journal of Clinical Investigation*, *111*(4), 469–478. Retrieved November 11, 2018, from http://www.ncbi.nlm.nih.gov/pubmed/12588885.

Yuan, J., Jiang, Y.-Y., Mayakonda, A., Huang, M., Ding, L.-W., Lin, H., et al. (2017). Super-enhancers promote transcriptional dysregulation in nasopharyngeal carcinoma. *Cancer Research*, 77(23), 6614–6626. Retrieved October 30, 2018, from http://www.ncbi.nlm.nih.gov/pubmed/28951465.

Zhang, B., Kirov, S., & Snoddy, J. (2005). WebGestalt: An integrated system for exploring gene sets in various biological contexts. *Nucleic Acids Research*, *33*, W741–W748.

(Web Server). Retrieved October 23, 2018, from https://academic.oup.com/nar/article-lookup/doi/10.1093/nar/gki475.

Zhang, Z., Kobayashi, S., Borczuk, A. C., Leidner, R. S., Laframboise, T., Levine, A. D., et al. (2010). Dual specificity phosphatase 6 (DUSP6) is an ETS-regulated negative feedback mediator of oncogenic ERK signaling in lung cancer cells. *Carcinogenesis*, *31*(4), 577–586. Retrieved October 24, 2018, from https://academic.oup.com/carcin/article-lookup/doi/10.1093/carcin/bgq020.

Zhou, T., Chou, J. W., Simpson, D. A., Zhou, Y., Mullen, T. E., Medeiros, M., et al. (2006). Profiles of global gene expression in ionizing-radiation-damaged human diploid fibroblasts reveal synchronization behind the G1 checkpoint in a G0-like state of quiescence. *Environmental Health Perspectives*, *114*(4), 553–559. Retrieved October 26, 2018, from https://ehp.niehs.nih.gov/doi/10.1289/ehp.8026.

Zhou, H., Guo, H., Zong, J., Dai, J., Yuan, Y., Bian, Z.-Y., et al. (2014). ATF3 regulates multiple targets and may play a dual role in cardiac hypertrophy and injury. *International Journal of Cardiology*, *174*(3), 838–839. Elsevier. Retrieved November 20, 2018, from http://www.ncbi.nlm.nih.gov/pubmed/24794959.

CHAPTER SIX

The Na,K-ATPase in vascular smooth muscle cells

Lin Zhang[a,b,†], Christian Staehr[a,†], Fanxing Zeng[b], Elena V. Bouzinova[a], Vladimir V. Matchkov[a,*]

[a]Department of Biomedicine, Health, Aarhus University, Aarhus, Denmark
[b]Department of Exercise Physiology, Beijing Sport University, Beijing, China
*Corresponding author: e-mail address: vvm@biomed.au.dk

Contents

Abstract

The Na,K-ATPase is an enzyme essential for ion homeostasis in all cells. Over the last decades, it has been well-established that in addition to the transport of Na^+/K^+ over the cell membrane, the Na,K-ATPase acts as a receptor transducing humoral signals intracellularly. It has been suggested that ouabain-like compounds serve as endogenous modulators of this Na,K-ATPase signal transduction. The molecular mechanisms underlying Na,K-ATPase signaling are complicated and suggest the confluence of divergent biological pathways. This review discusses recent updates on the Na,K-ATPase signaling pathways characterized or suggested in vascular smooth muscle cells. The conventional view on this signaling is based on a microdomain structure where the Na,K-ATPase controls the Na,Ca-exchanger activity via modulation of intracellular Na^+

† Contributed equally.

Current Topics in Membranes, Volume 83
ISSN 1063-5823
https://doi.org/10.1016/bs.ctm.2019.01.007

in the spatially restricted submembrane space. This, in turn, affects intracellular Ca^{2+} and Ca^{2+} load in the sarcoplasmic reticulum leading to modulation of contractility as well as gene expression. An ion-transport-independent signal transduction from the Na,K-ATPase is based on molecular interactions. This was primarily characterized in other cell types but recently also demonstrated in vascular smooth muscles. The downstream signaling from the Na,K-ATPase includes Src and phosphatidylinositol-4,5-bisphosphate 3 kinase signaling pathways and generation of reactive oxygen species. Moreover, in vascular smooth muscle cells the interaction between the Na,K-ATPase and proteins responsible for Ca^{2+} homeostasis, e.g., phospholipase C and inositol triphosphate receptors, contributes to an integration of the signaling pathways. Recent update on the Na,K-ATPase dependent intracellular signaling and the significance for physiological functions and pathophysiological changes are discussed in this review.

1. Introduction

The Na,K-ATPase is essential for all cells in the body. It is an ubiquitous membrane transport protein responsible for intracellular ion homeostasis critical for membrane potential and numerous cellular processes, including cell proliferation, differentiation, contraction and volume regulation (Matchkov & Krivoi, 2016). The Na,K-ATPase plays an important role in the cardiovascular system, as it is evident from its involvement in the pathology of hypertension, cardiomyopathies, arrhythmias and ischemia/reperfusion injury (for review see: Blaustein et al., 2016; Shattock et al., 2015). Currently, there are two major views on the regulatory role of the Na,K-ATPase in cell function and protein expression: The conventional view that is based on an ion-transport-dependent control of intracellular ion homeostasis (Blaustein & Lederer, 1999; Blaustein & Wier, 2007) and a mechanism that depends on intracellular signal transduction (Aizman & Aperia, 2003; Cui & Xie, 2017; Wu et al., 2013; Yu et al., 2018). These two different regulatory pathways dependent on the Na,K-ATPase seem to integrate with each other, although it remains a matter of debate (Cui & Xie, 2017; Weigand, Swarts, Fedosova, Russel, & Koenderink, 2012; Yu et al., 2018).

Cardiac glycosides, including ouabain and digoxin, are specific inhibitors of the Na,K-ATPase that is stabilized in the E2-P transition state after inhibition, so that Na^+ cannot be extruded (Noel, Fagoo, & Godfraind, 1990). Cardiac glycosides are originally derived from plants but endogenous ouabain-like compound(s) and other cardiac glycosides are suggested to circulate in plasma (Hamlyn et al., 1982). Although it still remains controversial

(Hamlyn & Blaustein, 2016), several facts point toward ouabain as being a hormone synthesized and secreted in the body. First, ouabain-like compound(s) have been found in high concentrations in the adrenal cortex (Hamlyn et al., 1991; Li, Eim, Kirch, Lang, & Schoner, 1998). Second, adrenocortical cells have been shown to secrete ouabain-like compound(s) in amounts greater than their storage capacity under in vitro conditions (Boulanger et al., 1993; Laredo, Hamilton, & Hamlyn, 1994, 1995). Third, the concentration of ouabain-like compounds in adrenal venous blood is significantly higher than in arterial plasma (Boulanger et al., 1993). Moreover, adrenal cortex tumors have been characterized by overproduction and secretion of ouabain-like compounds (Komiyama et al., 1999). Consistently, administration of anti-ouabain antibodies in rats produces adrenal cortex enlargement, further implicating the adrenal gland as a source of ouabain-like compounds (Nesher, Dvela, Igbokwe, Rosen, & Lichtstein, 2009). Endogenous-ouabain-like compound(s) are suggested synthesized in zona glomerulosa cells of the adrenal cortex, as other adrenal steroids (Hamlyn et al., 1998; Laredo et al., 1995). The exact mechanisms and precursors directly involved in the biosynthesis of ouabain-like compounds are, however, still unclear (Hamlyn & Blaustein, 2016; Hamlyn et al., 1998; Lichtstein et al., 1998). Importantly, hypothalamus has been suggested to synthesize ouabain-like compounds (Li et al., 1998) where they may play a central neuro-modulatory role leading to excitation of the central sympatho-excitatory pathway (Blaustein et al., 2012). Finally, the idea of endogenous ouabain-like compounds is strongly supported by the suggestion that the Na,K-ATPase acts as a membrane signal transduction receptor for ouabain (Aperia, 2007; Tian & Xie, 2008; Xie & Askari, 2002). The significance of this signaling for vascular function will be discussed in this review.

In accordance with increased peripheral resistance and normal cardiac output in hypertension, plasma ouabain level positively correlates with elevated peripheral resistance and left ventricular hypertrophy but not with cardiac output (Manunta et al., 1999; Pierdomenico et al., 2001). Almost 50% of patients with uncomplicated essential hypertension are reported to have elevated endogenous ouabain (Rossi et al., 1995). Chronic administration of ouabain increasing the plasma concentration to the level observed in essential hypertension produces hypertension in rats (Manunta, Hamilton, Rogowski, Hamilton, & Hamlyn, 2000; Yuan et al., 1993). Importantly, the plasma ouabain concentration is also elevated in several rodent models of hypertension, including DOCA-salt/reduced renal mass and adrenocorticotropic hormone (ACTH) induced hypertension, Milan hypertensive and Dahl S rats on high-salt diet (for review: Blaustein et al., 2012). Inhibition of endogenous ouabain by systemic administration of the antagonist rostafuroxin or an antibody against endogenous ouabain lowers blood pressure and prevents hypertension in the high-ouabain hypertension models (Dostanic-Larson, Van Huysse, Lorenz, & Lingrel, 2005; Kaide et al., 1999; Krep et al., 1995; Manunta, Ferrandi, Messaggio, & Ferrari, 2006).

In general, blood pressure can be elevated by either an increase in cardiac output or in vascular resistance. Since the Na,K-ATPase is expressed in both cardiac and vascular tissues both compartments have to be considered when discussing the pathology. The Na,K-ATPase is implicated in regulation of cardiac hypertrophy and contractility (for review: Shattock et al., 2015) as well as in the regulation of peripheral resistance via either sympathetic control (for review: Blaustein et al., 2012; Leenen, Blaustein, & Hamlyn, 2017) or via a direct modulation of smooth muscle contraction. This review primarily discusses the current mechanistic understanding of the role of the Na, K-ATPase in the regulation of vascular smooth muscle cell contraction. We will consider the conventional ion-transport-dependent mechanism and the intracellular signal transduction pathways in the regulation of vascular tone, with a major focus on the latter.

2. Na,K-ATPase in the vascular wall

The majority of cells in the body express the α1 isoform of the Na, K-ATPase and usually one more isoform is present as well (Blaustein et al., 2009; Juhaszova & Blaustein, 1997a; Matchkov & Krivoi, 2016). Skeletal, cardiac and smooth muscle cells co-express α1 and α2 isoforms (Blanco & Mercer, 1998; Matchkov, 2010). However, their distribution, expressional level and functional contribution remain uncertain and depend on numerous factors including species, organ and age (Matchkov & Krivoi, 2016). In general, expression ratio for vascular smooth muscle α1 and α2 isoforms was suggested approximately 7:3 (Shelly et al., 2004) but ionic current generated by the Na,K-ATPase α2 isoform is usually smaller than expected in relation to this ratio (Matchkov & Krivoi, 2016). This suggests that the α2 isoform is less active at resting conditions but may be activated during agonist stimulation (Mulvany, Aalkjaer, & Petersen, 1984).

Moreover, the α1 and α2 isoforms have different membrane localization. The α2 isoform is localized in spatially restricted areas in the plasma membrane in a close proximity to the Na^+,Ca^{2+}-exchanger (NCX), while the α1 isoform is homogeneously distributed in the membrane in smooth muscle cells (Juhaszova & Blaustein, 1997a, 1997b; Linde, Antos, Golovina, & Blaustein, 2012; Moore et al., 1993). In rodents, the functional contribution of the α1 and α2 isoforms is pharmacologically distinguishable since the α1 isoform is relatively insensitive to ouabain (Vilsen, 1999).

The Na,K-ATPase α2 isoform is considered most important for the regulation of vascular tone (Blaustein et al., 2016; Chen et al., 2015). Thus, knock-in of an ouabain-resistant mutation of the α2 isoform prevents ouabain-induced hypertension (Dostanic et al., 2005; Lorenz et al., 2008) suggesting the importance of this isoform in the pathogenesis of hypertension. Accordingly, heterozygote mice with global α2 isoform knockout (homozygote knockout is lethal (Shelly et al., 2004)) (Zhang et al., 2005) and mice with smooth-muscle-specific dominant-negative α2 isoform expression have elevated blood pressure (Chen et al., 2015), while mice overexpressing the α2 isoform are hypotensive (Chen et al., 2015; Pritchard, Bullard, Lynch, Lorenz, & Paul, 2007). In contrast, the α1 isoform heterozygote knockout mice have normal blood pressure (Zhang et al., 2005).

There is no generally accepted molecular mechanism that explains the pro-hypertensive action of the Na,K-ATPase α2 isoform. Thus, global heterozygote knockout of the α2 isoform have increased myogenic tone suggesting the mechanism underlying elevated peripheral resistance (Zhang et al., 2005). In contrast, the dominant-negative α2 isoform mice have reduced vascular tone whereas the α2 isoform overexpression is without any significance for myogenic contraction (Chen et al., 2015).

Furthermore, not all cardiac glycosides have similar effect on blood pressure. In contrast to ouabain, which elevates blood pressure (Manunta et al., 2000; Yuan et al., 1993), digoxin does not raise blood pressure and, conversely, seems to lower blood pressure in ouabain-dependent hypertension models (Huang, Kudlac, Kumarathasan, & Leenen, 1999; Kimura, Manunta, Hamilton, & Hamlyn, 2000; Manunta et al., 2000). Since both ouabain and digoxin inhibit the ion pump activity (Noel et al., 1990), this finding suggests that the pro-hypertensive action of ouabain is mediated, at least in part, through an ion-transport-independent mechanism.

3. Na,K-ATPase modulates local intracellular Ca^{2+} concentration ($[Ca^{2+}]_i$)

The spatially restricted Na,K-ATPase α2 isoform in vascular smooth muscle cells is located in close proximity to the sarcoplasmic reticulum (SR) forming a microdomain, which is able to control the local Na^+

concentration near the NCX and, thus, influence the intracellular calcium concentration ($[Ca^{2+}]_i$) in this restricted sub-membrane space (Blaustein & Wier, 2007). Elevation of local $[Ca^{2+}]_i$ increases the Ca^{2+} load in the SR (Blaustein et al., 2016; Blaustein & Hamlyn, 2010) contributing to greater Ca^{2+} release upon agonist-stimulation. This model (Fig. 1) suggests that reduction of the Na,K-ATPase activity either by knocking it down or by pharmacological inhibition is associated with elevated contraction of vascular smooth muscles (Blaustein & Lederer, 1999). Accordingly, micromolar ouabain, which inhibits only the α2 isoform in rodent tissue, elevates $[Ca^{2+}]_i$ in pressurized arteries and potentiates myogenic tone (Bouzinova et al., 2018; Iwamoto et al., 2004; Matchkov et al., 2012; Miriel, Mauban, Blaustein, & Wier, 1999; Pulina et al., 2010; Zhang et al., 2005). This is, however, not always the case. Thus, micromolar ouabain does not affect global intracellular Na^+ and Ca^{2+} in the vascular wall but potentiates the contraction (Aalkjaer & Mulvany, 1985; Arnon, Hamlyn, & Blaustein, 2000).

The ouabain-digoxin antagonism discussed above (Hamlyn & Blaustein, 2016) also argues against local intracellular Na^+/Ca^{2+} homeostasis as a main pathway in the Na,K-ATPase-dependent control of vascular contraction (Song, Karashima, Hamlyn, & Blaustein, 2014; Zulian et al., 2013). Furthermore, transient downregulation of the Na,K-ATPase α2 isoform in rat mesenteric arteries suppresses vasoconstriction in spite of higher $[Ca^{2+}]_i$ (Matchkov et al., 2012) suggesting reduced sensitivity of the contractile machinery to Ca^{2+} (Bouzinova et al., 2018; Zhang, Aalkjaer, & Matchkov, 2018). Chronic downregulation of the α2 isoform potentiates cerebral artery contraction in spite of smaller increase in $[Ca^{2+}]_i$ suggesting increased sensitization of smooth muscles to Ca^{2+} (Staehr et al., 2018). Altogether, these studies indicate that the modulatory action of the Na, K-ATPase on arterial tone and contractility cannot be entirely explained by modulation of the Na^+/Ca^{2+} homeostasis. Other signal transduction pathways involving the Na,K-ATPase in the control of arterial contraction are therefore likely to appear.

4. Scaffolding and signal transducing function of the Na,K-ATPase

Xie and colleagues demonstrated in rat cultured neonatal cardiomyocytes that, in addition to Ca^{2+}-dependent pathways (Peng, Huang, Xie, Huang, & Askari, 1996), ouabain stimulates mitogen-activated protein

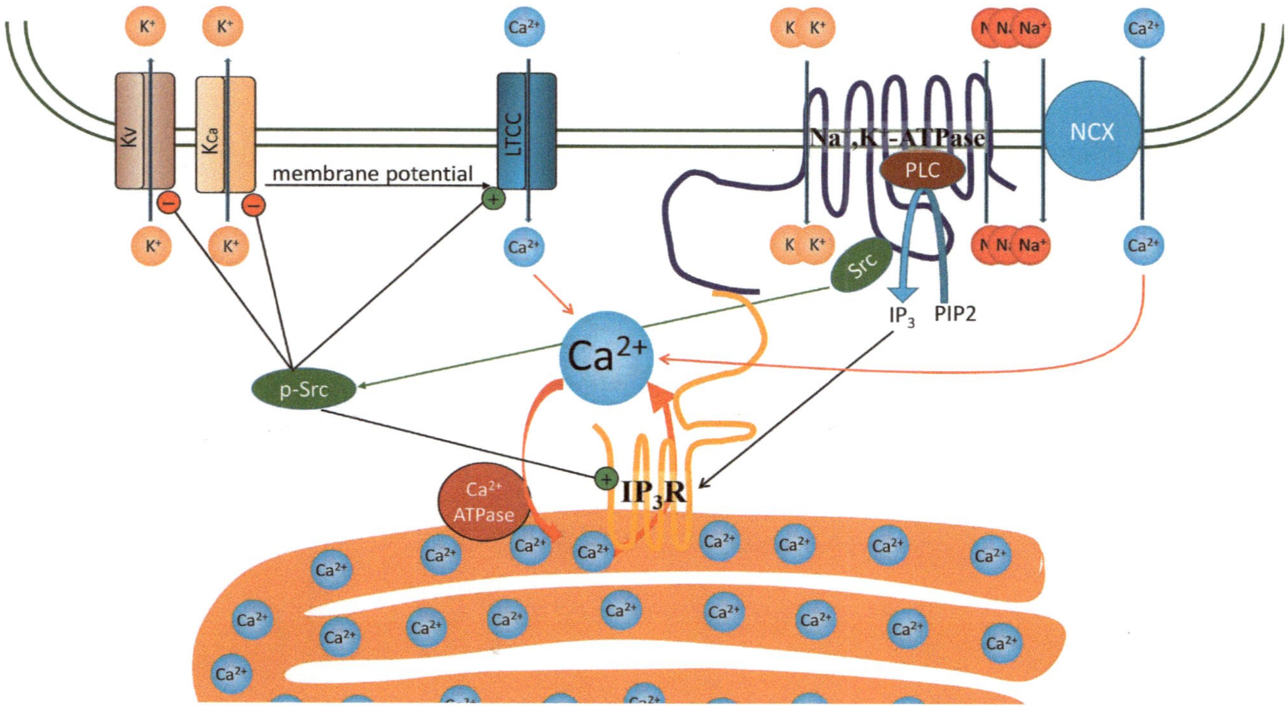

Fig. 1 A schematic presentation of putative Na,K-ATPase-dependent signaling pathways modulating intracellular Ca^{2+} in a caveolae-based signalosome of smooth muscle cell.

kinases and growth-related marker gene expression in a Ca^{2+}-independent manner (Liu et al., 2000). Binding of ouabain to the cardiac Na,K-ATPase initiates multiple signal transduction and generation of second messengers. One of the initial steps in this signaling is an activation of the membrane associated non-receptor tyrosine kinase, Src kinase. In vascular smooth muscles, this signaling mediates cell proliferation (Allen, Abramowitz, & Koksoy, 2003), affects Ca^{2+} homeostasis (Zulian et al., 2013) and increases arterial contraction (Bouzinova et al., 2018; Zhang et al., 2018). This Src kinase activation leads to tyrosine phosphorylation of the epidermal growth factor receptor (EGFR) and, possibly, other receptor tyrosine kinases, which transduce downstream signaling (Aydemir-Koksoy, Abramowitz, & Allen, 2001; Haas, Wang, Tian, & Xie, 2002). Ouabain initiates tyrosine phosphorylation in several cell types including smooth muscle cells in culture (Aydemir-Koksoy et al., 2001; Haas, Askari, & Xie, 2000) and in the vascular wall (Bouzinova et al., 2018; Hangaard et al., 2017; Zulian et al., 2013).

The trans-activation of EGFR is an important step in the signal amplification since it recruits adaptor protein Shc resulting in activation of the Ras/Raf/MEK/p42/44 mitogen-activated protein kinase (MAPK) cascade (Aydemir-Koksoy et al., 2001; Haas et al., 2002) and generation of reactive oxygen species (ROS) (for review: Yan & Shapiro, 2016). It is established that the MAPK signal pathway stimulates transcription and translation of a large range of genes that are involved in cell proliferation and migration (Ou, Pan, Zuo, & van der Hoorn, 2017). Not much, however, is known about this downstream signaling in the vascular wall. Thus, chronic exposure to ouabain upregulated NCX and the transient receptor potential cation channel 6 (TRPC6) expression in vascular smooth muscle cells (Zulian et al., 2013). Although this is associated with an activation of Src kinase, no increase (but a decrease) in MAPK phosphorylation is found in this study (Zulian et al., 2013) suggesting another pathway involved in the transcriptional regulation, possibly an elevation of $[Ca^{2+}]_i$ (see below) (Fig. 1).

5. Src kinase signaling pathway

Binding of ouabain to the Na,K-ATPase $\alpha 1$ isoform leads to phosphorylation of Src kinase, which in turn initiates signal transduction (Haas et al., 2000). Although this was elegantly shown by Xie and co-workers (Haas et al., 2000) and validated by several groups in different tissues including vasculature (Bouzinova et al., 2018; Hangaard et al., 2017; Wenceslau & Rossoni, 2014; Zhang et al., 2018; Zulian et al., 2013), the mechanism of Src activation is not generally accepted yet. Based on cell-free experiments, several groups suggest that Src autophosphorylation upon ouabain binding to the Na,K-ATPase is a result of changes in ATP consumption (Gable, Abdallah, Najjar, Liu, & Askari, 2014; Weigand et al., 2012; Yosef, Katz, Peleg, Mehlman, & Karlish, 2016). These studies failed to identify physical interaction between the Na,K-ATPase and Src kinase (Yosef et al., 2016) and specificity of this signaling for different Na,K-ATPase inhibitors (Gable et al., 2014; Weigand et al., 2012). They concluded that Src autophosphorylation depends on the ATP/ADP ratio (Gable et al., 2014; Weigand et al., 2012). Although the authors recognized the importance of membrane micro-environment, i.e., localization of putative Na, K-ATPase-Src complex in caveolae (Liu et al., 2003), this was not addressed in native cells (Yosef et al., 2016). In contrast, Xie and co-authors demonstrated an importance of the third cytosolic domain for this interaction (Tian et al., 2006) and characterized a specific amino acid sequence responsible for it (Li et al., 2009). The pull-down studies supporting (Li et al., 2009; Tian et al., 2006) and opposing (Yosef et al., 2016) the hypothesis of Src kinase association with the Na,K-ATPase are done in cell-free conditions and with similar experimental protocols. Clarification of the underlying reason of this discrepancy will be helpful for the mechanistic understanding of this signaling.

Nevertheless, ouabain increases phosphorylation of Src kinase in several cell types including vascular smooth muscle cells (Bouzinova et al., 2018; Hangaard et al., 2017; Wenceslau & Rossoni, 2014; Zhang et al., 2018; Zulian et al., 2013). In rodents, low concentrations of ouabain do not affect global intracellular ATP, Na^+ and Ca^{2+} concentrations (Aalkjaer & Mulvany, 1985; Hellstrand, Jorup, & Lydrup, 1984; Mulvany et al., 1984) but phosphorylate Src kinase (Bouzinova et al., 2018; Hangaard et al., 2017; Zhang et al., 2018). This supports the hypothesis of Na,K-ATPase/Src interaction although restricted ATP/ADP changes cannot be excluded (Glavind-Kristensen et al., 2004). However, micromolar ouabain does not affect the global ATP/ADP ratio in rodent skeletal muscles but phosphorylates Src kinase (Kotova et al., 2006). The autophosphorylation of Src kinase upon ouabain inhibition might still depend on ATP availability (Roskoski, 2004) and, therefore, on ATP consumption by the Na,K-ATPase. This needs, however, to be validated.

6. Src-dependent Ca^{2+} sensitization

We have reported that Src kinase activation in the vascular wall by ouabain application in vitro is associated with potentiation of contraction (Bouzinova et al., 2018; Hangaard et al., 2017; Zhang et al., 2018). However, we did not analyze the response for MAPK phosphorylation. In contrast to another study (Zulian et al., 2013), we did not find the effect of ouabain in terms of agonist induced $[Ca^{2+}]_i$ response and therefore suggested a sensitization of the smooth muscle contractile machinery to Ca^{2+} (Bouzinova et al., 2018). This conclusion is based on a steepening of the $[Ca^{2+}]_i$—contractile response relationship in the presence of ouabain that normalizes upon Src kinase inhibition.

We found that ouabain-induced Ca^{2+} sensitization is mediated by phosphorylation of myosin phosphatase target subunit 1 (MYPT1) (Bouzinova et al., 2018). MYPT1 modulates Ca^{2+}-dependent phosphorylation of myosin light chain by myosin light chain kinase and is therefore an essential modulator of smooth muscle contraction (MacKay & Knock, 2015). Phosphorylation of MYPT1 has been shown to be mediated via Rho kinase translocation and may significantly potentiate agonist-induced contraction. Although we did not measure Rho kinase translocation in this study, our findings are in line with a recent report demonstrating that Src kinase contributes to ROS-mediated Rho-kinase activation and vasoconstriction in response to agonist stimulation (MacKay et al., 2017). Neither RhoA nor Rho-kinase are regulated directly by tyrosine phosphorylation. Members of Rho guanine nucleotide exchange factor family (RhoGEF), which are involved in RhoA activation, can, however, be tyrosine phosphorylated (Guilluy et al., 2010; Ogita et al., 2003). Thus, ouabain-induced ROS and Src kinase signaling may stimulate Rho-kinase via activation of RhoA due to activation of PDZ-RhoGEF (MacKay et al., 2017). The role of this signaling pathway for ouabain-induced Ca^{2+} sensitization in the vascular wall needs to be validated since other mechanisms, e.g., a direct activation of Rho-kinase by ROS (Aghajanian, Wittchen, Campbell, & Burridge, 2009), are also possible.

Our hypothesis about ouabain-induced Ca^{2+} sensitization is supported by a previous report showing that orthovanadate constricts rat mesenteric arteries via cSrc activation that mediates Rho kinase associated MYPT1 phosphorylation (Ito, Matsuzaki, Sasahara, Shin, & Yayama, 2015). Moreover, Src signaling has been shown to occur upstream from Rho kinase activation during agonist-induced vasoconstriction (Lu et al., 2008).

Accordingly, our studies suggest that ouabain-induced Src activation is associated with MYPT1 phosphorylation and potentiates vasoconstriction independently of the type of agonist (Bouzinova et al., 2018; Staehr et al., 2018; Zhang et al., 2018). Moreover, myogenic tone is reduced after tyrosine kinase inhibition in rat cremaster arterioles (Murphy, Spurrell, & Hill, 2001, 2002) as well as after specific inhibition of the Na,K-ATPase-dependent Src phosphorylation in rat mesenteric arteries (Bouzinova et al., 2018).

In contrast to ouabain, digoxin has no pro-hypertensive action in rats and even antagonizes ouabain-induced hypertension (Huang et al., 1999; Kimura et al., 2000; Manunta et al., 2000). This is, at least in part, due to their different effects on arterial contraction (Song et al., 2014). It was shown that digoxin does not activate Src in the vascular wall (Bouzinova et al., 2018; Staehr et al., 2018; Zulian et al., 2013) suggesting a potential mechanism for the ouabain-digoxin antagonism (Hamlyn & Blaustein, 2016; Song et al., 2014). This is, however, in contrast with a report studying a human cancer cell line, where both cardiac glycosides are effective to activate Src kinase (Wang et al., 2009), suggesting a complex interaction between cardiac glycosides and elicited signal transduction.

A cross talk between the two major mechanistic views on the Na,K-ATPase-dependent signaling is complicated and requires further studies to clarify this issue (Song et al., 2014; Xie et al., 2015; Zulian et al., 2013). The contribution of this interaction seems to be tissue specific since digoxin has no effect on rat mesenteric artery contraction (Bouzinova et al., 2018; Song et al., 2014) but slightly potentiates contraction of cerebral arteries (Staehr et al., 2018). This potentiation of mouse cerebral artery contraction by digoxin is Src kinase independent and of less pronounced compared to the effect of ouabain. We suggested that this effect is mediated via $[Ca^{2+}]_i$. The variability in the contribution of different signaling pathways is further supported by our study showing that the molecular mechanism behind the pro-contractile action of ouabain changes with arterial diameter (Zhang et al., 2018). Using simultaneous measurements of $[Ca^{2+}]_i$ and the contractile response, we found that contribution of the ouabain-induced Ca^{2+} sensitization increases with the diameter of mesenteric arteries, while $[Ca^{2+}]_i$ responses are the same for different diameters (Zhang et al., 2018). Thus, in accordance with the current dual model for the pro-contractile action of cardiac glycosides in the vascular wall, both ouabain and digoxin can potentiate contraction via elevation of $[Ca^{2+}]_i$, while only ouabain is able to initiate "additional" Ca^{2+} sensitization.

7. Isoform specificity of Src kinase signaling

Another important concern regarding the Na,K-ATPase signal transduction in the vascular wall is its isoform specificity. A distinct role of the α1 isoform in this signaling is suggested based on experiments in the kidney epithelium and cell cultures derived from ovaries and human cancer cells (Banerjee et al., 2018; Banerjee, Duan, & Xie, 2015; Lai et al., 2013; Madan et al., 2017; Xie et al., 2015). It is therefore surprising that in the vascular wall the Na,K-ATPase α2 isoform is involved in ouabain-induced Src signaling (Bouzinova et al., 2018; Hangaard et al., 2017; Staehr et al., 2018; Zhang et al., 2018; Zulian et al., 2013). This suggestion is based on rodent studies, where signal transduction is elicited by ouabain concentrations affecting only the α2 isoform (Bouzinova et al., 2018; Hangaard et al., 2017). Moreover, transient downregulation of the Na,K-ATPase α2 isoform prevents ouabain-induced potentiation of arterial contraction in a Src-kinase-dependent manner (Bouzinova et al., 2018).

This is surprising since previous studies clearly indicate lack of a direct interaction between the Na,K-ATPase α2 isoform and Src kinase suggesting that this interaction only occurs for the α1 isoform (Xie et al., 2015; Yu et al., 2018). We suggest that this discrepancy is a result of either functional or expressional interaction between these two α isoforms. The Na,K-ATPase α1 isoform was shown to inhibit Src kinase preferentially in its E1 conformation (Ye et al., 2013). We, therefore, hypothesized that inhibition of the Na, K-ATPase α2 isoform could produce local changes in Na^+ leading to conformational changes in the α1 isoform and thus activation of Src kinase. This hypothesis remains to be validated but in accordance with findings from the vasculature (Bouzinova et al., 2018; Staehr et al., 2018), previous studies on human skeletal muscle cells demonstrated interaction of both α1 and α2 isoforms with Src kinase (Kotova et al., 2006). Importantly, the α1 isoform dependent Src signaling was characterized in cells expressing almost exclusively the α1 isoform (Banerjee et al., 2018, 2015; Lai et al., 2013; Madan et al., 2017; Xie et al., 2015), while skeletal and smooth muscle cells express both the α1 and α2 isoform (Matchkov & Krivoi, 2016).

8. pNaKtide

A pull-down assay identified the 20 amino acid sequence of the Na,K-ATPase α1 isoform, which is responsible for the interaction and inhibition of Src kinase (Li et al., 2009). A water-soluble and membrane-permeable peptide conjugate, pNaKtide, containing this amino acid sequence is capable to interact with Src kinase and inhibit its autophosphorylation (Li et al., 2009). pNaKtide was designed to target specifically the Na,K-ATPase-interacting pool of Src kinase and thereby prevent its phosphorylation by ouabain in an ATP-independent manner (Li et al., 2009). The specificity of pNaKtide has been tested and proven in several studies where pNaKtide antagonized ouabain-induced responses (Drummond et al., 2015; Lai et al., 2013; Li et al., 2009., 2011; Sodhi et al., 2015; Wang et al., 2014; Xie et al., 2015) including the vascular effects of ouabain, which are otherwise mediated via an phosphorylation of Src kinase (Bouzinova et al., 2018; Hangaard et al., 2017; Staehr et al., 2018).

Importantly, to interfere with the Src kinase signaling in the vascular wall, the supramaximal concentration of 2 μM pNaKtide was used to ensure effective inhibition (Bouzinova et al., 2018; Hangaard et al., 2017; Staehr et al., 2018), while half-maximal pNaKtide concentration for inhibition of Src kinase in cell-free system is suggested approximately 70 nM (Li et al., 2009). Whether this supramaximal pNaKtide concentration affects non-Na,K-ATPase related pools of Src kinase or other non-related targets remains to be tested. An evaluation of concentration-dependent effects is necessary for future implementation of pNaKtide in vascular intervention studies in vivo as it has been done for tumor growth (Li et al., 2011), adiposeness (Sodhi et al., 2015) and cardiomyopathy studies (Li et al., 2018; Liu et al., 2016).

9. Modulation of intercellular coupling

The Na,K-ATPase-dependent Src kinase signaling in the vascular wall is also important for modulation of intercellular communication (Hangaard et al., 2017). Intercellular coupling in the vascular wall is important for several functions including endothelium-dependent relaxation, arterial contraction and rhythmic oscillations in wall tension, i.e., vasomotion (Matchkov et al., 2007, 2012). Micromolar ouabain reduces electrical conductance through gap junctions—most likely formed by connexins 43 (Cx43)—and suppresses vasomotion (Matchkov, 2010; Matchkov et al., 2007, 2012). This is not associated with any changes in global $[Ca^{2+}]_i$ changes although spatially restricted Ca^{2+} events cannot be excluded.

Phosphorylation of multiple tyrosine and serine residues inhibits intercellular communication in tumor cells (Solan & Lampe, 2014) and an importance of Src signaling in endothelial cells is suggested (Suarez & Ballmer-Hofer, 2001). Moreover, the constitutively activated Src kinase in late-stage congestive heart failure is associated with Cx43 tyrosine phosphorylation and suppression of intercellular communication (Toyofuku et al., 1999). Accordingly, ouabain abolishes vasomotion in the vascular wall (Gustafsson, 1993; Gustafsson, Bülow, & Nilsson, 1994; Matchkov et al., 2007, 2012) via an activation of Src kinase and Cx43 phosphorylation (Hangaard et al., 2017). This uncoupling effect of ouabain is antagonized by pNaKtide where both Src kinase activation and Cx43 tyrosine phosphorylation are prevented and the rhythmic contractions in the vascular wall are preserved. This suggests a novel tool to re-establish vasomotion and, thus, tissue perfusion in pathological conditions (Aalkjaer, Boedtkjer, & Matchkov, 2011).

10. PI3K-Akt signaling pathway

Another important interaction partner of the Na,K-ATPase is phosphatidylinositol-4,5-bisphosphate 3-kinase (PI3K) that is suggested to initiate protein kinase B (Akt) kinase activation (Khundmiri et al., 2006; Liu, Zhao, Pierre, & Askari, 2007; Wang et al., 2017) independently of Src activation (Wu et al., 2013). Ouabain stimulates PI3K resulting in endocytosis of the activated Na,K-ATPase-based receptor complex (Liu et al., 2004). This signaling is important for termination of the ouabain-induced signal transduction. A recent study demonstrated that cardiac glycosides induce phosphorylation and activation of both MAPK and Akt kinases with opposite effects on the contractility of the heart (Buzaglo, Rosen, Ben Ami, Inbal, & Lichtstein, 2016). Interestingly, in this study the signaling was induced by both ouabain and digoxin. No study regarding the ouabain-induced PI3K signaling in smooth muscle cells have yet been reported but ouabain increases endothelial nitric oxide production and aorta relaxation via this pathway (Siman et al., 2015).

11. Na,K-ATPase scaffolds with several proteins important for $[Ca^{2+}]_i$ signaling

The Na,K-ATPase-dependent signal transduction is strongly integrated in $[Ca^{2+}]_i$ signaling (Fig. 1). Previous studies showed that Src can modulate $[Ca^{2+}]_i$ in isolated vascular smooth muscle cells via phosphorylation and activation of the L-type voltage-gated Ca^{2+} channels (Gui et al., 2006; Wijetunge, Lymn, & Hughes, 2000). We suggested that attenuation of the resting and agonist-induced $[Ca^{2+}]_i$ in the presence of tyrosine kinase inhibitors might be a result of suppressed tyrosine phosphorylation of voltage-gated Ca^{2+} channels (Bouzinova et al., 2018). However, this may also be an effect of membrane hyperpolarization due to activation of the K^+ channels (Alioua et al., 2002). This suggestion is in accordance with out finding that pNaKtide slightly hyperpolarizes smooth muscle cells (Hangaard et al., 2017).

Notably, well-described ouabain-induced elevation of $[Ca^{2+}]_i$ may be meditated not only through the conventional modulation of Na^+/Ca^{2+} homeostasis but also via physical interactions with several proteins involved in $[Ca^{2+}]_i$ signaling (Fig. 1). Thus, fluorescent resonance energy transfer (FRET) measurements in cultured epithelial cells demonstrated a close spatial proximity between the Na,K-ATPase in the plasma membrane and inositol trisphosphate receptor (IP_3R) on the SR. This interaction is further enhanced by ouabain, which induces $[Ca^{2+}]_i$ oscillations (Aizman, Uhlen, Lal, Brismar, & Aperia, 2001; Miyakawa-Naito et al., 2003). These oscillations elicit activation of the transcription factor NF-κB and are independent of IP_3 generation. The NH_2 terminal of the Na,K-ATPase is essential for this interaction with the IP_3R (Miyakawa-Naito et al., 2003) and a specific motif is identified (Zhang et al., 2006). It was proposed that binding of ouabain induces E1-to-E2 conformational changes leading to the Na,K-ATPase/ IP_3R complex formation.

Although these studies (Aizman et al., 2001; Miyakawa-Naito et al., 2003) were originally done in Na,K-ATPase α1 isoform expressing cells, the following co-immunoprecipitation study (Lencesova, O'Neill, Resneck, Bloch, & Blaustein, 2004) demonstrated isoform specificity of the Na-K-ATPase/IP_3R complex. This complex was formed with the α1 isoform in epithelial cells and with α3 and α2 isoforms in astrocytes and neurons, respectively. It suggests that proximity of specific isoforms of the Na, K-ATPase to IP_3R defines the cell-type-specific $[Ca^{2+}]_i$ signal. This protein-protein interaction was not shown in the vasculature. However, a close proximity between the α2 isoform and the SR was reported (Moore et al., 1993) and, therefore, it is likely that similar interactions appear in the vasculature (Lencesova et al., 2004).

A pull-down assay of epithelial cell lysate revealed that the central loop of the Na/K-ATPase α1 isoform interacts with phospholipase Cγ1 (PLCγ1) forming together with the IP_3R a Ca^{2+}-regulatory complex (Yuan et al., 2005).

This suggests that the Na,K-ATPase acts as an important scaffold bringing together IP_3Rs to their effector PLCγ1 to facilitate a Ca^{2+} response following conformational changes of the Na,K-ATPase. However, the interaction between PLC and the Na,K-ATPase was not shown for the vasculature. This is surprising taking into account the importance of PLC-dependent generation of IP_3 for smooth muscle contraction, gene transcription, proliferation and migration (Narayanan, Adebiyi, & Jaggar, 2012). Modulation of this signaling will therefore has important consequences for vascular tone and structure (Narayanan et al., 2012). Notably, both tyrosine kinase activity and ROS are shown to modulate PLC-IP_3R signaling in smooth muscle cells. Thus, inhibition of tyrosine kinase phosphorylation reduces PLCγ1 activation and IP_3 production in smooth muscle cells (Marrero, Paxton, Duff, Berk, & Bernstein, 1994). Moreover, ROS in smooth muscles suppresses IP_3 degradation (Suzuki & Ford, 1992) and increases IP_3R affinity in cultured smooth muscle cells (Bultynck et al., 2004) although this is not the case for rabbit mesenteric arteries (Wada & Okabe, 1997). Altogether, this indirectly suggests a possibility for potentiation of PLC-IP_3R signaling extending the Na-K-ATPase/IP_3R complex to the Na-K-ATPase → Src/ROS → PLC → IP_3 → IP_3R signal transduction. This hypothesis needs, however, to be validated.

12. Concluding remarks

Discovery of signal transduction dependent on the Na,K-ATPase-based receptor for endogenous ouabain-like compounds (Haas et al., 2000; Xie et al., 1999) suggested mechanistic explanations of several inconsistencies in the experimental findings regarding the vascular significance of the Na,K-ATPase. That is, an opposing action of ouabain and digoxin on blood pressure and arterial contractility (Song et al., 2014), an effect of ouabain on smooth muscle cell phenotype and arterial structure (Blaustein et al., 2016), and sensitization of the contractile machinery to $[Ca^{2+}]_i$ (Bouzinova et al., 2018). However, these suggestions are based on recent studies and need to be further validated. The detailed characterization of involved signaling pathways is necessary. Thus, although Src kinase is shown to be activated by ouabain and its phosphorylation depends on the Na,K-ATPase α2 isoform expression, the role of α1 and α2 isoforms in this signaling remains to be clarified. The pathway downstream from Src kinase to MYPT1 phosphorylation should also be elucidated, as well as divergent signaling from Src and PI3 kinases and their interaction with the conventional Ca^{2+} signaling should be identified. Taking into account the suggested significance of Na, K-ATPase in terms of vascular control of peripheral resistance and the putative role of endogenous ouabain-like compounds in cardiovascular pathologies, mechanistic understanding of the Na,K-ATPase signaling in vascular smooth muscle cells may suggest new therapeutic targets to improve cardiovascular function.

Acknowledgment

The study supported by the Novo Nordisk Foundation [nrs. NNF17OC0026198 and NNFOC0052021] and by the Independent Research Fund Denmark—Medical Sciences [nrs. 7025-00015B and 8020-00084A].

References

Aalkjaer, C., Boedtkjer, D., & Matchkov, V. (2011). Vasomotion—What is currently thought? *Acta Physiologica (Oxford, England)*, *202*(3), 253–269.

Aalkjaer, C., & Mulvany, M. J. (1985). Effect of ouabain on tone, membrane potential and sodium efflux compared with [^{3}H]ouabain binding in rat resistance vessels. *The Journal of Physiology*, *362*, 215–231.

Aghajanian, A., Wittchen, E. S., Campbell, S. L., & Burridge, K. (2009). Direct activation of RhoA by reactive oxygen species requires a redox-sensitive motif. *PLoS One*, *4*(11), e8045. https://doi.org/10.1371/journal.pone.0008045.

Aizman, O., & Aperia, A. (2003). Na,K-ATPase as a signal transducer. *Annals of the New York Academy of Sciences*, *986*, 489–496.

Aizman, O., Uhlen, P., Lal, M., Brismar, H., & Aperia, A. (2001). Ouabain, a steroid hormone that signals with slow calcium oscillations. *Proceedings of the National Academy of Sciences of the United States of America*, *98*(23), 13420–13424. https://doi.org/10.1073/pnas.221315298.

Alioua, A., Mahajan, A., Nishimaru, K., Zarei, M. M., Stefani, E., & Toro, L. (2002). Coupling of c-Src to large conductance voltage- and Ca^{2+}-activated K^+ channels as a new mechanism of agonist-induced vasoconstriction. *Proceedings of the National Academy of Sciences of the United States of America*, *99*(22), 14560–14565. https://doi.org/10.1073/pnas.222348099.

Allen, J. C., Abramowitz, J., & Koksoy, A. (2003). Low concentrations of ouabain activate vascular smooth muscle cell proliferation. *Annals of the New York Academy of Sciences*, *986*, 504–508.

Aperia, A. (2007). New roles for an old enzyme: Na,K-ATPase emerges as an interesting drug target. *Journal of Internal Medicine*, *261*(1), 44–52.

Arnon, A., Hamlyn, J. M., & Blaustein, M. P. (2000). Ouabain augments Ca^{2+} transients in arterial smooth muscle without raising cytosolic Na^+. *The American Journal of Physiology*, *279*(2), H679–H691.

Aydemir-Koksoy, A., Abramowitz, J., & Allen, J. C. (2001). Ouabain-induced signaling and vascular smooth muscle cell proliferation. *The Journal of Biological Chemistry*, *276*(49), 46605–46611.

Banerjee, M., Cui, X., Li, Z., Yu, H., Cai, L., Jia, X., et al. (2018). Na/K-ATPase Y260 phosphorylation-mediated Src regulation in control of aerobic glycolysis and tumor growth. *Scientific Reports*, *8*(1), 12322. https://doi.org/10.1038/s41598-018-29995-2.

Banerjee, M., Duan, Q., & Xie, Z. (2015). SH2 ligand-like effects of second cytosolic domain of Na/K-ATPase alpha1 subunit on Src kinase. *PLoS One*, *10*(11), e0142119. https://doi.org/10.1371/journal.pone.0142119.

Blanco, G., & Mercer, R. W. (1998). Isozymes of the Na-K-ATPase: Heterogeneity in structure, diversity in function. *The American Journal of Physiology*, *275*(5 Pt. 2), F633–F650.

Blaustein, M. P., Chen, L., Hamlyn, J. M., Leenen, F. H., Lingrel, J. B., Wier, W. G., et al. (2016). Pivotal role of alpha2 Na^+ pumps and their high affinity ouabain binding site in cardiovascular health and disease. *The Journal of Physiology*, *594*(21), 6079–6103. https://doi.org/10.1113/JP272419.

Blaustein, M. P., & Hamlyn, J. M. (2010). Signaling mechanisms that link salt retention to hypertension: Endogenous ouabain, the Na^+ pump, the Na^+/Ca^{2+} exchanger and TRPC proteins. *Biochimica et Biophysica Acta*, *1802*(12), 1219–1229.

Blaustein, M. P., & Lederer, W. J. (1999). Sodium/calcium exchange: Its physiological implications. *Physiological Reviews*, *79*(3), 763–854.

Blaustein, M. P., Leenen, F. H., Chen, L., Golovina, V. A., Hamlyn, J. M., Pallone, T. L., et al. (2012). How NaCl raises blood pressure: A new paradigm for the pathogenesis of salt-dependent hypertension. *The American Journal of Physiology*, *302*(5), H1031–H1049.

Blaustein, M. P., & Wier, W. G. (2007). Local sodium, global reach: Filling the gap between salt and hypertension. *Circulation Research*, *101*(10), 959–961.

Blaustein, M. P., Zhang, J., Chen, L., Song, H., Raina, H., Kinsey, S. P., et al. (2009). The pump, the exchanger, and endogenous ouabain: Signaling mechanisms that link salt retention to hypertension. *Hypertension*, *53*(2), 291–298.

Boulanger, B. R., Lilly, M. P., Hamlyn, J. M., Laredo, J., Shurtleff, D., & Gann, D. S. (1993). Ouabain is secreted by the adrenal gland in awake dogs. *The American Journal of Physiology*, *264*(3 Pt. 1), E413–E419.

Bouzinova, E. V., Hangaard, L., Staehr, C., Mazur, A., Ferreira, A., Chibalin, A., et al. (2018). The α2 isoform Na,K-ATPase modulates contraction of rat mesenteric small artery via cSrc–dependent Ca^{2+} sensitization. *Acta Physiologica (Oxford, England)*, e13059. https://doi.org/10.1111/apha.13059.

Bultynck, G., Szlufcik, K., Kasri, N. N., Assefa, Z., Callewaert, G., Missiaen, L., et al. (2004). Thimerosal stimulates Ca^{2+} flux through inositol 1,4,5-trisphosphate receptor type 1, but not type 3, via modulation of an isoform-specific Ca^{2+}—dependent intramolecular interaction. *The Biochemical Journal*, *381*(Pt. 1), 87–96. https://doi.org/10.1042/BJ20040072.

Buzaglo, N., Rosen, H., Ben Ami, H. C., Inbal, A., & Lichtstein, D. (2016). Essential opposite roles of ERK and Akt signaling in cardiac steroid-induced increase in heart contractility. *The Journal of Pharmacology and Experimental Therapeutics*, *357*(2), 345–356. https://doi.org/10.1124/jpet.115.230763.

Chen, L., Song, H., Wang, Y., Lee, J. C., Kotlikoff, M. I., Pritchard, T. J., et al. (2015). Arterial alpha2-Na^+ pump expression influences blood pressure: Lessons from novel, genetically engineered smooth muscle-specific alpha2 mice. *The American Journal of Physiology*, *309*(5), H958–H968. https://doi.org/10.1152/ajpheart.00430.2015.

Cui, X., & Xie, Z. (2017). Protein interaction and Na/K-ATPase-mediated signal transduction. *Molecules*, *22*(6). https://doi.org/10.3390/molecules22060990.

Dostanic, I., Paul, R. J., Lorenz, J. N., Theriault, S., Van Huysse, J. W., & Lingrel, J. B. (2005). The alpha2-isoform of Na-K-ATPase mediates ouabain-induced hypertension in mice and increased vascular contractility in vitro. *The American Journal of Physiology*, *288*(2), H477–H485.

Dostanic-Larson, I., Van Huysse, J. W., Lorenz, J. N., & Lingrel, J. B. (2005). The highly conserved cardiac glycoside binding site of Na,K-ATPase plays a role in blood pressure regulation. *Proceedings of the National Academy of Sciences of the United States of America*, *102*(44), 15845–15850.

Drummond, C. A., Hill, M. C., Shi, H., Fan, X., Xie, J. X., Haller, S. T., et al. (2015). Na/K-ATPase signaling regulates collagen synthesis through microRNA-29b-3p in cardiac fibroblasts. *Physiological Genomics*, *48*(3), 220–229. https://doi.org/10.1152/physiolgenomics.00116.2015.

Gable, M. E., Abdallah, S. L., Najjar, S. M., Liu, L., & Askari, A. (2014). Digitalis-induced cell signaling by the sodium pump: On the relation of Src to Na^+/K^+-ATPase. *Biochemical and Biophysical Research Communications*, *446*(4), 1151–1154.

Glavind-Kristensen, M., Matchkov, V., Hansen, V. B., Forman, A., Nilsson, H., & Aalkjaer, C. (2004). K_{ATP}-channel-induced vasodilation is modulated by the Na,K-pump activity in rabbit coronary small arteries. *British Journal of Pharmacology*, *143*, 872–880.

Gui, P., Wu, X., Ling, S., Stotz, S. C., Winkfein, R. J., Wilson, E., et al. (2006). Integrin receptor activation triggers converging regulation of Cav1.2 calcium channels by c-Src and protein kinase A pathways. *The Journal of Biological Chemistry*, *281*(20), 14015–14025. https://doi.org/10.1074/jbc.M600433200.

Guilluy, C., Bregeon, J., Toumaniantz, G., Rolli-Derkinderen, M., Retailleau, K., Loufrani, L., et al. (2010). The rho exchange factor Arhgef1 mediates the effects of angiotensin II on vascular tone and blood pressure. *Nature Medicine*, *16*(2), 183–190. https://doi.org/10.1038/nm.2079.

Gustafsson, H. (1993). Vasomotion and underlying mechanisms in small arteries. An in vitro study of rat blood vessels. *Acta Physiologica Scandinavica*, *149*(Suppl. 614), 1–44.

Gustafsson, H., Bülow, A., & Nilsson, H. (1994). Rhythmic contractions of isolated pressurized small arteries fom the rat. *Acta Physiologica Scandinavica*, *152*, 145–152.

Haas, M., Askari, A., & Xie, Z. (2000). Involvement of Src and epidermal growth factor receptor in the signal-transducing function of Na^+/K^+-ATPase. *The Journal of Biological Chemistry*, *275*(36), 27832–27837.

Haas, M., Wang, H., Tian, J., & Xie, Z. (2002). Src-mediated inter-receptor cross-talk between the Na^+/K^+-ATPase and the epidermal growth factor receptor relays the signal from ouabain to mitogen-activated protein kinases. *The Journal of Biological Chemistry*, *277*(21), 18694–18702. https://doi.org/10.1074/jbc.M111357200.

Hamlyn, J. M., & Blaustein, M. P. (2016). Endogenous ouabain: Recent advances and controversies. *Hypertension*, *68*(3), 526–532. https://doi.org/10.1161/HYPERTENSIONAHA.116.06599.

Hamlyn, J. M., Blaustein, M. P., Bova, S., DuCharme, D. W., Harris, D. W., Mandel, F., et al. (1991). Identification and characterization of a ouabain-like compound from human plasma. *Proceedings of the National Academy of Sciences of the United States of America*, *88*(14), 6259–6263.

Hamlyn, J. M., Lu, Z. R., Manunta, P., Ludens, J. H., Kimura, K., Shah, J. R., et al. (1998). Observations on the nature, biosynthesis, secretion and significance of endogenous ouabain. *Clinical and Experimental Hypertension*, *20*(5–6), 523–533.

Hamlyn, J. M., Ringel, R., Schaeffer, J., Levinson, P. D., Hamilton, B. P., Kowarski, A. A., et al. (1982). A circulating inhibitor of (Na+ + K+)ATPase associated with essential hypertension. *Nature*, *300*(5893), 650–652.

Hangaard, L., Bouzinova, E. V., Staehr, C., Dam, V. S., Kim, S., Xie, Z., et al. (2017). Na,K-ATPase regulates intercellular communication in the vascular wall via cSrc kinase dependent connexin43 phosphorylation. *The American Journal of Physiology*, *312*(4), C385–C397. https://doi.org/10.1152/ajpcell.00347.2016.

Hellstrand, P., Jorup, C., & Lydrup, M. L. (1984). O_2 consumption, aerobic glycolysis and tissue phosphagen content during activation of the Na^+/K^+ pump in rat portal vein. *Pflügers Archiv*, *401*(2), 119–124.

Huang, B. S., Kudlac, M., Kumarathasan, R., & Leenen, F. H. (1999). Digoxin prevents ouabain and high salt intake-induced hypertension in rats with sinoaortic denervation. *Hypertension*, *34*(4 Pt. 2), 733–738.

Ito, K., Matsuzaki, M., Sasahara, T., Shin, M., & Yayama, K. (2015). Orthovanadate-induced vasoconstriction of rat mesenteric arteries is mediated by rho kinase-dependent inhibition of myosin light chain phosphatase. *Biological & Pharmaceutical Bulletin*, *38*(11), 1809–1816. https://doi.org/10.1248/bpb.b15-00587.

Iwamoto, T., Kita, S., Zhang, J., Blaustein, M. P., Arai, Y., Yoshida, S., et al. (2004). Salt-sensitive hypertension is triggered by Ca^{2+} entry via Na^{+}/Ca^{2+} exchanger type-1 in vascular smooth muscle. *Nature Medicine*, *10*(11), 1193–1199.

Juhaszova, M., & Blaustein, M. P. (1997a). Distinct distribution of different Na^{+} pump alpha subunit isoforms in plasmalemma. Physiological implications. *Annals of the New York Academy of Sciences*, *834*, 524–536.

Juhaszova, M., & Blaustein, M. P. (1997b). Na^{+} pump low and high ouabain affinity alpha subunit isoforms are differently distributed in cells. *Proceedings of the National Academy of Sciences of the United States of America*, *94*(5), 1800–1805.

Kaide, J., Ura, N., Torii, T., Nakagawa, M., Takada, T., & Shimamoto, K. (1999). Effects of digoxin-specific antibody fab fragment (Digibind) on blood pressure and renal water-sodium metabolism in 5/6 reduced renal mass hypertensive rats. *American Journal of Hypertension*, *12*(6), 611–619.

Khundmiri, S. J., Metzler, M. A., Ameen, M., Amin, V., Rane, M. J., & Delamere, N. A. (2006). Ouabain induces cell proliferation through calcium-dependent phosphorylation of Akt (protein kinase B) in opossum kidney proximal tubule cells. *The American Journal of Physiology*, *291*(6), C1247–C1257.

Kimura, K., Manunta, P., Hamilton, B. P., & Hamlyn, J. M. (2000). Different effects of in vivo ouabain and digoxin on renal artery function and blood pressure in the rat. *Hypertension Research*, *23*(Suppl), S67–S76.

Komiyama, Y., Nishimura, N., Munakata, M., Okuda, K., Nishino, N., Kosaka, C., et al. (1999). Increases in plasma ouabainlike immunoreactivity during surgical extirpation of pheochromocytoma. *Hypertension Research*, *22*(2), 135–139.

Kotova, O., Al-Khalili, L., Talia, S., Hooke, C., Fedorova, O. V., Bagrov, A. Y., et al. (2006). Cardiotonic steroids stimulate glycogen synthesis in human skeletal muscle cells via a Src- and ERK1/2-dependent mechanism. *The Journal of Biological Chemistry*, *281*(29), 20085–20094. https://doi.org/10.1074/jbc.M601577200.

Krep, H., Price, D. A., Soszynski, P., Tao, Q. F., Graves, S. W., & Hollenberg, N. K. (1995). Volume sensitive hypertension and the digoxin-like factor. Reversal by a fab directed against digoxin in DOCA-salt hypertensive rats. *American Journal of Hypertension*, *8*(9), 921–927.

Lai, F., Madan, N., Ye, Q., Duan, Q., Li, Z., Wang, S., et al. (2013). Identification of a mutant alpha1 Na/K-ATPase that pumps but is defective in signal transduction. *The Journal of Biological Chemistry*, *288*(19), 13295–13304.

Laredo, J., Hamilton, B. P., & Hamlyn, J. M. (1994). Ouabain is secreted by bovine adrenocortical cells. *Endocrinology*, *135*(2), 794–797.

Laredo, J., Hamilton, B. P., & Hamlyn, J. M. (1995). Secretion of endogenous ouabain from bovine adrenocortical cells: Role of the zona glomerulosa and zona fasciculata. *Biochemical and Biophysical Research Communications*, *212*(2), 487–493.

Leenen, F., Blaustein, M. P., & Hamlyn, J. (2017). Update on angiotensin II: New endocrine connections between the brain, adrenal glands and the cardiovascular system. *Endocrine Connections*, *6*, R131–R145. https://doi.org/10.1530/EC-17-0161.

Lencesova, L., O'Neill, A., Resneck, W. G., Bloch, R. J., & Blaustein, M. P. (2004). Plasma membrane-cytoskeleton-endoplasmic reticulum complexes in neurons and astrocytes. *The Journal of Biological Chemistry*, *279*(4), 2885–2893.

Li, Z., Cai, T., Tian, J., Xie, J. X., Zhao, X., Liu, L., et al. (2009). NaKtide, a Na/K-ATPase-derived peptide Src inhibitor, antagonizes ouabain-activated signal transduction in cultured cells. *The Journal of Biological Chemistry*, *284*(31), 21066–21076.

Li, S., Eim, C., Kirch, U., Lang, R. E., & Schoner, W. (1998). Bovine adrenals and hypothalamus are a major source of proscillaridin A- and ouabain-immunoreactivities. *Life Sciences*, *62*(11), 1023–1033.

Li, H., Yin, A., Cheng, Z., Feng, M., Zhang, H., Xu, J., et al. (2018). Attenuation of Na/K-ATPase/Src/ROS amplification signal pathway with pNaktide ameliorates myocardial ischemia-reperfusion injury. *International Journal of Biological Macromolecules*, *118*(Pt. A), 1142–1148. https://doi.org/10.1016/j.ijbiomac.2018.07.001.

Li, Z., Zhang, Z., Xie, J. X., Li, X., Tian, J., Cai, T., et al. (2011). Na/K-ATPase mimetic pNaKtide peptide inhibits the growth of human cancer cells. *The Journal of Biological Chemistry*, *286*(37), 32394–32403.

Lichtstein, D., Steinitz, M., Gati, I., Samuelov, S., Deutsch, J., & Orly, J. (1998). Biosynthesis of digitalis-like compounds in rat adrenal cells: Hydroxycholesterol as possible precursor. *Life Sciences*, *62*(23), 2109–2126.

Linde, C. I., Antos, L. K., Golovina, V. A., & Blaustein, M. P. (2012). Nanomolar ouabain increases NCX1 expression and enhances Ca^{2+} signaling in human arterial myocytes: A mechanism that links salt to increased vascular resistance? *The American Journal of Physiology*, *303*(7), H784–H794. https://doi.org/10.1152/ajpheart.00399.2012.

Liu, J., Kesiry, R., Periyasamy, S. M., Malhotra, D., Xie, Z., & Shapiro, J. I. (2004). Ouabain induces endocytosis of plasmalemmal Na/K-ATPase in LLC-PK1 cells by a clathrin-dependent mechanism. *Kidney International*, *66*(1), 227–241. https://doi.org/10.1111/j.1523-1755.2004.00723.x.

Liu, L., Mohammadi, K., Aynafshar, B., Wang, H., Li, D., Liu, J., et al. (2003). Role of caveolae in signal-transducing function of cardiac Na^+/K^+-ATPase. *The American Journal of Physiology*, *284*(6), C1550–C1560.

Liu, J., Tian, J., Chaudhry, M., Maxwell, K., Yan, Y., Wang, X., et al. (2016). Attenuation of Na/K-ATPase mediated oxidant amplification with pNaKtide ameliorates experimental uremic cardiomyopathy. *Scientific Reports*, *6*, 34592. https://doi.org/10.1038/srep34592.

Liu, J., Tian, J., Haas, M., Shapiro, J. I., Askari, A., & Xie, Z. (2000). Ouabain interaction with cardiac Na^+/K^+-ATPase initiates signal cascades independent of changes in intracellular Na^+ and Ca^{2+} concentrations. *The Journal of Biological Chemistry*, *275*(36), 27838–27844.

Liu, L., Zhao, X., Pierre, S. V., & Askari, A. (2007). Association of PI3K-Akt signaling pathway with digitalis-induced hypertrophy of cardiac myocytes. *American Journal of Physiology. Cell Physiology*, *293*(5), C1489–C1497. https://doi.org/10.1152/ajpcell.00158.2007.

Lorenz, J. N., Loreaux, E. L., tanic-Larson, I., Lasko, V., Schnetzer, J. R., Paul, R. J., et al. (2008). ACTH-induced hypertension is dependent on the ouabain-binding site of the alpha2-Na^+-K^+-ATPase subunit. *The American Journal of Physiology*, *295*(1), H273–H280.

Lu, R., Alioua, A., Kumar, Y., Kundu, P., Eghbali, M., Weisstaub, N. V., et al. (2008). C-Src tyrosine kinase, a critical component for 5-HT2A receptor-mediated contraction in rat aorta. *The Journal of Physiology*, *586*(16), 3855–3869. https://doi.org/10.1113/jphysiol.2008.153593.

MacKay, C. E., & Knock, G. A. (2015). Control of vascular smooth muscle function by Src-family kinases and reactive oxygen species in health and disease. *The Journal of Physiology*, *593*(17), 3815–3828. https://doi.org/10.1113/jphysiol.2014.285304.

MacKay, C. E., Shaifta, Y., Snetkov, V. V., Francois, A. A., Ward, J. P. T., & Knock, G. A. (2017). ROS-dependent activation of RhoA/rho-kinase in pulmonary artery: Role of Src-family kinases and ARHGEF1. *Free Radical Biology & Medicine*, *110*, 316–331. https://doi.org/10.1016/j.freeradbiomed.2017.06.022.

Madan, N., Xu, Y., Duan, Q., Banerjee, M., Larre, I., Pierre, S. V., et al. (2017). Src-independent ERK signaling through the rat alpha3 isoform of Na/K-ATPase. *The American Journal of Physiology*, *312*(3), C222–C232. https://doi.org/10.1152/ajpcell.00199.2016.

Manunta, P., Ferrandi, M., Messaggio, E., & Ferrari, P. (2006). A new antihypertensive agent that antagonizes the prohypertensive effect of endogenous ouabain and adducin. *Cardiovascular & Hematological Agents in Medicinal Chemistry*, *4*(1), 61–66.

Manunta, P., Hamilton, J., Rogowski, A. C., Hamilton, B. P., & Hamlyn, J. M. (2000). Chronic hypertension induced by ouabain but not digoxin in the rat: Antihypertensive effect of digoxin and digitoxin. *Hypertension Research*, *23*, S77–S85.

Manunta, P., Stella, P., Rivera, R., Ciurlino, D., Cusi, D., Ferrandi, M., et al. (1999). Left ventricular mass, stroke volume, and ouabain-like factor in essential hypertension. *Hypertension*, *34*(3), 450–456.

Marrero, M. B., Paxton, W. G., Duff, J. L., Berk, B. C., & Bernstein, K. E. (1994). Angiotensin II stimulates tyrosine phosphorylation of phospholipase C-gamma 1 in vascular smooth muscle cells. *The Journal of Biological Chemistry*, *269*(14), 10935–10939.

Matchkov, V. V. (2010). Mechanisms of cellular synchronization in the vascular wall. Mechanisms of vasomotion. *Danish Medical Bulletin*, *57*(10), B4191.

Matchkov, V. V., Gustafsson, H., Rahman, A., Boedtkjer, D. M., Gorintin, S., Hansen, A. K., et al. (2007). Interaction between Na^+/K^+-pump and Na^+/Ca^{2+}-exchanger modulates intercellular communication. *Circulation Research*, *100*(7), 1026–1035.

Matchkov, V. V., & Krivoi, I. I. (2016). Specialized functional diversity and interactions of the Na,K-ATPase. *Frontiers in Physiology*, *7*, 179. https://doi.org/10.3389/fphys.2016.00179.

Matchkov, V. V., Moeller-Nielsen, N., Secher, D. V., Nourian, Z., Bodtkjer, D. M., & Aalkjaer, C. (2012). The alpha2 isoform of the Na,K-pump is important for intercellular communication, agonist-induced contraction and EDHF-like response in rat mesenteric arteries. *The American Journal of Physiology*, *303*(1), H36–H46.

Miriel, V. A., Mauban, J. R., Blaustein, M. P., & Wier, W. G. (1999). Local and cellular Ca^{2+} transients in smooth muscle of pressurized rat resistance arteries during myogenic and agonist stimulation. *The Journal of Physiology*, *518*(Pt. 3), 815–824.

Miyakawa-Naito, A., Uhlen, P., Lal, M., Aizman, O., Mikoshiba, K., Brismar, H., et al. (2003). Cell signaling microdomain with Na,K-ATPase and inositol 1,4,5-trisphosphate receptor generates calcium oscillations. *The Journal of Biological Chemistry*, *278*(50), 50355–50361. https://doi.org/10.1074/jbc.M305378200.

Moore, E. D., Etter, E. F., Philipson, K. D., Carrington, W. A., Fogarty, K. E., Lifshitz, L. M., et al. (1993). Coupling of the Na^+/Ca^{2+} exchanger, Na^+/K^+ pump and sarcoplasmic reticulum in smooth muscle. *Nature*, *365*(6447), 657–660.

Mulvany, M. J., Aalkjaer, C., & Petersen, T. T. (1984). Intracellular sodium, membrane potential, and contractility of rat mesenteric small arteries. *Circulation Research*, *54*(6), 740–749.

Murphy, T. V., Spurrell, B. E., & Hill, M. A. (2001). Tyrosine phosphorylation following alterations in arteriolar intraluminal pressure and wall tension. *The American Journal of Physiology*, *281*(3), H1047–H1056. https://doi.org/10.1152/ajpheart.2001.281.3.H1047.

Murphy, T. V., Spurrell, B. E., & Hill, M. A. (2002). Cellular signalling in arteriolar myogenic constriction: Involvement of tyrosine phosphorylation pathways. *Clinical and Experimental Pharmacology & Physiology*, *29*(7), 612–619.

Narayanan, D., Adebiyi, A., & Jaggar, J. H. (2012). Inositol trisphosphate receptors in smooth muscle cells. *American Journal of Physiology Heart and Circulatory Physiology*, *302*(11), H2190–H2210. https://doi.org/10.1152/ajpheart.01146.2011.

Nesher, M., Dvela, M., Igbokwe, V. U., Rosen, H., & Lichtstein, D. (2009). Physiological roles of endogenous ouabain in normal rats. *The American Journal of Physiology*, *297*(6), H2026–H2034.

Noel, F., Fagoo, M., & Godfraind, T. (1990). A comparison of the affinities of rat (Na+ + K+)-ATPase isozymes for cardioactive steroids, role of lactone ring, sugar moiety and KCl concentration. *Biochemical Pharmacology*, *40*(12), 2611–2616.

Ogita, H., Kunimoto, S., Kamioka, Y., Sawa, H., Masuda, M., & Mochizuki, N. (2003). EphA4-mediated rho activation via Vsm-RhoGEF expressed specifically in vascular smooth muscle cells. *Circulation Research, 93*(1), 23–31. https://doi.org/10.1161/01.RES.0000079310.81429.C8.

Ou, Y., Pan, C. X., Zuo, J., & van der Hoorn, F. A. (2017). Ouabain affects cell migration via Na,K-ATPase-p130cas and via nucleus-centrosome association. *PLoS One, 12*(8), e0183343. https://doi.org/10.1371/journal.pone.0183343.

Peng, M., Huang, L., Xie, Z., Huang, W. H., & Askari, A. (1996). Partial inhibition of Na+/K+-ATPase by ouabain induces the Ca^{2+}-dependent expressions of early-response genes in cardiac myocytes. *The Journal of Biological Chemistry, 271*(17), 10372–10378.

Pierdomenico, S. D., Bucci, A., Manunta, P., Rivera, R., Ferrandi, M., Hamlyn, J. M., et al. (2001). Endogenous ouabain and hemodynamic and left ventricular geometric patterns in essential hypertension. *American Journal of Hypertension, 14*(1), 44–50.

Pritchard, T. J., Bullard, D. P., Lynch, R. M., Lorenz, J. N., & Paul, R. J. (2007). Transgenic mice expressing Na^{+}-K^{+} ATPase in smooth muscle decreases blood pressure. *The American Journal of Physiology, 293*(2), H1172–H1182.

Pulina, M. V., Zulian, A., Berra-Romani, R., Beskina, O., Mazzocco-Spezzia, A., Baryshnikov, S. G., et al. (2010). Upregulation of Na^{+} and Ca^{2+} transporters in arterial smooth muscle from ouabain-induced hypertensive rats. *The American Journal of Physiology, 298*(1), H263–H274.

Roskoski, R., Jr. (2004). Src protein-tyrosine kinase structure and regulation. *Biochemical and Biophysical Research Communications, 324*(4), 1155–1164. https://doi.org/10.1016/j.bbrc.2004.09.171.

Rossi, G., Manunta, P., Hamlyn, J. M., Pavan, E., De, T. R., Semplicini, A., et al. (1995). Immunoreactive endogenous ouabain in primary aldosteronism and essential hypertension: Relationship with plasma renin, aldosterone and blood pressure levels. *Journal of Hypertension, 13*(10), 1181–1191.

Shattock, M. J., Ottolia, M., Bers, D. M., Blaustein, M. P., Boguslavskyi, A., Bossuyt, J., et al. (2015). Na^{+}/Ca^{2+} exchange and Na^{+}/K^{+}-ATPase in the heart. *The Journal of Physiology, 593*(6), 1361–1382. https://doi.org/10.1113/jphysiol.2014.282319.

Shelly, D. A., He, S., Moseley, A., Weber, C., Stegemeyer, M., Lynch, R. M., et al. (2004). Na^{+} pump alpha 2-isoform specifically couples to contractility in vascular smooth muscle: Evidence from gene-targeted neonatal mice. *The American Journal of Physiology, 286*(4), C813–C820.

Siman, F. D., Silveira, E. A., Fernandes, A. A., Stefanon, I., Vassallo, D. V., & Padilha, A. S. (2015). Ouabain induces nitric oxide release by a PI3K/Akt-dependent pathway in isolated aortic rings from rats with heart failure. *Journal of Cardiovascular Pharmacology, 65*(1), 28–38. https://doi.org/10.1097/FJC.0000000000000160.

Sodhi, K., Maxwell, K., Yan, Y., Liu, J., Chaudhry, M. A., Getty, M., et al. (2015). pNaKtide inhibits Na/K-ATPase reactive oxygen species amplification and attenuates adipogenesis. *Science Advances, 1*(9), e1500781. https://doi.org/10.1126/sciadv.1500781.

Solan, J. L., & Lampe, P. D. (2014). Specific Cx43 phosphorylation events regulate gap junction turnover in vivo. *FEBS Letters, 588*(8), 1423–1429.

Song, H., Karashima, E., Hamlyn, J. M., & Blaustein, M. P. (2014). Ouabain-digoxin antagonism in rat arteries and neurones. *The Journal of Physiology, 592*(Pt. 5), 941–969. https://doi.org/10.1113/jphysiol.2013.266866.

Staehr, C., Hangaard, L., Bouzinova, E. V., Kim, S., Rajanathan, R., Boegh Jessen, P., et al. (2018). Smooth muscle ca(2+) sensitization causes hypercontractility of middle cerebral arteries in mice bearing the familial hemiplegic migraine type 2 associated mutation. *Journal of Cerebral Blood Flow and Metabolism*. https://doi.org/10.1177/0271678X18761712.271678X18761712.

Suarez, S., & Ballmer-Hofer, K. (2001). VEGF transiently disrupts gap junctional communication in endothelial cells. *Journal of Cell Science, 114*(Pt. 6), 1229–1235.

Suzuki, Y. J., & Ford, G. D. (1992). Superoxide stimulates IP3-induced Ca^{2+} release from vascular smooth muscle sarcoplasmic reticulum. *The American Journal of Physiology, 262*(1 Pt. 2), H114–H116.

Tian, J., Cai, T., Yuan, Z., Wang, H., Liu, L., Haas, M., et al. (2006). Binding of Src to Na^+/K^+-ATPase forms a functional signaling complex. *Molecular Biology of the Cell, 17*(1), 317–326.

Tian, J., & Xie, Z. J. (2008). The Na-K-ATPase and calcium-signaling microdomains. *Physiology (Bethesda), 23*, 205–211.

Toyofuku, T., Yabuki, M., Otsu, K., Kuzuya, T., Tada, M., & Hori, M. (1999). Functional role of c-Src in gap junctions of the cardiomyopathic heart. *Circulation Research, 85*(8), 672–681.

Vilsen, B. (1999). Mutant Phe788 → Leu of the Na,K-ATPase is inhibited by micromolar concentrations of potassium and exhibits high Na-ATPase activity at low sodium concentrations. *Biochemistry, 38*(35), 11389–11400.

Wada, S., & Okabe, E. (1997). Susceptibility of caffeine- and ins(1,4,5)P3-induced contractions to oxidants in permeabilized vascular smooth muscle. *European Journal of Pharmacology, 320*(1), 51–59.

Wang, L., Ren, J., Li, G., Moorman, J. P., Yao, Z. Q., & Ning, S. (2017). LMP1 signaling pathway activates IRF4 in latent EBV infection and a positive circuit between PI3K and Src is required. *Oncogene, 36*(16), 2265–2274. https://doi.org/10.1038/onc.2016.380.

Wang, Y., Ye, Q., Liu, C., Xie, J. X., Yan, Y., Lai, F., et al. (2014). Involvement of Na/K-ATPase in hydrogen peroxide-induced activation of the Src/ERK pathway in LLC-PK1 cells. *Free Radical Biology & Medicine, 71*, 415–426. https://doi.org/10.1016/j.freeradbiomed.2014.03.036.

Wang, Z., Zheng, M., Li, Z., Li, R., Jia, L., Xiong, X., et al. (2009). Cardiac glycosides inhibit p53 synthesis by a mechanism relieved by Src or MAPK inhibition. *Cancer Research, 69*(16), 6556–6564. https://doi.org/10.1158/0008-5472.CAN-09-0891.

Weigand, K. M., Swarts, H. G., Fedosova, N. U., Russel, F. G., & Koenderink, J. B. (2012). Na,K-ATPase activity modulates Src activation: A role for ATP/ADP ratio. *Biochimica et Biophysica Acta, 1818*(5), 1269–1273.

Wenceslau, C. F., & Rossoni, L. V. (2014). Rostafuroxin ameliorates endothelial dysfunction and oxidative stress in resistance arteries from deoxycorticosterone acetate-salt hypertensive rats: The role of Na^+K^+-ATPase/cSRC pathway. *Journal of Hypertension, 32*(3), 542–554. https://doi.org/10.1097/HJH.0000000000000059.

Wijetunge, S., Lymn, J. S., & Hughes, A. D. (2000). Effects of protein tyrosine kinase inhibitors on voltage-operated calcium channel currents in vascular smooth muscle cells and pp60(c-src) kinase activity. *British Journal of Pharmacology, 129*(7), 1347–1354. https://doi.org/10.1038/sj.bjp.0703186.

Wu, J., Akkuratov, E. E., Bai, Y., Gaskill, C. M., Askari, A., & Liu, L. (2013). Cell signaling associated with Na^+/K^+-ATPase: Activation of phosphatidylinositide 3-kinase IA/Akt by ouabain is independent of Src. *Biochemistry, 52*(50), 9059–9067. https://doi.org/10.1021/bi4011804.

Xie, Z., & Askari, A. (2002). Na^+/K^+-ATPase as a signal transducer. *European Journal of Biochemistry, 269*(10), 2434–2439.

Xie, Z., Kometiani, P., Liu, J., Li, J., Shapiro, J. I., & Askari, A. (1999). Intracellular reactive oxygen species mediate the linkage of Na+/K+-ATPase to hypertrophy and its marker genes in cardiac myocytes. *The Journal of Biological Chemistry, 274*(27), 19323–19328.

Xie, J., Ye, Q., Cui, X., Madan, N., Yi, Q., Pierre, S. V., et al. (2015). Expression of rat Na/K-ATPase alpha2 enables ion pumping but not Ouabain-induced signaling in alpha1-deficient porcine renal epithelial cells. *The American Journal of Physiology, 309*(6), C373–C382. https://doi.org/10.1152/ajpcell.00103.2015.

Yan, Y., & Shapiro, J. I. (2016). The physiological and clinical importance of sodium potassium ATPase in cardiovascular diseases. *Current Opinion in Pharmacology*, *27*, 43–49. https://doi.org/10.1016/j.coph.2016.01.009.

Ye, Q., Lai, F., Banerjee, M., Duan, Q., Li, Z., Si, S., et al. (2013). Expression of mutant alpha1 Na/K-ATPase defective in conformational transition attenuates Src-mediated signal transduction. *The Journal of Biological Chemistry*, *288*(8), 5803–5814.

Yosef, E., Katz, A., Peleg, Y., Mehlman, T., & Karlish, S. J. (2016). Do Src kinase and Caveolin interact directly with Na,K-ATPase? *The Journal of Biological Chemistry*, *291*(22), 11736–11750. https://doi.org/10.1074/jbc.M116.721084.

Yu, H., Cui, X., Zhang, J., Xie, J. X., Banerjee, M., Pierre, S. V., et al. (2018). Heterogeneity of signal transduction by Na-K-ATPase alpha-isoforms: Role of Src interaction. *American Journal of Physiology. Cell Physiology*, *314*(2), C202–C210. https://doi.org/10.1152/ajpcell.00124.2017.

Yuan, Z., Cai, T., Tian, J., Ivanov, A. V., Giovannucci, D. R., & Xie, Z. (2005). Na/K-ATPase tethers phospholipase C and IP_3 receptor into a calcium-regulatory complex. *Molecular Biology of the Cell*, *16*(9), 4034–4045.

Yuan, C. M., Manunta, P., Hamlyn, J. M., Chen, S., Bohen, E., Yeun, J., et al. (1993). Long-term ouabain administration produces hypertension in rats. *Hypertension*, *22*(2), 178–187.

Zhang, L., Aalkjaer, C., & Matchkov, V. V. (2018). The Na,K-ATPase-dependent Src kinase signaling changes with mesenteric artery diameter. *International Journal of Molecular Sciences*, *19*(9). https://doi.org/10.3390/ijms19092489.

Zhang, J., Lee, M. Y., Cavalli, M., Chen, L., Berra-Romani, R., Balke, C. W., et al. (2005). Sodium pump alpha2 subunits control myogenic tone and blood pressure in mice. *The Journal of Physiology*, *569*(Pt. 1), 243–256.

Zhang, S., Malmersjo, S., Li, J., Ando, H., Aizman, O., Uhlen, P., et al. (2006). Distinct role of the N-terminal tail of the Na,K-ATPase catalytic subunit as a signal transducer. *The Journal of Biological Chemistry*, *281*(31), 21954–21962.

Zulian, A., Linde, C. I., Pulina, M. V., Baryshnikov, S. G., Papparella, I., Hamlyn, J. M., et al. (2013). Activation of c-SRC underlies the differential effects of ouabain and digoxin on Ca^{2+} signaling in arterial smooth muscle cells. *The American Journal of Physiology*, *304*(4), C324–C333.

CHAPTER SEVEN

Structure-function relationships in the renal NaCl cotransporter (NCC)

Erika Moreno[a], Paola de los Heros[b], Consuelo Plata[a], Christopher Cutler[c], Antonio Vega-Mateos[a], Norma Vázquez[a,d], Gerardo Gamba[a,d,e,*]

[a]Department of Nephrology and Mineral Metabolism, Instituto Nacional de Ciencias Médicas y Nutrición Salvador Zubirán, México City, Mexico
[b]Unidad de Investigación Traslacional UNAM-INCICH, División de Investigación, Facultad de Medicina, Universidad Nacional Autónoma de México, México City, Mexico
[c]Biology Department, Georgia Southern University, Statesboro, GA, United States
[d]Molecular Physiology Unit, Instituto de Investigaciones Biomédicas, Universidad Nacional Autónoma de México, México City, Mexico
[e]Tecnológico de Monterrey, Escuela de Medicina y Ciencias de la Salud, Monterrey, NL, Mexico
*Corresponding author: e-mail address: gamba@biomedicas.unam.mx

Contents

Abstract

The thiazide-sensitive Na^{+}-Cl^{-} cotransporter (NCC) is the major pathway for salt reabsorption in the distal convoluted tubule, serves as a receptor for thiazide-type diuretics, and is involved in inherited diseases associated with abnormal blood pressure. The functional and structural characterization of NCC from different species has led us to gain insights into the structure-function relationships of the cotransporter. Here we present an overview of different studies that had described these properties. Additionally, we report the cloning and characterization of the NCC from the spiny dogfish (*Squalus acanthias*) kidney (sNCC). The purpose of the present study was to determine the main

Current Topics in Membranes, Volume 83
ISSN 1063-5823
https://doi.org/10.1016/bs.ctm.2019.01.003

functional, pharmacological and regulatory properties of sNCC to make a direct comparison with other NCC orthologous. The sNCC cRNA encodes a 1033 amino acid membrane protein, when expressed in *Xenopus oocytes*, functions as a thiazide-sensitive Na–Cl cotransporter with NCC regulation and thiazide-inhibition properties similar to mammals, rather than to teleosts. However, the K_m values for ion transport kinetics are significantly higher than those observed in the mammal species. In summary, we present a review on NCC structure-function relationships with the addition of the sNCC information in order to enrich the NCC cotransporter knowledge.

Abbreviations

AHT	arterial hypertension
COOH	carboxy terminal loop
Cul3	cullin-3
DCT	distal convoluted tubule
FHHt	familial hyperkalemic hypertension
GS	Gitelman Syndrome
KCC	K^+-Cl^- cotransporter
KLHL3	kelch like 3
NCC	Na^+-Cl^- cotransporter
NH_2	amino terminal loop
NKCC	Na^+-K^+-$2Cl^-$ cotransporter
OSR1	oxidative stress-responsive kinase 1
PHAll	pseudo-hypoaldosteronism type ll
SeSAME	seizures, sensorineural deafness, ataxia, mental retardation and the electrolyte imbalance syndrome
SLC12A	solute carrier cotransporter 12 A
SNP	single nucleotide polymorphism
SPAK	Ste20-related proline alanine-rich kinase
TM	transmembrane regions
WNK	with no lysine kinases

The thiazide-sensitive Na–Cl cotransporter (NCC) is located at the apical membrane of the renal distal convoluted tubule (DCT), where the reabsorption of Na–Cl constitutes its primary function (Gamba, 2005). NCC belongs to the SLC12A family of Cation-Chloride Cotransporters family and it is the main target of thiazide diuretics, widely used in the treatment of arterial hypertension (AHT) (James et al., 2014). NCC loss of function is associated with hereditary diseases such as Gitelman syndrome (Cruz et al., 2001; Gitelman, Graham, & Welt, 1966) and the (SeSAME) syndrome in which mutations in the *KCNJ10* gene encoding the basolateral potassium channel Kir4.1 reduces the activity of NCC, resulting in a Gitelman-like electrolyte disorder and in addition, produce a complex neurological syndrome exhibiting seizures, sensorineural deafness, ataxia, and mental retardation (Bockenhauer et al., 2009; Cuevas et al., 2017; Scholl et al., 2009; Zhang et al., 2014). Conversely, NCC activation is associated with the development of the Familial Hyperkalemic Hypertension or Pseudohypoaldosteronism type II (PHAII); an inherited disease featuring arterial hypertension together with hypokalemic metabolic acidosis and hypocalciuria, constituting thus a mirror image of Gitelman disease. Mutations in the With-No-lysine Kinases (WNK1 and WNK4) genes and ubiquitin ligases Cul3 (Cullin-3) and KLHL3 have been identified to be responsible for this disease (Boyden et al., 2012; Louis-Dit-Picard et al., 2012; Wilson et al., 2001).

NCC has been identified and cloned from many different species; including winter flounder (Gamba et al., 1993), rabbit (Velázquez et al., 1998), rat (Gamba et al., 1994), mouse (Kunchaparty et al., 1999), European and Japanese eel (Cutler & Cramb, 2008; Watanabe, Mekuchi, Ideuchi, Kim, & Kaneko, 2011) and the human (Mastroianni et al., 1996; Simon et al., 1996); this variety has allowed us to study and evaluate diverse NCC functional and structural properties. In addition, we have recently cloned and analyzed the functional properties of NCC from the Squalus kidney and will present these data in the present work.

NCC primary structure analysis (Gamba, 2005) has shown that mammalian species possess up to 90% of identity between them, while between winter flounder and mammalian NCC is about 61% and with the European and Japanese eel NCC only ~45% (Cutler & Cramb, 2008).

Despite NCC importance in renal physiology, pathology and pharmacology, little is known about specific amino acid residues or structural domains where transported ions and/or thiazide diuretics bind to the cotransporter. Studies from our lab in the last decades have focused in the analysis of NCC structure-function in order to find out structural motifs or particular amino acid residues that could eventually increase the understanding of structure and function relationship and eventually to the development of better and stronger drugs. Here we present a review from different studies concerned with NCC structure-function analysis.

1. Molecular properties of the NaCl cotransporter

NCC was first cloned and identified in 1993 from the urinary bladder of the winter flounder teleost fish (flNCC) by means of a functional expression strategy in *Xenopus laevis* oocytes (Gamba et al., 1994). Hydropathy analysis of the flNCC amino acid sequence by the Kyte and Doolitle scale (Kyte & Doolittle, 1982) revealed a hydrophobic central pattern corresponding to 12 putative transmembrane segments (TM) and an extracellular loop between TM segments 7 and 8 with putative N-glycosylation sites. TM segments are flanked by the predominantly hydrophilic amino (NH_2) and carboxy (COOH) terminal intracellular loops, respectively.

The molecular identification of flNCC opened the possibility to find and clone the other SLC12A family members in different animal species by means of sequence homology-based methodologies. Gamba et al. analyzed a rat kidney cRNA gene library with a ^{32}P radiolabeled probe synthetized from flNCC. Two clones were isolated from this assay, first a ~4.4 Kb clone that encodes for the thiazide sensitive Na–Cl cotransporter from renal rat cortex (rNCC); and second, a ~4.7 Kb clone that encodes the bumetanide sensitive Na^+-K^+-$2Cl^-$ cotransporter (rNKCC2). rNCC sequence analysis revealed a very similar protein to that from the winter flounder, with a codifying region of 3006 DNA bp that predicts a 1002 amino acid residues protein, that shares a 61% identity with flNCC (Gamba et al., 1994).

2. Cation chloride cotransporters family

NCC belongs to the SLC12A cation-chloride cotransporters family (Human Genome Organization). This family receives its name because all members transport one or two chloride ions coupled with a cation (Na^+ and/or K^+) following a stoichiometry of 1:1. Up to date, there are nine known genes of the SLC12A family known as SLC12A1-SLC12A9; and they encode for NKCC2, NKCC1, NCC, KCC1, KCC2, KCC3, KCC4, CCC9 and CIP, respectively, localized in different chromosomes in the human. As shown in Fig. 1, the different members of the SLC12A share an important percentage of identity in their amino acid sequence and topology. At the functional level the differences between them are the number and type of ions transported, the sensibility to specific inhibitors (diuretics), the tissue expression and in epithelial cells, the apical or basolateral localization. All members present potential N-glycosylation sites. The SLC12A phylogenetic tree presented in Fig. 1 shows the family is divided in two main branches: one that contains the cotransporters that move Na^+ coupled to Cl^- (NCC, NKKC1 and 2) and another one that includes the K^+-Cl^- cotransporters (KCC1–KCC4). There is a 20% of identity between these branches. There are two more members of the family: CIP and CCC9, they both share approximately

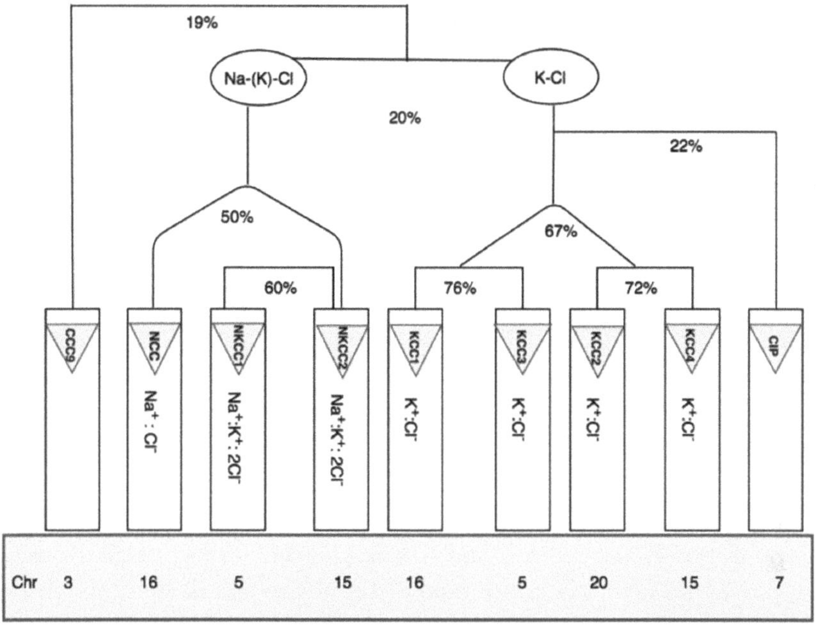

Fig. 1 Phylogenetic tree of SLC12A cotransporter family. The solute carrier family 12 is composed by nine genes located in different chromosomes (Chr). Two main branches can be distinguished in the family by structural and functional characteristics. One is the Na^+-coupled chloride cotransporters, that include the Na-K-2Cl cotransporters 1 and 2: NKCC1 and NKCC2, respectively, and the Na–Cl– cotransporter: NCC, with identity degree of about 50%. The other branch is the K^+-coupled chloride cotransporters that encompasses four members known as KCC1, KCC2, KCC3 and KCC4. The percentage of identity among them is 67% and about 20% with the NKCCs. Chromosome localization is shown.

20% identity with respect to the Na^+ and K^+ cotransporters but their functional properties have not yet been clarified (Gamba, 2005).

The gene codifying for NCC in humans (SLC12A3) is situated in chromosome 16q13 (Mastroianni et al., 1996; Simon et al., 1996), 19p12–14 in the rat (Taniyama et al., 2001) and chromosome 8 in the mouse (Pathak et al., 1996). Human SLC12A3 is a 55 Kb gene contained in 26 exons (Simon et al., 1996). Once transcribed and depending on the specie, NCC is a membrane protein with an amino acid range from 1002 to 1028 residues. Since it's cloning it was established that NCC is expressed exclusively in the kidney (Gamba et al., 1993). However, recent studies showed NCC expression in the intestine (Bazzini et al., 2005; Cutler & Cramb, 2008; Watanabe et al., 2011) and bone (Dvorak et al., 2007). Expression of the cotransporter has been suggested in brain (Deng & Chen, 2003), endothelium

(Calder, Schachter, & Sever, 1992), pancreas (33), peripheral blood monocytes (Abuladze et al., 1998), gall bladder (Cremaschi et al., 2000) and heart (Drewnowska & Baumgarten, 1991), but NCC molecular identification in these tissues is still inconclusive.

3. Functional properties of NCC

The first studies related to structural and functional aspects of NCC were published in the 90's decade. In these studies, functional properties of rNCC (Monroy, Plata, Hebert, & Gamba, 2000) and flNCC (Vázquez, Monroy, Dorantes, Muñoz-Clares, & Gamba, 2002) were evaluated, showing important differences in the sensibility to thiazides and the transport kinetics for Na^+ and Cl^- ions. By means of a functional heterologous expression system in *Xenopus laevis* oocytes, K_m values in flNCC were determined for Na^+ in 58.2 ± 7.1 mM and for Cl^- in 22.1 ± 4.2 mM, compared to rNCC K_m values of 7.26 ± 2.4 mM and 6.46 ± 1.7 mM, for Na^+ and Cl^- ions, respectively. These values are below the physiological concentration of Na^+ and Cl^- in both, the flounder urinary bladder or the mammalian distal convoluted tubule. Thus, in physiological conditions NCC activity is not limited by the ions concentrations. These results showed that rNCC has a higher affinity for the transported ions than flNCC, and it can also be seen that K_m values for both ions in rNCC are very close, while flNCC has higher affinity for the Cl^- than the Na^+ ion.

When thiazide affinity was analyzed, it showed important differences between species; rNCC was 95% inhibited in presence of 10^{-4} M chlortalidone or trichlormethiazide; whereas flNCC activity was only partially reduced under this concentration (68% and 46%, respectively) (Monroy et al., 2000; Vázquez et al., 2002).

Based on these studies, Vázquez et al. (2002) analyzed rNCC and flNCC primary structure in detail. They observed that the percentage of identity between the species is 62%. The transmembrane segments are the regions where greater similarity can be found, with an identity percent up to 80%. In contrast, the NH_2 and $COOH^-$ regions, proved to be the less conserved, sharing 20% and 55% identity, respectively.

Important differences were also found on the extracellular loop between TM 7 and TM 8, where for flNCC three potential N-glycosylation sites were observed, while for rNCC only two glycosylation sites are present (Fig. 2B). A further study not only supported the presence of the two-glycosylation sites in rNCC, but it also showed that they are necessary for

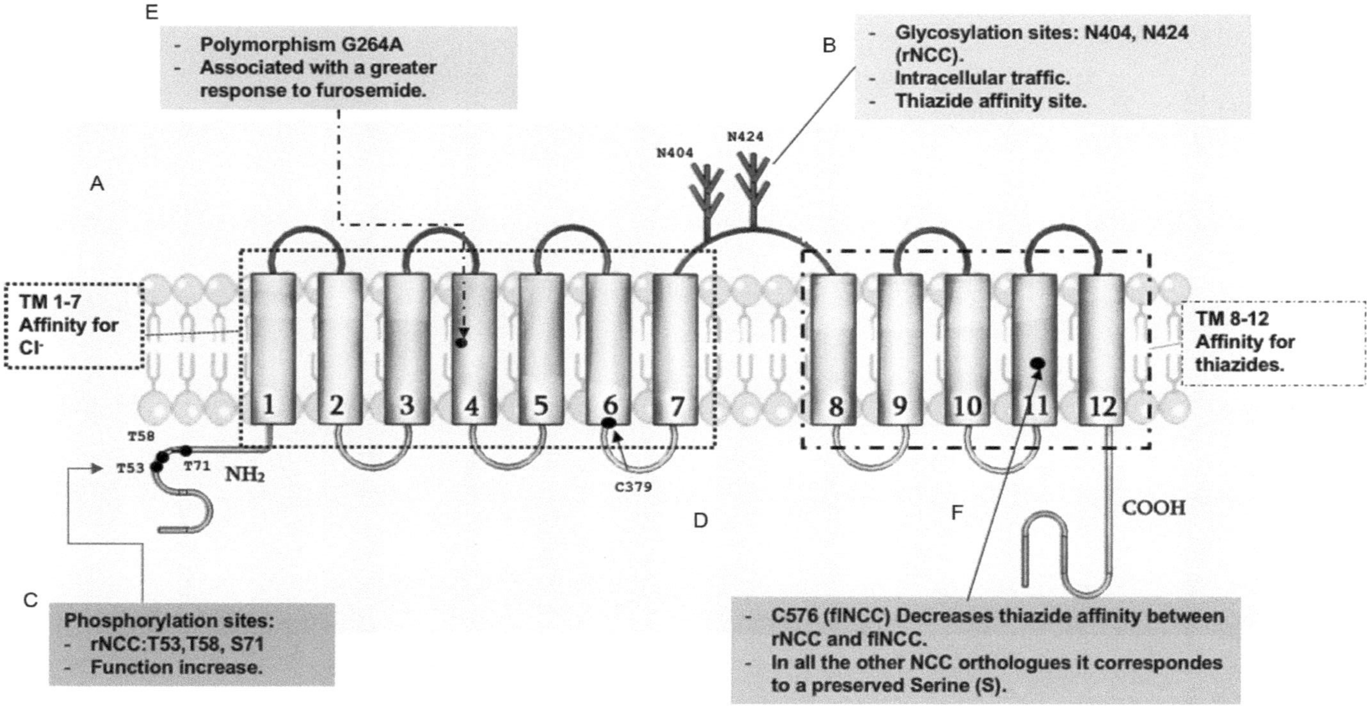

Fig. 2 NCC structure and function relationship. Important regions and amino acids are depicted as follows: (A) Proposed secondary structure with a central hydrophobic domain containing 12 putative transmembrane segments (TM) and an extracellular glycosylated loop located between transmembrane segments 7 and 8. TM are flanked by an NH_2 and a COOH-terminal region. Studies have shown that TM regions 1–7 are involved in chloride (Cl^-) affinity, whereas TM 8–12 are involved in affinity for thiazides. rNCC and the other NCC orthologues share a degree of identity of ~78% in the TM regions and ~39% and ~65% at the amino and carboxyl terminal segments, respectively. (B) Glycosylation sites in the rNCC are found at the connecting loop between TM regions 7–8 and correspond to the amino acids N404 and N424. (C) Black circles shown the potential phosphorylation sites in the rNCC corresponding to AA T53, T58, S71. (D) eNCCβ C379 is indicated with an arrow at the end of TM 6, where it plays a key role in the NCC thiazide-inhibition. (E) Polymorphism G264A is indicated in the TM4. Glycine 264 is a highly preserved AA in all the NCCs. (F) Residue C576 is involved in the affinity for thiazides between rNCC and flNCC. In all the other cloned NCCs, it corresponds to a Serine (S).

proper function, synthesis and processing of the cotransporter (Hoover et al., 2003). Elimination of any of the sites reduced its membrane localization in 50% and consequently its function. Removal of both sites reduced rNCC activity and membrane localization in 95%. Nevertheless, the few amounts of protein that managed to get to the membrane, showed an increased affinity for thiazide, which suggested that N-glycosylation in the extracellular loop might interfere with thiazides binding to the cotransporter.

Years later, Moreno et al. were able to determine specific regions of the NCC protein where ions and thiazides bind. This was possible by means of chimeric proteins constructed between rNCC and flNCC using genetic engineering. That is, proteins constructed with fragments of both species. In this study it was established that all 12 TM regions are involved in Na^+ ion transport, while TM 1–7 are the regions that include the affinity defining residues por Cl^- binding (Fig. 2A). In contrast, the residues defining the affinity for thiazides were found in TM 8–12 (Moreno et al., 2006). The finding that Cl^- affinity was defined by residues located in TM 1–7 was supported by another study by Moreno et al. (2004) where a single nucleotide polymorphism (SNP) that changes a single residue in NCC was analyzed. This residue corresponds to a well conserved glycine located in the TM 4 where it plays an important role in the defining the activity of the cotransporter, as well as the Cl^- binding (Fig. 2E). The glycine/alanine SNP located in position 264 resulted in a 50% loss of cotransporter function, associated with an increased affinity for Cl^- ions of one order of magnitude. This observation was later translated to pharmacogenomics, given that subjects presenting G264A SNP have a greater diuretic response to furosemide (Vormfelde et al., 2007). Reduction in the activity of NCC-G264A resulted in elevated chloride and potassium excretion 24h after a loop diuretic administration. Thus, given that NCC-G264A polymorphism reduces the activity of NCC, it might be possible that the DCT cannot increase salt reabsorption secondary to furosemide administration, increasing in this way the natriuretic response to the loop diuretic.

Castañeda-Bueno et al. (2010) analyzed the amino acids of TM segments 8–12 likely to be responsible for the differences in thiazides affinity between rNCC and flNCC. A detailed amino acid sequence analysis was done in order to find conserved residues in all mammalian species but different in flNCC. This analysis revealed six amino acid residues localized as follows: two in TM 9, three in TM 11 and one in TM 12. Three rNCC mutants were generated where the conserved amino acid was changed to the residue found in flNCC. Mutant rNCC activity was evaluated in the *Xenopus laevis* oocytes functional expression system. Oocytes were micro-injected with wild type (WT) or mutant cRNA and cotransporter function was determined by $^{22}Na^{+}$ uptake in presence of increasing thiazide concentrations. TM 9 and TM 12 mutants showed the same metolazone inhibitory kinetics as WT rNCC, suggesting that TM 9 and 12 residues in flNCC are not responsible for the differences in thiazide affinity between the species. Based on these observations, the following experiments were focused on the three different residues found on TM 11: alanine (ala) 568, isoleucine (iso) 574 and serine (ser) 575 of rNCC. Experiments showed that substitution of serine 575 to cysteine (cys) was enough to confer rNCC a similar thiazide affinity to the one observed in flNCC. When the reverse mutation was performed in flNCC, that is substitution of cys to ser, flNCC presented a thiazide affinity similar to rNCC. It was concluded then, that a single residue in TM segment 11 is enough for determining the differences in thiazide affinity between these two cotransporters.

4. Two genes encoding NCC isoforms in the European eel

Cutler and Cramb (2008) reported the molecular identification and cloning of two NCC isoforms found in different genes of the European eel (eNCC), expressed on the kidney and intestine. These two variants were named as alpha (eNCCα) and beta (eNCCβ). eNCCα is expressed in the kidney and is 65–70% identical to mammal and flounder NCC, respectively. eNCCβ is not expressed in mammals and shares a 55% identity with mammalian NCC. It is worth noticing that the identity of eNCCα; the presumably mammalian orthologue is only about 70% identical with flNCC, despite that they are both teleost fishes. Because eNCCβ seems to be distantly related to mammalian species and teleosts, we decided to perform the molecular characterization of this isoform in order to continue the analysis of NCC structure-function and possibly obtain new useful information. We thus recently reported the functional properties of eNCCβ evaluated in *Xenopus laevis* oocytes (Moreno et al., 2016). We observed that eNCCβ cRNA injection into oocytes induced the appearance of a tracer $^{22}Na^{+}$ uptake in an extracellular Cl^{-} concentration dependent manner, that surprisingly was not inhibited by thiazide diuretics. eNCCβ was exposed to metolazone, hydrochlorothiazide, bendroflumethiazide, trichlormethiazide and chlorthalidone, in concentrations ranging from 10^{-9} up to 10^{-4}, but none of them had inhibitory effect over its function, while any of these diuretics were able to inhibit more than 90% of hNCC, rNCC and flNCC transport in 10^{-4} M concentration (Vázquez et al., 2002). Watanabe et al. (2011) cloned and studied the β isoform from the Japanese eel (jNCCβ). They observed that the jNCCβ is expressed in the apical membrane of the rectal epithelium of the fish and that its expression is higher in fresh water eels than in sea water eels, suggesting that the jNCCβ plays a major role in ion absorption where their concentration is low. In this study they showed that in rectal sacs preparations, 1 mM hydrochlorothiazide reduced significantly water and NaCl absorption, suggesting jNCCβ exhibits diuretic sensitivity. The alignment analysis of the NCC sequences reported by Cutler and Cramb (2008)) and Watanabe et al. (2011) showed that the identity between these two eel orthologues is 98%. The 2% difference is given by six amino acids located in the NH_2 terminal domain, three amino acid residues of the TM regions and four residues of the COOH terminal domain. Specifically, for the TM regions, the three divergent amino acids are: serine 194 placed in the middle of the TM 2 in eNCCβ, corresponding to a phenylalanine in the jNCCβ; cysteine 379 located at the end of the TM 6, in place of a serine in the jNCCβ, and alanine 482 found in the middle of TM 8, that in the jNCCβ corresponds to a glycine residue. In order to prove if these amino acid residues could account for the putative differences observed between eNCCβ and jNCCβ. Moreno et al. (2016) evaluated the effect of switching these residues between the species. A482G had no effect on thiazide response over eNCCβ. On the contrary, C379S and S194F substitutions had a positive, but marginal effect over thiazide inhibition of about 30%. Nevertheless, neither a double (C379S/S194F) nor triple mutation (C379S/S194F/A482G) had an additive or further effect, suggesting only single mutations C379S (See Fig. 3) and S194F can induce structural modifications in eNCCβ that allow it to be slightly inhibited by thiazide diuretics.

Finally, little is known regarding the domains or residues involved in the inhibition by loop diuretics or thiazides in all the SLC12A family. Studies with chimeric constructions between NCC and NKCC2 indicated that the key residues must be contained in the TM regions, given that NH_2 and COOH-terminal residues can be interchanged between NCC and NKCC2 without affecting the sensibility to thiazides or loop diuretics, respectively (Tovar-Palacio et al., 2004).

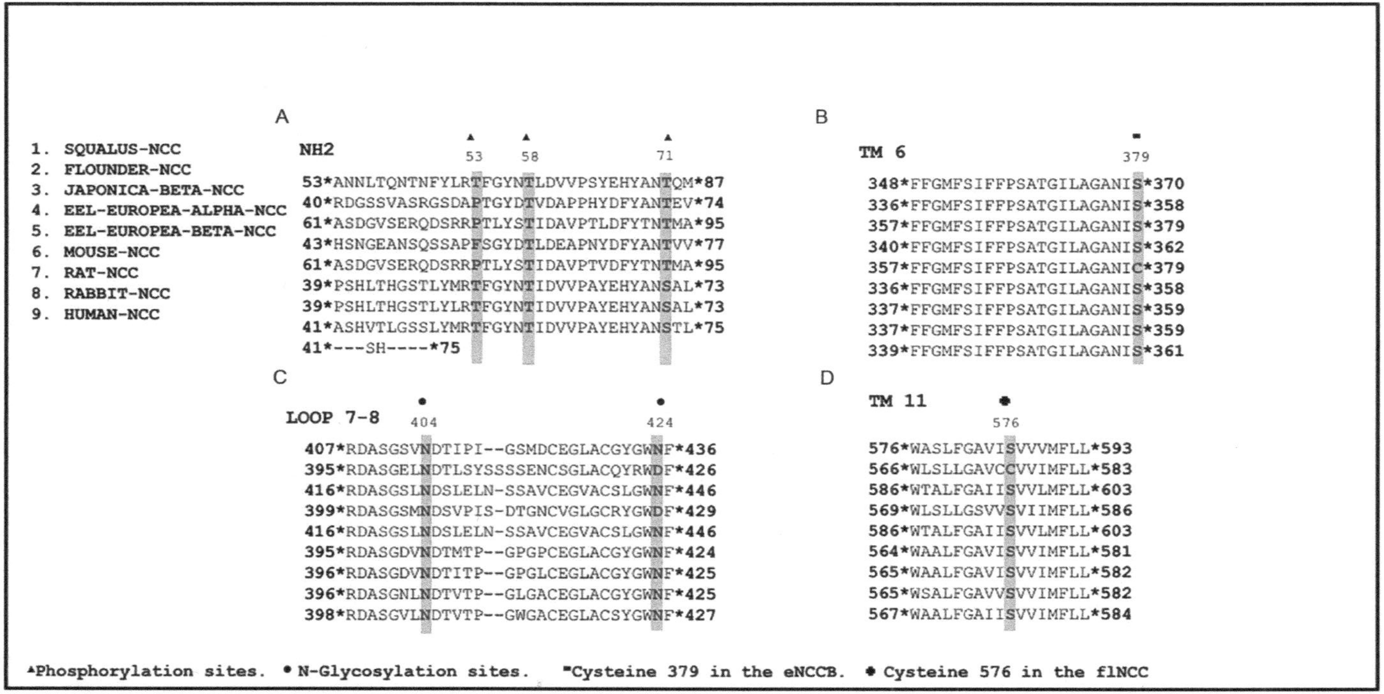

Fig. 3 Alignment analysis of the certain regions of NCC from different species, highlighting key conserved amino acids. (A) Phosphorylation sites in the rNCC (AA T53, T58, T71) are marked and highlighted with the corresponding AA in the other NCC orthologues. (B) Highly preserved Serine in TM6, except in eNCCβ. (C) Glycosylation sites in the rNCC (AA N404 and N423) indicated with a circle. Corresponding AA are highlighted in the different NCC orthologues. (D) Cysteine 576 in TM11 of flNCC, is indicated in the alignment. Corresponding AA in the NCC orthologues are highlighted.

5. NCC regulation by intracellular chloride concentration $[Cl^-]i$

Another interesting feature of eNCCβ is that regulation of its function is different from other NCCs. Previous studies from our group (Pacheco-Alvarez et al., 2006) have shown that rNCC function is clearly influenced by intracellular chloride concentration $[Cl^-]i$. Phosphorylation of the cotransporter in the NH_2 terminal residues threonines 53 (T53) and 58 (T58) and serine 71 (S71) is influenced by low $[Cl^-]i$ (Fig. 2C). Mutation of these residues to alanine produced a decrease in cotransporter function, with phosphorylation of T58 proving to be the critical regulation residue for NCC basal activity (Richardson et al., 2008). Single mutations T53A, T58A and S71A were still able to respond to low $[Cl^-]i$, while the triple mutation, lost sensitivity to $[Cl^-]i$. Joined to this effect, it was observed that cell surface expression of the triple mutant is not different from that of the WT. In summary, this data shows that residues T53, T58 and S71 are critical for rNCC regulation (Pacheco-Alvarez et al., 2006) (see Fig. 3).

Interestingly, a sequence analysis of eNCCβ revealed the presence in this cotransporter of phosphorylation sites T80 and T93, equivalent to rNCC T58 and S71, as well as the SPAK (Ste20-related proline alanine-rich kinase) binding motif nearby in the amino terminal domain. However, in contrast to what have been observed for mammalian NCC and other SLC12A members, eNCCβ is not regulated by changes in $[Cl^-]i$. Thus, further studies comparing eNCCβ with mammalian NCC could give insight into other sequence requirements for NCC to be phosphorylated at the key residues of the amino terminal domain.

6. NCC regulation by WNK kinases

The WNK family of serine/threonine kinases is composed of four genes known as WNK1, WNK2, WNK3 and WNK4, localized in human chromosomes 12, 9, X and 17, respectively (Wilson et al., 2001). These kinases contain a conserved ser/thr kinase domain that lacks a characteristic catalytic lysine in subdomain II. A short NH_2-terminal domain and a long COOH-terminal region that varies in length among the four members flank the kinase domain. They all present an autoinhibitory domain, two coiled-coil regions and a highly conserved 10 amino acids acidic region, where most of the WNK4-PHAII mutations occur. WNK1, WNK3 and WNK4 are expressed in various tissues, including Cl^- absorbing or secreting epithelia (Delaloy et al., 2008; Kahle et al., 2004, 2005; Rinehart et al., 2005; Vitari, Deak, Morrice, & Alessi, 2005; Xu et al., 2000). WNK2 is highly expressed in brain neurons (Xu et al., 2000) and unlike the other members it is not expressed in the kidney. It has been shown that WNK1 and WNK3 are expressed in all nephron segments (Delaloy et al., 2008; Kahle et al., 2005; O'Reilly, Marshall, Speirs, & Brown, 2003; Vidal-Petiot et al., 2012), while WNK4 can only be found along the aldosterone sensitive nephron region (Ohno et al., 2011).

In 2001, it was discovered that mutations in WNK1 and WNK4 cause the monogenic disease of Familial Hyperkalemic Hypertension (FHHt), also known as Pseudohypoaldosteronism type II or Gordon syndrome. Given that the phenotype reported for this disease is associated with abnormal renal Na^+, Cl^- and K^+ electrolyte equilibrium and handling, results obtained from the study of this signaling pathway established an important role for WNK kinases in the regulation of solute transport and blood pressure (Wilson et al., 2001).

Different studies have shown that under osmotic stress or a low $[Cl^-]$i, WNK kinases become autophosphorylated in the highly conserved residue serine 382, this in turn drives the activation and phosphorylation of their downstream effectors, the SPAK and Oxidative stress-responsive kinase 1 (OSR1) (Gagnon & Delpire, 2012). WNKs phosphorylate residue threonine 233 of human SPAK and threonine 185 in human OSR1 (Vitari et al., 2005), once phosphorylated and activated, SPAK/OSR1 will induce directly NCC T53, T58 and S71 phosphorylation, as previously described (Pacheco-Alvarez et al., 2006).

7. NCC mutations in Gitelman syndrome

Gitelman Syndrome (GS) [OMIM 263800] is a hereditary salt wasting tubule pathology characterized by the presence of metabolic alkalosis, hypokalemia, hypomagnesemia and hypocalciuric, described in 1966 by Gitelman et al. (Cruz et al., 2001; Gitelman et al., 1966). GS is caused by mutations that reduce SLC12A3 gene function, the gene that codifies for NCC, inherited in an autosomic recessive fashion (Melander et al., 2000). GS was considered an uncommon disease, with a reported incidence of approximately 1/40,000 cases. The frequency of heterozygote carriers is approximately 1/100 (Melander et al., 2000; Riveira-Munoz et al., 2007). Recent advances in genetic technology have helped to diagnose hypokalemic patients as GS. Nowadays this disease has turned in the most common inherited renal tubular disease (Riveira-Munoz et al., 2007). By means of a database search of the human genome (HGMD, professional 2018.3), we were able to find 426 mutations in the SLC12A3 gene related to GS, 268 of them correspond to missense/nonsense mutations, 57 are splicing events, 47 deletions, 22 insertions, 5 small indels, 22 gross deletions, 4 gross insertions and 1 complex rearrangement (Sabath et al., 2004).

To understand how these mutations activate or reduce NCC activity, at least five mechanisms have been described: (1) a mutation can affect the synthesis of the protein; (2) it can affect protein processing; (3) it can affect membrane protein insertion; (4) it can change the functional properties of the cotransporter; and (5) it can accelerate protein degradation (Sabath et al., 2004). GS mutations in NCC can be found all along the protein from the NH_2 terminus up to the COOH terminal region, with no preference of any region.

8. Cloning and analysis from NCC *Squalus acanthias* (sNCC)

As shown in previous paragraphs, the functional and structural characterization of NCC from different species has led us to gain insights into the structure-function relationships of the cotransporter. We have just recently cloned and analyzed the NCC from the spiny dogfish (*Squalus acanthias*) kidney, one of the best-known members of the family of sharks. Being a cartilaginous fish (*Elasmobranchii*), the dogfish possess no bones and thus is older in evolution than the teleost fishes, like the flounder or the eel. Here, we present our new and original data on the characterization of the dogfish NCC (sNCC). The more the NCC orthologues that are characterized at the functional level, the greater the possibility to recognize highly or poorly conserved regions involved in defining the cotransporter functional properties along the evolution and thus, to eventually define the critical amino acid residues involved in defining those properties.

8.1 Methods

Structure-function analysis of the sNCC was performed following methodologies previously described by our group (Castañeda-Bueno et al., 2010; Monroy et al., 2000; Moreno et al., 2006, 2016; Vázquez et al., 2002), summarized as follows.

sNCC Cloning. The full-length cDNA clone was obtained by RT-PCR amplified fragments of the *Squalus acanthias* NCC (sNCC) kidney mRNA. Five original fragments were linked together by PCR until one complete cDNA was obtained. The complete sequence of both strands from the resultant cDNA clone revealed that it contains 3437 bp exhibiting an open reading frame of 3099 bp, that encodes a 1033 amino acid residue protein (accession number MK256760). The sNCC sequence was inserted into the PGEM *Xenopus* expression vector.

In Vitro *cRNA Translation.* rNCC and sNCC cDNA were digested at the 3′-end using *Not* I from New England Biolabs. cRNA was transcribed in vitro using the T7 RNA polymerase mMESSAGE mMACHINE (Ambion) transcription system. cRNA product integrity was confirmed on agarose gels, and the concentration was determined using an absorbance of 260 nm (DU 640, Beckman, Fullerton, CA). The cRNA was stored frozen in aliquots at −80 °C until ready for use.

X. laevis Oocyte Preparation. Oocytes were harvested surgically from adult female *Xenopus laevis* frogs (Nasco) under 0.17% tricaine anesthesia and incubated in ND96 (96 mM NaCl, 2 mM KCl, 1.8 mM CaCl2, 1 mM MgCl2, and 5 mM HEPES-Tris (pH 7.4)) in the presence of collagenase B (2 mg/mL) for 1 h. After four washes in ND96, the oocytes were manually defolliculated and incubated overnight at 16 °C in ND96 supplemented with 2.5 mM sodium pyruvate and 5 mg/100 mL gentamicin. The next day, stage V-VI oocytes were injected with 50 nL of water or 10–20 ng of rNCC or sNCC cRNA. Next, the oocytes were incubated for 2 or 3 days in ND96 supplemented with sodium pyruvate and gentamicin, which was changed every 24 h (Castañeda-Bueno et al., 2010; Monroy et al., 2000; Moreno et al., 2006, 2004; Pacheco-Alvarez et al., 2006; Tovar-Palacio et al., 2004; Vázquez et al., 2002).

Assessment of sNCC Cotransporter Function. The function of sNCC was determined by assessing tracer $^{22}Na^+$, $^{86}Rb^+$ and $^{36}Cl^-$ influx in groups of 10 to 15 oocytes following our protocol. To determine the ion transport kinetics of sNCC, we performed experiments varying the concentrations of Na^+ and Cl^-. To maintain osmolarity and ionic strength, gluconate was used as a Cl^- substitute, and *N*-methyl-D-glucamine was used as Na^+ substitute. sNCC was subjected to at least four different ion transport kinetic experiments with each set of solutions. All the influx experiments were performed at 32 °C. At the end of the influx period, oocytes were washed five times in ice-cold influx solution without isotope to remove extracellular fluid tracer. After the oocytes were dissolved in 10% sodium dodecyl sulfate, tracer activity was determined for each oocyte by β-scintillation counting (Castañeda-Bueno et al., 2010; Monroy et al., 2000; Moreno et al., 2006, 2016; Pacheco-Alvarez et al., 2006; Tovar-Palacio et al., 2004; Vázquez et al., 2002).

Statistical Analysis. The results are presented as mean S.E., and significance was defined as two-tailed $P < 0.05$. The significance of the differences between the groups was tested using Student's *t*-test. For three or more groups, one-way analysis of variance with multiple comparisons using Bonferroni correction was applied.

8.2 Results

Squalus acanthias NCC Molecular Identification. sNCC cDNA presented and open reading frame of 1033 amino acids, with a 5′UTR of 36 bp and a 3′UTR of 302 bp. According to the topology prediction (See Fig. 2), sNCC is made up of 12 TM segments flanked by 2 intracellular regions that correspond to the NH_2 and COOH-terminal domains, present in all SLC12A members (Gamba, 2005). Comparative NCC sequence analysis from different species displayed that sNCC has a greater identity degree with the mammalian species than with the marine one. Interestingly, sNCC shares more than 70% of its sequence with human, rat, mouse and rabbit NCC and approximately only 60% with winter flounder and both European and Japanese eels (Fig. 4).

sNCC encodes for a thiazide sensitive Na–Cl cotransporter. In order to study sNCC functional properties, the tracer $^{22}Na^+$ uptake was evaluated in presence or absence of extracellular Cl^- in *Xenopus laevis* oocytes 2 or 3 days after they were microinjected with sNCC cRNA. As shown in Fig. 5A, microinjection of sNCC induced the appearance of a significant $^{22}Na^+$ uptake when compared with water injected oocytes. The uptake induced by sNCC was chloride dependent since it was completely prevented by elimination of Cl^- from the uptake medium and it was sensitive to thiazide diuretics since it was prevented by the addition 10^{-4} M hydrochlorothiazide. The uptake media used in these experiments was K^+ free. Nevertheless, we decided to analyze if sNCC can as well transport K^+ ions (Fig. 5B). Tracer $^{86}Rb^+$ uptake as a K^+ surrogate was evaluated in sNCC and in control rat NKCC2 (rNKCC2) micro-injected oocytes. Functional assays showed that even though rNKCC2 oocytes significantly increased furosemide sensitive $^{86}Rb^+$ uptake, sNCC did not induced any $^{86}Rb^+$ uptake under the same conditions. Therefore, we concluded sNCC is a K^+ independent cotransporter. Additionally, Fig. 5C displays sNCC $^{36}Cl^-$ uptake in presence or absence

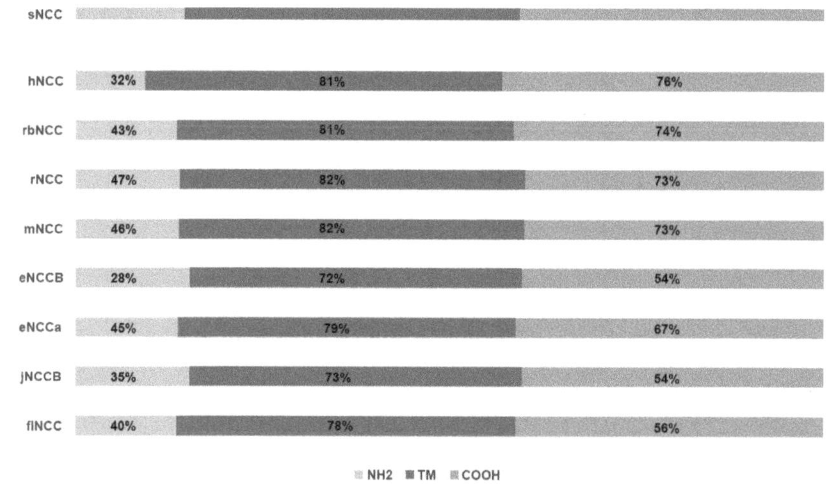

Fig. 4 Degree of identity of sNCC amino terminal, transmembrane and carboxyl terminal segments with NCC orthologues. Comparison was done using BLAST platform (Altschul et al., 1997). The degree of identity of every segment compared to sNCC is annotated in percentage. As noted, sNCC is more similar to mammalian NCC in every segment than to other aquatic species.

of extracellular Na^+. The $^{36}Cl^-$ uptake observed in control conditions was completely prevented in the absence of extracellular Na^+. These results together indicate that sNCC perform as a thiazide-sensitive NaCl cotransporter.

sNCC Thiazide Kinetics. One of the characteristic features of the different reported NCCs is the inhibition by thiazide type diuretics (Castañeda-Bueno et al., 2010; Monroy et al., 2000; Moreno et al., 2006; Sabath et al., 2004; Vázquez et al., 2002). Nevertheless, as previously discussed eNCCβ did not showed sensitivity to these compounds (Moreno et al., 2016). sNCC proved to be sensitive to metolazone (Fig. 5A and C). We evaluated the effect of different concentrations of metolazone in order to define the IC_{50} for sNCC inhibition in order to compare sNCC thiazide affinity with other species (Moreno et al., 2006; Vázquez et al., 2002). Fig. 6 presents sNCC metholazone dose-response curve, compared to rNCC microinjected oocytes. The IC_{50} for sNCC was 10^{-6}, which is similar to that previously shown for rNCC (Monroy et al., 2000; Moreno et al., 2006; Vázquez et al., 2002). These data show that, consistent with the higher degree of identity with mammalian NCC, the sensibility of the dogfish NCC to thiazides resembles the mammalian rather than the flounder NCC.

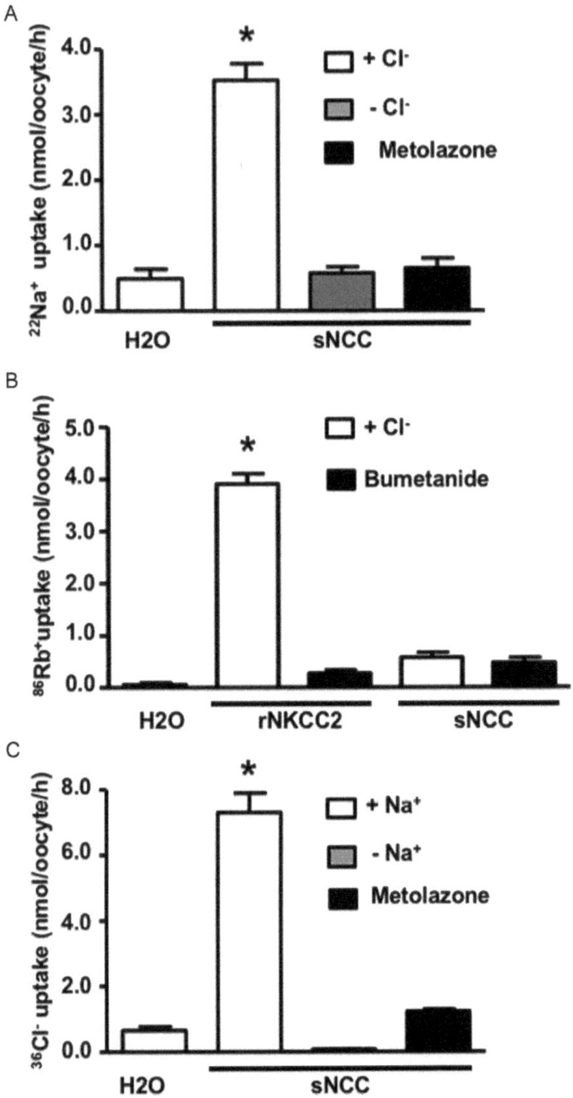

Fig. 5 sNCC encodes a thiazide sensitive Na–Cl cotransporter. (A) $^{22}Na^+$ uptake in oocytes injected with water or with sNCC cRNA (as stated) incubated in the presence (white columns) or absence of extracellular chloride (gray columns) or in the presence of thiazide (black columns). (B) $^{86}Rb^+$ uptake in oocytes injected with water, with rNKCC2 cRNA or with sNCC cRNA (as stated) in absence (white columns) or presence (black columns) of bumetanide. (C) $^{36}Cl^-$ uptake of oocytes injected with water or with sNCC cRNA (as stated) incubated in the presence (white columns) or absence of extracellular sodium (gray columns) or in the presence of thiazide (black columns). *P 0.0001 versus water-injected control or versus absence of chloride or sodium, or versus in presence of bumetanide or thiazide of the same group.

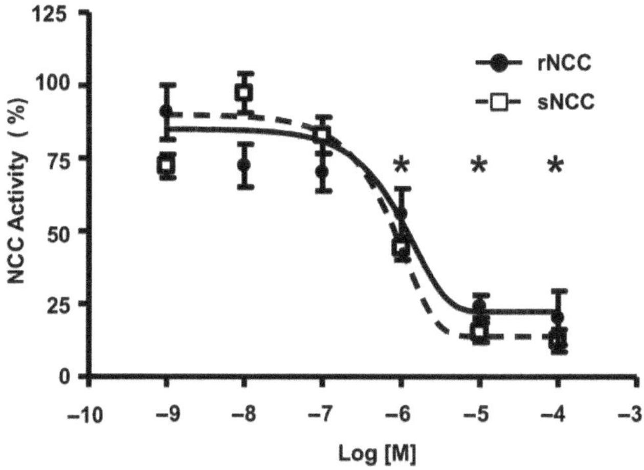

Fig. 6 Thiazide inhibitory kinetics. *Xenopus* oocytes were injected with rat NCC cRNA (black circles, continuous line) or sNCC cRNA (white boxes, discontinuous line), 3 days later a dose-response curve for metolazone was obtained by assessing the $^{22}Na^{+}$ uptake in the absence of the diuretic or in its presence from 10^{-9} to 10^{-4} M concentration. An inhibitory effect of metolazone was observed in both NCCs, with similar IC_{50} of about 10^{-6} M. *$P < 0.001$ versus control group.

sNCC is regulated by low Cl^- hypotonic stress. Previous research from our group have shown that rNCC activity is stimulated under low $[Cl^-]$i. by amino terminal T53, T58 and S71 residues direct phosphorylation via WNK-SPAK signaling pathway (Pacheco-Alvarez et al., 2006; Richardson et al., 2008). However, as discussed above eNCCβ is not (Moreno et al., 2016). We thus analyzed the tracer $^{22}Na^{+}$ uptake in *Xenopus laevis* oocytes expressing sNCC and rNCC that were incubated in control conditions or under low chloride hypotonic stress that consist of exposing oocytes to slight hypotonic conditions of 170 mOsm/kg in the absence of extracellular chloride for 16h before the $^{22}Na^{+}$ uptake assay. Fig. 7 shows that both cotransporters activity was significantly increased under low chloride hypotonic stress when compared to uptake observed in control isotonic conditions.

NCC is regulated by WNK1 and WNK3 kinases. As earlier discussed in this chapter, kinases WNK1 and WNK3 have been reported as important NCC function regulators (Chávez-Canales et al., 2014; Rinehart et al., 2005). Therefore, we evaluated if sNCC can be also affected by these kinases. Oocytes were co-injected with sNCC and WNK1 or WNK3 cRNA, or co-injected with rNCC plus WNK1 or WNK3 cRNA as a control. As Fig. 8 shows, both kinases were able to stimulate sNCC activity. The same

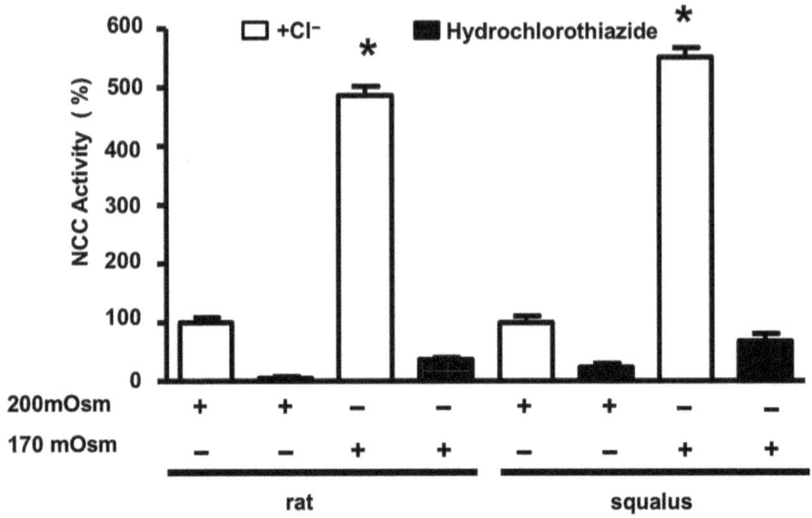

Fig. 7 Effect of low-chloride hypotonic stress on the activity of rat NCC and sNCC. Two days after microinjection with corresponding cRNAs (as stated), oocytes were incubated overnight in regular ND96 (200 mosmol/kg) (as stated with +), or a slightly hypotonic medium (170 mosmol/kg) (as stated with +) without chloride (substituted with isethionate). The next day, the $^{22}Na^+$ influx was assessed in influx media containing NaCl with a similar osmolarity as that used overnight in absence (white boxes) or in presence (black columns) of thiazide. The data observed for each control group was taken as 100%, and the hypotonic medium groups were normalized accordingly. The uptake in the absence of thiazides for each group was similar. $^*P < 0.05$ versus control group.

effect takes place in rNCC expressing ooctyes. This data demonstrates that sNCC, as the mammalian NCCs can be regulated by WNKs kinases.

sNCC Na^+ and Cl^- transport kinetics. We have previously reported significant different ion transport kinetics for mammalian and European eel NCCs (Castañeda-Bueno et al., 2010; Monroy et al., 2000; Moreno et al., 2006, 2016; Vázquez et al., 2002). Our findings showed higher affinity for Na^+ and Cl^- in mammalian orthologues than in the fish versions. In the present work, we determined sNCC ion transport kinetics by measuring $^{22}Na^+$ uptake in sNCC expressing *Xenopus laevis* oocytes. Results from these experiments are presented in Fig. 9. Uptake assays were performed fixing the concentration of either Na^+ or Cl^- ions in 96 mM and increasing the counter ion concentration from 0 to 80 or 96 mM. Uptake average from water injected oocytes (data not shown) was subtracted from the sNCC oocytes groups.

sNCC K_m and V_{max} values for extracellular Na^+ are 19.88 ± 4.75 mM and 6.39 ± 0.50 mM nmol/oocyte/h, respectively (Fig. 9A). Cl^- kinetics K_m and V_{max} values are 13.44 ± 8.2 mM and 1.32 ± 0.24 mM nmol/oocyte/h,

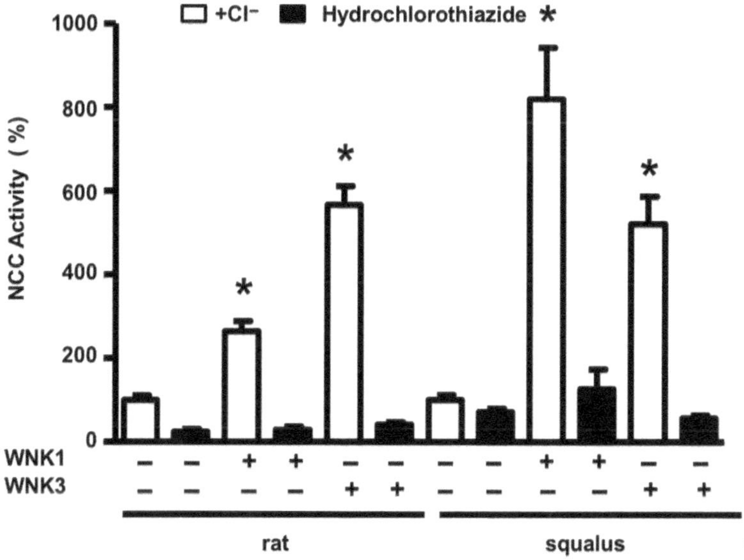

Fig. 8 WNK1 and WNK3 increase the sNCC activity. The $^{22}Na^{+}$ uptake data was normalized to 100% of the sNCC activity or rNCC activity with a previous subtraction of background ($^{22}Na^{+}$ uptake of water oocytes group) and was considered control groups. The percentage of sNCC and rNCC activity in oocytes injected with sNCC or rNCC cRNA alone (as stated) or co-injected WNK1 or WNK3 cRNA (as stated with +) incubated in the absence (white columns) or in presence of thiazide (black columns) was assessed. **P* 0.0001 versus control or versus in presence thiazide of the same group.

respectively (Fig. 9C). Hill coefficient transport kinetics value for Na^{+} ion is 0.86 (Fig. 9B), and 1.3 for Cl^{-} (Fig. 9D), suggesting electroneutrality. In summary, sNCC transport kinetics data, demonstrates it corresponds to a Na–Cl electroneutral cotransporter with a similar affinity to previously characterized NCCs (Castañeda-Bueno et al., 2010; Monroy et al., 2000; Moreno et al., 2006, 2016; Vázquez et al., 2002).

sNCC was cloned and identified from the *Squalus* renal tissue. On a first step we evaluated the functional properties of this cotransporter in order to identify similarities and differences between sNCC and other reported NCCs. As expected, sNCC behaves as a typical NaCl cotransporter that does not transports K^{+} as all the NCCs studied up to date (Castañeda-Bueno et al., 2010; Monroy et al., 2000; Moreno et al., 2006; Vázquez et al., 2002). sNCC function is inhibited in presence of thiazides and its sensitivity to these diuretics is similar to the mammalian orthologues, rather than to the winter flounder fish (Vázquez et al., 2002). Consistent with this, sNCC contains a serine at position 586 of TM segment 11 (equivalent to

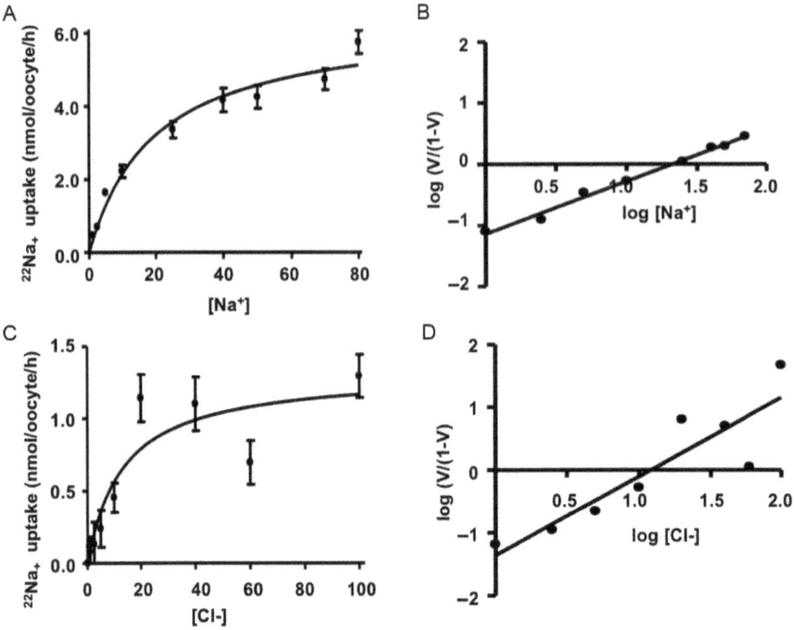

Fig. 9 Chloride and sodium transport kinetic analysis. The $^{22}Na^+$ uptake in the oocytes injected with sNCC cRNA were exposed to (A) different Na^+ extracellular concentrations from 0 to 80mM and in (B) Hill coefficient for Na^+ was calculated. Similarly, sNCC cRNA injected oocytes were exposed to (C) different Cl^- extracellular concentrations from 0 to 100mM and in (D) Hill coefficient for Cl^- was calculated. The slope of the Hill plot for Na^+ is 1.3 ± 0.25 and for Cl^- is 0.86 ± 0.03 (D).

serine 575 in rNCC or cysteine 576 in flNCC discussed above) that is known to confer a mammalian like affinity for thiazides. When it comes to ions affinity, we showed that sNCC is more attracted to the Cl^- ion (K_m 13.44 ± 8.270), than to the Na^+ ion, (K_m 19.88 ± 4.755). We analyzed sNCC TM 4, that has proved to be an important region for Cl^- transport in rNCC and flNCC chimeras, in this analysis we could see that this region keeps an identity percentage of approximately 83% when it is compared with hNCC or rNCC. We believe this amino acid conservation between the species can account for its Cl^- affinity similarity.

Another shared feature between sNCC and rNCC is the increase in activity by decreasing the $[Cl^-]$i and/or when it is expressed with WNK1 or WNK3 kinases. Consistent with this, the sNCC primary structure contains the conserved phosphorylation sites reported in rNCC (Pacheco-Alvarez et al., 2006). The equivalent sNCC phosphorylation sites are T67, T72 and T85 suggesting sNCC activity can be regulated like other

NCCs, with the exception of eNCCβ, that in spite of sharing two equivalent phosphorylation sites, T80 and T93, is not affected by changes in $[Cl^-]i$ (Moreno et al., 2016).

Sequence analysis also revealed putative N-glycosylation sites in sNCC. They are present in asparagine N415 and N435 that in rNCC correspond to residues N404 and N424. It is worth mentioning that in all NCCs sequences identified these sites are conserved, except in the flNCC, that presents three glycosylation sites: N403, N414 and N432 and, in contrast to what was observed for rNCC, elimination of the glycosylation sites in flNCC did not affect its thiazide sensitivity (Moreno et al., 2006).

9. Concluding remarks

Since the cloning of NCC from flounder urinary bladder 25 years ago, tremendous amount of information has been generated regarding many aspects of the NCC molecular biology, biochemistry, regulation, physiology and pathophysiology. This review focused in the information regarding the structure and function relationship of the cotransporter. Most of the information comes from studies performed in *Xenopus laevis* oocytes that constitute a robust system for NCC expression (Castañeda-Bueno et al., 2010; Monroy et al., 2000; Moreno et al., 2006, 2016, 2004; Tovar-Palacio et al., 2004; Vázquez et al., 2002). Functional expression of NCC in mammalian cells has been unsuccessful (de Jong, Willems, van den Heuvel, Knoers, & Bindels, 2003). So far it has been demonstrated that NCC is a glycoprotein and that glycosylation in specific residues is required for proper folding and expression of the transporter. Differences in chloride kinetics between species can be attributed to amino acid residues located in TM segments 1–7, where a specific polymorphism of a single residue affects the renal response to furosemide. A single residue in the TM segment 11 can explain the difference in thiazide sensibility between mammalian and flounder NCC. Three specific phosphor-residues modulating the activity of NCC have been localized to the amino terminal domain. Finally, eNCCβ an isoform identified from eel intestine behave as Na–Cl cotransporter that is nevertheless not sensitive to thiazide diuretics and in contrast to other NCCs studied, is not affected by WNK kinases or intracellular chloride concentration, suggesting that proper study and comparison between this variant and mammalian NCC can eventually guide to the discovery of the amino acid residues in charge of binding the thiazide diuretics, a knowledge that could in turn lead to design new and more potent diuretics, and the sequence requirements for WNK-SPAK regulation of the cotransporter.

Acknowledgment

This work was supported by the Consejo Nacional de Ciencia y Tecnología (CONACYT) Grants 222918 (to E.M.) and 23 (to G.G.).

Conflict of interest

The authors declare that they have no conflicts of interest with the contents of this article.

References

Abuladze, N., Yanagawa, N., Lee, I., Jo, O. D., Newman, D., Hwang, J., et al. (1998). Peripheral blood mononuclear cells express mutated NCCT mRNA in Gitelman's syndrome: Evidence for abnormal thiazide-sensitive NaCl cotransport. *Journal of the American Society of Nephrology: JASN*, *9*(5), 819–826. Retrieved from http://www.ncbi.nlm.nih.gov/pubmed/9596079.

Altschul, S. F., Madden, T. L., Schäffer, A. A., Zhang, J., Zhang, Z., Miller, W., et al. (1997). Gapped BLAST and PSI-BLAST: A new generation of protein database search programs. *Nucleic Acids Research*, *25*(17), 3389–3402. Retrieved from http://www.ncbi.nlm.nih.gov/pubmed/9254694.

Bazzini, C., Vezzoli, V., Sironi, C., Dossena, S., Ravasio, A., De Biasi, S., et al. (2005). Thiazide-sensitive NaCl-cotransporter in the intestine: Possible role of hydrochlorothiazide in the intestinal Ca^{2+} uptake. *The Journal of Biological Chemistry*, *280*(20), 19902–19910. https://doi.org/10.1074/jbc.M411961200.

Bockenhauer, D., Feather, S., Stanescu, H. C., Bandulik, S., Zdebik, A. A., Reichold, M., et al. (2009). Epilepsy, ataxia, sensorineural deafness, tubulopathy, and KCNJ10 mutations. *The New England Journal of Medicine*, *360*(19), 1960–1970. https://doi.org/10.1056/NEJMoa0810276.

Boyden, L. M., Choi, M., Choate, K. A., Nelson-Williams, C. J., Farhi, A., Toka, H. R., et al. (2012). Mutations in kelch-like 3 and cullin 3 cause hypertension and electrolyte abnormalities. *Nature*, *482*(7383), 98–102. https://doi.org/10.1038/nature10814.

Calder, J. A., Schachter, M., & Sever, P. S. (1992). Direct vascular actions of hydrochlorothiazide and indapamide in isolated small vessels. *European Journal of Pharmacology*, *220*(1), 19–26. Retrieved from http://www.ncbi.nlm.nih.gov/pubmed/1425977.

Castañeda-Bueno, M., Vázquez, N., Bustos-Jaimes, I., Hernández, D., Rodríguez-Lobato, E., Pacheco-Alvarez, D., et al. (2010). A single residue in transmembrane domain 11 defines the different affinity for thiazides between the mammalian and flounder NaCl transporters. *American Journal of Physiology—Renal Physiology*, *299*(5), F1111–F1119. https://doi.org/10.1152/ajprenal.00412.2010.

Chávez-Canales, M., Zhang, C., Soukaseum, C., Moreno, E., Pacheco-Alvarez, D., Vidal-Petiot, E., et al. (2014). WNK-SPAK-NCC cascade revisited: WNK1 stimulates the activity of the Na-Cl cotransporter via SPAK, an effect antagonized by WNK4. *Hypertension*, *64*(5), 1047–1053. https://doi.org/10.1161/HYPERTENSIONAHA.114.04036.

Cremaschi, D., Porta, C., Bottà, G., Bazzini, C., Baroni, M. D., & Garavaglia, M. (2000). Apical Na+-Cl-symport in rabbit gallbladder epithelium: A thiazide-sensitive cotransporter (TSC). *The Journal of Membrane Biology, 176*(1), 53–65. Retrieved from http://www.ncbi.nlm.nih.gov/pubmed/10882428.

Cruz, D. N., Simon, D. B., Nelson-Williams, C., Farhi, A., Finberg, K., Burleson, L., et al. (2001). Mutations in the Na-Cl cotransporter reduce blood pressure in humans. *Hypertension, 37*(6), 1458–1464. Retrieved from http://www.ncbi.nlm.nih.gov/pubmed/11408395.

Cuevas, C. A., Su, X.-T., Wang, M.-X., Terker, A. S., Lin, D.-H., McCormick, J. A., et al. (2017). Potassium sensing by renal distal tubules requires Kir4.1. *Journal of the American Society of Nephrology, 28*(6), 1814–1825. https://doi.org/10.1681/ASN.2016090935.

Cutler, C. P., & Cramb, G. (2008). Differential expression of absorptive cation-chloride-cotransporters in the intestinal and renal tissues of the European eel (*Anguilla anguilla*). *Comparative Biochemistry and Physiology Part B: Biochemistry and Molecular Biology, 149*(1), 63–73. https://doi.org/10.1016/j.cbpb.2007.08.007.

de Jong, J. C., Willems, P. H. G. M., van den Heuvel, L. P. W. J., Knoers, N. V. A. M., & Bindels, R. J. M. (2003). Functional expression of the human thiazide-sensitive NaCl cotransporter in Madin-Darby canine kidney cells. *Journal of the American Society of Nephrology: JASN, 14*(10), 2428–2435. Retrieved from http://www.ncbi.nlm.nih.gov/pubmed/14514720.

Delaloy, C., Elvira-Matelot, E., Clemessy, M., Zhou, X., Imbert-Teboul, M., Houot, A.-M., et al. (2008). Deletion of *WNK1* first intron results in misregulation of both isoforms in renal and extrarenal tissues. *Hypertension, 52*(6), 1149–1154. https://doi.org/10.1161/HYPERTENSIONAHA.108.120899.

Deng, L., & Chen, G. (2003). Cyclothiazide potently inhibits gamma-aminobutyric acid type A receptors in addition to enhancing glutamate responses. *Proceedings of the National Academy of Sciences of the United States of America, 100*(22), 13025–13029. https://doi.org/10.1073/pnas.2133370100.

Drewnowska, K., & Baumgarten, C. M. (1991). Regulation of cellular volume in rabbit ventricular myocytes: Bumetanide, chlorothiazide, and ouabain. *American Journal of Physiology-Cell Physiology, 260*(1), C122–C131. https://doi.org/10.1152/ajpcell.1991.260.1.C122.

Dvorak, M. M., De Joussineau, C., Carter, D. H., Pisitkun, T., Knepper, M. A., Gamba, G., et al. (2007). Thiazide diuretics directly induce osteoblast differentiation and mineralized nodule formation by interacting with a sodium chloride co-transporter in bone. *Journal of the American Society of Nephrology: JASN, 18*(9), 2509–2516. https://doi.org/10.1681/ASN.2007030348.

Gagnon, K. B., & Delpire, E. (2012). Molecular physiology of SPAK and OSR1: Two Ste20-related protein kinases regulating ion transport. *Physiological Reviews, 92*(4), 1577–1617. https://doi.org/10.1152/physrev.00009.2012.

Gamba, G. (2005). Molecular physiology and pathophysiology of electroneutral cation-chloride cotransporters. *Physiological Reviews, 85*(2), 423–493. https://doi.org/10.1152/physrev.00011.2004.

Gamba, G., Miyanoshita, A., Lombardi, M., Lytton, J., Lee, W. S., Hediger, M. A., et al. (1994). Molecular cloning, primary structure, and characterization of two members of the mammalian electroneutral sodium-(potassium)-chloride cotransporter family expressed in kidney. *The Journal of Biological Chemistry, 269*(26), 17713–17722. Retrieved from http://www.ncbi.nlm.nih.gov/pubmed/8021284.

Gamba, G., Saltzberg, S. N., Lombardi, M., Miyanoshita, A., Lytton, J., Hediger, M. A., et al. (1993). Primary structure and functional expression of a cDNA encoding the thiazide-sensitive, electroneutral sodium-chloride cotransporter. *Proceedings of the National Academy of Sciences of the United States of America, 90*(7), 2749–2753. Retrieved from http://www.ncbi.nlm.nih.gov/pubmed/8464884.

Gitelman, H. J., Graham, J. B., & Welt, L. G. (1966). A new familial disorder characterized by hypokalemia and hypomagnesemia. *Transactions of the Association of American Physicians*, *79*, 221–235. Retrieved from http://www.ncbi.nlm.nih.gov/pubmed/5929460.

Hoover, R. S., Poch, E., Monroy, A., Vázquez, N., Nishio, T., Gamba, G., et al. (2003). N-glycosylation at two sites critically alters thiazide binding and activity of the rat thiazide-sensitive Na(+):Cl(−) cotransporter. *Journal of the American Society of Nephrology: JASN*, *14*(2), 271–282. Retrieved from http://www.ncbi.nlm.nih.gov/pubmed/12538726.

James, P. A., Oparil, S., Carter, B. L., Cushman, W. C., Dennison-Himmelfarb, C., Handler, J., et al. (2014). 2014 evidence-based guideline for the management of high blood pressure in adults. *JAMA*, *311*(5), 507. https://doi.org/10.1001/jama.2013.284427.

Kahle, K. T., Gimenez, I., Hassan, H., Wilson, F. H., Wong, R. D., Forbush, B., et al. (2004). WNK4 regulates apical and basolateral Cl- flux in extrarenal epithelia. *Proceedings of the National Academy of Sciences of the United States of America*, *101*(7), 2064–2069. https://doi.org/10.1073/pnas.0308434100.

Kahle, K. T., Rinehart, J., de los Heros, P., Louvi, A., Meade, P., Vazquez, N., et al. (2005). WNK3 modulates transport of Cl^- in and out of cells: Implications for control of cell volume and neuronal excitability. *Proceedings of the National Academy of Sciences of the United States of America*, *102*(46), 16783–16788. https://doi.org/10.1073/pnas.0508307102.

Kunchaparty, S., Palcso, M., Berkman, J., Velázquez, H., Desir, G. V., Bernstein, P., et al. (1999). Defective processing and expression of thiazide-sensitive Na-Cl cotransporter as a cause of Gitelman's syndrome. *American Journal of Physiology—Renal Physiology*, *277*(4), F643–F649. https://doi.org/10.1152/ajprenal.1999.277.4.F643.

Kyte, J., & Doolittle, R. F. (1982). A simple method for displaying the hydropathic character of a protein. *Journal of Molecular Biology*, *157*(1), 105–132. https://doi.org/10.1016/0022-2836(82)90515-0.

Louis-Dit-Picard, H., Barc, J., Trujillano, D., Miserey-Lenkei, S., Bouatia-Naji, N., Pylypenko, O., et al. (2012). KLHL3 mutations cause familial hyperkalemic hypertension by impairing ion transport in the distal nephron. *Nature Genetics*, *44*(4), 456–460. https://doi.org/10.1038/ng.2218.

Mastroianni, N., De Fusco, M., Zollo, M., Arrigo, G., Zuffardi, O., Bettinelli, A., et al. (1996). Molecular cloning, expression pattern, and chromosomal localization of the human Na-Cl thiazide-sensitive cotransporter (SLC12A3). *Genomics*, *35*(3), 486–493. Retrieved from http://www.ncbi.nlm.nih.gov/pubmed/8812482.

Melander, O., Orho-Melander, M., Bengtsson, K., Lindblad, U., Râstam, L., Groop, L., et al. (2000). Genetic variants of thiazide-sensitive NaCl-cotransporter in Gitelman's syndrome and primary hypertension. *Hypertension*, *36*(3), 389–394. Retrieved from http://www.ncbi.nlm.nih.gov/pubmed/10988270.

Monroy, A., Plata, C., Hebert, S. C., & Gamba, G. (2000). Characterization of the thiazide-sensitive Na^+-Cl^- cotransporter: A new model for ions and diuretics interaction. *American Journal of Physiology—Renal Physiology*, *279*(1), F161–F169. https://doi.org/10.1152/ajprenal.2000.279.1.F161.

Moreno, E., Cristóbal, P. S., Rivera, M., Vázquez, N., Bobadilla, N. A., & Gamba, G. (2006). Affinity-defining domains in the Na-Cl cotransporter. *Journal of Biological Chemistry*, *281*(25), 17266–17275. https://doi.org/10.1074/jbc.M602614200.

Moreno, E., Plata, C., Rodríguez-Gama, A., Argaiz, E. R., Vázquez, N., Leyva-Ríos, K., et al. (2016). The European eel NCCβ gene encodes a thiazide-resistant Na-Cl cotransporter. *The Journal of Biological Chemistry*, *291*(43), 22472–22481. https://doi.org/10.1074/jbc.M116.742783.

Moreno, E., Tovar-Palacio, C., de los Heros, P., Guzmán, B., Bobadilla, N. A., Vázquez, N., et al. (2004). A single nucleotide polymorphism alters the activity of the renal Na^{+}:Cl^{-} cotransporter and reveals a role for transmembrane segment 4 in chloride and thiazide affinity. *Journal of Biological Chemistry*, *279*(16), 16553–16560. https://doi.org/10.1074/jbc.M400602200.

Ohno, M., Uchida, K., Ohashi, T., Nitta, K., Ohta, A., Chiga, M., et al. (2011). Immunolocalization of WNK4 in mouse kidney. *Histochemistry and Cell Biology*, *136*(1), 25–35. https://doi.org/10.1007/s00418-011-0827-x.

O'Reilly, M., Marshall, E., Speirs, H. J. L., & Brown, R. W. (2003). WNK1, a gene within a novel blood pressure control pathway, tissue-specifically generates radically different isoforms with and without a kinase domain. *Journal of the American Society of Nephrology: JASN*, *14*(10), 2447–2456. Retrieved from http://www.ncbi.nlm.nih.gov/pubmed/14514722.

Pacheco-Alvarez, D., Cristóbal, P. S., Meade, P., Moreno, E., Vazquez, N., Muñoz, E., et al. (2006). The Na^{+}:Cl^{-} cotransporter is activated and phosphorylated at the amino-terminal domain upon intracellular chloride depletion. *Journal of Biological Chemistry*, *281*(39), 28755–28763. https://doi.org/10.1074/jbc.M603773200.

Pathak, B. G., Shaughnessy, J. D., Meneton, P., Greeb, J., Shull, G. E., Jenkins, N. A., et al. (1996). Mouse chromosomal location of three epithelial sodium channel subunit genes and an apical sodium chloride cotransporter gene. *Genomics*, *33*(1), 124–127. Retrieved from http://www.ncbi.nlm.nih.gov/pubmed/8617496.

Richardson, C., Rafiqi, F. H., Karlsson, H. K. R., Moleleki, N., Vandewalle, A., Campbell, D. G., et al. (2008). Activation of the thiazide-sensitive Na^{+}-Cl^{-} cotransporter by the WNK-regulated kinases SPAK and OSR1. *Journal of Cell Science*, *121*(Pt. 5), 675–684. https://doi.org/10.1242/jcs.025312.

Rinehart, J., Kahle, K. T., de los Heros, P., Vazquez, N., Meade, P., Wilson, F. H., et al. (2005). WNK3 kinase is a positive regulator of NKCC2 and NCC, renal cation-Cl-cotransporters required for normal blood pressure homeostasis. *Proceedings of the National Academy of Sciences of the United States of America*, *102*(46), 16777–16782. https://doi.org/10.1073/pnas.0508303102.

Riveira-Munoz, E., Chang, Q., Godefroid, N., Hoenderop, J. G., Bindels, R. J., Dahan, K., et al. (2007). Transcriptional and functional analyses of SLC12A3 mutations: New clues for the pathogenesis of Gitelman syndrome. *Journal of the American Society of Nephrology: JASN*, *18*(4), 1271–1283. https://doi.org/10.1681/ASN.2006101095.

Sabath, E., Meade, P., Berkman, J., de los Heros, P., Moreno, E., Bobadilla, N. A., et al. (2004). Pathophysiology of functional mutations of the thiazide-sensitive Na-Cl cotransporter in Gitelman disease. *American Journal of Physiology. Renal Physiology*, *287*(2), F195–F203. https://doi.org/10.1152/ajprenal.00044.2004.

Scholl, U. I., Choi, M., Liu, T., Ramaekers, V. T., Hausler, M. G., Grimmer, J., et al. (2009). Seizures, sensorineural deafness, ataxia, mental retardation, and electrolyte imbalance (SeSAME syndrome) caused by mutations in KCNJ10. *Proceedings of the National Academy of Sciences of the United States of America*, *106*(14), 5842–5847. https://doi.org/10.1073/pnas.0901749106.

Simon, D. B., Nelson-Williams, C., Johnson Bia, M., Ellison, D., Karet, F. E., Morey Molina, A., et al. (1996). Gitelman's variant of Barter's syndrome, inherited hypokalaemic alkalosis, is caused by mutations in the thiazide-sensitive Na–Cl cotransporter. *Nature Genetics*, *12*(1), 24–30. https://doi.org/10.1038/ng0196-24.

Taniyama, Y., Sato, K., Sugawara, A., Uruno, A., Ikeda, Y., Kudo, M., et al. (2001). Renal tubule-specific transcription and chromosomal localization of rat thiazide-sensitive Na-Cl cotransporter gene. *The Journal of Biological Chemistry*, *276*(28), 26260–26268. https://doi.org/10.1074/jbc.M101614200.

Tovar-Palacio, C., Bobadilla, N. A., Cortés, P., Plata, C., de los Heros, P., Vázquez, N., et al. (2004). Ion and diuretic specificity of chimeric proteins between apical Na^{+}-K^{+}-$2Cl^{-}$ and Na^{+}-Cl^{-} cotransporters. *American Journal of Physiology—Renal Physiology, 287*(3), F570–F577. https://doi.org/10.1152/ajprenal.00124.2004.

Vázquez, N., Monroy, A., Dorantes, E., Muñoz-Clares, R. A., & Gamba, G. (2002). Functional differences between flounder and rat thiazide-sensitive Na-Cl cotransporter. *American Journal of Physiology—Renal Physiology, 282*(4), F599–F607. https://doi.org/10.1152/ajprenal.00284.2001.

Velázquez, H., Náray-Fejes-Tóth, A., Silva, T., Andújar, E., Reilly, R. F., Desir, G. V., et al. (1998). Rabbit distal convoluted tubule coexpresses NaCl cotransporter and 11β-hydroxysteroid dehydrogenase II mRNA. *Kidney International, 54*(2), 464–472. https://doi.org/10.1046/j.1523-1755.1998.00036.x.

Vidal-Petiot, E., Cheval, L., Faugeroux, J., Malard, T., Doucet, A., Jeunemaitre, X., et al. (2012). A new methodology for quantification of alternatively spliced exons reveals a highly tissue-specific expression pattern of WNK1 isoforms. *PLoS One*, 7(5), e37751. https://doi.org/10.1371/journal.pone.0037751.

Vitari, A. C., Deak, M., Morrice, N. A., & Alessi, D. R. (2005). The WNK1 and WNK4 protein kinases that are mutated in Gordon's hypertension syndrome phosphorylate and activate SPAK and OSR1 protein kinases. *Biochemical Journal, 391*(1), 17–24. https://doi.org/10.1042/BJ20051180.

Vormfelde, S. V., Sehrt, D., Toliat, M. R., Schirmer, M., Meineke, I., Tzvetkov, M., et al. (2007). Genetic variation in the renal sodium transporters NKCC2, NCC, and ENaC in relation to the effects of loop diuretic drugs. *Clinical Pharmacology & Therapeutics, 82*(3), 300–309. https://doi.org/10.1038/sj.clpt.6100131.

Watanabe, S., Mekuchi, M., Ideuchi, H., Kim, Y. K., & Kaneko, T. (2011). Electroneutral cation-Cl− cotransporters NKCC2β and NCCβ expressed in the intestinal tract of Japanese eel *Anguilla japonica. Comparative Biochemistry and Physiology Part A: Molecular & Integrative Physiology, 159*(4), 427–435. https://doi.org/10.1016/j.cbpa.2011.04.009.

Wilson, F. H., Disse-Nicodème, S., Choate, K. A., Ishikawa, K., Nelson-Williams, C., Desitter, I., et al. (2001). Human hypertension caused by mutations in WNK kinases. *Science, 293*(5532), 1107–1112. https://doi.org/10.1126/science.1062844.

Xu, B., English, J. M., Wilsbacher, J. L., Stippec, S., Goldsmith, E. J., & Cobb, M. H. (2000). WNK1, a novel mammalian serine/threonine protein kinase lacking the catalytic lysine in subdomain II. *The Journal of Biological Chemistry, 275*(22), 16795–16801. Retrieved from http://www.ncbi.nlm.nih.gov/pubmed/10828064.

Zhang, C., Wang, L., Zhang, J., Su, X.-T., Lin, D.-H., Scholl, U. I., et al. (2014). KCNJ10 determines the expression of the apical Na-Cl cotransporter (NCC) in the early distal convoluted tubule (DCT1). *Proceedings of the National Academy of Sciences of the United States of America, 111*(32), 11864–11869. https://doi.org/10.1073/pnas.1411705111.

CHAPTER EIGHT

Roles of volume-regulatory anion channels, VSOR and Maxi-Cl, in apoptosis, cisplatin resistance, necrosis, ischemic cell death, stroke and myocardial infarction

Yasunobu Okada[a,b,*], Tomohiro Numata[c], Kaori Sato-Numata[d], Ravshan Z. Sabirov[e], Hongtao Liu[b,†], Shin-ichiro Mori[b], Shigeru Morishima[b]

[a]Department of Physiology and Systems Bioscience, Kyoto Prefectural University of Medicine, Kyoto, Japan
[b]Division of Cell Signaling, National Institute for Physiological Sciences (NIPS), Okazaki, Japan
[c]Department of Physiology, School of Medicine, Fukuoka University, Fukuoka, Japan
[d]Japan Society for the Promotion of Science, Tokyo, Japan
[e]Laboratory of Molecular Physiology, Institute of Biophysics and Biochemistry, National University of Uzbekistan, Tashkent, Uzbekistan
*Corresponding author: e-mail address: okada@nips.ac.jp

Contents

† Present address: Department of Anesthesiology, Shengjing Hospital, China Medical University, Shenyang 110004, China.

Current Topics in Membranes, Volume 83
ISSN 1063-5823
https://doi.org/10.1016/bs.ctm.2019.03.001

Abstract

Two types of anion channels are directly activated by osmotic swelling and are involved in the regulatory volume decrease (RVD) in most types of mammalian cells, and they include the volume-sensitive outwardly rectifying anion channel (VSOR), also called the volume-regulated anion channel (VRAC), and the large-conductance maxi-anion channel (Maxi-Cl). In cardiomyocytes, a splice variant of cystic fibrosis transmembrane conductance regulator anion channel (cardiac CFTR) participates in the RVD mechanism under β-adrenergic stimulation. VSOR and Maxi-Cl are also involved in facilitation of the RVD process by releasing extracellular autocrine/paracrine signals, glutamate and ATP. Apoptotic cell death starts with cell shrinkage, called the apoptotic volume decrease (AVD), which is also caused by activation of VSOR. Since VSOR is implicated not only in the AVD induction but also in the uptake of an anti-cancer drug, cisplatin, downregulation of VSOR activity is causatively involved in acquisition of cisplatin resistance in cancer cells. Necrotic cell death exhibits persistent cell swelling, called the necrotic volume increase (NVI), which is coupled to RVD dysfunction due to impaired VSOR function. Acidotoxic and lactacidosis-induced necrotic cell death is induced both by glutamate release mediated by astroglial VSOR and Maxi-Cl and by exaggerated Cl^- influx mediated by neuronal VSOR under prolonged depolarization caused by activation of ionotropic glutamate receptor (iGluR) cation channels. Both VSOR and Maxi-Cl are elaborately involved, in a manner as double-edged swords, in ischemia- and ischemia-reperfusion-induced apoptotic or necrotic cell death in cerebral and myocardial cells by mediating not only Cl^- transport but also release of glutamate and/or ATP. Cardiac CFTR exerts a protective action against ischemia(-reperfusion)-induced cardiac injury, called myocardial infarction (MI), which is largely necrotic. Cardiac Maxi-Cl activity may participate in protection against ischemia(-reperfusion) injury by mediating ATP release.

Abbreviations

ABC ATP-binding cassette
ACTN4 α-actinin-4
AE Cl^-/HCO_3^- anion exchanger
AQP aquaporin
ASIC acid-sensing ion channel

ASOR	acid-sensitive outwardly rectifying Cl^- channel
AVD	apoptotic volume decrease
AVP	arginine vasopressin
BK	bradykinin
BK2	bradykinin type 2 receptor
CaR	Ca^{2+}/polyvalent cation-sensing receptor
CFTR	cystic fibrosis transmembrane conductance regulator
CVD	cooling volume decrease
CVR	cell volume regulation
DAMPs	danger/damage-associated molecular patterns
DCPIB	4-[(2-butyl-6,7-dichloro-2-cyclopentyl-2,3-dihydro-1-oxo-1*H*-inden-5-yl) oxy]butanoic acid
DND	delayed neuronal death
EGFP	enhanced green fluorescent protein
EGFR	epidermal growth factor receptor
ENaC	epithelial Na^+ channel
FasL	Fas ligand
FVD	freezing volume decrease
GABA	gamma-aminobutyric acid
$GABA_AR$	$GABA_A$ receptors
GPCR	G protein-coupled receptors
GSDND	gasderin D
HDAC	histone deacetylase
HICC	hypertonicity-induced cation channel
HMGB1	high mobility group box 1
IAA-94	indanyloxyacetic acid 94
IPC	ischemic preconditioning
iGluR	ionotropic glutamate receptor
KCC	K^+-Cl^- co-transporter
LDH	lactate dehydrogenase
LRRC8	leucine-rich repeat-containing protein 8
Maxi-Cl	large-conductance maxi-anion channel
MCN	magnocellular neurons
MCT	monocarboxylate transporter
MDR	multidrug resistance
mGluR	metabotropic glutamate receptor
MI	myocardial infarction
MLKL	mixed lineage kinase domain-like
NHE	Na^+/H^+ exchanger
NKCC	Na^+-K^+-$2Cl^-$ cotransporter
NMDA	*N*-methyl-D-aspartate
NPPB	5-nitro-2-(3-phenylpropylamino)benzoic acid
NVI	necrotic volume increase
OGD	oxygen-glucose deprivation
OVD	osmotic volume decrease
OVI	osmotic volume increase
Panx1	pannexin 1

PI	propidium iodide
PLM	plasmalemmal membrane
PMCA	plasmalemmal Ca^{2+}-ATPase
RIP	receptor-interacting serine/threonine-protein kinase
ROS	reactive oxygen species
RVD	regulatory volume decrease
RVI	regulatory volume increase
SLCO2A1	solute carrier organic anion transporter family member 2A1
SON	supraoptic nucleus
STS	staurosporine
SVD	secretory volume decrease
TMZ	temozolomide
TNF	tumor necrosis factor
TRP	transient receptor potential cation channel
TSA	trichostatin A
VRAC	volume-regulated anion channel
VSOR	volume-sensitive outwardly rectifying anion channel

1. Introduction

Animal cells incessantly encounter osmotic perturbations, which are associated with their living activities under physiological conditions and is caused by a variety of insults under pathological conditions. Among a variety of anion channel types, two types of anion channel are directly activated by the cell volume increase and thereby involved in the regulatory volume decrease (RVD). One of these volume-regulatory anion channel is the volume-sensitive outwardly rectifying anion channel (VSOR), also called the volume-regulated anion channel (VRAC), which is also known to serve as the pathway for glutamate release, and another one is the large-conductance maxi-anion channel (Maxi-Cl), which is known to serve as the release pathway for ATP and glutamate from swollen cells. In cardiomyocytes, a splice variant of the cAMP-activated cystic fibrosis transmembrane conductance regulator (CFTR) anion channel also participates in RVD when intracellular cAMP is increased. The purpose of this review article is to comprehensively describe the essential roles of VSOR and Maxi-Cl as well as cardiac CFTR first in the cell volume regulation, second in the induction of or protection against necrotic and apoptotic cell death as well as the acquisition of cisplatin resistance, and third in the induction of ischemia(-reperfusion) injury in mammalian cells with mainly focusing to cerebral and cardiac cells. In this article, special attention is also drawn to the roles of VSOR, Maxi-Cl and cardiac CFTR in the induction of cell death associated with excitotoxicity, acidotoxicity and lactacidosis, and in delayed neuronal death (DND) as well as in myocardial infarction (MI).

2. Cell volume regulation and volume-regulatory anion channels

2.1 Osmotic perturbation and cell volume changes under physiological and pathological conditions

Body fluid osmolarity is maintained constant within 1–2% at around 300 mOsm through the thirst-antidiuretic hormone mechanisms. Even in physiological situations, however, large changes in local osmolarity are produced within the intestinal villi exposed to a hypotonic interstitial fluid during excessive water intake or to a hypertonic interstitial fluid (reaching 1200 mOsm: Hallback, Jodal, & Lundgren, 1979; Jodal, Hallback, & Lundgren, 1978) during absorption of ions and nutrients as well as within the kidney medulla during diuretic and antidiuretic (reaching 1400 mOsm: Kurbel, Dodig, & Radic, 2002) states. In the lamina propria mucosae of intestine and the renal medulla, thus, intestinal and renal epithelial and fibroblastic cells therein as well as blood and lymph cells traversing capillaries normally encounter large osmotic perturbation. Also, changes in intracellular osmolarity are produced in neurons during their intense excitation (Andrew & MacVicar, 1994; Churchwell, Wright, Emma, Rosenberg, & Strange, 1996; Fields & Ni, 2010; Iwasa, Tasaki, & Gibbons, 1980; Ransom, Yamate, & Connors, 1985; Tasaki & Byrne, 1990; Uckermann et al., 2004), in liver cells in response to hormonal stimulation of glycogen synthesis/degradation (Lang, Busch, et al., 1998), and in kidney macula densa cells in response to alteration of the tubular NaCl concentration (Bell et al., 2003; Peti-Peterdi, Morishima, Bell, & Okada, 2002). It is known that sperms, which have been standing by in the hypertonic fluid (around 340 mOsm) within the vas deferens (Hinton, Pryor, Hirsh, & Setchell, 1981; Rossato, Di Virgilio, & Foresta, 1996), are subjected to a hypotonic challenge upon ejaculation into the hypotonic fluid (280–290 mOsm) within the cervical mucus and the uterine fluid (Casslen & Nilsson, 1984). Under pathological conditions, sizable changes in the plasma osmolarity are brought about in association with a variety of diseases and iatrogenic outcomes causing hypernatremia, hyponatremia, hyperglycemia and azotemia, as summarized in Table 1.

Table 1 Plasma osmolarity disorder coupled to human diseases.

Increased plasma osmolarity		
Hypernatremia	Na^+ overload	Iatrogenesis
	Increased Na reabsorption	Primary aldosteronism
	Decreased water reabsorption	Diabetes insipidus
	Water loss	Dehydration/diarrhea
Hyperglycemia	Polyuria	Ketoacidosis
Azotemia	Increased BUN[a]	Uremic acidosis
Decreased plasma osmolarity		
Hyponatremia	Decreased Na^+ reabsorption	Cardiac/hepatic/renal insufficiency
	Extrarenal Na^+ loss	Vomiting/diarrhea/excessive sweating
	Renal Na^+ loss	Adverse effects of diuretics/ NSAIDs[b]
	Increased water reabsorption	SIADH[c]/adverse effects of amiodarone
	Excessive water intake	Water intoxication/polydipsia

[a]BUN, blood urea nitrogen.
[b]NSAID, non-steroidal anti-inflammatory drug.
[c]SIADH, syndrome of inappropriate antidiuretic hormone secretion.

Under such anisotonic conditions, the osmotic volume increase (OVI) and decrease (OVD) are to be forcibly induced upon exposure to hypotonic and hypertonic extracellular solutions, respectively (Fig. 1), because animal cells, unlike plant cells, are not surrounded by the rigid cell wall and also because their plasma membrane exhibits relatively high water permeability. Even under normotonic conditions, cell volume changes are induced by water movements driven by uptake and release of inorganic and organic osmolytes that are coupled to cell functions (for example, see Table 1 in Okada, 2004). A number of secretory cells are known to show shrinkage, called the secretory volume decrease (SVD: Manabe, Shimizu, Morishima, & Okada, 2004), in response to stimulation with secretagogues (Bachmann, Heinzmann, Mack, Manns, & Seidler, 2007; Dissing, Hansen, Unden, & Nauntofte, 1990; Foskett, 1990; Lee & Foskett, 2010; Manabe

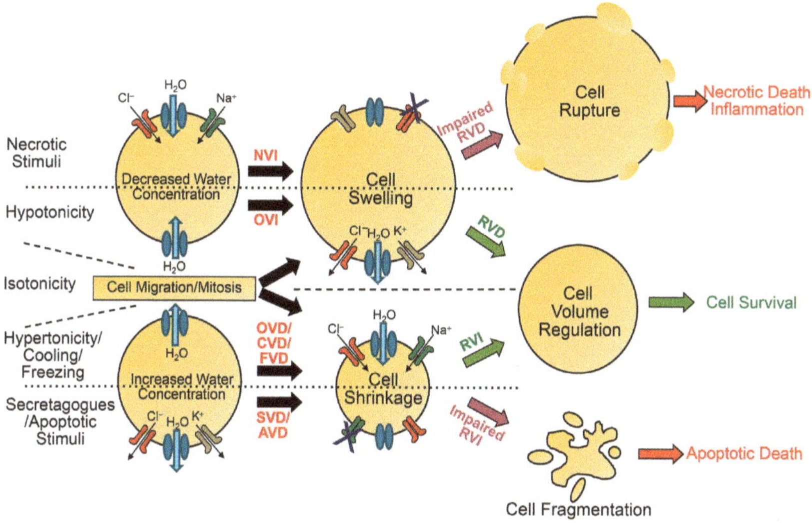

Fig. 1 Schematic illustration of the cell volume changes in response to osmotic perturbation and a variety of stimuli or insults, of the cell volume regulation in response to volume alterations, and of the cell death-survival switching mediated by ion and water channels. NVI, necrotic volume increase; OVI, osmotic volume increase; OVD, osmotic volume decrease; CVD, cooling volume decrease; FVD, freezing volume decrease; SVD, secretory volume decrease; AVD, apoptotic volume decrease; RVD, regulatory volume decrease; RVI, regulatory volume increase. (See the text for details.)

et al., 2004; Nakahari et al., 1991; Numata, Sato-Numata, Okada, & Inoue, 2018; Pedersen, Hoffmann, & Novak, 2013). Sizable cell volume decreases (Fig. 1), called the freezing and cooling volume decrease (FVD and CVD: Okada, 2016), are also observed upon freezing (Griffiths, Cox, Beadle, Hunt, & Reid, 1979) and cooling (Christmann et al., 2016), respectively, because a standing gradient of chemical potential for water is produced between "supercooled" intracellular and ice-containing extracellular fluids (Okada, 2016). Isoosmotic whole-cell volume fluctuations take place during cell mitosis and proliferation (see Review: Delpire & Gagnon, 2018). In the process of cell migration, isotonic local cell volume changes are associated with formation of lamellipodia and ruffles at the leading edge and uropodial retraction at the lagging edge of migrating cells (Schwab, Nechyporuk-Zloy, Fabian, & Stock, 2007). Normotonic cell volume changes are also coupled to cell differentiation (Delpire & Gullans, 1994; Lauf & Bauer, 1987; Marks & Rifkind, 1978). The apoptotic volume decrease (AVD: Maeno, Ishizaki, Kanaseki, Hazama, & Okada, 2000) and necrotic volume increase (NVI: Okada et al., 2001) are also attained under normotonic conditions upon exposure to apoptotic and necrotic stimuli (Fig. 1), as described in detail in Sections 3.1 and 4.1.

2.2 Steady-state and non-steady-state cell volume regulation

The ability of cell volume regulation (CVR) had evolved as essential mechanisms for cells in the primitive age to survive in environments with varying osmolarity. The CVR mechanisms are now utilized in living cells to cope with osmotic perturbation and normotonic cell volume changes in response to a variety of situations under physiological and pathological conditions, as described in Section 2.1. Unlike the protozoa cells, the metazoan cells devoid of rigid cell walls and contractile vacuoles must cope with anisotonic or normotonic volume changes by compensatory flux of inorganic and organic osmolytes as a result of activation of channels and/or transporters to drive volume-regulatory water flux.

Cell wall-lacking animal cells must ceaselessly regulate their cell volume even in the steady state under normotonic conditions, because a large number of polyvalently anionic macromolecules, which are impermeable to the plasma membrane, exist within the cells and result in the Gibbs-Donnan equilibrium of membrane-permeable inorganic cations and anions between the intracellular and extracellular solutions. This physicochemical effect inevitably leads to an increase in the intracellular osmolarity due to intracellular accumulation of inorganic cations (see Reviews: Dick, 1959; Okada, 2004). Animal cells counteract such an oncotic pressure by actively egesting Na^+ through continuous operation of the Na^+-K^+ pump. Such a steady-state CVR process is called the double-Donnan mechanism (Leaf, 1956) or the pump-leak balance mechanism (Tosteson & Hoffman, 1960).

The non-steady-state CVR processes elicited after osmotic cell swelling (OVI) and shrinkage (OVD) are termed the regulatory volume decrease (RVD) and increase (RVI), respectively (Fig. 1). Such non-steady-state CVR abilities are fundamental to the survival and functions of animal cells under both physiological and pathological conditions (Hoffmann, Lambert, & Pedersen, 2009; Hoffmann & Pedersen, 2006; Lang, Busch, et al., 1998; Okada, 1997, 1998, 2004). The failure of CVR is a major issue especially for brain cells in a variety of pathological situations, because the brain is encased in the rigid skull and the total volume of its extracellular fluid is very small, as discussed by Wilson & Mongin (2018).

The RVI event is attained by the uptake of most abundant extracellular osmolytes, mainly Na^+ and Cl^-, which drives influx of osmotically obliged water. The volume-regulatory NaCl uptake is mediated by operation of Na^+-K^+-$2Cl^-$ cotransporter (NKCC) and Na^+/H^+ exchanger (NHE) coupled to Cl^-/HCO_3^- anion exchanger (AE) (see Reviews: Hoffmann et al., 2009; Lang, Busch, et al., 1998; Mongin & Orlov, 2001; Wehner, Olsen, Tinel, Kinne-Saffran, & Kinne, 2003) as well as activation of the Na^+-permeable hypertonicity-induced cation channel (HICC) (see Review: Wehner et al., 2006). HICC activity was first functionally observed in human epithelial Intestine 407 cells upon a hypertonic challenge (Okada & Hazama, 1989) and molecularly identified as the epithelial Na^+ channel α, β and γ subunits (α, β and γENaC) in rat hepatocytes (Plettenberg, Weiss, Lemor, & Wehner, 2008), the ΔC-splice variant of a type of transient receptor potential cation channel (TRP), TRPM2, in HeLa cells (Numata et al., 2012), and δENaC plus TRPM2 and/or TRPM5 in HepG2 cells (Koos et al., 2018). HICC activity was found to be involved also in the RVI process after CVD or FVD (Christmann et al., 2016). To attain RVI by electrogenic HICC-mediated Na^+ inflow, electrogenic Cl^- inflow may be accompanied by activating some type of anion channel (Fig. 1, near the center), tentatively called the hypertonicity-induced anion channel (Okada, 2016), although this activity was not proved as yet.

The neuronal RVI event was observed in immature rat neurons (Schomberg et al., 2003) but was deemed to be lacking in mature neurons (Aitken et al., 1998; Andrew, Lobinowich, & Osehobo, 1997; Andrew & MacVicar, 1994; Pasantes-Morales, Maar, & Moran, 1993) including rat osmosensing magnocellular neurons (MCNs), that consist of arginine vasopressin (AVP) and oxytocin neurons, in the supraoptic nucleus (SON) (Zhang & Bourque, 2003). However, the potential RVI capability was recently evidenced in mature AVP neurons in rat SON (Sato-Numata, Ueta, & Okada, 2014). The glial RVI response to hypertonic stress was shown not only in glioma cells (Ernest & Sontheimer, 2007; McManus, Fischbarg, Sun, Hebert, & Strange, 1993; Strange, 1991) but also in brain astrocytes (Eriksson, Nilsson, Wagberg, Ronnback, & Hansson, 1992; Isaacks, Bender, Kim, Prieto, & Norenberg, 1994). The cardiac RVI ability was, though not strong, observed in rabbit ventricular myocytes (Drewnowska & Baumgarten, 1991), whereas a cardiomyocyte cell line (HL-1) can exhibit a clear RVI response to a hypertonic challenge (Cacace et al., 2014).

The RVD process involves release of most abundant intracellular osmolytes, such as K^+ and Cl^- as well as organic solutes, which drives the efflux of osmotically obliged water. The volume-regulatory osmolyte release is mediated by activation of ion channels and transporters. The first direct evidence for essential involvements of parallel activation of K^+ and Cl^- channels was provided by electrophysiological studies in human epithelial cells (Hazama & Okada, 1988). Most types of mammalian cells are now known to accomplish RVD by parallel activation of K^+ and Cl^- channels (Fig. 1, near the center) (for Reviews: Hoffmann & Simonsen, 1989; Okada, 1997; Sarkadi & Parker, 1991), whereas a few cell types such as erythrocytes attain RVD by operation of electroneutral transporters such as K^+-Cl^- cotransporter (KCC) (for Reviews: Gibson, Ellory, Adragna, & Lauf, 2009; Hoffmann & Mills, 1999; Lauf & Adragna, 2000).

The first observation of neuronal RVD was made in mouse dorsal root ganglion neurons by Horie, Ikuta, Takenaka, and Ito (1989). In contrast, in brain neurons, the early years of observations failed to detect the RVD ability (Aitken et al., 1998; Andrew et al., 1997; Andrew & MacVicar, 1994). However, now ample evidence for neuronal RVD capability has been accumulated (Chebabo, Hester, Jing, Aitken, & Somjen, 1995; Churchwell et al., 1996; Hernandez-Benitez, Sedeno-Cortes, Ramos-Mandujano, & Pasantes-Morales, 2014; Inoue, Mori, Morishima, & Okada, 2005; Inoue, Ohtaki, Nakamachi, Shioda, & Okada, 2007; Kreisman & Olson, 2003; Leaney, Marsh, & Brown, 1997; Loo & McNamara, 2006; Pasantes-Morales, Alavez, Sanchez Olea, & Moran, 1993; Pasantes-Morales, Chacon, Sanchez-Olea, & Moran, 1994; Pasantes-Morales, Maar, et al., 1993; Patel, Lauritzen, Lazdunski, & Honore, 1998; Quinn & Pierce, 1992; Viana, de la Pena, Pecson, Schmidt, & Belmonte, 2001). Although MCNs including AVP and oxytocin neurons were reported to lack the RVD ability (Zhang & Bourque, 2003), recently a slow but clear RVD response to a hypotonic challenge was demonstrated in AVP neurons (Sato, Numata, Saito, Ueta, & Okada, 2011). Since the first observation of RVD in astrocytes (Kimelberg & Frangakis, 1985), a large number of laboratories have confirmed the glial RVD ability (Akita, Fedorovich, & Okada, 2011; Bender, Neary, Blicharska, Norenberg, & Norenberg, 1992; Benfenati et al., 2011; Eriksson et al., 1992; Formaggio et al., 2019; Fraser & Swanson, 1994; Kimelberg, 1987; Liu, Akita, Shimizu, Sabirov, & Okada, 2009; O'Connor, Kimelberg, Keese, & Giaever, 1993; Olson, Sankar, Holtzman, James, & Fleischhacker, 1986; Parkerson & Sontheimer, 2003; Parsons, Cole, & Kimelberg, 1989; Pasantes-Morales, Murray, Lilja, & Moran, 1994; Pasantes Morales & Schousboe, 1988; Sanchez-Olea, Pena, Moran, & Pasantes-Morales, 1993; Sarfaraz & Fraser, 1999). In cardiomyocytes, the RVD event was well confirmed not only in chick embryo heart cells (Hall, Zhang, & Lieberman, 1995; Piwnica-Worms, Jacob, Horres, & Lieberman, 1985; Rasmusson, Davis, & Lieberman, 1993; Souza, Boyle, & Lieberman, 2000) but also in rabbit ventricular myocytes (Clemo & Baumgarten, 1991, 1997; Suleymanian & Baumgarten, 1996; Suleymanian, Clemo, Cohen, & Baumgarten, 1995) and neonatal rat cardiomyocytes (Rojas-Rivera et al., 2009; Shi et al., 2009; Taouil, Giancola, Morel, & Hannaert, 1998; Taouil & Hannaert, 1999) as well as in guinea pig ventricular myocytes (Wang, Mitsuiye, Rees, & Noma, 1997; Yamamoto, Ishihara, Ehara, & Shioya, 2004). Interestingly, the RVD capability was lost in hypertrophic guinea pig ventricular cells (Yamamoto, Kita, Iyoda, Yamada, & Iwamoto, 2011).

To rapidly cope with cell swelling emergency, the RVD process is facilitated by a number of cell signals. For example, as schematically depicted in Fig. 2, epithelial and glial cells make full use of cytosolic Ca^{2+}, cAMP and reactive oxygen species (ROS) as well as extracellular ATP and glutamate. The pathway for volume-regulatory K^+ efflux in human epithelial cells was shown to be activated by cytosolic Ca^{2+} rise, and therefore functionally classified as a Ca^{2+}-activated K^+ channel (Hazama & Okada, 1988), and then molecularly identified as the intermediate-conductance Ca^{2+}-activated K^+ channel, IK1 (Wang, Morishima, & Okada, 2003) (Fig. 2: left half). In human epithelial cells, the intracellular Ca^{2+} rise in response to a hypotonic challenge is produced both by Ca^{2+} influx via swelling-activated Ca^{2+}-permeable cation channels (Okada, Hazama, & Yuan, 1990), the molecular identity of which is TRPM7 (Numata, Shimizu, & Okada, 2007a, 2007b), and by Ca^{2+} release from intracellular Ca^{2+} stores (Hazama & Okada, 1990). Ca^{2+} mobilization may also be induced by stimulation of G protein-coupled receptors (GPCRs) by extracellular ATP in human epithelial cells (Dezaki, Tsumura, Maeno, & Okada, 2000) and glutamate in rat astrocytes (Akita et al., 2011). In many cell types, one of major pathways for swelling-induced ATP release is known to be provided by the large-conductance anion channel called Maxi-Cl (see Reviews: Okada, Okada, Islam, & Sabirov, 2018; Sabirov & Okada, 2005), the molecular identity of which has recently been identified as the solute carrier organic anion transporter family member 2A1 (SLCO2A1) by applying unbiased genome-wide approaches (Sabirov et al., 2017). Swelling-induced glutamate release from astrocytes may be mediated by both types of volume-regulatory anion channels, Maxi-Cl and VSOR (Liu, Tashmukhamedov, Inoue, Okada, & Sabirov, 2006). On the other hand, the pathway for volume-regulatory Cl^- efflux in most cell types was

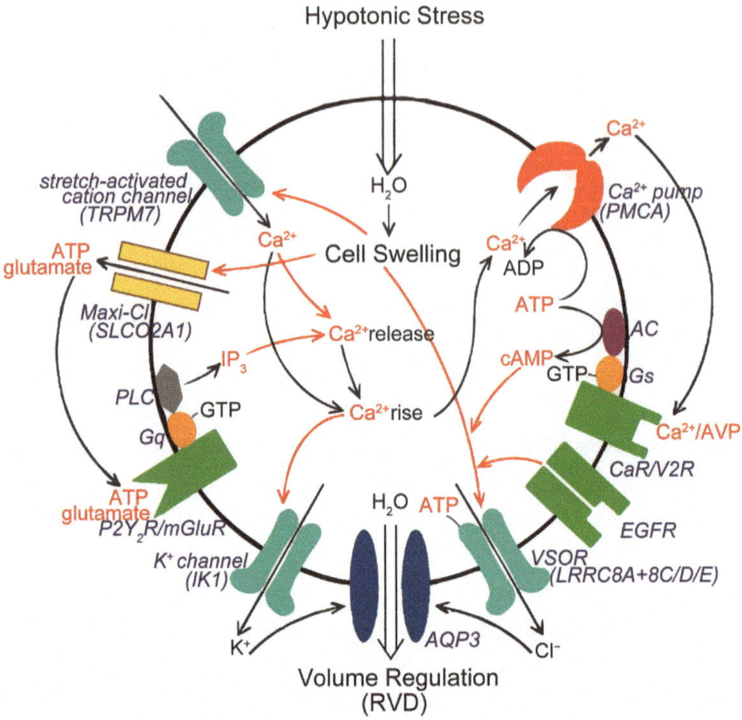

Fig. 2 Schematic model of the ionic and signaling mechanisms of RVD in epithelial and glial cells. $P2Y_2R$, mGluR, CaR, V2R and EGFR represent purinergic P2Y2-type, metabotropic glutamate, Ca^{2+}/polyvalent cation-sensing, arginine vasopressin type-2 and epidermal growth factor (EGF) receptors, respectively. (See the text for details.)

functionally identified as VSOR (Hazama & Okada, 1988; Kubo & Okada, 1992; Okada, 1997), which is also called the volume-regulated anion channel (VRAC: Nilius et al., 1997) or the volume-sensitive organic osmolyte/anion channel (VSOAC: Strange & Jackson, 1995). After a long time of history with numerous false positive candidates (see Reviews: Okada et al., 2018; Pedersen, Okada, & Nilius, 2016), the core molecule of VSOR was eventually identified as leucine-rich repeat-containing protein 8A (LRRC8A) (Formaggio et al., 2019; Planells-Cases et al., 2015; Qiu et al., 2014; Sato-Numata, Numata, Inoue, & Okada, 2016; Voss et al., 2014). As a natural consequence, RVD was found to be inhibited or eliminated by knockdown or knockout of LRRC8A (Qiu et al., 2014; Sorensen, Dam, Sturup, & Lambert, 2016; Sorensen, Thorsteinsdottir, & Lambert, 2014; Voss et al., 2014). Now, it is established that VSOR activity requires not only LRRC8A but also at least one of its paralogs, LRRC8C, LRRC8D and LRRC8E

(Gaitan-Penas et al., 2016; Lutter, Ullrich, Lueck, Kempa, & Jentsch, 2017; Sato-Numata, Numata, Inoue, Sabirov, & Okada, 2017; Voss et al., 2014). VSOR was demonstrated to be dually activated by cell swelling and by production of reactive oxygen species (ROS) (see Review: Akita & Okada, 2014). Both activation mechanisms have recently been clarified to be mediated by the interaction between a cytosolic member of the ATP-binding cassette (ABC) transporter superfamily, ABCF2, and α-actinin-4 (ACTN4) via cytoplasmic nanodomain Ca^{2+} rise (Okada et al., 2019). As summarized in Fig. 2 (right half), VSOR activity is also upregulated by cAMP produced in response to stimulation of Ca^{2+}/polyvalent cation-sensing receptor (CaR) by extracellular Ca^{2+} which is pumped out of increased cytosolic Ca^{2+} via plasmalemmal Ca^{2+}-ATPase (PMCA) and also in response to activation of the V2 type of AVP receptor (Sato et al., 2011). Stimulation of epidermal growth factor receptor (EGFR) is known to upregulate VSOR activity by enhancing volume sensitivity of VSOR (Abdullaev, Sabirov, & Okada, 2003) presumably via modulation of the ABCF2-ACTN4 interaction. On balance, swelling-induced parallel activation of K^+ channels and Cl^- channels induces KCl efflux according to their electrochemical potential gradients especially under hyperpolarized conditions predominantly elicited by increased K^+ conductance. KCl efflux drives the efflux of osmotically obliged water via water channel aquaporins (AQPs) (Fig. 1, near the center), mainly AQP3 in the case of human epithelial cells (Kida et al., 2005). In astrocytes, recently, AQP4 and VSOR as well as the glial swelling-activated Ca^{2+} permeable cation channel, TRPV4, are found to be essential (Benfenati et al., 2011) probably through the direct interaction of TRPV4 and AQP4 with the VSOR molecule (Benfenati et al., 2011, 2007).

2.3 Cell volume regulation and volume-regulatory anion channels

Among a large number of types of anion channels, only VSOR and Maxi-Cl are directly activated by cell swelling and involved in CVR after cell swelling (RVD) and thus collectively termed volume-regulatory anion channels (Okada et al., 2018). A splice variant of the CFTR Cl^- channel expressed in mammalian cardiomyocytes (Levesque, Hart, Hume, Kenyon, & Horowitz, 1992), here called cardiac CFTR, is known to be involved in the RVD process under the restraint condition with increased intracellular cAMP (Wang et al., 1997). Thus, cardiac CFTR may also be classified into the volume-regulatory anion channel in a broad sense. Biophysical and pharmacological properties are phenotypically different between VSOR, Maxi-Cl and CFTR (see Reviews: Okada et al., 2018, 2019).

The VSOR currents are characterized by an intermediate single-channel conductance, mild outward rectification, inactivation kinetics only at large positive potentials, sensitivity to cell shrinkage and to 4-[(2-butyl-6,7-dichloro-2-cyclopentyl-2,3-dihydro-1-oxo-1*H*-inden-5-yl)oxy]butanoic acid (DCPIB) as well as dependence on intracellular ATP and on ROS generation (Nilius et al., 1997; Okada, 1997; Okada et al., 2018, 2019; Strange, Emma, & Jackson, 1996).

VSOR activity has so far been shown to be functionally expressed in a variety of neurons (Carpenter & Peers, 1997; Inoue et al., 2005, 2007; Leaney et al., 1997; Li & Olson, 2004; Patel et al., 1998; Sato et al., 2011; Wang, Lu, et al., 2017; Zhang, Cao, Kimelberg, & Zhou, 2011). Neuronal VSOR currents were confirmed to be molecularly associated with LRRC8A (Wang, Lu, et al., 2017). Functional expression of VSOR has also been demonstrated in astrocytes (Akita et al., 2011; Akita & Okada, 2011; Benfenati et al., 2009, 2007; Crepel, Panenka, Kelly, & MacVicar, 1998; Darby, Kuzmiski, Panenka, Feighan, & MacVicar, 2003; Formaggio et al., 2019; Kimelberg, Macvicar, & Sontheimer, 2006; Lascola & Kraig, 1996; Lascola, Nelson, & Kraig, 1998; Liu et al., 2006; Minieri, Pivonkova, Harantova, Anderova, & Ferroni, 2015; Parkerson & Sontheimer, 2004; Ye, Oberheim, Kettenmann, & Ransom, 2009). Glial VSOR currents were also shown to be molecularly associated with LRRC8A (Elorza-Vidal et al., 2018; Formaggio et al., 2019). Even without cell swelling, astroglial VSOR currents can be activated upon GPCR stimulation by extracellular ATP (Akita et al., 2011; Darby et al., 2003), bradykinin (Akita & Okada, 2011; Liu et al., 2009) and glutamate (Akita & Okada, 2014) via nanodomain Ca^{2+} mobilization (Akita et al., 2011; Akita & Okada, 2011, 2014). Since the first direct evidence for VSOR-mediated conduction of charged small amino acids such as glutamate was provided by Banderali & Roy (1992), actual amino acid currents via VSOR have been observed in glial cells (Abdullaev, Rudkouskaya, Schools, Kimelberg, & Mongin, 2006; Jackson & Strange, 1993; Liu et al., 2006; Olson & Li, 1997; Roy, 1995). In fact, VSOR is a major pathway for glutamate release from astrocytes (see Review: Wilson & Mongin, 2018). Essential roles of LRRC8A and its paralogs in release of excitatory amino acids from astrocytes have recently been demonstrated (Hyzinski-Garcia, Rudkouskaya, & Mongin, 2014; Schober, Wilson, & Mongin, 2017). Cardiac VSOR activity has been demonstrated in ventricular myocytes (Browe & Baumgarten, 2003, 2004, 2006; Chiang, Luk, & Wang, 2004; Clemo & Baumgarten, 1997; Deng, Baki, &

Baumgarten, 2010; Deng, Baki, Yin, Zhou, & Baumgarten, 2010; Diaz et al., 1999; Gong, et al., 2004; Heusch, Liu, Rose, Cohen, & Downey, 2000; Okada, et al., 2006; Raucci, Wijesinghe, Chalfant, & Baumgarten, 2010; Ren & Baumgarten, 2005; Ren, Raucci, Browe, & Baumgarten, 2008; Shuba, Ogura, & McDonald, 1996; Tseng, 1992; Vandenberg, Yoshida, Kirk, & Powell, 1994; Wang, Takahashi, Uramoto, & Okada, 2005; Yamamoto, Ichishima, & Ehara, 2008, 2009) and in atrial myocytes (Deng, Baki, & Baumgarten, 2010; Du & Sorota, 1997, 1999, 2000; Duan, Fermini, & Nattel, 1995; Duan, Hume, & Nattel, 1997; Hagiwara, Masuda, Shoda, & Irisawa, 1992; Ichishima, Yamamoto, Iwamoto, & Ehara, 2010; Nagasaki, Ye, Duan, Horowitz, & Hume, 2000; Sakaguchi, Matsuura, & Ehara, 1997; Sorota, 1992, 1994, 1995; Sorota & Du, 1998; Xu, et al., 1997; Yamamoto-Mizuma, et al., 2004; Yamazaki & Hume, 1997) as well as in sinoatrial cells (Hagiwara, et al., 1992). Human cardiac VSOR activity was found in ventricular cells (Oz & Sorota, 1995) and atrial cells (Oz & Sorota, 1995; Sakai, Hagiwara, Kasanuki, & Hosoda, 1995). In relation to mammalian heart diseases, VSOR activity was found to be persistently activated in dog ventricular myocytes with congenital heart failure (Clemo, Stambler, & Baumgarten, 1999) and rat ventricular cells in hypertrophic hearts (de Jonge, Dekkers, Tilly, & Lamers, 2002), whereas VSOR activity was reported to be reduced in rabbit ventricular cells in hypertrophic hearts (van Borren et al., 2002) and mouse ventricular cells in long-term hypertrophic hearts (Yamamoto et al., 2011) as well as in diabetic mouse ventricular cells (Yamamoto et al., 2009).

Maxi-Cl activity was first found in cultured rat muscle cells (Blatz & Magleby, 1983) and is known to be functionally expressed in a large variety of cell types (see Reviews: Okada et al., 2018; Sabirov, Merzlyak, Islam, Okada, & Okada, 2016; Sabirov & Okada, 2009). The Maxi-Cl currents are characterized by a very large single-channel conductance (over 300 pS), an ohmic current-voltage (I–V) relationship, inactivation kinetics at both positive and negative potentials, and sensitivity to Gd^{3+} ions (Okada et al., 2019; Sabirov et al., 2016; Sabirov & Okada, 2009; Strange et al., 1996). Maxi-Cl currents are activated by cell swelling (Coulombe & Coraboeuf, 1992; Dutta et al., 2008; Dutta, Okada, & Sabirov, 2002; Dutta, Sabirov, Uramoto, & Okada, 2004; Falke & Misler, 1989; Jalonen, 1993; Kurbannazarova, Bessonova, Okada, & Sabirov, 2011; Liu et al., 2006; Sabirov, Dutta, & Okada, 2001). Thus, Maxi-Cl activity has been deemed to be directly involved in the RVD process by providing a volume-regulatory Cl^- efflux pathway (Falke & Misler, 1989; Jalonen,

1993; Light et al., 1990; Mitchell, Wang, & Jacob, 1997; Schlichter, Grygorczyk, Pahapill, & Grygorczyk, 1990). Also, Maxi-Cl is indirectly involved in the RVD process by releasing ATP which stimulates purinergic P2Y receptors in an autocrine fashion (Islam, Uramoto, Okada, Sabirov, & Okada, 2012; Kurbannazarova et al., 2011) (see Fig. 2).

Neuronal Maxi-Cl activity was found in neuroblastoma cell lines (Bettendorff, Kolb, & Schoffeniels, 1993; Bolotina, Borecky, Vlachova, Baudysova, & Vyskocil, 1987; Falke & Misler, 1989; Forshaw, Lister, & Ray, 1993, 2000) and a hippocampal cell line (Elinder et al., 2005). Glial Maxi-Cl activity was found in neonatal rat astrocytes (Jalonen, 1993; Jalonen, Johansson, Holopainen, Oja, & Arhem, 1989), newborn mouse astrocytes (Liu, Sabirov, & Okada, 2008; Liu et al., 2006) and adult human Schwann cells (McLarnon & Kim, 1991). Cardiac Maxi-Cl currents were observed in cardiomyocytes from newborn rats (Coulombe & Coraboeuf, 1992; Coulombe, Duclohier, Coraboeuf, & Touzet, 1987; Dutta et al., 2004) and those from adult rats (Coulombe & Coraboeuf, 1992; Dutta et al., 2008).

Not only vesicular (exocytotic) but also conductive (channel-mediated) mechanisms are known to be involved in the release of nucleotides which convey local autocrine and/or paracrine signaling (see Reviews: Lazarowski, 2012; Praetorius & Leipziger, 2009; Sabirov & Okada, 2005). Maxi-Cl has been suggested to represent one of the conductive ATP release pathways, because Gd^{3+}, which is a Maxi-Cl-selective blocker (see Review: Okada et al., 2019), inhibited ATP release from a variety of cell types that express Maxi-Cl activity, including mammary C127 cells (Hazama et al., 2000; Sabirov et al., 2001), astrocytes (Liu, Toychiev, Takahashi, Sabirov, & Okada, 2008; Stout, Costantin, Naus, & Charles, 2002; Zhao, Gu, Liu, Zhang, & Liu, 2017), DRG neurons (Fields & Ni, 2010), and cardiomyocytes (Dutta et al., 2008, 2004; Kunugi, Iwabuchi, Matsuyama, Okajima, & Kawahara, 2011). Consistently, the functional pore size of Maxi-Cl (around 2.6 nm in diameter: Sabirov & Okada, 2004) was found to be larger enough than the effective size of ATP and UTP (around 1.1 nm in diameter: Sabirov & Okada, 2005). The direct evidence for conductivity for anionic ATP/UTP of Maxi-Cl was obtained by measuring ATP currents via Maxi-Cl channel pores not only in C127 cells (the nucleotide/chloride permeability ratio $P_{ATP}/P_{Cl}=0.10$ and $P_{UTP}/P_{Cl}=0.09$: Sabirov & Okada, 2004) but also in cardiomyocytes ($P_{ATP}/P_{Cl}=0.12$: Dutta et al., 2004) and astrocytes ($P_{ATP}/P_{Cl}=0.11$: Liu, Toychiev, et al., 2008). Eventually, gene silencing of SLCO2A1 was found to suppress swelling-induced ATP release from C127 cells and ischemia-reperfusion-induced ATP release from Langendorff-perfused hearts (Sabirov et al., 2017).

3. Apoptotic cell death and volume-regulatory anion channels

3.1 Roles of the apoptotic volume decrease (AVD) in apoptotic cell death

Apoptosis is the physiological process by which living bodies eliminate unwanted cells, thus playing essential roles in maintenance of homeostasis in ceaselessly renewing tissues and organs and also in removal of useless, damaged and inflammatory cells. Two major cell signaling pathways were found to be associated with the apoptosis induction in mammalian cells: One is the intrinsic pathway mediated by mitochondrial stimulation and an initiator caspase, caspase-9 (Adams & Cory, 1998), and another is the extrinsic pathway mediated by CD95 death receptor stimulation and another initiator caspase, caspase-8 (Ashkenazi & Dixit, 1998). Endoplasmic reticulum (ER) stress induces apoptosis via mitochondrion-mediated intrinsic pathway (Bhat et al., 2017), and DNA damage also converges at the mitochondrion-mediated intrinsic apoptotic pathway (Kroemer & Reed, 2000; Whelan, Kaplinskiy, & Kitsis, 2010). Intrinsic stimuli include a non-selective blocker for protein kinase C, staurosporine (STS); a sarco/endoplasmic reticulum Ca^{2+}-ATPase (SERCA) inhibitor, thapsigargin; and anti-inflammatory glucocorticoid, dexamethasone, and radiation. Fas ligand (FasL) and tumor necrosis factor (TNF) family members represent extrinsic stimuli.

Persistent cell shrinkage is ubiquitously observed in apoptotic cells (Bortner & Cidlowski, 1998; Kerr, Wyllie, & Currie, 1972; Wyllie, Kerr, & Currie, 1980) and was termed AVD (Maeno et al., 2000), whereas persistent cell swelling termed NVI (Okada & Maeno, 2001) is associated with the early phase of necrotic cell death, as schematically depicted in Fig. 3. Induction of AVD and NVI is triggered by KCl release (Step-*1*) and NaCl uptake (Step-*a*), respectively, together with the flux of osmotically obliged water (Maeno et al., 2000; Okada & Maeno, 2001; Poulsen et al., 2010). After sustained AVD, cells undergo further shrinkage with caspase activation (Step-*2*), cell condensation and fragmentation (Step-*3*), and formation of apoptotic bodies (Step-*4*) that are to be cleaned up by phagocytes (Step-*5*). Since apoptotic cells release ATP/UTP, thereby sending a find-me

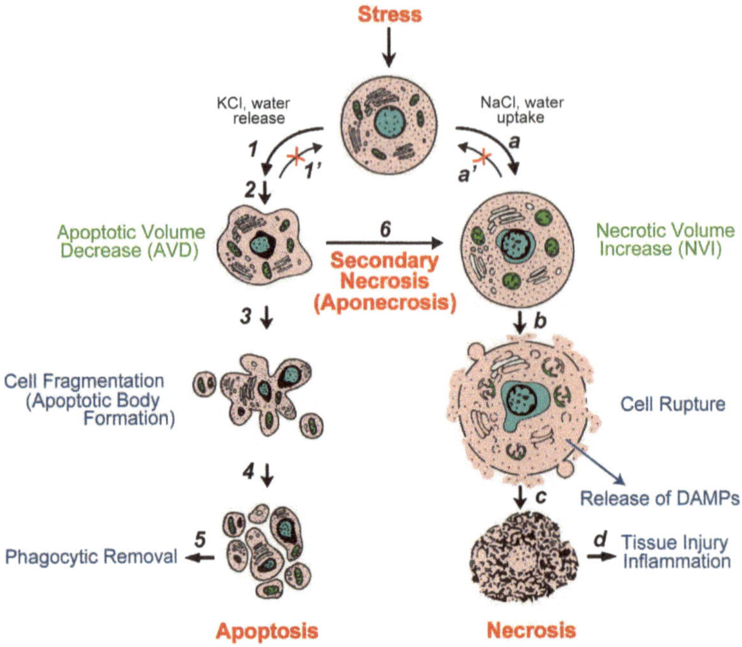

Fig. 3 Schematic illustration of the roles of AVD and NVI in stepwise processes for cell death in apoptosis, necrosis and secondary necrosis or aponecrosis. (See the text for details.)

signal to phagocytes (Elliott et al., 2009), there is a possibility of the involvement of activation of some conductive ATP release pathway(s) (see Reviews: Okada et al., 2018; Taruno, 2018) including Maxi-Cl, calcium homeostasis modulator 1 (CALHM1) and pannexin 1 (Panx1) at the last stage of apoptosis (Step-*5*). After sustained NVI, on the other hand, rupture of the cell membrane is elicited (Step-*b*) and thereby giving rise to cytolytic release of harmful substances, called danger/damage-associated molecular patterns (DAMPs), to the extracellular space (Step-*c*), thereby resulting in tissue injury and inflammation (Step-*d*).

The AVD time course is divided into two different phases (Maeno, Tsubata, & Okada, 2012). The early phase (Step-*1* in Fig. 3) starts as early as 5–10 min and persists for 1–3 h after apoptotic stimulation (Hasegawa, Shimizu, Takahashi, & Okada, 2012) in a manner insensitive to pancaspase inhibitors (Bortner & Cidlowski, 1998; Maeno, et al., 2000, 2012; Okada & Maeno, 2001; Vu, Bortner, & Cidlowski, 2001). Thus, the early-phase AVD was shown to precede caspase activation, as clearly demonstrated in Fig. 4 (Maeno et al., 2012), in a variety of cell types (Bortner &

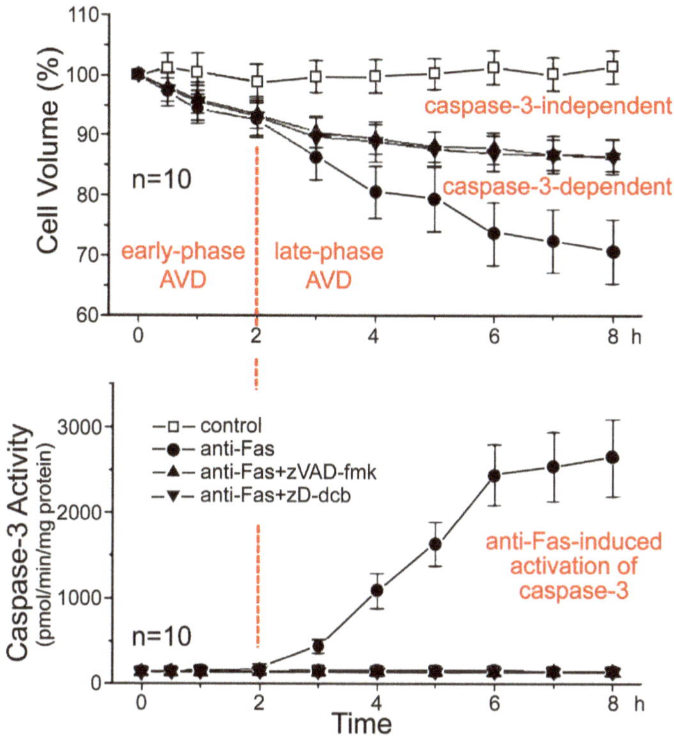

Fig. 4 The early-phase and late-phase of AVD that are independent of and dependent on activation of caspase-3. The presented data represent the effects of pan-caspase inhibitors, zVAD-fmk (50 μM) and zD-dcb (100 μM), on the AVD induction (top) and caspase-3 activation (bottom) in HeLa cells treated with anti-Fas antibody (anti-Fas). *The data per se were previously reported by Maeno, E., Tsubata, T., & Okada, Y. (2012). Apoptotic volume decrease (AVD) is independent of mitochondrial dysfunction and initiator caspase activation.* Cells, 1, *1156–1167.*

Cidlowski, 1998; Maeno et al., 2000, 2012; Vu et al., 2001) including neuronal (Xiao, Wei, Xia, Rothman, & Yu, 2002), glial (Ernest, Habela, & Sontheimer, 2008) and cardiac (d'Anglemont de Tassigny, Souktani, Henry, Ghaleh, & Berdeaux, 2004) cells. The early-phase AVD was shown to be associated with facilitation of RVD in response to a hypotonic challenge (Maeno et al., 2000; Shimizu, Numata, & Okada, 2004), decreases in the intracellular K^+ and Cl^- concentrations as well as with a rise of the intracellular free Ca^{2+} concentration (Dezaki, Maeno, Sato, Akita, & Okada, 2012) and a rise of the intracellular ATP concentration (Zamaraeva, et al., 2005). Furthermore, the AVD onset was shown to precede most other apoptotic morphological changes (Klassen, Walker, Ross, Cygler, & Lach, 1993), phosphatidylserine (PS) exposure caused by scramblase activation

(Elliott & Higgins, 2003), DNA laddering characterized by formation of multiple 180- to 200-base pair fragments (Bortner, Hughes, & Cidlowski, 1997; Maeno, et al., 2000; Wesselborg & Kabelitz, 1993; Xiao, et al., 2002), chromatin condensation (Orlov, Dam, Tremblay, & Hamet, 1996), and cell fragmentation (Benson, Heer, Dive, & Watson, 1996). The early-phase AVD is independent of mitochondrial signalings and dysfunction, because Bcl-2 overexpression failed to inhibit the AVD event and also because ligation of the death receptor Fas induced AVD in a cell line which can undergo Fas-mediated apoptosis without involving mitochondria (Maeno et al., 2012). The early-phase AVD event is also upstream of apoptotic activation of MAP kinases, JNK and p38 (Hasegawa et al., 2012). In contrast, the late phase of AVD (Step-*2* in Fig. 3) starts over 3 h later with exhibiting massive cell shrinkage (Bortner & Cidlowski, 1998; Maeno et al., 2000, 2012) and then blebbing followed by apoptotic body formation (Nunez, Sancho-Martinez, Novoa, & Lopez-Hernandez, 2010). This late-phase AVD was observed to be dependent on caspase activation (Bortner & Cidlowski, 1996; Hughes & Cidlowski, 1998; Maeno et al., 2000, 2012; Okada & Maeno, 2001; Schrantz, et al., 1999; Wolf, Reynolds, Morana, & Eastman, 1997; Zhivotovsky, Gahm, Ankarcrona, Nicotera, & Orrenius, 1995), as clearly shown in Fig. 4.

In addition to the AVD induction, RVI dysfunction (Step-*1′* in Fig. 3) is associated with persistent cell shrinkage which enables the cells proceed from the early-phase to the late-phase AVD and eventually to apoptotic cell death. Both intrinsic and extrinsic apoptotic stimuli inhibit the RVI process in HeLa cells (Maeno, Takahashi, & Okada, 2006). Since the RVI event is known to be attained by shrinkage-induced NaCl uptake by activation of NHE, NKCC and/or HICC (see Section 2.3), it is likely that apoptotic stimulation may suppress these NaCl uptake pathways. Actually, NHE activity was shown to be inhibited by stimulation with FasL (Lang et al., 2000) and STS (Wu et al., 2003). HICC activity was also found to be inhibited (Fig. 1, bottom cross mark) by stimulation with STS (Numata, Sato, Okada, & Wehner, 2008). Also, suppression of NKCC activity by bumetanide may augment cisplatin-induced apoptosis in the presence of K^+ ionophore (Marklund et al., 2004; Marklund, Henriksson, & Grankvist, 2001). In addition, suppression of inward operation of NKCC and/or NHE/AE by extracellular Na^+ deprivation (Nukui, Shimizu, & Okada, 2006) or Cl^- depletion (Maeno, Shimizu, & Okada, 2006) per se was shown to result in apoptotic cell death. The mechanism of apoptotic RVI dysfunction was shown to be caused by Akt inhibition due to

ROS-mediated ASK1 phosphorylation induced by apoptosis stimulation with STS, FasL, TNFα or H_2O_2 (Subramanyam, Takahashi, Hasegawa, Mohri, & Okada, 2010).

It is known that hypertonicity-induced osmotic cell shrinkage may not always induce apoptosis, and its induction is dependent on cell types (Bortner & Cidlowski, 1996; Orlov et al., 2000) and on the degree of hyperosmolarity (Bortner & Cidlowski, 1996; Gulbins, Welsch, Lepple-Wienhuis, Heinle, & Lang, 1997). Osmotic cell shrinkage itself resulted in apoptotic cell death in the cells lacking the RVI ability but not in the cells bearing the RVI capability (Bortner & Cidlowski, 1996; Shimizu, Maeno, & Okada, 2007). Furthermore, after repetitive exposures to hypertonic stress, RVI-lacking lymphocytes were found not only to acquire RVI ability but also to become resistant to hypertonicity-induced apoptosis (Bortner, Scoltock, Sifre, & Cidlowski, 2012). However, it must be noted that strong hypertonic stimulation can induce apoptosis even in the RVI-exhibiting cells (Ernest et al., 2008; Yurinskaya et al., 2012), when the level of volume recovery is not enough, because there is a volume threshold for apoptosis induction (Ernest et al., 2008). Hypertonicity could also cause apoptosis even in the cells possessing the RVI ability when RVI was pharmacologically inhibited by blockers of NHE and AE (Maeno, Takahashi, et al., 2006) or a HICC blocker (Shimizu et al., 2007). Apoptosis induction was caused even by normotonic cell shrinkage elicited by exposure of cells to Na^+-free isotonic solution which brings about facilitation of Na^+ efflux (Nukui et al., 2006) or to Cl^--deficient isotonic solution, which enhances Cl^- efflux (Maeno, Shimizu, et al., 2006), and also by inhibition of RVI presumably due to abolition of inward operation of NKCC and/or NHE/AE under such a Na^+-free or Cl^--deficient condition. Moreover, STS-induced apoptosis was found to be rescued by pre-activation of HICC by a hypertonic challenge given prior to STS stimulation (Numata et al., 2008). Thus, it is concluded that RVI dysfunction is indispensable for persistence of AVD which eventually leads to apoptotic cell death.

3.2 Roles of volume-regulatory anion channels in apoptotic cell death

The early phase of AVD is triggered by KCl release (Step-*1* in Fig. 3). The KCl release associated with AVD induction was assumed to be caused by parallel activation of K^+ and Cl^- channels (Fig. 1, bottom left) based on the following observations: First, cells undergoing AVD exhibited facilitation of the RVD event, which is known to be attained by parallel activation of K^+ and Cl^- channels, in response to a hypotonic challenge (Maeno et al., 2000); second, a number of blockers for K^+ and Cl^- channels prevented AVD initiation upon apoptotic stimulation (Maeno et al., 2000); and third, an apoptosis inducer STS shortly (within 10 min) led to decreases in the intracellular K^+ and Cl^- concentrations in a manner sensitive to blockers for K^+ and Cl^- channels (Dezaki et al., 2012). Actually, K^+ channel currents were shown to be activated by apoptosis inducers in a wide variety of cell types (see Reviews: Burg, Remillard, & Yuan, 2006; Okada, 2004; Shimizu, Lee, Ise, & Okada, 2008). In non-excitable cells, Cl^- channel currents were demonstrated to be activated by a large variety of apoptosis inducers, including STS (Coca-Prados, Anguita, Chalfant, & Civan, 1995; Kumagai et al., 2016; l'Hoste et al., 2010; Okada, 2004; Okada et al., 2006; Planells-Cases et al., 2015; Porcelli et al., 2003, 2004; Shimizu et al., 2004; Simoes et al., 2018), FasL (Okada et al., 2006; Shimizu et al., 2004; Szabo et al., 1998), TNFα (Nietsch, Roe, Fiekers, Moore, & Lidofsky, 2000; Schumann, Gardner, & Raffin, 1993; Shimizu et al., 2004; Yang et al., 2012), lipopolysaccharide (Meng, Carruth, & Weinman, 1997), sphingolipids (Souktani et al., 2000), an anti-cancer drug doxorubicin (d'Anglemont de Tassigny et al., 2004) and H_2O_2 (Shimizu et al., 2004; Varela, Simon, Riveros, Jorgensen, & Stutzin, 2004). Apoptosis inducers could activate Cl^- channel currents also in excitable cells such as a mouse hippocampal neuronal cell line (Akanda & Elinder, 2006; Elinder et al., 2005) and cardiomyocytes (Browe & Baumgarten, 2004; d'Anglemont de Tassigny, et al., 2004; Liu, Cao, et al., 2008; Okada, et al., 2006; Shen, Wang, Wang, et al., 2014; Shen, Wang, Yang, et al., 2014; Wang, Shen, et al., 2017). Among these apoptotic Cl^- channel currents, most ones are judged to be mediated by VSOR based on their mild outward rectification, sensitivity to osmotic shrinkage, blocking by a VSOR-selective blocker DCPIB, and dependence on ROS and intracellular ATP (Browe & Baumgarten, 2004; Kumagai et al., 2016; l'Hoste et al., 2010; Meng et al., 1997; Nietsch et al., 2000; Shimizu et al., 2004; Varela et al., 2004), whereas some are judged to be mediated by Maxi-Cl based on their large single-channel conductance, ohmic current-voltage relationship and Gd^{3+} sensitivity (Akanda & Elinder, 2006; Elinder et al., 2005). Thus, it is noted that the mechanisms of RVD induction are alternatively used as those for AVD initiation without subjecting to swelling stress but even during persistent apoptotic shrinkage. Normotonic VSOR activation is triggered by ROS-induced oxidation for the AVD initiation, whereas hypotonic VSOR activation is primarily induced by cell swelling (or membrane unfolding) for the RVD induction, as described in detail in our previous review article (Okada et al., 2019).

Numbers of Cl^- channel blockers have been shown to eliminate the AVD generation in a large variety of cell types (see Reviews: Okada, Maeno, & Mori, 2005; Okada et al., 2006; Shimizu et al., 2007), including neuronal cell lines, PC12 (Maeno et al., 2000) and NG108-15 (Okada et al., 2006), mouse cortical neurons (Wei et al., 2004) and rabbit ventricular myocytes (d'Anglemont de Tassigny et al., 2004). Also, apoptotic cell death has been reported to be prevented by a variety of Cl^- channel blockers in many cell types (see Review: Okada et al., 2006), including neuronal cell lines, PC12 (Maeno et al., 2000), NIE115 (O'Reilly, Xia, Fiander, Tauskela, & Small, 2002) and NG108-15 (Okada et al., 2006), and cortical neurons (Small, Tauskela, & Xia, 2002; Wei et al., 2004) as well as rat ventricular myocytes (d'Anglemont de Tassigny et al., 2004; Liu, Cao, et al., 2008; Okada et al., 2006; Tanabe, Wang, Takahashi, Uramoto, & Okada, 2005; Wang, Shen, et al., 2017) and mouse ventricular myocytes (Takahashi et al., 2005; Wang et al., 2005). These facts led us to conclude that AVD mediated by VSOR activity is an early prerequisite to apoptotic events leading to cell death by taking the fact that persistent cell shrinkage per se often cause apoptotic cell death into consideration. However, this concept may have been challenged by several reports: First, the NO-induced apoptotic cell death was observed to proceed without exhibiting AVD in a macrophage cell line (Hortelano, Zeini, Castrillo, Alvarez, & Bosca, 2002). However, the observed cell death looked largely necrotic cell death, because the smear-like DNA degradation and minimal activation of caspase-3. Second, in cortical neurons, Cl^- channel blockers eliminated AVD but only partially inhibited cell death monitored by release of lactate dehydrogenase (LDH) (Wei et al., 2004). However, LDH release does not always represent apoptotic cell death but does often represent necrotic one, especially when observed after long-time exposures to apoptotic stimuli. Third, FasL induced swelling in Jurkat T cells after 6-h stimulation with exhibiting a number of apoptotic characteristics under low Na^+ conditions (Bortner & Cidlowski, 2003). Also, serum deprivation induced cell swelling, though STS in contrast induced cell shrinkage, within 1–2h followed by apoptotic cell death characterized by caspase activation and chromatin condensation in vascular smooth muscle cells transfected with E1A-adenoviral protein (Platonova, Koltsova, Hamet, Grygorczyk, & Orlov, 2012). However, the early-phase AVD event is to be observed much earlier (usually starting within 10min after apoptosis stimulation). It must be pointed out that on the way of or a long time (often hours) after the early-phase AVD event, cells may turn, in some situations, to show swelling (Step-*6* in Fig. 3), because apoptosis becomes aborted and switches to secondary necrosis (Majno & Joris, 1995) due to no available neighboring phagocytes or to aponecrosis (Formigli et al., 2000) due to insufficient ATP production (Nicotera, Leist, & Ferrando-May, 1998), albeit the apoptosis processes require a rise of intracellular ATP (Zamaraeva et al., 2005). Fourth, Cl^- channel blockers and surprisingly caspase-3 inhibitor as well failed to prevent chromatin condensation in actinomycin D-stimulated HeLa cells (Kasim, Kuzelova, Holoubek, & Model, 2013). In this study, however, cells attached to coverslips were mounted on slides, and cell morphology observations typically started 12h after addition of actinomycin D and continued for up to additional 8–10h. As pointed out by the authors themselves, secondary necrosis and membrane rupture were often observed under the experimental conditions employed in the study. Thus, swelling-exhibiting cell death after apoptosis stimulation must be carefully distinguished from aponecrotic or secondary necrotic cells in the in vitro experiments where no phagocytes present and O_2 supply is limited, especially for cells situated for a long time under some non-physiological conditions (such as incubation in Na^+-deficient or serum-free solution and/or within a narrow-spaced chamber).

Apoptotic cells are known to release ATP (Aguirre, Shoji, Saez, Henriquez, & Quest, 2013; Elliott et al., 2009) and UTP (Elliott et al., 2009). Released ATP was deemed to enhance the apoptotic execution processes (Aguirre et al., 2013) and provide a find-me signal to phagocytes (Elliott ct al., 2009) via activation of purinergic receptors. Although Panx1 was hypothesized to mediate the ATP release, no direct evidence has been presented as yet. Thus, further studies are required to clarify the apoptotic ATP release pathway by examining a number of the candidates proposed for ATP release pathway, including Maxi-Cl, so far (Okada et al., 2018).

3.3 Roles of volume-regulatory anion channels in cisplatin resistance in cancer cells

The first generation platinum-based anticancer drug cisplatin (*cis*-diamminedichloroplatinum: CDDP) is widely used in chemotherapy for human cancers (Rozencweig, von Hoff, Slavik, & Muggia, 1977; Todd & Lippard, 2009). Cisplatin is taken up by cancer cells and forms DNA adducts eventually leading to apoptosis in a variety of cancer cells (see Reviews: Chu, 1994; Jamieson & Lippard, 1999) as well as dorsal root ganglion (DRG) neurons and PC12 cells (Gill & Windebank, 1998). Currently, cisplatin and second- and third-generation platinum-based drugs including carboplatin and oxaliplatin are the most effective and widely used anticancer drugs (Kelland, 2007).

Our group found that pretreatment with cisplatin enhanced hypotonicity-induced VSOR currents in human epidermoid cancer KB cells (Ise et al., 2005). Cisplatin itself was then found to activate VSOR currents and thereby induce AVD in a number of cancer cells such as human lung adenocarcinoma A549 cells (Min et al., 2011), osteosarcoma MG-63 cells (Cai, Zhang, Zhang, Qiu, & Liu, 2015) and nasopharyngeal carcinoma CNE-2Z cells (Yang et al., 2015). Carboplatin also induced VSOR activation and AVD induction in human adenocarcinoma A549 cells (He et al., 2010). Another type of anticancer drug doxorubicin was also shown to activate VSOR currents, thereby leading to AVD and apoptotic cell death in rabbit articular chondrocytes (Kumagai et al., 2012) and in adult rabbit ventricular myocytes (d'Anglemont de Tassigny et al., 2004). Doxorubicin-induced AVD and apoptosis was blocked by prior activation of VSOR by hypotonic stimulation (d'Anglemont de Tassigny et al., 2004) or by VSOR blockers (d'Anglemont de Tassigny, Berdeaux, Souktani, Henry, & Ghaleh, 2008) in the cardiomyocytes. Cisplatin-induced VSOR activation in CNE-2Z cells was found to be highly dependent on P2Y receptor stimulation by extracellular ATP (Yang et al., 2015). This observation may suggest first that VSOR activation is mediated by ROS production controlled by Ca^{2+} nanodomain, as previously observed in astrocytes (Akita et al., 2011), and second that ATP was released out of the cancer cells via a number of pathways including Maxi-Cl upon cisplatin stimulation.

The fact that cisplatin-induced apoptosis is mediated by VSOR activation has been molecularly confirmed by recent following studies: First, cisplatin-induced caspase-3 activation was found to be reduced by gene disruption of LRRC8A or/and of all the paralogs in human colonic cancer HCT116 cells (Planells-Cases et al., 2015). Second, LRRC8A knockdown and overexpression in human ovarian cancer A2780 cells rescued and exaggerated, respectively, cisplatin-induced apoptosis (Sorensen, Nielsen, Thorsteinsdottir, Hoffmann, & Lambert, 2016). Similarly, sensitivity to another anticancer drug temozolomide (TMZ) in human glioma U87 cells was enhanced and suppressed by overexpression and knockdown, respectively, of LRRC8A (Yang, He, Chen, Ning, & Xia, 2019). Third, overnight treatment with cisplatin resulted in activation of VSOR currents under isotonic conditions in *Xenopus* oocytes co-transfected with LRRC8A and LRRC8D or LRRC8E (Gradogna, Gaitan-Penas, Boccaccio, Estevez, & Pusch, 2017). Fourth, the treatment with cisplatin was found to increase the LRRC8A expression in human cancer cells (Sorensen et al., 2014; Thorsteinsdottir, Thorsteinsdottir, & Lambert,

2016). Thus, it is concluded that LRRC8A-dependent VSOR activity plays an essential role in cisplatin-induced apoptotic death in cancer cells.

The acquisition of resistance to anticancer drugs by cancer cells is the major limitation for cancer chemotherapy. The mechanisms of acquisition of anticancer drug resistance can be mainly classified into four categories, including (1) reduction in the uptake or in the accumulation of anticancer drugs by cancer cells, (2) enhancement of detoxification of anticancer drugs or prevention of their actions to DNA such as the formation of DNA adducts and crosslinks, (3) suppression of intracellular reactions to anticancer drugs leading to apoptosis induction, and (4) promotion of active extrusion of anticancer drugs from the cancer cells by multidrug resistance (MDR) drug pumps such as MDR1 or P-glycoprotein.

MDR1 is not involved in cisplatin resistance, because cisplatin is not the substrate of MDR1 (Chu, 1994), and because cisplatin-resistant KCP-4 cells do not express MDR1 (Fujii et al., 1994). Reduction of Cl^- conductance sensitive to 4,4′-diisothiocyanato-2,2′-stilbenedisulfonic acid (DIDS) was found in multidrug-resistant leukemic HL60/AR cells (Gollapudi, McDonald, Gardner, Kang, & Gupta, 1992). This observation may suggest that cisplatin resistance is coupled to reduction of VSOR activity. In fact, we identified this DIDS-sensitive Cl^- conductance as VSOR in cisplatin-resistant human epidermoid cancer KCP-4 cells (Lee et al., 2007). Hypotonicity-induced VSOR activity was largely deficient in this cisplatin-resistant cell line, whereas its parental KB cell line exhibited markedly prominent VSOR currents (Lee et al., 2007). Similar downregulation of VSOR activity was also observed in multidrug-resistant Ehrlich ascites tumor cells (Poulsen et al., 2010) and human lung adenocarcinoma A549/CDDP cells (Min et al., 2011). Furthermore, we found that not only VSOR activity but also cisplatin-induced apoptosis were partially restored by the treatment of KCP-4 cells with a histone deacetylase (HDAC) inhibitor, trichostatin A (TSA) (Lee et al., 2007). HDACs are known to play a role in regulation of gene expression and restoration of silenced genes (Johnstone, 2002; Marks, Richon, & Rifkind, 2000) and also to enhance cisplatin sensitivity in cancer cells (Kim et al., 2003). We also found that restoration of VSOR activity was partially induced by another HDAC inhibitor, acidipin, which is structurally unrelated to TSA in cisplatin-resistant KCP-4 cells (Shimizu et al., 2008). Thereafter, similar partial restoration of VSOR activity by TSA was found in cisplatin-resistant A549/CDDP cells (Min et al., 2011). Cisplatin resistance has been found to be correlated with reduced expression of a core component of VSOR, LRRC8A, in human cancer cells (Planells-Cases et al., 2015; Sorensen, Dam, et al., 2016;

Sorensen, Nielsen, et al., 2016). Similarly, reduction of LRRC8A expression was found to be associated with TMZ resistance, whereas TMZ sensitivity was induced by overexpression of LRRC8A in human glioblastoma cells (Yang et al., 2019). Taken together, it is concluded that downregulation of LRRC8A-dependent VSOR activity is causatively involved in cisplatin resistance and TMZ resistance.

Interestingly, gene disruption of LRRC8A was found to reduce cisplatin uptake in human epithelial HEK and cancer KBM7 cells (Planells-Cases et al., 2015). Similarly, reduced cisplatin uptake was induced by LRRC8A knockdown in human ovarian cancer A2780 cells and was found to be associated with cisplatin-resistant A2780CisR cells (Sorensen, Dam, et al., 2016). Moreover, acquisition of carboplatin resistance of human cancer KBM7 cells was found to show an enrichment of disruption/insertion not only in LRRC8A but also in LRRC8D genes, and LRRC8D-deficient KBM7 clones were resistant to cisplatin and carboplatin but not to oxaliplatin (Planells-Cases et al., 2015). Thus, it appears that VSOR channels composed of both LRRC8A and LRRC8D provide a pathway for cisplatin and carboplatin uptake. Actually, the radius of narrowest constriction of VSOR pore was determined as 0.63 nm by our nonelectrolyte partition studies (Ternovsky, Okada, & Sabirov, 2004) and is larger enough to be permeated by cisplatin and carboplatin, the mean radii of which are 0.30 and 0.40 nm, respectively, as shown in Fig. 5. The mean radius of oxaliplatin (0.40 nm: Fig. 5) suggests that LRRC8D-containing VSOR pores can transport oxaliplatin, because LRRC8D is known to be essentially involved in the import pathway for a nucleoside antibiotic blasticidin S (Lee, Freinkman, Sabatini, & Ploegh, 2014), the size of which is much bigger (0.59 nm: Fig. 5) than oxaliplatin. Since the drug import is, in general, mediated not only by facilitated diffusion via channels/transporters but also by simple diffusion via lipid bilayers of the plasma membrane, we estimated hydrophobicity of these anticancer and antibiotic drugs by calculating the octanol/water partitioning coefficient $P_{o/w}$ values using the method reported by Moriguchi, Hirono, Liu, Nakagome, & Matsushita (1992). These $P_{o/w}$ values given in Fig. 5 indicate that oxaliplatin is most hydrophobic (that is, least water-soluble) among these four drugs and the solubility in the organic phase (thus permeability through the lipid bilayer membrane) of oxaliplatin is expected to be four and 33 times higher than cisplatin and carboplatin, respectively, under the same conditions. The higher lipid solubility of oxaliplatin may explain why LRRC8D-deficient cells were still sensitive to oxaliplatin (Planells-Cases et al., 2015). Taken together, in any case, it is clear that downregulation of VSOR activity is dually involved in cisplatin

Molecule	*Shape*	*R (nm)*	$P_{o/w}$
Cisplatin	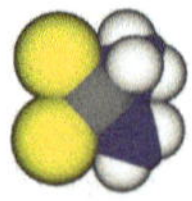	0.30	0.089
Carboplatin		0.40	0.011
Oxaliplatin	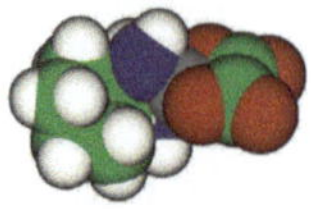	0.40	0.363
Blasticidin S	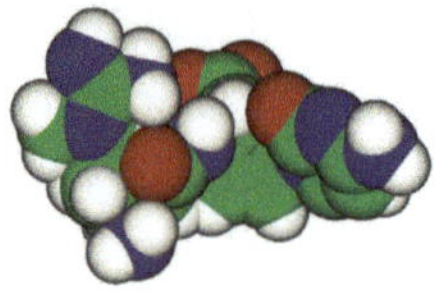	0.59	0.331

Fig. 5 Molecular shapes, sizes and hydrophobicity of platinum-based anti-cancer drugs, cisplatin, carboplatin and oxaliplatin, as well as of an antibiotic drug, blasticidin S. Each unhydrated radius (R) was calculated as a geometric mean of three dimensions according to the formula: $R = (1/2)(l_1 l_2 l_3)^{1/3}$, where ls are molecular dimensions estimated from space-filling models. Hydrophobicity of these drugs was estimated by the octanol/water partitioning coefficient $P_{o/w}$ value. The dimension and $P_{o/w}$ values were calculated by using Molecular Modeling Pro computer software (Norgwyn Montgomery Software Inc., North Wales, PA) after energy minimization by bond rotation and by using the method reported by Moriguchi et al. (1992). Color coding: C, dark green; H, white; O, red; N, blue; Pt, grey; Cl, yellow.

resistance by escaping from AVD induction in a manner independent of LRRC8D and by preventing cisplatin entry via VSOR pores in a manner essentially dependent on LRRC8D, as pointed out by Planells-Cases et al. (Planells-Cases et al., 2015).

4. Necrotic cell death and volume-regulatory anion channels

4.1 Roles of the necrotic volume increase (NVI) in necrotic cell death

Necrotic cell death is distinct from apoptotic cell death in morphological and biochemical characteristics, although both types of cell death can often occur simultaneously in tissues. Necrosis is induced by a variety of insults upon exposure to many toxins and poisons, lytic viral infection, complement attack, hyperthermia, acidosis, lactacidosis, excitotoxicity, hypoxia and ischemia. However, necrotic cell death commonly shows quintessential features, as follows: Necrosis starts with swelling of mitochondria and cell body, called NVI (Step-*a* in Fig. 3), and exhibits depletion of intracellular ATP, intracellular Ca^{2+} overload, ROS generation and disruption of the plasmalemmal membrane (PLM) integrity (Step-*b* in Fig. 3), thereby leading to cell death with showing a smear, but not laddering, pattern of DNA degradation in a manner independent of activation of apoptotic caspases. Cell membrane rupture gives rise to cytolytic release (Step-*c* in Fig. 3) of harmful substances, called DAMPs or alarmin, especially proteins associated with nuclear materials such as HMGB1 (high mobility group box 1) (Scaffidi, Misteli, & Bianchi, 2002), resulting in tissue injury and inflammation (Step-*d* in Fig. 3). Interestingly, cyclophilin D-dependent mitochondrial permeability transition caused by ROS and Ca^{2+} overload was found to be critical in some forms of necrotic, but not apoptotic, cell death (Nakagawa et al., 2005).

Necrosis is classified into primary necrosis and secondary necrosis. Secondary necrosis or aponecrosis is caused by an incomplete execution of the apoptotic program and represents necrosis generated after abortion of apoptotic processes on the way (Step-*6* in Fig. 3) due to the loss of cellular energy and/or of engulfment by phagocytes. Primary necrosis is further classified into accidental necrosis caused by extreme physicochemical stress and regulated necrosis which includes necroptosis and pyroptosis (Kaczmarek, Vandenabeele, & Krysko, 2013). Necroptosis is initiated by receptor-interacting serine/threonine-protein kinase 1 (RIP1) and the downstream kinase RIP3 in response to activation of death receptors by FasL, TNFα or tumor necrosis factor-related apoptosis-inducing ligand (TRAIL) under the conditions where caspase-8 activity is inhibited or depleted (Chan et al., 2003; Degterev et al., 2005; Holler et al., 2000; Vercammen et al., 1998). RIP3 phosphorylates mixed lineage kinase domain-like (MLKL), promotes homotrimer formation of MLKL and translocates the homotrimers to PLM, thereby activating TRPM7 cation channels which causes Ca^{2+} influx leading to PLM rupture (Cai et al., 2014). Excessive intracellular cholesterol accumulation also induces necroptosis in neuronal cells (Funakoshi, Aki, Tajiri, Unuma, & Uemura, 2016), the fact suggesting that necroptosis is involved in neuronal lyposomal storage disorder such as Nieman-Pick disease Type C. Pyroptosis is an inflammatory form of programmed necrosis dependent on caspase-1 activation (Cookson & Brennan, 2001). Pyroptosis is also characterized with cell swelling, cell lysis and release of

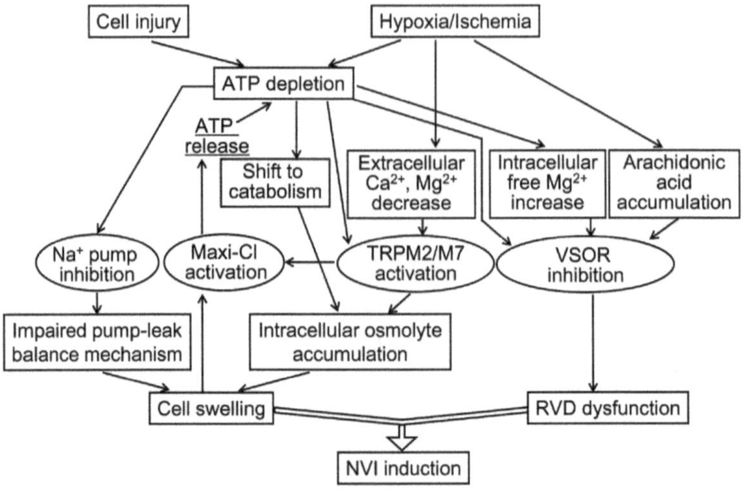

Fig. 6 Schematic diagram for the mechanisms of NVI induction due to persistent cell swelling coupled to RVI dysfunction under the conditions where cells are traumatized or subjected to hypoxia/ischemia. (See the text for details.)

proinflammatory cytokines (Fink & Cookson, 2006). Cleavage of gasderin D (GSDND) by inflammatory caspases, such as caspase-1 and -11 in mice or caspase-1, -4 and -5 in humans, promotes the membrane pore formation (He et al., 2015; Kayagaki et al., 2015; Shi et al., 2015).

Necrotic cell death, irrespective of its types, is paralleled by normotonic persistent cell swelling (Duvall & Wyllie, 1986; Kroemer, Dallaporta, & Resche-Rigon, 1998), which is thus called NVI (Okada & Maeno, 2001) due to the inflows of Na^+, Cl^- and water mainly via their channels (Fig. 1, upper left), although cell swelling itself may not necessarily lead to necrotic cell death. Accidental cell injury and hypoxic or ischemic insults bring about NVI due to intracellular ATP depletion, as schematically diagrammed in Fig. 6. Especially cerebral ischemia induces rapid ATP depletion (Kristian & Siesjo, 1998; Lipton, 1999). Cell injury-induced ATP depletion is caused not only by ATP leakage out of the cells but also by increased ATP consumption due to Na^+-K^+ pump stimulation by Na^+ leakage into the cells. ATP depletion may be exaggerated by ATP release via Maxi-Cl activated by hypoxia/ischemia-induced cell swelling. Hypoxia/ischemia-induced ATP depletion is caused by decreased mitochondrial ATP production via oxidative phosphorylation. ATP depletion results not only in an impairment of the pump-leak balance mechanism preventing oncotic pressure-induced steady-state cell swelling due to reduction of Na^+-K^+ pump activity but also in accumulation of intracellular osmolytes due to a metabolic shift to catabolism, thereby causing cell swelling. ATP depletion thus induced is known to activate Na^+- and Ca^{2+}-permeable TRPM2 channels (Liu & Schnellmann, 2003). In addition, reduction in extracellular divalent cations is known to be induced by ischemia in the brain (Hansen & Zeuthen, 1981; Siemkowicz & Hansen, 1981; Silver & Erecinska, 1990), and to activate Na^+- and Ca^{2+}-permeable TRPM7 channels in the brain (Wei et al., 2007). ROS generated under ischemia and ischemia-reperfusion situations may also cause activation of TRPM7 (Aarts et al., 2003; McNulty & Fonfria, 2005) and TRPM2 (Hara et al., 2002; Kaneko et al., 2006; McNulty & Fonfria, 2005). Therefore, Na^+ inflow via TRPM2/M7 results in additional intracellular osmolyte accumulation. In excitable cells including cardiac cells (Ju, Saint, & Gage, 1996) and hypocampal cells (Hammarstrom & Gage, 1998), hypoxia induces persistent activation of voltage-gated Na^+ channels, thereby leading to Na^+ inflow which induces Ca^{2+} entry via Na^+/Ca^{2+} exchanger (NCX) and also to membrane depolarization which in turn triggers activation of voltage-gated Ca^{2+} channels, thereby leading to the intracellular Ca^{2+} rise (Hammarstrom & Gage, 2000). On balance, these events may exaggerate intracellular osmolyte accumulation in excitable cells. It is known that excessive cell swelling can result in necrotic cell death, if not compensated by the inherent RVD mechanism (Cain, Langlais, Sun, Brown, & Cohen, 2001; Nieminen, 2003). Under ATP-depleted conditions, the RVD event must be impaired by inhibition of volume-regulatory VSOR activity (Fig. 1, upper cross mark) which is essentially dependent on intracellular ATP (Jackson, Morrison, & Strange, 1994; Oike, Droogmans, & Nilius, 1994; Oiki, Kubo, & Okada, 1994). Since intracellular ATP depletion inevitably leads to liberation of Mg^{2+} from MgATP, VSOR activity is also to be suppressed by the increase in the cytosolic free Mg^{2+} level (Oiki et al., 1994; Okada et al., 2019). Cerebral ischemia also induces accumulation of arachidonic acid (Bazan, 1970) which is one of the most potent blockers of VSOR (Kubo & Okada, 1992). Both cell swelling and RVD dysfunction together, though not either one alone, can provide a sufficient conditions for NVI induction. That is why, for example, ouabain-induced Na^+-K^+ pump inhibition alone induced cell swelling but failed to affect survival in a number of, though not all, cell types (see Review: Orlov, Platonova, Hamet, & Grygorczyk, 2013).

It must be stressed that persistent cell swelling per se is a necessary factor but not a sufficient factor for induction of PLM rupture. Physical cell swelling was found to provoke PLM rupture only after attaining a five-fold volume increase under hypoosmotic stress, whereas ATP depletion caused by oxidative stress and mitochondrial inhibition brought about gradual cell swelling and finally the PLM rupture already after reaching a 1.3- to 1.4-fold volume increase in rat hepatocytes (Carini, Autelli, Bellomo, & Albano, 1999). Similarly, PLM rupture was observed after reaching a 30 to 40% volume increase in MDCK cells treated with ouabain (Platonova, Koltsova, Maksimov, Grygorczyk, & Orlov, 2011). Also, simple physicochemical mechanisms cannot explain how necrotic PLM rupture takes place after a 30–40% volume increase, because the cell membrane surface area does not change even by a three-fold volume increase due to unfolding of plenty of membrane folds (Groulx, Boudreault, Orlov, & Grygorczyk, 2006; Levitan & Garber, 1997; Okada, 1997, 2004; Pangrsic, Potokar, Haydon, Zorec, & Kreft, 2006). The additional factor(s) to eventually accomplish PLM rupture together with persistent cell swelling is missing as yet. However, it seems likely that the intracellular rise of Na^+ and Ca^{2+} caused by ATP depletion-induced TRPM7 activation (Cai et al., 2014) and voltage-gated Na^+ channel activation induced by hypoxia may be additional prerequisite factors for necrotic cell death presumably by promoting PLM rupture, because it was reported that necrosis requires a combination of reduction of intracellular ATP and elevation of intracellular Na^+ (Barros, Hermosilla, & Castro, 2001). Also, we need to remind that insertion of MLKL homotrimers to PLM was shown to play a role in induction of PLM rupture during necroptosis (Cai et al., 2014), and that GSDND cleavage was reported to promote PLM rupture during pyroptosis (He et al., 2015; Kayagaki et al., 2015; Shi et al., 2015). In addition, in order to release HMGB1, which is one of DAMPs, from necrotic cells, liberation of this nuclear factor protein from the nucleus must be, in advance, induced by histone acetylation of chromatin (Scaffidi et al., 2002).

4.2 Roles of volume-regulatory anion channels in excitotoxicity

Excitotoxicity, a term coined by Olney (Olney, 1969), is defined as degeneration of neurons caused by exposure to excessive excitatory amino acids such as glutamate and aspartate, and is associated with stroke, cerebral ischemia, brain trauma, spinal injuries, ocular disorder such as glaucoma and a number of neurogenic diseases or neurodegenerative disorders including epilepsy and Huntington's, Alzheimer's and Parkinson's diseases (Hasbani, Hyrc, Faddis, Romano, & Goldberg, 1998). In the brain in vivo, neurons rapidly undergo somatic swelling together with dendritic focal swelling, called varicosities (Hori & Carpenter, 1994; Hsu & Buzsaki, 1993; Matesic & Lin, 1994).

The most predominant excitatory amino acid is glutamate which is massively released from astrocytes upon traumatic, ischemic and inflammatory brain injuries (Olney, Adamo, & Ratner, 1971). A number of putative pathways for astrocytic glutamate release have been proposed, including exocytotic vesicular transport, reverse operation of glutamate transporters, cysteine/glutamate exchangers, gap junction hemichannels and purinergic P2X7 receptor cation channels (see Liu et al., 2009). However, swelling-, ischemia- or ischemia-reperfusion-induced glial release of glutamate and/or aspartate is most predominantly mediated by VSOR and Maxi-Cl, both of which are activated by cell swelling, as schematically diagrammed in the left part of Fig. 7. Actually, glutamate release from astrocytes subjected to hypotonic, ischemic or ischemia-reperfusion-induced stress was shown to be markedly suppressed by a relatively selective VSOR blocker phloretin (Abdullaev et al., 2006; Liu et al., 2006) and a most selective VSOR blocker DCPIB (Abdullaev et al., 2006; Bowens, Dohare, Kuo, & Mongin, 2013;

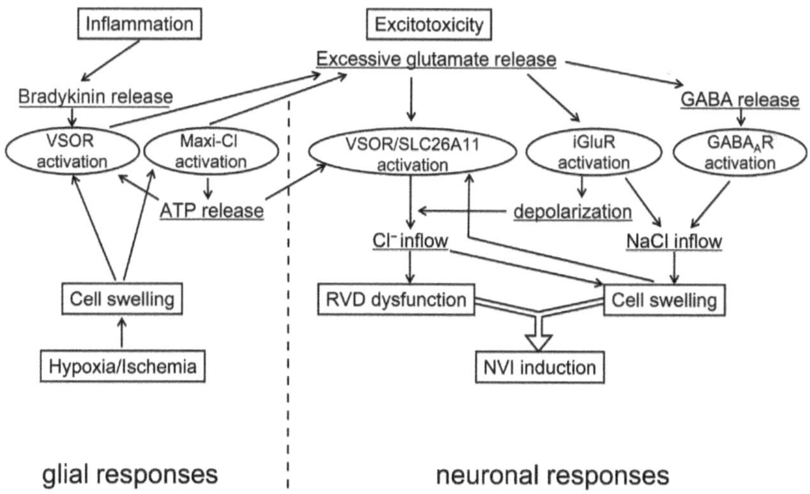

Fig. 7 Schematic diagram for the mechanisms of NVI induction due to persistent cell swelling coupled to RVD dysfunction under excitotoxic conditions where excessive glutamate release from glial cells is induced in response to inflammation and/or hypoxia/ischemia. (See the text for details.)

Zhang, Zhang, Feustel, & Kimelberg, 2008). Also, swelling- or ischemia-induced glutamate release from mouse astrocytes was shown to be very sensitive to a selective Maxi-Cl blocker Gd^{3+} in a manner additive to phloretin (Liu et al., 2006). Glutamate is well known to be released from astrocytes in response to bradykinin (BK) (Jeftinija, Jeftinija, Stefanovic, & Liu, 1996; Liu et al., 2009; Parpura et al., 1994; Parpura, Fang, Basarsky, Jahn, & Haydon, 1995; Parpura, Liu, et al., 1995), which is an initial mediator of brain inflammation (Wahl et al., 1996; Zausinger et al., 2002) released upon ischemia (Groger et al., 2005). Stimulation of BK type 2 receptor (BK2) was shown to cause ROS production, thereby activating VSOR under normotonic conditions and thus causing glutamate release from mouse astrocytes (Liu et al., 2009). Glial VSOR-mediated glutamate release was also shown to be triggered by ROS in rat astrocytes (Haskew-Layton, Mongin, & Kimelberg, 2005) and microglia (Harrigan, Abdullaev, Jourd'heuil, & Mongin, 2008). Recent studies demonstrated that VSOR channels must be composed of, at least, LRRC8A and LRRC8D to exhibit conductivity or permeability to charged or non-charged organic osmolytes (Gaitan-Penas et al., 2016; Lutter et al., 2017; Schober et al., 2017).

The ionic mechanism of excitotoxic neuronal NVI (Fig. 1, upper left) consists of a number of pathways for Na^+ and Cl^- inflows (Choi, 1987; Dessi, Charriaut-Marlangue, & Ben-Ari, 1994; Goldberg & Choi, 1993), and the excessive NaCl entry is responsible for induction of cell swelling in neurons subjected to exitotoxic insults, as summarized in the right part of Fig. 7. Excessive Na^+ inflow is primarily mediated by ionotropic glutamate receptor (iGluR) cation channels, that include *N*-methyl-D-aspartate (NMDA), kainate and 3-amino-hydroxy-5-methyl-4-isoxazolepropionicacid (AMPA) receptors, activated by glutamate released from astrocytes. Especially, stimulation of extrasynaptic iGluRs is deleterious by causing aberrant entry of Na^+ and membrane depolarization (Tymianski, 2011). On the other hand, glutamate is well known to induce release of gamma-aminobutyric acid (GABA) from GABAergic neurons (Belhage, Hansen, & Schousboe, 1993; Erdo, Cai, & Lakics, 1993; Harris & Miller, 1989; Pin & Bockaert, 1989; Weiss, 1988). GABA release was also observed upon ischemia in hippocampal slices containing not only neurons but also numerous astrocytes (Inglefield & Schwartz-Bloom, 1998). Released GABA stimulates $GABA_A$ receptors ($GABA_AR$), thereby inducing Cl^- flux mediated by $GABA_AR$-associated Cl^- channels in neurons. Membrane depolarization produced by iGluR hyperactivity drives $GABA_AR$-mediated neuronal Cl^- influx, thereby producing net NaCl uptake leading to neuronal cell swelling and varicosity formation. Actually, $GABA_AR$ blockers were shown to suppress NMDA-induced varicosity formation in cortical neurons (Hasbani et al., 1998; Inoue et al., 2007). An additional Cl^- influx pathway might be involved in excitotoxic Cl^- uptake, because the Na^+-independent sulfate anion transporter, SLC26A11, was reported to act as a voltage-dependent Cl^- channel under excessively Na^+-inflowing and depolarizing conditions produced by activation of NMDA receptors in rat hypocampal neurons in slices, thereby causing neuronal swelling induced by intracellular NaCl accumulation which drives water entry (Rungta et al., 2015). After such excitotoxic swelling, neuronal cells were found to fail to exhibit RVD (Churchwell et al., 1996; Loo & McNamara, 2006). This is not due to an inhibition of VSOR activity, but instead we observed very energetic VSOR activity not only at the whole-cell level in mouse cortical neurons subjected to NMDA-induced swelling but also at the single-channel level on their varicosities (Inoue et al., 2007). Neuronal VSOR activation is also induced by glutamate released from astrocytes via activation of the metabotropic glutamate receptor (mGluR) (Akita & Okada, 2014), by ATP released through astrocytic Maxi-Cl via stimulation of purinergic P2Y receptors (Akita et al., 2011) and by BK released upon inflammation via BK2 (Liu et al., 2009). Under excitotoxic conditions, VSOR was shown to serve as a pathway not for volume-regulatory Cl^- efflux but rather for swelling-exaggerating Cl^- influx, because iGluR-induced depolarization drives Cl^- flux to the inward, but not outward, direction, thereby resulting in the RVD dysfunction. Although some exitotoxic cells showed both necrotic and apoptotic signs (Hirashima et al., 1999), especially in the proximity of the lesion border (Ayata et al., 1997), most excitotoxic neuronal cell death was shown to be largely necrotic (Beck, Lenart, Kintner, & Sun, 2003; Choi, 1987; Rothman, Thurston, & Hauhart, 1987; Sun, Sombati, & DeLorenzo, 2001; Sztriha, Joo, & Szerdahelyi, 1986). In fact, after exhibiting massive somata swelling and dendritic beadings with a number of varicosities, neurons isolated from the brain cortex and in situ cortical neurons in brain slices eventually fell into necrotic death (Inoue et al., 2007). Not only varicosity formation but also necrotic cell death in cortical neurons exposed to NMDA was markedly prevented by potent VSOR blockers such as phloretin, 5-nitro-2-(3-phenylpropylamino)benzoic acid (NPPB), and indanyloxyacetic acid 94 (IAA-94) (Inoue et al., 2007). Thus, it is clear that massive cell swelling and RVD dysfunction are both prerequisite to induction of excitotoxic neuronal necrosis. Excessive Ca^{2+} entry via iGluR may not directly be involved in necrotic death (Chen, Olney, Lukasiewicz, Almli, & Romano, 1998; Price, Olney, Samson, & Labruyere, 1985) but may be implicated in the dendritic varicosity formation, because varicosities were observed to be often formed at the Ca^{2+}-rising points in dendrites (Bindokas & Miller, 1995). Also, it is possible that Ca^{2+} entry was shown to trigger NADPH oxidase (NOX) activation and generate ROS (Brennan et al., 2009), thereby presumably leading to VSOR activation in neurons, as observed in non-neuronal cells (Browe & Baumgarten, 2004; Shimizu et al., 2004; Varela et al., 2004).

4.3 Roles of volume-regulatory anion channels in acidotoxicity and lactacidosis-induced cell death

Cell acidosis is a common feature of a number of pathological states in which blood supply to tissues is reduced. Extracellular pH may fall below 6 in many tissues, especially in brain, during injury due to ischemia, trauma, inflammation, infection, seizures and hyperglycemia as well as malignant tumors and abscesses. Severe ischemia in the brain and heart results in strong acidosis, because of H^+ liberation by hydrolysis of ATP and creatine phosphate at a rate greater than ATP synthesis and by predominant anaerobic ATP synthesis by glycolysis.

Intracellular acidosis may induce persistent cell swelling, eventually leading to NVI, by the mechanisms, as schematically diagrammed in the left part of Fig. 8. Acidosis promotes H_2CO_3 production followed by release of H^+ and HCO_3^- within the cells, thereby leading to Na^+ and Cl^- uptake via stimulation of NHE and AE, respectively. In addition, acidosis induces reduction of negative charge valency of intracellular polyvalent anion macromolecules, which in turn enhances intracellular Cl^- accumulation. Furthermore, acidosis induces simultaneous activation of the acid-sensitive outwardly rectifying anion channel (ASOR) and TRPM7 cation channels, thereby inducing net NaCl uptake. The half-maximal pH values for activation of ASOR and TRPM7 are around 5.4 and 5.0 in HeLa and HEK293T cells, respectively, at room temperature, as shown in Fig. 9 (open symbols), and as reported previously (for ASOR: Sato-Numata, Numata, Okada, & Okada, 2013) (for TRPM7: Li, Jiang, & Yue, 2006; Numata & Okada, 2008). ASOR is expressed in a variety of cell types (see Review: Okada et al., 2019) including cortical neurons (Sato-Numata, Numata, & Okada, 2014), hippocampal astrocytes (Lambert & Oberwinkler, 2005) and cardiac myocytes (Yamamoto & Ehara, 2006). TRPM7 is also ubiquitously expressed in a wide variety of tissues including the brain (see Review: Minke & Cook, 2002) and the heart (see Review: Yue, Zhang, Xie, Jiang, & Yue, 2013). Moreover, extracellular pH reduction due to proton accumulation also induces activation of the acid-sensing

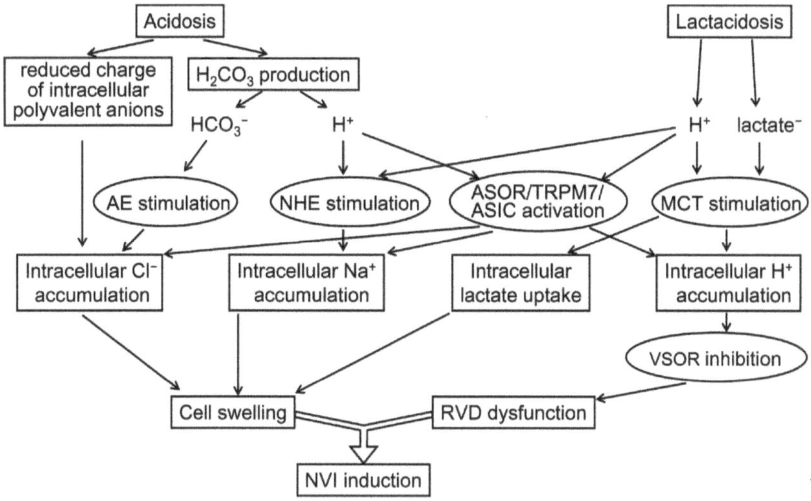

Fig. 8 Schematic diagram for the mechanisms of NVI induction due to persistent cell swelling coupled to RVD dysfunction under acidotoxic or lactacidosis conditions. (See the text for details.)

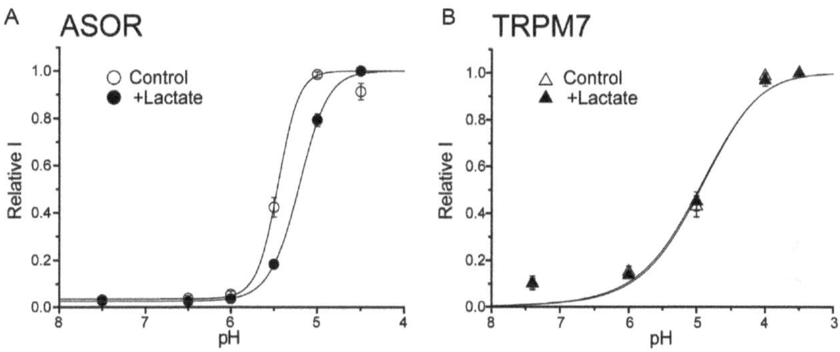

Fig. 9 Effects of lactate on the pH dependence of ASOR (A) and TRPM7 (B) activities. The mean whole-cell ASOR and TRPM7 currents were recorded at +100 and −140 mV, respectively, at room temperature. Each symbol represents the normalized mean current of seven (A) or five (B) different cells with the S.E. value (vertical bar) in the absence (open symbol) or presence (filled symbol) of 7.5 mM lactate added to the extracellular solution. Experimental conditions for (A) and (B) are the same as reported by Sato-Numata et al. (2013) and Numata and Okada (2008), respectively.

ion channel (ASIC), which belongs to a Na^+ channel superfamily that includes ENaC and is expressed in the nervous (see Review: Vick & Askwith, 2015) and cardiovascular (see Review: Abboud & Benson, 2015) tissues, thereby inducing Na^+ entry into the cells subjected to acidosis.

On balance, cell acidosis results in persistent cell swelling due to intracellular NaCl accumulation caused by the inflows of Na^+ and Cl^- mediated by transporters and/or channels (Fig. 1, upper left). Since severe intracellular acidosis was shown to inhibit VSOR activity (Fig. 1, upper cross mark) (Gerard, Rouzaire-Dubois, Dilda, & Dubois, 1998; Sabirov, Prenen, Droogmans, & Nilius, 2000), RVD is to be retarded or prevented in the brain and heart subjected to severe acidosis, as actually observed in astrocytes (Pasantes-Morales, Murray, et al., 1994) and epithelial cells (Li, De Smet, Jans, Simaels, & Van Driessche, 1998; Smets, Ameloot, Steels, & Van Driessche, 2002). Such acidotoxic conditions where persistent cell swelling is coupled to RVD dysfunction should result in NVI induction. Acidotoxic cell death may not be apoptotic, because anti-apoptotic signals such as Bcl-2 and Bcl-xL are stimulated by acidosis (Ryder, McColl, Zhong, & Distelhorst, 2012). Actually, ASOR blockers were found to prevent not only persistent swelling but also necrotic cell death in epithelial cells (Wang, Shimizu, Numata, & Okada, 2007) and cortical neurons (Sato-Numata, Numata, et al., 2014). Also, necrotic cell death was observed in neurons under acidotoxic conditions and was found to be prevented by blockers of ASIC (Xiong et al., 2004; Yermolaieva, Leonard, Schnizler, Abboud, & Welsh, 2004).

Lactacidosis represents tissue acidosis (see above) coupled to lactate accumulation due to augmented anaerobic glycolysis-fermentation reactions and occurs in association with cerebral ischemia, hypoxia, seizure, spreading depression and trauma. Lactacidosis causes neuronal and glial cell swelling eventually leading to irreversible cytotoxic brain edema and necrotic death of neurons and glial cells. The ionic mechanisms are schematically diagrammed in the right part of Fig. 8. Lactate is a strong acid (with pK of 3.9) and thus negatively charged under physiological and even acidosis conditions at $pH > 3.9$. $Lactate^-$ is taken up together with H^+ by a facilitated cotransport mechanism mediated by the monocarboxylate transporter (MCT) in neurons and glial cells both of which are endogenously expressing MCT (Mori, Morishima, Takasaki, & Okada, 2002; Nabekura et al., 2003; Nedergaard & Goldman, 1993), thereby causing intracellular lactate accumulation as well as Na^+ accumulation via H^+-induced stimulation of NHE (see Reviews: Okada, 2004; Okada et al., 2005). Here, we first show that even under lactacidosis, protons activate anionic ASOR and cationic TRPM7 channels, as shown in Fig. 9 (filled symbols), thereby causing intracellular accumulation of Cl^- and Na^+ mediated by their channels

(Fig. 1, upper left). Since proton-activated ASIC currents were found to be enhanced by lactate, Na^+ accumulation is further augmented in the presence of lactate (Allen & Attwell, 2002; Immke & McCleskey, 2001). Intracellular accumulation of Na^+, Cl^- and lactate together then leads to swelling both in neuronal cells (Alojado, Morimoto, Morimoto, & Kemmotsu, 1996; Mori et al., 2002; Staub, Mackert, Kempski, Peters, & Baethmann, 1993) and in astrocytes (Jakubovicz, Grinstein, & Klip, 1987; Jakubovicz & Klip, 1989; Lomneth, Medrano, & Gruenstein, 1990; Morihata, Nakamura, Tsutada, & Kuno, 2000; Nabekura et al., 2003; Plesnila et al., 1999; Ringel, Chang, Staub, Baethmann, & Plesnila, 2000; Ringel et al., 1997; Staub, Baethmann, Peters, Weigt, & Kempski, 1990; Staub et al., 1996). However, accumulated protons inhibit VSOR activity (Fig. 1, upper cross mark), thereby provoking RVD dysfunction in neuronal cells (Mori et al., 2002) and glial astrocytes (Nabekura et al., 2003). Both lactacidosis-induced persistent cell swelling and RVD dysfunction together then bring about NVI induction, thereby finally resulting in necrotic cell death, as actually observed in neurons (Goldman, Pulsinelli, Clarke, Kraig, & Plum, 1989; Nedergaard, Goldman, Desai, & Pulsinelli, 1991; Staub et al., 1993) and glial cells (Goldman et al., 1989; Nedergaard et al., 1991; Okada et al., 2004; Staub et al., 1990). When exogenous acid-resistant anion channels were introduced into the cells by pretreatment with vacuolating cytotoxin A (VacA) purified from *Helicobacter pylori*, however, even after lactacidosis-induced swelling, C6 glioma cells became undergoing RVD (Nabekura et al., 2003), thereby being largely rescued from necrotic cell death (Okada et al., 2004). These facts indicates that not only cell swelling but also RVD dysfunction due to impaired VSOR activity are prerequisite to induction of NVI and necrotic cell death.

5. Ischemic cell death and volume-regulatory anion channels

5.1 Effects of ischemia and ischemia-reperfusion on volume-regulatory anion channels

Ischemia/hypoxia induces cell swelling (Fig. 6) which should trigger VSOR activation if the conditions are normal. Under severe ischemia/hypoxia conditions, however, intracellular ATP depletion and increased intracellular free Mg^{2+} may prevent VSOR activation, thereby causing NVI induction (Fig. 6). Actually, VSOR activity was found to be suppressed in mouse cerebellar granule neurons under hypoxic conditions (Patel et al., 1998). Acidosis frequently associated with ischemia should markedly affect VSOR activity, because its single-channel conductance is enhanced by mildly acidic intracellular pH (~6.5) but its overall activity evaluated by the nP_o values, where n and P_o represent the channel number and open probability, respectively, is suppressed at pH below 6.5 but enhanced at pH 7 to 8 (Sabirov et al., 2000). Although brief acid exposure was observed to activate LRRC8A-dependent VSOR currents by ROS generation in mouse nodose ganglia neurons, the intracellular pH was rather slightly increased in the acid-preconditioned neurons (Wang, Lu, et al., 2017). Severe acidosis produced by a lactacidosis insult abolished VSOR activity in both neuronal cells (Mori et al., 2002) and astrocytes (Nabekura et al., 2003), thereby causing NVI induction (Fig. 8). As far as intracellular ATP, free Mg^{2+} and pH levels are maintained near normal, on the other hand, VSOR can be activated not only by cell swelling but also by ROS (see Reviews: Akita & Okada, 2014; Okada et al., 2019). Neuronal and glial cell swelling is caused by ischemia, excitotoxicity and acidotoxicity (Figs. 6–8). Since ROS generation is elicited by ischemia, ischemia-reperfusion, excitotoxicity, β-amyloid peptides and oxidative burst as part of immune response (Abramov, Canevari, & Duchen, 2004; Banasiak, Xia, & Haddad, 2000; Dubinsky, Kristal, & Elizondo-Fournier, 1995; Jenner, 2003; Lipton, 1999), VSOR is to be often activated under these ROS-generating conditions in neurons, astrocytes and cardiomyocytes. Glial cell swelling induced by ischemia/hypoxia may initially activate VSOR until lapsing into ATP depletion, and activated VSOR then resulted in glutamate release (Liu et al., 2006). Released glutamate may in turn activate VSOR in a positive feedback manner and furthermore trigger excitotoxic neuronal cell death (Fig. 7).

Ischemia/hypoxia-induced cell swelling may elicit Maxi-Cl activation in glial cells, and this activation becomes more prominent after ATP depletion, because Maxi-Cl activation is promoted by protein tyrosine dephosphorylation (Toychiev et al., 2009). Actually, Maxi-Cl was shown to be activated by ischemia/hypoxia stress in isolated astrocytes (Liu, Sabirov, & Okada, 2008; Liu et al., 2006). Here, we provide first evidence for actual activation of Maxi-Cl by oxygen-glucose deprivation (OGD) in the cell-attached patch on the enhanced green fluorescent protein (EGFP)-expressing astrocyte in a hypocampal slice prepared from an EGFP transgenic mouse (Fig. 10). Astroglial Maxi-Cl activation brings about massive release of glutamate (Liu et al., 2006), thereby triggering excitotoxic neuronal cell death (Fig. 7). Maxi-Cl activity also induces ATP release from astrocytes to brain tissues upon osmotic swelling (Liu, Toychiev, et al., 2008) and

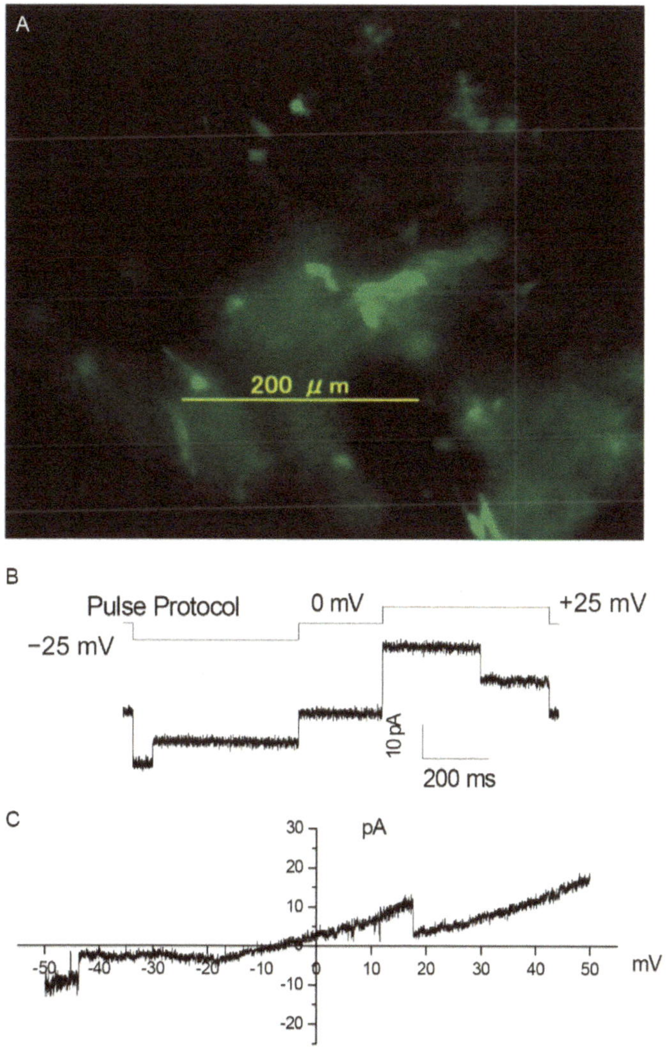

Fig. 10 OGD-induced activation of Maxi-Cl currents in the cell-attached patch on astrocytes in hippocampal slices prepared from EGFP-transgenic mice. (A) EGFP-expressing astrocytes (seen in green) in a hippocampal slice observed under a fluorescence microscope. (B) Maxi-Cl unitary events recorded in an on-cell patch upon application of step pulses (the protocol shown at the top of trace). (C) Maxi-Cl unitary events recorded in another on-cell patch upon application of a ramp pulse from −50 to +50 mV. Experimental conditions are the same as reported by Liu, Toychiev, et al. (2008).

ischemic stress (Liu, Sabirov, & Okada, 2008). Released ATP may activate VSOR in glial and neuronal cells (Fig. 7). Activation of Ca^{2+}-permeable TRPM2 and TRPM7 channels induced by ischemia or ischemia-reperfusion (see above) may cause intracellular Ca^{2+} load and thereby induce activation of Maxi-Cl, because Maxi-Cl was shown to be activated by a Ca^{2+} ionophore (Bajnath, Groot, de Jonge, Kansen, & Bijman, 1993; Kawahara & Takuwa, 1991; Light et al., 1990; Sun, Supplisson, & Mayer, 1993).

Maxi-Cl is functionally expressed in cardiomyocytes and activated in response to hypotonic, hypoxic and ischemic stress not only in neonatal ventricular myocytes (Coulombe & Coraboeuf, 1992; Dutta et al., 2004) but also in adult ventricular cells (Dutta et al., 2008). Swelling-induced Maxi-Cl activation was shown to cause ATP release from ventricular cardiomyocytes of newborn and adult rats (Dutta et al., 2008, 2004). Ischemia-induced ATP release was also observed and shown to be mediated by Maxi-Cl in neonatal cardiomyocytes (Dutta et al., 2004). Moreover, Langendorff-perfused hearts of adult mice subjected to ischemia-reperfusion were found to respond with massive ATP release in a manner sensitive to gene knockdown of SLCO2A1, which is the core molecule of Maxi-Cl (Sabirov et al., 2017).

5.2 Roles of volume-regulatory anion channels in ischemia- and ischemia-reperfusion-induced death in cerebral cells

In injured cerebral tissues under ischemia or ischemia-reperfusion stress, cell death can be coexisting as apoptosis, necrosis and their hybrid form, aponecrosis (Martin et al., 1998; Portera-Cailliau, Price, & Martin, 1997a, 1997b). Necrosis occurs mainly at the ischemic core in the early period, and apoptosis does at the peri-infarct zone or in the ischemic penumbra region several days later in the brain following transient or permanent ischemia (stroke) (see Reviews: Choi, 1996; Dirnagl, Iadecola, & Moskowitz, 1999; Lipton, 1999) as well as in the retina upon increased intraocular pressure (Joo et al., 1999). Correspondingly, activation of VSOR and Maxi-Cl plays roles in induction of both types of cell death, apoptosis and necrosis, depending on the situations. Also, both VSOR and Maxi-Cl activities are double-edged swords for ischemia- or ischemia-reperfusion injury, with exerting physiological and pathological functions. Their physiological functions include promotion of cell survival by their volume-regulatory actions and cell-to-cell signal transduction (including glia-neuron communication) by extracellular signals such as glutamate and ATP, whereas pathological functions include induction of cell death by causing AVD and NVI as well as facilitation of cell death by mediating release of intracellular glutamate and ATP, as summarized in Figs. 6 and 8.

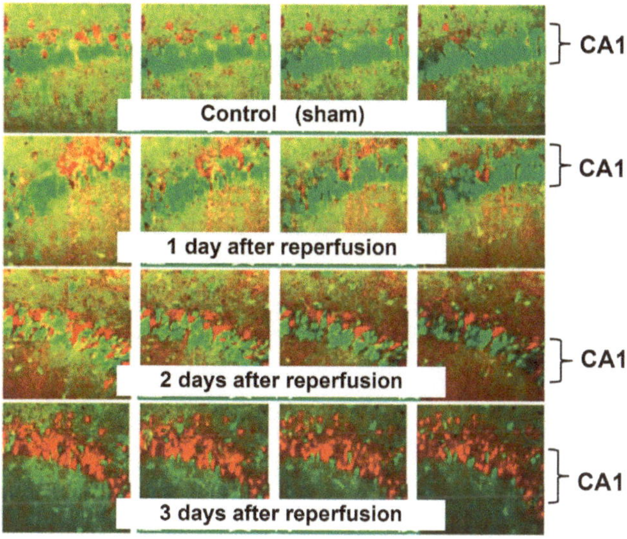

Fig. 11 Two-photon laser microscopic images of delayed neuronal death (DND) in the CA1 region of mouse hippocampal slices subjected to transient forebrain ischemia. Experimental conditions are the same as reported by Inoue et al. (2007). Green- and red-colored cells represent survived and dead cells, respectively.

CA1 pyramidal neurons in the hippocampus are known to undergo delayed neuronal death (DND) several days after transient forebrain ischemia in vivo (Pulsinelli, Brierley, & Plum, 1982). DND occurs largely through apoptosis (Love, 2003). When mice were subjected to transient common carotid artery occlusion, actually, neuronal cell death was induced in the CA1 region 4 days after reperfusion with releasing cytochrome *c* (Inoue et al., 2007). In Fig. 11, we present additional morphological evidence for the apoptotic sign, namely AVD induction, observed under a two-photon laser microscope in a hippocampal slice subjected to transient forebrain ischemia in vivo. Although neuronal cell death was observed only in the CA1 region 3 days after reperfusion (red-colored cells), most CA1 neurons were still survived but became shrunken (green-colored cells) 2 days after reperfusion. When VSOR blockers were administered via the jugular vein prior to the ischemic treatment and into the peritoneal cavity after reperfusion, neuronal cell death and cytochrome *c* release were both strongly abrogated (Inoue et al., 2007), indicating involvements of VSOR activity in AVD and DND induction.

An involvement of VSOR activity in necrotic cell death due to excitotoxicity caused by ischemia or ischemia-reperfusion was indirectly shown by inhibitory effects of VSOR blockers, as follows: A relatively VSOR-selective blocker, phloretin, inhibited NVI induction and necrotic cell death induced by NMDA exposure of mouse cortical neurons (Inoue et al., 2007). Similarly, a VSOR-specific blocker, DCPIB, markedly attenuated neuronal damage caused by transient OGD in rat hippocampal neurons (Han et al., 2014). DCPIB administered intracisternally reduced glutamate release to the extracellular fluid and the volume of infarct induced by reversible middle cerebral artery occlusion in rat brain cortex in vivo (Zhang et al., 2008). Intraperitoneal administration of DCPIB diminished the size of infarct observed 24 h after the right common carotid artery occlusion in vivo in the neonatal mice in which the blood-brain barrier is less developed (Alibrahim et al., 2013).

OGD was shown to induce ATP release in rat hippocampal slices (Juranyi, Sperlagh, & Vizi, 1999). Since Maxi-Cl is known to mediate ATP release from astrocytes subjected to an ischemic insult (Liu, Sabirov, & Okada, 2008), ischemia-reperfusion-induced cerebral ATP release may be largely, though not solely, mediated by Maxi-Cl. Both detrimental (cytotoxic) and beneficial (protective) roles have been suggested for astroglial ATP release upon stroke (Franke, Verkhratsky, Burnstock, & Illes, 2012; Rossi, Brady, & Mohr, 2007). ATP released from astrocytes may, in a paracrine fashion, activates VSOR in nearby neurons via their purinergic receptors (Akita & Okada, 2014), thereby causing facilitation of ATP depletion in astrocytes under hypoxia/ischemia stress (Fig. 6) and/or leading to excessive Cl^- uptake by neurons under excitotoxic depolarized conditions (Fig. 7). Also, it is possible that Maxi-Cl-mediated ATP release from glial cells is involved in DND induction, because stimulation of purinergic receptors in neuronal and/or microglial cells was suggested to participate in ischemia-reperfusion-induced neurotoxicity (Bargiotas et al., 2011; Webster et al., 2013; Yu et al., 2013). On the other hand, there is a possibility that adenosine delivered from the hydrolysis of released ATP plays a protective role during brain ischemia, because the tissue adenosine level was found to increase in the dog and rat brains and exerts a vasodilator effect on the cerebral vessels (Berne & Rubio, 1974).

5.3 Roles of volume-regulatory anion channels in ischemia- and ischemia-reperfusion-induced death in cardiac myocytes

Myocardial infarction (MI), commonly known as a heart attack, is caused by cardiac myocyte death during coronary occlusion and/or reperfusion thereafter. MI-associated cell death occurs by apoptosis mainly in the early period and necrosis mostly in the late stage in the animal model (Kajstura et al., 1996; Kurrelmeyer et al., 2000; Li et al., 1997; Whelan et al., 2010) and in the human patients (Saraste et al., 1997).

Cardiac cell swelling was observed after acute regional MI in the porcine heart (Tranum-Jensen et al., 1981) but was inhibited by ischemic preconditioning in rabbit ventricular myocytes (Diaz et al., 2014). Therefore, swelling-induced cardiac VSOR activity is possibly involved in protection against ischemia- or ischemia-reperfusion-induced injury. Actually, a VSOR blocker, IAA-94, was found to suppress the protective action of ischemic preconditioning (Batthish, Diaz, Zeng, Backx, & Wilson, 2002; Diaz et al., 1999) with inhibiting the RVD response (Batthish et al., 2002).

Apoptotic cardiac cell death was often observed during ischemia-reperfusion in hearts in vivo (Bardales, Hailey, Xie, Schaefer, & Hsu, 1996; Bialik et al., 1997; Fliss & Gattinger, 1996; Ji, Yue, Wu, & He, 2004). Apoptosis induced by ischemia-reperfusion was shown to be inhibited by VSOR blockers including phloretin, DIDS and NPPB and by ROS scavengers in mouse cardiomyocytes (Wang et al., 2005), suggesting an involvement of ROS-activated VSOR activity in apoptosis induction elicited by ischemia-reperfusion. Actually, massive ROS production was found to be induced in mouse cardiomyocytes upon reperfusion (Wang et al., 2005). Consistently, it has been recognized that ROS are involved in the pathogenesis of MI (Curello et al., 1995; Jeroudi, Hartley, & Bolli, 1994; Lucchesi, 1990).

During ischemia or ischemia-reperfusion, cardiac cell swelling has been reported to be induced due to an accumulation of anaerobic metabolites before the onset of cell damage (Ganote & Armstrong, 2003; Jennings & Ganote, 1974; Macknight & Leaf, 1977; Tranum-Jensen et al., 1981; Vandenberg, Rees, Wright, & Powell, 1996; Whalen, Hamilton, Ganote, & Jennings, 1974). Ischemia-induced cardiac cell swelling persists, because the counteracting RVD mechanism is impaired by inhibition of VSOR activity (Fig. 6), thereby causing necrotic cell death. Actually, ischemia-reperfusion was shown to induce cardiac injury, namely MI, largely due to necrosis (Nakagawa et al., 2005; Uramoto, Okada, & Okada, 2012; Uramoto et al., 2003). An isoflavonoid, genistein, was found to markedly decrease the size of MI and necrotic release of myocardial-specific isoenzyme of creatine kinase (CK-MB) in the mouse hearts subjected to ischemia-reperfusion in vivo and also to largely ameliorate OGD-induced necrotic cell injury in vitro which was monitored by propidium iodide (PI) uptake in cultured neonatal rat ventricular cardiomyocytes (Uramoto et al., 2012). Since genestein is known not only as a potent blocker of VSOR activity (Bryan-Sisneros, Sabanov, Thoroed, & Doroshenko, 2000; Inoue et al., 2007; Shuba et al., 2000) but also as a best-studied potentiator of cardiac CFTR activity (Chiang, Chen, Chang, Lin, & Luk, 1996; Hool, Middleton, & Harvey, 1998; Obayashi, Horie, Washizuka, Nishimoto, & Sasayama, 1999; Shuba, Asai, Pelzer, & McDonald, 1996; Shuba & McDonald, 1997; Tilly et al., 1996; Zhou, Hazama, & Okada, 1998), the above genistein-induced protective effect for ischemia-reperfusion-induced necrosis cannot be explained by its action to VSOR. Instead, cardiac CFTR is likely to be involved in the protective effect presumably by facilitating the RVD process for the following three reasons. First, under ischemic/hypoxic conditions, catecholamines are known to be massively released (Karwatowska-Krynska & Beresewicz, 1983; Videbaek, Christensen, & Sterndorff, 1972). Second, stimulation of β-adrenergic receptors leads to activation of cardiac CFTR (Bahinski, Nairn, Greengard, & Gadsby, 1989; Ehara & Ishihara, 1990; Harvey & Hume, 1989; Matsuoka, Ehara, & Noma, 1990). Third, cardiac CFTR activity is involved in the RVD process in ventricular myocytes under β-adrenergic stimulation (Wang et al., 1997; Yamamoto, Ehara, & Shioya, 2001). In fact, molecular expression of CFTR was found to be upregulated 3h after ischemic treatment in rat ventricular myocytes in vitro (Uramoto et al., 2003) and in mouse ventricles in vivo (Uramoto et al., 2012). Also, CFTR currents were shown to be increased after the ischemic treatment for 3–4h in rat ventricular myocytes in culture (Uramoto et al., 2003), Finally, the mean size of ischemia-reperfusion-induced infarct was demonstrated to be aggravated by administration of known CFTR blockers and was found to be larger in CFTR-deficient mice compared to wild-type mice (Uramoto et al., 2012). Furthermore, hypoxia-reoxygenation-induced PI-positive necrosis was ameliorated by over-expression of CFTR gene but aggravated by CFTR gene silencing in cultured neonatal rat ventricular myocytes (Uramoto et al., 2012). Taken together, it is concluded that activation of endogenous cardiac CFTR is involved in protection against necrotic myocardial injury induced by ischemia-reperfusion in vivo and in vitro.

Ischemic preconditioning (IPC) is the most potent intervention whereby repeated short episodes of ischemia-reperfusion can protect the heart against a subsequent more prolonged ischemic insult. Although the precise mechanism of IPC still remains obscure, Cl^- efflux mediated by some anion channels was suggested to be a key mechanism for IPC (Diaz et al., 2003; Diaz, Hinek, & Wilson, 2010). Since IPC was found to reduce ischemic cell swelling in isolated rabbit cardiomyocytes (Ganote & Armstrong, 2003), there raises a possibility that volume-regulatory anion channels are involved in the IPC mechanism. Actually, IPC-conferred cardioprotection was abolished by administration of NPPB and IAA-94 (Diaz et al., 1999), both of which are known blockers of VSOR and CFTR. Finally, by using CFTR-knockout mice, cardiac CFTR was shown to be essential for the IPC-induced protection against MI in vitro (Chen et al., 2004) and in vivo (Xiang et al., 2011). Taken together, it appears that the volume-regulatory function of cardiac CFTR plays an essential role in protection against and rescue from necrotic myocardial injury elicited by ischemia and ischemia-reperfusion insults.

Ischemia/hypoxia or ischemia-reperfusion stress was observed to induce ATP release from the hearts (Borst & Schrader, 1991; Clemens & Forrester, 1981; Erlinge et al., 2005; Kuzmin, Gourine, Molosh, Lakomkin, & Vassort, 2000; Kuzmin, Lakomkin, Kapelko, & Vassort, 1998; Paddle & Burnstock, 1974; Sabirov et al., 2017; Vial, Owen, Opie, & Posel, 1987; Wee, Peart, & Headrick, 2007) and from isolated cardiomyocytes (Dutta et al., 2004; Forrester & Williams, 1977). Adenosine release from the hearts was also induced by ischemia/hypoxia insults (Borst & Schrader, 1991; Van Belle, Goossens, & Wynants, 1987) especially following the ATP release episode (Kuzmin et al., 1998; Vial et al., 1987). Also, adenosine was shown to appear in ischemic cardiac muscles (Berne, Rubio, Dobson, & Curnish, 1971) and suggested as a possible mediator of the vasodilation when the oxygen supply to the tissue is reduced (Berne, 1963; Rubio, Berne, & Katori, 1969). In addition, UTP was released together with ATP from the hearts during ischemia and early after reperfusion (Erlinge et al., 2005; Wee et al., 2007). Released ATP, adenosine and UTP were reported to exert the protective or preconditioning action against ischemic and ischemia-reperfusion injuries via purinergic receptors (Armstrong & Ganote, 1994; Cohen et al., 2011; Golan et al., 2011; Hochhauser et al., 2013; Ninomiya et al., 2002; Thornton, Liu, Olsson, & Downey, 1992; Vessey, Li, & Kelley, 2011; Wee et al., 2007; Yitzhaki, Shneyvays, Jacobson, & Shainberg, 2005). Since ischemia-induced cardiac ATP release was suppressed by Gd^{3+} and NPPB (Dutta et al., 2004; Kunugi et al., 2011), it is highly likely that the ATP release is mediated by Maxi-Cl. Consistently, the spatial distribution of Maxi-Cl in rat cardiomyocytes was shown to coincide with the spatial distribution of the ATP-releasing site (Dutta et al., 2008). Also, ischemia-reperfusion induced ATP release from Langendorff-perfused adult mouse hearts was recently demonstrated to be significantly suppressed when the Maxi-Cl-forming SLCO2A1 gene was partially silenced by injection of its siRNA in vivo (Sabirov et al., 2017). Thus, it is concluded that cardiac Maxi-Cl activity exerts some protective action by mediating ATP and UTP release.

6. Conclusions and future direction

Animal cells, that lack the cell wall, must cope with cell swelling induced by osmotic perturbation under physiological and pathological conditions by the volume regulation mechanism, called RVD, which is mainly attained by activation of K^+ and Cl^- channels. In most mammalian cells, the RVD process involves two types of the volume-activated and volume-regulatory anion channels, VSOR and Maxi-Cl, the core molecules of which have been recently identified as LRRC8A *plus* LRRC8C/D/E and SLCO2A1, respectively. In cardiomyocytes, a cardiac type of CFTR isoform is also involved in the RVD process under β-adrenergic stimulation. VSOR and Maxi-Cl are also involved in release of glutamate and ATP, that are autocrine/paracrine signals, and augment VSOR activity via GPCRs.

Persistent cell shrinkage, called AVD, is an essential hallmark of apoptotic cell death. The early phase of AVD occurring prior to caspase activation is triggered by KCl release due to activation of K^+ and Cl^- channels. Since VSOR activity is responsible for the AVD-causing Cl^- efflux, pharmacological blocking or genetic disruption of VSOR can provide a most effective way to prevent or rescue apoptotic cell death. Downregulation of VSOR activity is involved in acquisition of cisplatin resistance in cancer cells not only by escaping from cisplatin-induced AVD induction but also by preventing the entry of cisplatin largely mediated via the VSOR pore. Not surprisingly therefore, it is expected that the development of interventions targeting the VSOR activity and/or the genes of VSOR-forming core molecules may provide most promising approaches for restoration of cisplatin sensitivity in cancer cells.

Necrotic cell death starts with cell swelling coupled to RVD dysfunction. Under excitotoxicity due to excessive glutamate release in the brain, activation of VSOR in neurons is induced by swelling initiated by the cation influx via iGluR and anion influx via $GABA_AR$. However, this VSOR activity does not serve as a pathway for volume-regulatory anion efflux but rather mediates swelling-exaggerating anion influx, thereby causing

NVI. Excitotoxicity-associated glutamate release is mediated by VSOR and Maxi-Cl in swollen astrocytes. Under ischemic/hypoxic, acidotoxic and lactacidosis conditions, NVI is caused not only by cell swelling but also by RVD dysfunction in animal cells including neuronal, glial and cardiac cells, because their VSOR activity is inhibited by ATP depletion together with increased intracellular free Mg^{2+} and H^{+}. Therefore, restoration of endogenous VSOR activity or introduction/overexpression of some exogenous ATP-independent and proton-insensitive anion channels may provide the effective approach for the rescue from ischemic/hypoxic, acidotoxic and lactacidosis-induced necrosis in cerebral and cardiac cells. In the case of myocardial infarction caused by ischemia(-reperfusion), one additional therapeutic target would be cardiac CFTR, because cardiac CFTR, in place of VSOR, plays a protective role by serving as a pathway of volume-regulatory Cl^{-} efflux, under the conditions where cAMP production is increased by ischemic β-adrenergic stimulation.

In a word, volume-regulatory anion channels including VSOR, Maxi-Cl and cardiac CFTR are essentially and elaborately involved in the induction of cell death associated with ischemia(-reperfusion) injury, excitotoxicity, acidotoxicity, lactacidosis, delayed neuronal death and myocardial infarction. Thus, it appears that these anion channels are promising targets for the development of neuroprotective and cardioprotective drugs in the future.

Acknowledgment

This work was supported by JSPS KAKENHI Grants #17K19517 (to Y.O.).

References

Aarts, M., Iihara, K., Wei, W. L., Xiong, Z. G., Arundine, M., Cerwinski, W., et al. (2003). A key role for TRPM7 channels in anoxic neuronal death. *Cell*, *115*, 863–877.

Abboud, F. M., & Benson, C. J. (2015). ASICs and cardiovascular homeostasis. *Neuropharmacology*, *94*, 87–98.

Abdullaev, I. F., Rudkouskaya, A., Schools, G. P., Kimelberg, H. K., & Mongin, A. A. (2006). Pharmacological comparison of swelling-activated excitatory amino acid release and Cl^{-} currents in cultured rat astrocytes. *The Journal of Physiology*, *572*, 677–689.

Abdullaev, I. F., Sabirov, R. Z., & Okada, Y. (2003). Upregulation of swelling-activated Cl^{-} channel sensitivity to cell volume by activation of EGF receptors in murine mammary cells. *The Journal of Physiology*, *549*, 749–758.

Abramov, A. Y., Canevari, L., & Duchen, M. R. (2004). Beta-amyloid peptides induce mitochondrial dysfunction and oxidative stress in astrocytes and death of neurons through activation of NADPH oxidase. *The Journal of Neuroscience*, *24*, 565–575.

Adams, J. M., & Cory, S. (1998). The Bcl-2 protein family: Arbiters of cell survival. *Science*, *281*, 1322–1326.

Aguirre, A., Shoji, K. F., Saez, J. C., Henriquez, M., & Quest, A. F. (2013). FasL-triggered death of Jurkat cells requires caspase 8-induced, ATP-dependent cross-talk between Fas and the purinergic receptor $P2X_7$. *Journal of Cellular Physiology*, *228*, 485–493.

Aitken, P. G., Borgdorff, A. J., Juta, A. J., Kiehart, D. P., Somjen, G. G., & Wadman, W. J. (1998). Volume changes induced by osmotic stress in freshly isolated rat hippocampal neurons. *Pflügers Archiv*, *436*, 991–998.

Akanda, N., & Elinder, F. (2006). Biophysical properties of the apoptosis-inducing plasma membrane voltage-dependent anion channel. *Biophysical Journal*, *90*, 4405–4417.

Akita, T., Fedorovich, S. V., & Okada, Y. (2011). Ca^{2+} nanodomain-mediated component of swelling-induced volume-sensitive outwardly rectifying anion current triggered by autocrine action of ATP in mouse astrocytes. *Cellular Physiology and Biochemistry*, *28*, 1181–1190.

Akita, T., & Okada, Y. (2011). Regulation of bradykinin-induced activation of volume-sensitive outwardly rectifying anion channels by Ca^{2+} nanodomains in mouse astrocytes. *The Journal of Physiology*, *589*, 3909–3927.

Akita, T., & Okada, Y. (2014). Characteristics and roles of the volume-sensitive outwardly rectifying (VSOR) anion channel in the central nervous system. *Neuroscience*, *275*, 211–231.

Alibrahim, A., Zhao, L. Y., Bae, C. Y., Barszczyk, A., Sun, C. L., Wang, G. L., et al. (2013). Neuroprotective effects of volume-regulated anion channel blocker DCPIB on neonatal hypoxic-ischemic injury. *Acta Pharmacologica Sinica*, *34*, 113–118.

Allen, N. J., & Attwell, D. (2002). Modulation of ASIC channels in rat cerebellar Purkinje neurons by ischaemia-related signals. *The Journal of Physiology*, *543*, 521–529.

Alojado, M. E., Morimoto, Y., Morimoto, Y., & Kemmotsu, O. (1996). Mechanism of cellular swelling induced by extracellular lactic acidosis in neuroblastoma-glioma hybrid (NG108-15) cells. *Anesthesia and Analgesia*, *83*, 1002–1008.

Andrew, R. D., Lobinowich, M. E., & Osehobo, E. P. (1997). Evidence against volume regulation by cortical brain cells during acute osmotic stress. *Experimental Neurology*, *143*, 300–312.

Andrew, R. D., & MacVicar, B. A. (1994). Imaging cell volume changes and neuronal excitation in the hippocampal slice. *Neuroscience*, *62*, 371–383.

Armstrong, S., & Ganote, C. E. (1994). Adenosine receptor specificity in preconditioning of isolated rabbit cardiomyocytes: Evidence of A3 receptor involvement. *Cardiovascular Research*, *28*, 1049–1056.

Ashkenazi, A., & Dixit, V. M. (1998). Death receptors: Signaling and modulation. *Science*, *281*, 1305–1308.

Ayata, C., Ayata, G., Hara, H., Matthews, R. T., Beal, M. F., Ferrante, R. J., et al. (1997). Mechanisms of reduced striatal NMDA excitotoxicity in type I nitric oxide synthase knock-out mice. *The Journal of Neuroscience*, *17*, 6908–6917.

Bachmann, O., Heinzmann, A., Mack, A., Manns, M. P., & Seidler, U. (2007). Mechanisms of secretion-associated shrinkage and volume recovery in cultured rabbit parietal cells. *American Journal of Physiology. Gastrointestinal and Liver Physiology*, *292*, G711–G717.

Bahinski, A., Nairn, A. C., Greengard, P., & Gadsby, D. C. (1989). Chloride conductance regulated by cyclic AMP-dependent protein kinase in cardiac myocytes. *Nature*, *340*, 718–721.

Bajnath, R. B., Groot, J. A., de Jonge, H. R., Kansen, M., & Bijman, J. (1993). Calcium ionophore plus excision induce a large conductance chloride channel in membrane patches of human colon carcinoma cells HT-29cl.19A. *Experientia*, *49*, 313–316.

Banasiak, K. J., Xia, Y., & Haddad, G. G. (2000). Mechanisms underlying hypoxia-induced neuronal apoptosis. *Progress in Neurobiology*, *62*, 215–249.

Banderali, U., & Roy, G. (1992). Anion channels for amino acids in MDCK cells. *The American Journal of Physiology*, *263*, C1200–C1207.

Bardales, R. H., Hailey, L. S., Xie, S. S., Schaefer, R. F., & Hsu, S. M. (1996). In situ apoptosis assay for the detection of early acute myocardial infarction. *The American Journal of Pathology*, *149*, 821–829.
Bargiotas, P., Krenz, A., Hormuzdi, S. G., Ridder, D. A., Herb, A., Barakat, W., et al. (2011). Pannexins in ischemia-induced neurodegeneration. *Proceedings of the National Academy of Sciences of the United States of America*, *108*, 20772–20777.
Barros, L. F., Hermosilla, T., & Castro, J. (2001). Necrotic volume increase and the early physiology of necrosis. *Comparative Biochemistry and Physiology. Part A, Molecular & Integrative Physiology*, *130*, 401–409.
Batthish, M., Diaz, R. J., Zeng, H. P., Backx, P. H., & Wilson, G. J. (2002). Pharmacological preconditioning in rabbit myocardium is blocked by chloride channel inhibition. *Cardiovascular Research*, *55*, 660–671.
Bazan, N. G., Jr. (1970). Effects of ischemia and electroconvulsive shock on free fatty acid pool in the brain. *Biochimica et Biophysica Acta*, *218*, 1–10.
Beck, J., Lenart, B., Kintner, D. B., & Sun, D. (2003). Na-K-Cl cotransporter contributes to glutamate-mediated excitotoxicity. *The Journal of Neuroscience*, *23*, 5061–5068.
Belhage, B., Hansen, G. H., & Schousboe, A. (1993). Depolarization by K^+ and glutamate activates different neurotransmitter release mechanisms in GABAergic neurons: Vesicular versus non-vesicular release of GABA. *Neuroscience*, *54*, 1019–1034.
Bell, P. D., Lapointe, J. Y., Sabirov, R., Hayashi, S., Peti-Peterdi, J., Manabe, K., et al. (2003). Macula densa cell signaling involves ATP release through a maxi anion channel. *Proceedings of the National Academy of Sciences of the United States of America*, *100*, 4322–4327.
Bender, A. S., Neary, J. T., Blicharska, J., Norenberg, L. O., & Norenberg, M. D. (1992). Role of calmodulin and protein kinase C in astrocytic cell volume regulation. *Journal of Neurochemistry*, *58*, 1874–1882.
Benfenati, V., Caprini, M., Dovizio, M., Mylonakou, M. N., Ferroni, S., Ottersen, O. P., et al. (2011). An aquaporin-4/transient receptor potential vanilloid 4 (AQP4/TRPV4) complex is essential for cell-volume control in astrocytes. *Proceedings of the National Academy of Sciences of the United States of America*, *108*, 2563–2568.
Benfenati, V., Caprini, M., Nicchia, G. P., Rossi, A., Dovizio, M., Cervetto, C., et al. (2009). Carbenoxolone inhibits volume-regulated anion conductance in cultured rat cortical astroglia. *Channels (Austin, Tex.)*, *3*, 323–336.
Benfenati, V., Nicchia, G. P., Svelto, M., Rapisarda, C., Frigeri, A., & Ferroni, S. (2007). Functional down-regulation of volume-regulated anion channels in AQP4 knockdown cultured rat cortical astrocytes. *Journal of Neurochemistry*, *100*, 87–104.
Benson, R. S., Heer, S., Dive, C., & Watson, A. J. (1996). Characterization of cell volume loss in CEM-C7A cells during dexamethasone-induced apoptosis. *The American Journal of Physiology*, *270*, C1190–C1203.
Berne, R. M. (1963). Cardiac nucleotides in hypoxia: Possible role in regulation of coronary blood flow. *The American Journal of Physiology*, *204*, 317–322.
Berne, R. M., & Rubio, R. (1974). Adenine nucleotide metabolism in the heart. *Circulation Research*, *35*(Suppl. 3), 109–120.
Berne, R. M., Rubio, R., Dobson, J. G., Jr., & Curnish, R. R. (1971). Adenosine and adenine nucleotides as possible mediators of cardiac and skeletal muscle blood flow regulation. *Circulation Research*, *28*(Suppl. 1), 115+.
Bettendorff, L., Kolb, H. A., & Schoffeniels, E. (1993). Thiamine triphosphate activates an anion channel of large unit conductance in neuroblastoma cells. *The Journal of Membrane Biology*, *136*, 281–288.
Bhat, T. A., Chaudhary, A. K., Kumar, S., O'Malley, J., Inigo, J. R., Kumar, R., et al. (2017). Endoplasmic reticulum-mediated unfolded protein response and mitochondrial apoptosis in cancer. *Biochimica et Biophysica Acta, Reviews on Cancer*, *1867*, 58–66.

Bialik, S., Geenen, D. L., Sasson, I. E., Cheng, R., Horner, J. W., Evans, S. M., et al. (1997). Myocyte apoptosis during acute myocardial infarction in the mouse localizes to hypoxic regions but occurs independently of p53. *The Journal of Clinical Investigation, 100*, 1363–1372.

Bindokas, V. P., & Miller, R. J. (1995). Excitotoxic degeneration is initiated at non-random sites in cultured rat cerebellar neurons. *The Journal of Neuroscience, 15*, 6999–7011.

Blatz, A. L., & Magleby, K. L. (1983). Single voltage-dependent chloride-selective channels of large conductance in cultured rat muscle. *Biophysical Journal, 43*, 237–241.

Bolotina, V., Borecky, J., Vlachova, V., Baudysova, M., & Vyskocil, F. (1987). Voltage-dependent chloride channels with several substates in excised patches from mouse neuroblastoma cells. *Neuroscience Letters, 77*, 298–302.

Borst, M. M., & Schrader, J. (1991). Adenine nucleotide release from isolated perfused guinea pig hearts and extracellular formation of adenosine. *Circulation Research, 68*, 797–806.

Bortner, C. D., & Cidlowski, J. A. (1996). Absence of volume regulatory mechanisms contributes to the rapid activation of apoptosis in thymocytes. *The American Journal of Physiology, 271*, C950–C961.

Bortner, C. D., & Cidlowski, J. A. (1998). A necessary role for cell shrinkage in apoptosis. *Biochemical Pharmacology, 56*, 1549–1559.

Bortner, C. D., & Cidlowski, J. A. (2003). Uncoupling cell shrinkage from apoptosis reveals that Na^+ influx is required for volume loss during programmed cell death. *The Journal of Biological Chemistry, 278*, 39176–39184.

Bortner, C. D., Hughes, F. M., Jr., & Cidlowski, J. A. (1997). A primary role for K^+ and Na^+ efflux in the activation of apoptosis. *The Journal of Biological Chemistry, 272*, 32436–32442.

Bortner, C. D., Scoltock, A. B., Sifre, M. I., & Cidlowski, J. A. (2012). Osmotic stress resistance imparts acquired anti-apoptotic mechanisms in lymphocytes. *The Journal of Biological Chemistry, 287*, 6284–6295.

Bowens, N. H., Dohare, P., Kuo, Y. H., & Mongin, A. A. (2013). DCPIB, the proposed selective blocker of volume-regulated anion channels, inhibits several glutamate transport pathways in glial cells. *Molecular Pharmacology, 83*, 22–32.

Brennan, A. M., Suh, S. W., Won, S. J., Narasimhan, P., Kauppinen, T. M., Lee, H., et al. (2009). NADPH oxidase is the primary source of superoxide induced by NMDA receptor activation. *Nature Neuroscience, 12*, 857–863.

Browe, D. M., & Baumgarten, C. M. (2003). Stretch of beta 1 integrin activates an outwardly rectifying chloride current via FAK and Src in rabbit ventricular myocytes. *The Journal of General Physiology, 122*, 689–702.

Browe, D. M., & Baumgarten, C. M. (2004). Angiotensin II (AT1) receptors and NADPH oxidase regulate Cl^- current elicited by beta1 integrin stretch in rabbit ventricular myocytes. *The Journal of General Physiology, 124*, 273–287.

Browe, D. M., & Baumgarten, C. M. (2006). EGFR kinase regulates volume-sensitive chloride current elicited by integrin stretch via PI-3K and NADPH oxidase in ventricular myocytes. *The Journal of General Physiology, 127*, 237–251.

Bryan-Sisneros, A., Sabanov, V., Thoroed, S. M., & Doroshenko, P. (2000). Dual role of ATP in supporting volume-regulated chloride channels in mouse fibroblasts. *Biochimica et Biophysica Acta, 1468*, 63–72.

Burg, E. D., Remillard, C. V., & Yuan, J. X. (2006). K^+ channels in apoptosis. *The Journal of Membrane Biology, 209*, 3–20.

Cacace, V. I., Finkelsteyn, A. G., Tasso, L. M., Kusnier, C. F., Gomez, K. A., & Fischbarg, J. (2014). Regulatory volume increase and regulatory volume decrease responses in HL-1 atrial myocytes. *Cellular Physiology and Biochemistry, 33*, 1745–1757.

Cai, Z., Jitkaew, S., Zhao, J., Chiang, H. C., Choksi, S., Liu, J., et al. (2014). Plasma membrane translocation of trimerized MLKL protein is required for TNF-induced necroptosis. *Nature Cell Biology, 16*, 55–65.

Cai, S., Zhang, T., Zhang, D., Qiu, G., & Liu, Y. (2015). Volume-sensitive chloride channels are involved in cisplatin treatment of osteosarcoma. *Molecular Medicine Reports*, *11*, 2465–2470.

Cain, K., Langlais, C., Sun, X. M., Brown, D. G., & Cohen, G. M. (2001). Physiological concentrations of K^+ inhibit cytochrome c-dependent formation of the apoptosome. *The Journal of Biological Chemistry*, *276*, 41985–41990.

Carini, R., Autelli, R., Bellomo, G., & Albano, E. (1999). Alterations of cell volume regulation in the development of hepatocyte necrosis. *Experimental Cell Research*, *248*, 280–293.

Carpenter, E., & Peers, C. (1997). Swelling- and cAMP-activated Cl^- currents in isolated rat carotid body type I cells. *The Journal of Physiology*, *503*(Pt. 3), 497–511.

Casslen, B., & Nilsson, B. (1984). Human uterine fluid, examined in undiluted samples for osmolarity and the concentrations of inorganic ions, albumin, glucose, and urea. *American Journal of Obstetrics and Gynecology*, *150*, 877–881.

Chan, F. K., Shisler, J., Bixby, J. G., Felices, M., Zheng, L., Appel, M., et al. (2003). A role for tumor necrosis factor receptor-2 and receptor-interacting protein in programmed necrosis and antiviral responses. *The Journal of Biological Chemistry*, *278*, 51613–51621.

Chebabo, S. R., Hester, M. A., Jing, J., Aitken, P. G., & Somjen, G. G. (1995). Interstitial space, electrical resistance and ion concentrations during hypotonia of rat hippocampal slices. *The Journal of Physiology*, *487*(Pt. 3), 685–697.

Chen, H., Liu, L. L., Ye, L. L., McGuckin, C., Tamowski, S., Scowen, P., et al. (2004). Targeted inactivation of cystic fibrosis transmembrane conductance regulator chloride channel gene prevents ischemic preconditioning in isolated mouse heart. *Circulation*, *110*, 700–704.

Chen, Q., Olney, J. W., Lukasiewicz, P. D., Almli, T., & Romano, C. (1998). Ca^{2+}-independent excitotoxic neurodegeneration in isolated retina, an intact neural net: A role for Cl^- and inhibitory transmitters. *Molecular Pharmacology*, *53*, 564–572.

Chiang, C. E., Chen, S. A., Chang, M. S., Lin, C. I., & Luk, H. N. (1996). Genistein directly inhibits L-type calcium currents but potentiates cAMP-dependent chloride currents in cardiomyocytes. *Biochemical and Biophysical Research Communications*, *223*, 598–603.

Chiang, C. E., Luk, H. N., & Wang, T. M. (2004). Swelling-activated chloride current is activated in guinea pig cardiomyocytes from endotoxic shock. *Cardiovascular Research*, *62*, 96–104.

Choi, D. W. (1987). Ionic dependence of glutamate neurotoxicity. *The Journal of Neuroscience*, *7*, 369–379.

Choi, D. W. (1996). Ischemia-induced neuronal apoptosis. *Current Opinion in Neurobiology*, *6*, 667–672.

Christmann, J., Azer, L., Dorr, D., Fuhr, G. R., Bastiaens, P. I., & Wehner, F. (2016). Adaptive responses of cell hydration to a low temperature arrest. *The Journal of Physiology*, *594*, 1663–1676.

Chu, G. (1994). Cellular responses to cisplatin. The roles of DNA-binding proteins and DNA repair. *The Journal of Biological Chemistry*, *269*, 787–790.

Churchwell, K. B., Wright, S. H., Emma, F., Rosenberg, P. A., & Strange, K. (1996). NMDA receptor activation inhibits neuronal volume regulation after swelling induced by veratridine-stimulated Na^+ influx in rat cortical cultures. *The Journal of Neuroscience*, *16*, 7447–7457.

Clemens, M. G., & Forrester, T. (1981). Appearance of adenosine triphosphate in the coronary sinus effluent from isolated working rat heart in response to hypoxia. *The Journal of Physiology*, *312*, 143–158.

Clemo, H. F., & Baumgarten, C. M. (1991). Atrial natriuretic factor decreases cell volume of rabbit atrial and ventricular myocytes. *The American Journal of Physiology*, *260*, C681–C690.

Clemo, H. F., & Baumgarten, C. M. (1997). Swelling-activated Gd^{3+}-sensitive cation current and cell volume regulation in rabbit ventricular myocytes. *The Journal of General Physiology, 110*, 297–312.
Clemo, H. F., Stambler, B. S., & Baumgarten, C. M. (1999). Swelling-activated chloride current is persistently activated in ventricular myocytes from dogs with tachycardia-induced congestive heart failure. *Circulation Research, 84*, 157–165.
Coca-Prados, M., Anguita, J., Chalfant, M. L., & Civan, M. M. (1995). PKC-sensitive Cl^- channels associated with ciliary epithelial homologue of pICln. *The American Journal of Physiology, 268*, C572–C579.
Cohen, R., Shainberg, A., Hochhauser, E., Cheporko, Y., Tobar, A., Birk, E., et al. (2011). UTP reduces infarct size and improves mice heart function after myocardial infarct via P2Y2 receptor. *Biochemical Pharmacology, 82*, 1126–1133.
Cookson, B. T., & Brennan, M. A. (2001). Pro-inflammatory programmed cell death. *Trends in Microbiology, 9*, 113–114.
Coulombe, A., & Coraboeuf, E. (1992). Large-conductance chloride channels of new-born rat cardiac myocytes are activated by hypotonic media. *Pflügers Archiv, 422*, 143–150.
Coulombe, A., Duclohier, H., Coraboeuf, E., & Touzet, N. (1987). Single chloride-permeable channels of large conductance in cultured cardiac cells of new-born rats. *European Biophysics Journal, 14*, 155–162.
Crepel, V., Panenka, W., Kelly, M. E., & MacVicar, B. A. (1998). Mitogen-activated protein and tyrosine kinases in the activation of astrocyte volume-activated chloride current. *The Journal of Neuroscience, 18*, 1196–1206.
Curello, S., Ceconi, C., de Giuli, F., Panzali, A. F., Milanesi, B., Calarco, M., et al. (1995). Oxidative stress during reperfusion of human hearts: Potential sources of oxygen free radicals. *Cardiovascular Research, 29*, 118–125.
d'Anglemont de Tassigny, A., Berdeaux, A., Souktani, R., Henry, P., & Ghaleh, B. (2008). The volume-sensitive chloride channel inhibitors prevent both contractile dysfunction and apoptosis induced by doxorubicin through PI3kinase, Akt and Erk 1/2. *European Journal of Heart Failure, 10*, 39–46.
d'Anglemont de Tassigny, A., Souktani, R., Henry, P., Ghaleh, B., & Berdeaux, A. (2004). Volume-sensitive chloride channels (ICl,vol) mediate doxorubicin induced apoptosis through apoptotic volume decrease in cardiomyocytes. *Fundamental & Clinical Pharmacology, 18*, 531–538.
Darby, M., Kuzmiski, J. B., Panenka, W., Feighan, D., & MacVicar, B. A. (2003). ATP released from astrocytes during swelling activates chloride channels. *Journal of Neurophysiology, 89*, 1870–1877.
de Jonge, H. W., Dekkers, D. H., Tilly, B. C., & Lamers, J. M. (2002). Cyclic stretch and endothelin-1 mediated activation of chloride channels in cultured neonatal rat ventricular myocytes. *Clinical Science (London, England: 1979), 103*(Suppl. 48), 148s–151s.
Degterev, A., Huang, Z., Boyce, M., Li, Y., Jagtap, P., Mizushima, N., et al. (2005). Chemical inhibitor of nonapoptotic cell death with therapeutic potential for ischemic brain injury. *Nature Chemical Biology, 1*, 112–119.
Delpire, E., & Gagnon, K. B. (2018). Water homeostasis and cell volume maintenance and regulation. *Current Topics in Membranes, 81*, 3–52.
Delpire, E., & Gullans, S. R. (1994). Cell volume and K^+ transport during differentiation of mouse erythroleukemia cells. *The American Journal of Physiology, 266*, C515–C523.
Deng, W., Baki, L., & Baumgarten, C. M. (2010). Endothelin signalling regulates volume-sensitive Cl^- current via NADPH oxidase and mitochondrial reactive oxygen species. *Cardiovascular Research, 88*, 93–100.
Deng, W., Baki, L., Yin, J., Zhou, H., & Baumgarten, C. M. (2010). HIV protease inhibitors elicit volume-sensitive Cl^- current in cardiac myocytes via mitochondrial ROS. *Journal of Molecular and Cellular Cardiology, 49*, 746–752.

Dessi, F., Charriaut-Marlangue, C., & Ben-Ari, Y. (1994). Glutamate-induced neuronal death in cerebellar culture is mediated by two distinct components: A sodium-chloride component and a calcium component. *Brain Research*, *650*, 49–55.

Dezaki, K., Maeno, E., Sato, K., Akita, T., & Okada, Y. (2012). Early-phase occurrence of K^+ and Cl^- efflux in addition to Ca^{2+} mobilization is a prerequisite to apoptosis in HeLa cells. *Apoptosis*, *17*, 821–831.

Dezaki, K., Tsumura, T., Maeno, E., & Okada, Y. (2000). Receptor-mediated facilitation of cell volume regulation by swelling-induced ATP release in human epithelial cells. *The Japanese Journal of Physiology*, *50*, 235–241.

Diaz, R. J., Armstrong, S. C., Batthish, M., Backx, P. H., Ganote, C. E., & Wilson, G. J. (2003). Enhanced cell volume regulation: A key protective mechanism of ischemic preconditioning in rabbit ventricular myocytes. *Journal of Molecular and Cellular Cardiology*, *35*, 45–58.

Diaz, R. J., Harvey, K., Boloorchi, A., Hossain, T., Hinek, A., Backx, P. H., et al. (2014). Enhanced cell volume regulation: A key mechanism in local and remote ischemic preconditioning. *American Journal of Physiology. Cell Physiology*, *306*, C1191–C1199.

Diaz, R. J., Hinek, A., & Wilson, G. J. (2010). Direct evidence of chloride ion efflux in ischaemic and pharmacological preconditioning of cultured cardiomyocytes. *Cardiovascular Research*, *87*, 545–551.

Diaz, R. J., Losito, V. A., Mao, G. D., Ford, M. K., Backx, P. H., & Wilson, G. J. (1999). Chloride channel inhibition blocks the protection of ischemic preconditioning and hypo-osmotic stress in rabbit ventricular myocardium. *Circulation Research*, *84*, 763–775.

Dick, D. A. (1959). Osmotic properties of living cells. *International Review of Cytology*, *8*, 387–448.

Dirnagl, U., Iadecola, C., & Moskowitz, M. A. (1999). Pathobiology of ischaemic stroke: An integrated view. *Trends in Neurosciences*, *22*, 391–397.

Dissing, S., Hansen, H. J., Unden, M., & Nauntofte, B. (1990). Inhibitory effects of amitriptyline on the stimulation-induced Ca^{2+} increase in parotid acini. *European Journal of Pharmacology*, *177*, 43–54.

Drewnowska, K., & Baumgarten, C. M. (1991). Regulation of cellular volume in rabbit ventricular myocytes: Bumetanide, chlorothiazide, and ouabain. *The American Journal of Physiology*, *260*, C122–C131.

Du, X. Y., & Sorota, S. (1997). Modulation of dog atrial swelling-induced chloride current by cAMP: Protein kinase A-dependent and -independent pathways. *The Journal of Physiology*, *500*(Pt. 1), 111–122.

Du, X. Y., & Sorota, S. (1999). Protein kinase C stimulates swelling-induced chloride current in canine atrial cells. *Pflügers Archiv*, *437*, 227–234.

Du, X. Y., & Sorota, S. (2000). Cardiac swelling-induced chloride current is enhanced by endothelin. *Journal of Cardiovascular Pharmacology*, *35*, 769–776.

Duan, D., Fermini, B., & Nattel, S. (1995). Alpha-adrenergic control of volume-regulated Cl^- currents in rabbit atrial myocytes. Characterization of a novel ionic regulatory mechanism. *Circulation Research*, *77*, 379–393.

Duan, D., Hume, J. R., & Nattel, S. (1997). Evidence that outwardly rectifying Cl^- channels underlie volume-regulated Cl- currents in heart. *Circulation Research*, *80*, 103–113.

Dubinsky, J. M., Kristal, B. S., & Elizondo-Fournier, M. (1995). An obligate role for oxygen in the early stages of glutamate-induced, delayed neuronal death. *The Journal of Neuroscience*, *15*, 7071–7078.

Dutta, A. K., Korchev, Y. E., Shevchuk, A. I., Hayashi, S., Okada, Y., & Sabirov, R. Z. (2008). Spatial distribution of maxi-anion channel on cardiomyocytes detected by smart-patch technique. *Biophysical Journal*, *94*, 1646–1655.

Dutta, A. K., Okada, Y., & Sabirov, R. Z. (2002). Regulation of an ATP-conductive large-conductance anion channel and swelling-induced ATP release by arachidonic acid. *The Journal of Physiology*, *542*, 803–816.

Dutta, A. K., Sabirov, R. Z., Uramoto, H., & Okada, Y. (2004). Role of ATP-conductive anion channel in ATP release from neonatal rat cardiomyocytes in ischaemic or hypoxic conditions. *The Journal of Physiology, 559*, 799–812.

Duvall, E., & Wyllie, A. H. (1986). Death and the cell. *Immunology Today, 7*, 115–119.

Ehara, T., & Ishihara, K. (1990). Anion channels activated by adrenaline in cardiac myocytes. *Nature, 347*, 284–286.

Elinder, F., Akanda, N., Tofighi, R., Shimizu, S., Tsujimoto, Y., Orrenius, S., et al. (2005). Opening of plasma membrane voltage-dependent anion channels (VDAC) precedes caspase activation in neuronal apoptosis induced by toxic stimuli. *Cell Death and Differentiation, 12*, 1134–1140.

Elliott, M. R., Chekeni, F. B., Trampont, P. C., Lazarowski, E. R., Kadl, A., Walk, S. F., et al. (2009). Nucleotides released by apoptotic cells act as a find-me signal to promote phagocytic clearance. *Nature, 461*, 282–286.

Elliott, J. I., & Higgins, C. F. (2003). IKCa1 activity is required for cell shrinkage, phosphatidylserine translocation and death in T lymphocyte apoptosis. *EMBO Reports, 4*, 189–194.

Elorza-Vidal, X., Sirisi, S., Gaitan-Penas, H., Perez-Rius, C., Alonso-Gardon, M., Armand-Ugon, M., et al. (2018). GlialCAM/MLC1 modulates LRRC8/VRAC currents in an indirect manner: Implications for megalencephalic leukoencephalopathy. *Neurobiology of Disease, 119*, 88–99.

Erdo, S. L., Cai, N., & Lakics, V. (1993). Medium-dependent dissociation of cytotoxic and GABA-releasing effects of N-methyl-D-aspartate (NMDA), alpha-amino-3-hydroxy-5-methyl-4-isoxazolepropionate (AMPA) and kainate in rat cortical cultures. *Neuroscience Letters, 152*, 84–86.

Eriksson, P. S., Nilsson, M., Wagberg, M., Ronnback, L., & Hansson, E. (1992). Volume regulation of single astroglial cells in primary culture. *Neuroscience Letters, 143*, 195–199.

Erlinge, D., Harnek, J., van Heusden, C., Olivecrona, G., Jern, S., & Lazarowski, E. (2005). Uridine triphosphate (UTP) is released during cardiac ischemia. *International Journal of Cardiology, 100*, 427–433.

Ernest, N. J., Habela, C. W., & Sontheimer, H. (2008). Cytoplasmic condensation is both necessary and sufficient to induce apoptotic cell death. *Journal of Cell Science, 121*, 290–297.

Ernest, N. J., & Sontheimer, H. (2007). Extracellular glutamine is a critical modulator for regulatory volume increase in human glioma cells. *Brain Research, 1144*, 231–238.

Falke, L. C., & Misler, S. (1989). Activity of ion channels during volume regulation by clonal N1E115 neuroblastoma cells. *Proceedings of the National Academy of Sciences of the United States of America, 86*, 3919–3923.

Fields, R. D., & Ni, Y. (2010). Nonsynaptic communication through ATP release from volume-activated anion channels in axons. *Science Signaling, 3*, ra73.

Fink, S. L., & Cookson, B. T. (2006). Caspase-1-dependent pore formation during pyroptosis leads to osmotic lysis of infected host macrophages. *Cellular Microbiology, 8*, 1812–1825.

Fliss, H., & Gattinger, D. (1996). Apoptosis in ischemic and reperfused rat myocardium. *Circulation Research, 79*, 949–956.

Formaggio, F., Saracino, E., Mola, M. G., Rao, S. B., Amiry-Moghaddam, M., Muccini, M., et al. (2019). LRRC8A is essential for swelling-activated chloride current and for regulatory volume decrease in astrocytes. *The FASEB Journal, 33*, 101–113.

Formigli, L., Papucci, L., Tani, A., Schiavone, N., Tempestini, A., Orlandini, G. E., et al. (2000). Aponecrosis: Morphological and biochemical exploration of a syncretic process of cell death sharing apoptosis and necrosis. *Journal of Cellular Physiology, 182*, 41–49.

Forrester, T., & Williams, C. A. (1977). Release of adenosine triphosphate from isolated adult heart cells in response to hypoxia. *The Journal of Physiology, 268*, 371–390.

Forshaw, P. J., Lister, T., & Ray, D. E. (1993). Inhibition of a neuronal voltage-dependent chloride channel by the type II pyrethroid, deltamethrin. *Neuropharmacology, 32*, 105–111.

Forshaw, P. J., Lister, T., & Ray, D. E. (2000). The role of voltage-gated chloride channels in type II pyrethroid insecticide poisoning. *Toxicology and Applied Pharmacology, 163*, 1–8.

Foskett, J. K. (1990). $[Ca^{2+}]_i$ modulation of Cl^- content controls cell volume in single salivary acinar cells during fluid secretion. *The American Journal of Physiology, 259*, C998–1004.

Franke, H., Verkhratsky, A., Burnstock, G., & Illes, P. (2012). Pathophysiology of astroglial purinergic signalling. *Purinergic Signal, 8*, 629–657.

Fraser, C. L., & Swanson, R. A. (1994). Female sex hormones inhibit volume regulation in rat brain astrocyte culture. *The American Journal of Physiology, 267*, C909–C914.

Fujii, R., Mutoh, M., Niwa, K., Yamada, K., Aikou, T., Nakagawa, M., et al. (1994). Active efflux system for cisplatin in cisplatin-resistant human KB cells. *Japanese Journal of Cancer Research, 85*, 426–433.

Funakoshi, T., Aki, T., Tajiri, M., Unuma, K., & Uemura, K. (2016). Necroptosis-like neuronal cell death caused by cellular cholesterol accumulation. *The Journal of Biological Chemistry, 291*, 25050–25065.

Gaitan-Penas, H., Gradogna, A., Laparra-Cuervo, L., Solsona, C., Fernandez-Duenas, V., Barrallo-Gimeno, A., et al. (2016). Investigation of LRRC8-mediated volume-regulated anion currents in Xenopus oocytes. *Biophysical Journal, 111*, 1429–1443.

Ganote, C. E., & Armstrong, S. C. (2003). Effects of CCCP-induced mitochondrial uncoupling and cyclosporin A on cell volume, cell injury and preconditioning protection of isolated rabbit cardiomyocytes. *Journal of Molecular and Cellular Cardiology, 35*, 749–759.

Gerard, V., Rouzaire-Dubois, B., Dilda, P., & Dubois, J. M. (1998). Alterations of ionic membrane permeabilities in multidrug-resistant neuroblastoma x glioma hybrid cells. *The Journal of Experimental Biology, 201*, 21–31.

Gibson, J. S., Ellory, J. C., Adragna, N. C., & Lauf, P. K. (2009). In F. J. Alvarez-Leefmans & E. Delpire (Eds.), *Physiology and pathology of chloride transporters and channels in the nervous system* (pp. 27–42). San Diego: Elsevier.

Gill, J. S., & Windebank, A. J. (1998). Cisplatin-induced apoptosis in rat dorsal root ganglion neurons is associated with attempted entry into the cell cycle. *The Journal of Clinical Investigation, 101*, 2842–2850.

Golan, O., Issan, Y., Isak, A., Leipziger, J., Robaye, B., & Shainberg, A. (2011). Extracellular nucleotide derivatives protect cardiomyocytes against hypoxic stress. *Biochemical Pharmacology, 81*, 1219–1227.

Goldberg, M. P., & Choi, D. W. (1993). Combined oxygen and glucose deprivation in cortical cell culture: Calcium-dependent and calcium-independent mechanisms of neuronal injury. *The Journal of Neuroscience, 13*, 3510–3524.

Goldman, S. A., Pulsinelli, W. A., Clarke, W. Y., Kraig, R. P., & Plum, F. (1989). The effects of extracellular acidosis on neurons and glia in vitro. *Journal of Cerebral Blood Flow and Metabolism, 9*, 471–477.

Gollapudi, S., McDonald, T., Gardner, P., Kang, N., & Gupta, S. (1992). Abnormal chloride conductance in multidrug resistant HL60/AR cells. *Cancer Letters, 66*, 83–89.

Gong, W., Xu, H., Shimizu, T., Morishima, S., Tanabe, S., Tachibe, T., et al. (2004). ClC-3-independent, PKC-dependent activity of volume-sensitive Cl channel in mouse ventricular cardiomyocytes. *Cellular Physiology and Biochemistry, 14*, 213–224.

Gradogna, A., Gaitan-Penas, H., Boccaccio, A., Estevez, R., & Pusch, M. (2017). Cisplatin activates volume sensitive LRRC8 channel mediated currents in Xenopus oocytes. *Channels (Austin, Tex.), 11*, 254–260.

Griffiths, J. B., Cox, C. S., Beadle, D. J., Hunt, C. J., & Reid, D. S. (1979). Changes in cell size during the cooling, warming and post-thawing periods of the freeze-thaw cycle. *Cryobiology*, *16*, 141–151.

Groger, M., Lebesgue, D., Pruneau, D., Relton, J., Kim, S. W., Nussberger, J., et al. (2005). Release of bradykinin and expression of kinin B2 receptors in the brain: Role for cell death and brain edema formation after focal cerebral ischemia in mice. *Journal of Cerebral Blood Flow and Metabolism*, *25*, 978–989.

Groulx, N., Boudreault, F., Orlov, S. N., & Grygorczyk, R. (2006). Membrane reserves and hypotonic cell swelling. *The Journal of Membrane Biology*, *214*, 43–56.

Gulbins, E., Welsch, J., Lepple-Wienhuis, A., Heinle, H., & Lang, F. (1997). Inhibition of Fas-induced apoptotic cell death by osmotic cell shrinkage. *Biochemical and Biophysical Research Communications*, *236*, 517–521.

Hagiwara, N., Masuda, H., Shoda, M., & Irisawa, H. (1992). Stretch-activated anion currents of rabbit cardiac myocytes. *The Journal of Physiology*, *456*, 285–302.

Hall, S. K., Zhang, J., & Lieberman, M. (1995). Cyclic AMP prevents activation of a swelling-induced chloride-sensitive conductance in chick heart cells. *The Journal of Physiology*, *488*(Pt. 2), 359–369.

Hallback, D. A., Jodal, M., & Lundgren, O. (1979). Importance of sodium and glucose for the establishment of a villous tissue hyperosmolality by the intestinal countercurrent multiplier. *Acta Physiologica Scandinavica*, *107*, 89–96.

Hammarstrom, A. K., & Gage, P. W. (1998). Inhibition of oxidative metabolism increases persistent sodium current in rat CA1 hippocampal neurons. *The Journal of Physiology*, *510*(Pt. 3), 735–741.

Hammarstrom, A. K., & Gage, P. W. (2000). Oxygen-sensing persistent sodium channels in rat hippocampus. *The Journal of Physiology*, *529*(Pt. 1), 107–118.

Han, Q., Liu, S., Li, Z., Hu, F., Zhang, Q., Zhou, M., et al. (2014). DCPIB, a potent volume-regulated anion channel antagonist, attenuates microglia-mediated inflammatory response and neuronal injury following focal cerebral ischemia. *Brain Research*, *1542*, 176–185.

Hansen, A. J., & Zeuthen, T. (1981). Extracellular ion concentrations during spreading depression and ischemia in the rat brain cortex. *Acta Physiologica Scandinavica*, *113*, 437–445.

Hara, Y., Wakamori, M., Ishii, M., Maeno, E., Nishida, M., Yoshida, T., et al. (2002). LTRPC2 Ca^{2+}-permeable channel activated by changes in redox status confers susceptibility to cell death. *Molecular Cell*, *9*, 163–173.

Harrigan, T. J., Abdullaev, I. F., Jourd'heuil, D., & Mongin, A. A. (2008). Activation of microglia with zymosan promotes excitatory amino acid release via volume-regulated anion channels: The role of NADPH oxidases. *Journal of Neurochemistry*, *106*, 2449–2462.

Harris, K. M., & Miller, R. J. (1989). Excitatory amino acid-evoked release of [3H]GABA from hippocampal neurons in primary culture. *Brain Research*, *482*, 23–33.

Harvey, R. D., & Hume, J. R. (1989). Autonomic regulation of a chloride current in heart. *Science*, *244*, 983–985.

Hasbani, M. J., Hyrc, K. L., Faddis, B. T., Romano, C., & Goldberg, M. P. (1998). Distinct roles for sodium, chloride, and calcium in excitotoxic dendritic injury and recovery. *Experimental Neurology*, *154*, 241–258.

Hasegawa, Y., Shimizu, T., Takahashi, N., & Okada, Y. (2012). The apoptotic volume decrease is an upstream event of MAP kinase activation during Staurosporine-induced apoptosis in HeLa cells. *International Journal of Molecular Sciences*, *13*, 9363–9379.

Haskew-Layton, R. E., Mongin, A. A., & Kimelberg, H. K. (2005). Hydrogen peroxide potentiates volume-sensitive excitatory amino acid release via a mechanism involving Ca^{2+}/calmodulin-dependent protein kinase II. *The Journal of Biological Chemistry*, *280*, 3548–3554.

Hazama, A., Fan, H. T., Abdullaev, I., Maeno, E., Tanaka, S., Ando-Akatsuka, Y., et al. (2000). Swelling-activated, cystic fibrosis transmembrane conductance regulator-augmented ATP release and Cl^- conductances in murine C127 cells. *The Journal of Physiology, 523*(Pt. 1), 1–11.
Hazama, A., & Okada, Y. (1988). Ca^{2+} sensitivity of volume-regulatory K^+ and Cl^- channels in cultured human epithelial cells. *The Journal of Physiology, 402*, 687–702.
Hazama, A., & Okada, Y. (1990). Involvement of Ca^{2+}-induced Ca^{2+} release in the volume regulation of human epithelial cells exposed to a hypotonic medium. *Biochemical and Biophysical Research Communications, 167*, 287–293.
He, W., Li, H., Min, X., Liu, J., Hu, B., Hou, S., et al. (2010). Activation of volume-sensitive Cl- channel is involved in carboplatin-induced apoptosis in human lung adenocarcinoma cells. *Cancer Biology & Therapy, 9*, 885–891.
He, W. T., Wan, H., Hu, L., Chen, P., Wang, X., Huang, Z., et al. (2015). Gasdermin D is an executor of pyroptosis and required for interleukin-1beta secretion. *Cell Research, 25*, 1285–1298.
Hernandez-Benitez, R., Sedeno-Cortes, A., Ramos-Mandujano, G., & Pasantes-Morales,-H. (2014). Regulatory volume decrease in neural precursor cells: Taurine efflux and gene microarray analysis. *Cellular Physiology and Biochemistry, 34*, 2038–2048.
Heusch, G., Liu, G. S., Rose, J., Cohen, M. V., & Downey, J. M. (2000). No confirmation for a causal role of volume-regulated chloride channels in ischemic preconditioning in rabbits. *Journal of Molecular and Cellular Cardiology, 32*, 2279–2285.
Hinton, B. T., Pryor, J. P., Hirsh, A. V., & Setchell, B. P. (1981). The concentration of some inorganic ions and organic compounds in the luminal fluid of the human ductus deferens. *International Journal of Andrology, 4*, 457–461.
Hirashima, Y., Kurimoto, M., Nogami, K., Endo, S., Saitoh, M., Ohtani, O., et al. (1999). Correlation of glutamate-induced apoptosis with caspase activities in cultured rat cerebral cortical neurons. *Brain Research, 849*, 109–118.
Hochhauser, E., Cohen, R., Waldman, M., Maksin, A., Isak, A., Aravot, D., et al. (2013). P2Y2 receptor agonist with enhanced stability protects the heart from ischemic damage in vitro and in vivo. *Purinergic Signal, 9*, 633–642.
Hoffmann, E. K., Lambert, I. H., & Pedersen, S. F. (2009). Physiology of cell volume regulation in vertebrates. *Physiological Reviews, 89*, 193–277.
Hoffmann, E. K., & Mills, J. W. (1999). Membrane events involved in volume regulation. *Current Topics in Membranes, 48*, 123–196.
Hoffmann, E. K., & Pedersen, S. F. (2006). Sensors and signal transduction pathways in vertebrate cell volume regulation. *Contributions to Nephrology, 152*, 54–104.
Hoffmann, E. K., & Simonsen, L. O. (1989). Membrane mechanisms in volume and pH regulation in vertebrate cells. *Physiological Reviews, 69*, 315–382.
Holler, N., Zaru, R., Micheau, O., Thome, M., Attinger, A., Valitutti, S., et al. (2000). Fas triggers an alternative, caspase-8-independent cell death pathway using the kinase RIP as effector molecule. *Nature Immunology, 1*, 489–495.
Hool, L. C., Middleton, L. M., & Harvey, R. D. (1998). Genistein increases the sensitivity of cardiac ion channels to beta-adrenergic receptor stimulation. *Circulation Research, 83*, 33–42.
Hori, N., & Carpenter, D. O. (1994). Functional and morphological changes induced by transient in vivo ischemia. *Experimental Neurology, 129*, 279–289.
Horie, H., Ikuta, S., Takenaka, T., & Ito, S. (1989). Adaptation of cultured mammalian neurons to a hypotonic environment with age-related response. *Brain Research, 477*, 233–240.
Hortelano, S., Zeini, M., Castrillo, A., Alvarez, A. M., & Bosca, L. (2002). Induction of apoptosis by nitric oxide in macrophages is independent of apoptotic volume decrease. *Cell Death and Differentiation, 9*, 643–650.

Hsu, M., & Buzsaki, G. (1993). Vulnerability of mossy fiber targets in the rat hippocampus to forebrain ischemia. *The Journal of Neuroscience, 13*, 3964–3979.

Hughes, F. M., Jr., & Cidlowski, J. A. (1998). Glucocorticoid-induced thymocyte apoptosis: Protease-dependent activation of cell shrinkage and DNA degradation. *The Journal of Steroid Biochemistry and Molecular Biology, 65*, 207–217.

Hyzinski-Garcia, M. C., Rudkouskaya, A., & Mongin, A. A. (2014). LRRC8A protein is indispensable for swelling-activated and ATP-induced release of excitatory amino acids in rat astrocytes. *The Journal of Physiology, 592*, 4855–4862.

Ichishima, K., Yamamoto, S., Iwamoto, T., & Ehara, T. (2010). Alpha-adrenoceptor-mediated depletion of phosphatidylinositol 4, 5-bisphosphate inhibits activation of volume-regulated anion channels in mouse ventricular myocytes. *British Journal of Pharmacology, 161*, 193–206.

Immke, D. C., & McCleskey, E. W. (2001). Lactate enhances the acid-sensing Na+ channel on ischemia-sensing neurons. *Nature Neuroscience, 4*, 869–870.

Inglefield, J. R., & Schwartz-Bloom, R. D. (1998). Optical imaging of hippocampal neurons with a chloride-sensitive dye: Early effects of in vitro ischemia. *Journal of Neurochemistry, 70*, 2500–2509.

Inoue, H., Mori, S., Morishima, S., & Okada, Y. (2005). Volume-sensitive chloride channels in mouse cortical neurons: Characterization and role in volume regulation. *The European Journal of Neuroscience, 21*, 1648–1658.

Inoue, H., Ohtaki, H., Nakamachi, T., Shioda, S., & Okada, Y. (2007). Anion channel blockers attenuate delayed neuronal cell death induced by transient forebrain ischemia. *Journal of Neuroscience Research, 85*, 1427–1435.

Isaacks, R. E., Bender, A. S., Kim, C. Y., Prieto, N. M., & Norenberg, M. D. (1994). Osmotic regulation of myo-inositol uptake in primary astrocyte cultures. *Neurochemical Research, 19*, 331–338.

Ise, T., Shimizu, T., Lee, E. L., Inoue, H., Kohno, K., & Okada, Y. (2005). Roles of volume-sensitive Cl- channel in cisplatin-induced apoptosis in human epidermoid cancer cells. *The Journal of Membrane Biology, 205*, 139–145.

Islam, M. R., Uramoto, H., Okada, T., Sabirov, R. Z., & Okada, Y. (2012). Maxi-anion channel and pannexin 1 hemichannel constitute separate pathways for swelling-induced ATP release in murine L929 fibrosarcoma cells. *American Journal of Physiology Cell Physiology, 303*, C924–C935.

Iwasa, K., Tasaki, I., & Gibbons, R. C. (1980). Swelling of nerve fibers associated with action potentials. *Science, 210*, 338–339.

Jackson, P. S., Morrison, R., & Strange, K. (1994). The volume-sensitive organic osmolyte-anion channel VSOAC is regulated by nonhydrolytic ATP binding. *The American Journal of Physiology, 267*, C1203–C1209.

Jackson, P. S., & Strange, K. (1993). Volume-sensitive anion channels mediate swelling-activated inositol and taurine efflux. *The American Journal of Physiology, 265*, C1489–C1500.

Jakubovicz, D. E., Grinstein, S., & Klip, A. (1987). Cell swelling following recovery from acidification in C6 glioma cells: An in vitro model of postischemic brain edema. *Brain Research, 435*, 138–146.

Jakubovicz, D. E., & Klip, A. (1989). Lactic acid-induced swelling in C6 glial cells via Na^+/H^+ exchange. *Brain Research, 485*, 215–224.

Jalonen, T. (1993). Single-channel characteristics of the large-conductance anion channel in rat cortical astrocytes in primary culture. *Glia, 9*, 227–237.

Jalonen, T., Johansson, S., Holopainen, I., Oja, S. S., & Arhem, P. (1989). A high-conductance multi-state anion channel in cultured rat astrocytes. *Acta Physiologica Scandinavica, 136*, 611–612.

Jamieson, E. R., & Lippard, S. J. (1999). Structure, recognition, and processing of cisplatin-DNA adducts. *Chemical Reviews*, *99*, 2467–2498.

Jeftinija, S. D., Jeftinija, K. V., Stefanovic, G., & Liu, F. (1996). Neuroligand-evoked calcium-dependent release of excitatory amino acids from cultured astrocytes. *Journal of Neurochemistry*, *66*, 676–684.

Jenner, P. (2003). Oxidative stress in Parkinson's disease. *Annals of Neurology*, *53*(Suppl. 3), S26–S36. discussion S36-28.

Jennings, R. B., & Ganote, C. E. (1974). Structural changes in myocardium during acute ischemia. *Circulation Research*, *35*(Suppl. 3), 156–172.

Jeroudi, M. O., Hartley, C. J., & Bolli, R. (1994). Myocardial reperfusion injury: Role of oxygen radicals and potential therapy with antioxidants. *The American Journal of Cardiology*, *73*, 2b–7b.

Ji, E. S., Yue, H., Wu, Y. M., & He, R. R. (2004). Effects of phytoestrogen genistein on myocardial ischemia/reperfusion injury and apoptosis in rabbits. *Acta Pharmacologica Sinica*, *25*, 306–312.

Jodal, M., Hallback, D. A., & Lundgren, O. (1978). Tissue osmolality in intestinal villi during luminal perfusion with isotonic electrolyte solutions. *Acta Physiologica Scandinavica*, *102*, 94–107.

Johnstone, R. W. (2002). Histone-deacetylase inhibitors: Novel drugs for the treatment of cancer. *Nature Reviews. Drug Discovery*, *1*, 287–299.

Joo, C. K., Choi, J. S., Ko, H. W., Park, K. Y., Sohn, S., Chun, M. H., et al. (1999). Necrosis and apoptosis after retinal ischemia: Involvement of NMDA-mediated excitotoxicity and p53. *Investigative Ophthalmology & Visual Science*, *40*, 713–720.

Ju, Y. K., Saint, D. A., & Gage, P. W. (1996). Hypoxia increases persistent sodium current in rat ventricular myocytes. *The Journal of Physiology*, *497*(Pt. 2), 337–347.

Juranyi, Z., Sperlagh, B., & Vizi, E. S. (1999). Involvement of P2 purinoceptors and the nitric oxide pathway in [3H]purine outflow evoked by short-term hypoxia and hypoglycemia in rat hippocampal slices. *Brain Research*, *823*, 183–190.

Kaczmarek, A., Vandenabeele, P., & Krysko, D. V. (2013). Necroptosis: The release of damage-associated molecular patterns and its physiological relevance. *Immunity*, *38*, 209–223.

Kajstura, J., Cheng, W., Reiss, K., Clark, W. A., Sonnenblick, E. H., Krajewski, S., et al. (1996). Apoptotic and necrotic myocyte cell deaths are independent contributing variables of infarct size in rats. *Laboratory Investigation*, *74*, 86–107.

Kaneko, S., Kawakami, S., Hara, Y., Wakamori, M., Itoh, E., Minami, T., et al. (2006). A critical role of TRPM2 in neuronal cell death by hydrogen peroxide. *Journal of Pharmacological Sciences*, *101*, 66–76.

Karwatowska-Krynska, E., & Beresewicz, A. (1983). Effect of locally released catecholamines on lipolysis and injury of the hypoxic isolated rabbit heart. *Journal of Molecular and Cellular Cardiology*, *15*, 523–536.

Kasim, N. R., Kuzelova, K., Holoubek, A., & Model, M. A. (2013). Live fluorescence and transmission-through-dye microscopic study of actinomycin D-induced apoptosis and apoptotic volume decrease. *Apoptosis*, *18*, 521–532.

Kawahara, K., & Takuwa, N. (1991). Bombesin activates large-conductance chloride channels in swiss 3T3 fibroblasts. *Biochemical and Biophysical Research Communications*, *177*, 292–298.

Kayagaki, N., Stowe, I. B., Lee, B. L., O'Rourke, K., Anderson, K., Warming, S., et al. (2015). Caspase-11 cleaves gasdermin D for non-canonical inflammasome signalling. *Nature*, *526*, 666–671.

Kelland, L. (2007). The resurgence of platinum-based cancer chemotherapy. *Nature Reviews. Cancer*, *7*, 573–584.

Kerr, J. F., Wyllie, A. H., & Currie, A. R. (1972). Apoptosis: A basic biological phenomenon with wide-ranging implications in tissue kinetics. *British Journal of Cancer*, *26*, 239–257.

Kida, H., Miyoshi, T., Manabe, K., Takahashi, N., Konno, T., Ueda, S., et al. (2005). Roles of aquaporin-3 water channels in volume-regulatory water flow in a human epithelial cell line. *The Journal of Membrane Biology*, *208*, 55–64.

Kim, M. S., Blake, M., Baek, J. H., Kohlhagen, G., Pommier, Y., & Carrier, F. (2003). Inhibition of histone deacetylase increases cytotoxicity to anticancer drugs targeting DNA. *Cancer Research*, *63*, 7291–7300.

Kimelberg, H. K. (1987). Anisotonic media and glutamate-induced ion transport and volume responses in primary astrocyte cultures. *Journal of Physiology, Paris*, *82*, 294–303.

Kimelberg, H. K., & Frangakis, M. V. (1985). Furosemide- and bumetanide-sensitive ion transport and volume control in primary astrocyte cultures from rat brain. *Brain Research*, *361*, 125–134.

Kimelberg, H. K., Macvicar, B. A., & Sontheimer, H. (2006). Anion channels in astrocytes: Biophysics, pharmacology, and function. *Glia*, *54*, 747–757.

Klassen, N. V., Walker, P. R., Ross, C. K., Cygler, J., & Lach, B. (1993). Two-stage cell shrinkage and the OER for radiation-induced apoptosis of rat thymocytes. *International Journal of Radiation Biology*, *64*, 571–581.

Koos, B., Christmann, J., Plettenberg, S., Kading, D., Becker, J., Keteku, M., et al. (2018). Hypertonicity-induced cation channels in HepG2 cells: Architecture and role in proliferation vs. apoptosis. *The Journal of Physiology*, *596*, 1227–1241.

Kreisman, N. R., & Olson, J. E. (2003). Taurine enhances volume regulation in hippocampal slices swollen osmotically. *Neuroscience*, *120*, 635–642.

Kristian, T., & Siesjo, B. K. (1998). Calcium in ischemic cell death. *Stroke*, *29*, 705–718.

Kroemer, G., Dallaporta, B., & Resche-Rigon, M. (1998). The mitochondrial death/life regulator in apoptosis and necrosis. *Annual Review of Physiology*, *60*, 619–642.

Kroemer, G., & Reed, J. C. (2000). Mitochondrial control of cell death. *Nature Medicine*, *6*, 513–519.

Kubo, M., & Okada, Y. (1992). Volume-regulatory Cl- channel currents in cultured human epithelial cells. *The Journal of Physiology*, *456*, 351–371.

Kumagai, K., Imai, S., Toyoda, F., Okumura, N., Isoya, E., Matsuura, H., et al. (2012). 17beta-Oestradiol inhibits doxorubicin-induced apoptosis via block of the volume-sensitive Cl^- current in rabbit articular chondrocytes. *British Journal of Pharmacology*, *166*, 702–720.

Kumagai, K., Toyoda, F., Staunton, C. A., Maeda, T., Okumura, N., Matsuura, H., et al. (2016). Activation of a chondrocyte volume-sensitive Cl^- conductance prior to macroscopic cartilage lesion formation in the rabbit knee anterior cruciate ligament transection osteoarthritis model. *Osteoarthritis and Cartilage*, *24*, 1786–1794.

Kunugi, S., Iwabuchi, S., Matsuyama, D., Okajima, T., & Kawahara, K. (2011). Negative-feedback regulation of ATP release: ATP release from cardiomyocytes is strictly regulated during ischemia. *Biochemical and Biophysical Research Communications*, *416*, 409–415.

Kurbannazarova, R. S., Bessonova, S. V., Okada, Y., & Sabirov, R. Z. (2011). Swelling-activated anion channels are essential for volume regulation of mouse thymocytes. *International Journal of Molecular Sciences*, *12*, 9125–9137.

Kurbel, S., Dodig, K., & Radic, R. (2002). The osmotic gradient in kidney medulla: A retold story. *Advances in Physiology Education*, *26*, 278–281.

Kurrelmeyer, K. M., Michael, L. H., Baumgarten, G., Taffet, G. E., Peschon, J. J., Sivasubramanian, N., et al. (2000). Endogenous tumor necrosis factor protects the adult cardiac myocyte against ischemic-induced apoptosis in a murine model of acute myocardial infarction. *Proceedings of the National Academy of Sciences of the United States of America*, *97*, 5456–5461.

Kuzmin, A. I., Gourine, A. V., Molosh, A. I., Lakomkin, V. L., & Vassort, G. (2000). Effects of preconditioning on myocardial interstitial levels of ATP and its catabolites during regional ischemia and reperfusion in the rat. *Basic Research in Cardiology, 95*, 127–136.

Kuzmin, A. I., Lakomkin, V. L., Kapelko, V. I., & Vassort, G. (1998). Interstitial ATP level and degradation in control and postmyocardial infarcted rats. *The American Journal of Physiology, 275*, C766–C771.

Lambert, S., & Oberwinkler, J. (2005). Characterization of a proton-activated, outwardly rectifying anion channel. *The Journal of Physiology, 567*, 191–213.

Lang, F., Busch, G. L., Ritter, M., Volkl, H., Waldegger, S., Gulbins, E., et al. (1998). Functional significance of cell volume regulatory mechanisms. *Physiological Reviews, 78*, 247–306.

Lang, F., Madlung, J., Bock, J., Lukewille, U., Kaltenbach, S., Lang, K. S., et al. (2000). Inhibition of jurkat-T-lymphocyte Na^+/H^+-exchanger by CD95(Fas/Apo-1)-receptor stimulation. *Pflügers Archiv, 440*, 902–907.

Lascola, C. D., & Kraig, R. P. (1996). Whole-cell chloride currents in rat astrocytes accompany changes in cell morphology. *The Journal of Neuroscience, 16*, 2532–2545.

Lascola, C. D., Nelson, D. J., & Kraig, R. P. (1998). Cytoskeletal actin gates a Cl^- channel in neocortical astrocytes. *The Journal of Neuroscience, 18*, 1679–1692.

Lauf, P. K., & Adragna, N. C. (2000). K-Cl cotransport: Properties and molecular mechanism. *Cellular Physiology and Biochemistry, 10*, 341–354.

Lauf, P. K., & Bauer, J. (1987). Direct evidence for chloride-dependent volume reduction in macrocytic sheep reticulocytes. *Biochemical and Biophysical Research Communications, 144*, 849–855.

Lazarowski, E. R. (2012). Vesicular and conductive mechanisms of nucleotide release. *Purinergic Signal, 8*, 359–373.

Leaf, A. (1956). On the mechanism of fluid exchange of tissues in vitro. *The Biochemical Journal, 62*, 241–248.

Leaney, J. L., Marsh, S. J., & Brown, D. A. (1997). A swelling-activated chloride current in rat sympathetic neurones. *The Journal of Physiology, 501*(Pt. 3), 555–564.

Lee, R. J., & Foskett, J. K. (2010). Mechanisms of Ca^{2+}-stimulated fluid secretion by porcine bronchial submucosal gland serous acinar cells. *American Journal of Physiology. Lung Cellular and Molecular Physiology, 298*, L210–L231.

Lee, C. C., Freinkman, E., Sabatini, D. M., & Ploegh, H. L. (2014). The protein synthesis inhibitor blasticidin s enters mammalian cells via leucine-rich repeat-containing protein 8D. *The Journal of Biological Chemistry, 289*, 17124–17131.

Lee, E. L., Shimizu, T., Ise, T., Numata, T., Kohno, K., & Okada, Y. (2007). Impaired activity of volume-sensitive Cl- channel is involved in cisplatin resistance of cancer cells. *Journal of Cellular Physiology, 211*, 513–521.

Levesque, P. C., Hart, P. J., Hume, J. R., Kenyon, J. L., & Horowitz, B. (1992). Expression of cystic fibrosis transmembrane regulator Cl^- channels in heart. *Circulation Research, 71*, 1002–1007.

Levitan, I., & Garber, S. S. (1997). Volume regulated anion channels and cytoskeletal interactions. In R. Latorre & J. C. Saez (Eds.), *Ion channels to cell-to-cell conversations* (pp. 245–267). New York: Plenum Press.

l'Hoste, S., Chargui, A., Belfodil, R., Corcelle, E., Duranton, C., Rubera, I., et al. (2010). CFTR mediates apoptotic volume decrease and cell death by controlling glutathione efflux and ROS production in cultured mice proximal tubules. *American Journal of Physiology Renal Physiology, 298*, F435–F453.

Li, J., De Smet, P., Jans, D., Simaels, J., & Van Driessche, W. (1998). Swelling-activated cation-selective channels in A6 epithelia are permeable to large cations. *The American Journal of Physiology, 275*, C358–C366.

Li, M., Jiang, J., & Yue, L. (2006). Functional characterization of homo- and heteromeric channel kinases TRPM6 and TRPM7. *The Journal of General Physiology*, *127*, 525–537.

Li, Q., Li, B., Wang, X., Leri, A., Jana, K. P., Liu, Y., et al. (1997). Overexpression of insulin-like growth factor-1 in mice protects from myocyte death after infarction, attenuating ventricular dilation, wall stress, and cardiac hypertrophy. *The Journal of Clinical Investigation*, *100*, 1991–1999.

Li, G., & Olson, J. E. (2004). Extracellular ATP activates chloride and taurine conductances in cultured hippocampal neurons. *Neurochemical Research*, *29*, 239–246.

Light, D. B., Schwiebert, E. M., Fejes-Toth, G., Naray-Fejes-Toth, A., Karlson, K. H., McCann, F. V., et al. (1990). Chloride channels in the apical membrane of cortical collecting duct cells. *The American Journal of Physiology*, *258*, F273–F280.

Lipton, P. (1999). Ischemic cell death in brain neurons. *Physiological Reviews*, *79*, 1431–1568.

Liu, H. T., Akita, T., Shimizu, T., Sabirov, R. Z., & Okada, Y. (2009). Bradykinin-induced astrocyte-neuron signalling: Glutamate release is mediated by ROS-activated volume-sensitive outwardly rectifying anion channels. *The Journal of Physiology*, *587*, 2197–2209.

Liu, A. H., Cao, Y. N., Liu, H. T., Zhang, W. W., Liu, Y., Shi, T. W., et al. (2008). DIDS attenuates staurosporine-induced cardiomyocyte apoptosis by PI3K/Akt signaling pathway: Activation of eNOS/NO and inhibition of Bax translocation. *Cellular Physiology and Biochemistry*, *22*, 177–186.

Liu, H. T., Sabirov, R. Z., & Okada, Y. (2008). Oxygen-glucose deprivation induces ATP release via maxi-anion channels in astrocytes. *Purinergic Signal*, *4*, 147–154.

Liu, X., & Schnellmann, R. G. (2003). Calpain mediates progressive plasma membrane permeability and proteolysis of cytoskeleton-associated paxillin, talin, and vinculin during renal cell death. *The Journal of Pharmacology and Experimental Therapeutics*, *304*, 63–70.

Liu, H. T., Tashmukhamedov, B. A., Inoue, H., Okada, Y., & Sabirov, R. Z. (2006). Roles of two types of anion channels in glutamate release from mouse astrocytes under ischemic or osmotic stress. *Glia*, *54*, 343–357.

Liu, H. T., Toychiev, A. H., Takahashi, N., Sabirov, R. Z., & Okada, Y. (2008). Maxi-anion channel as a candidate pathway for osmosensitive ATP release from mouse astrocytes in primary culture. *Cell Research*, *18*, 558–565.

Lomneth, R., Medrano, S., & Gruenstein, E. I. (1990). The role of transmembrane pH gradients in the lactic acid induced swelling of astrocytes. *Brain Research*, *523*, 69–77.

Loo, L. S., & McNamara, J. O. (2006). Impaired volume regulation is the mechanism of excitotoxic sensitization to complement. *The Journal of Neuroscience*, *26*, 10177–10187.

Love, S. (2003). Apoptosis and brain ischaemia. *Progress in Neuro-Psychopharmacology & Biological Psychiatry*, *27*, 267–282.

Lucchesi, B. R. (1990). Myocardial ischemia, reperfusion and free radical injury. *The American Journal of Cardiology*, *65*, 14i–23i.

Lutter, D., Ullrich, F., Lueck, J. C., Kempa, S., & Jentsch, T. J. (2017). Selective transport of neurotransmitters and modulators by distinct volume-regulated LRRC8 anion channels. *Journal of Cell Science*, *130*, 1122–1133.

Macknight, A. D., & Leaf, A. (1977). Regulation of cellular volume. *Physiological Reviews*, *57*, 510–573.

Maeno, E., Ishizaki, Y., Kanaseki, T., Hazama, A., & Okada, Y. (2000). Normotonic cell shrinkage because of disordered volume regulation is an early prerequisite to apoptosis. *Proceedings of the National Academy of Sciences of the United States of America*, *97*, 9487–9492.

Maeno, E., Shimizu, T., & Okada, Y. (2006). Normotonic cell shrinkage induces apoptosis under extracellular low Cl conditions in human lymphoid and epithelial cells. *Acta Physiologica (Oxford, England)*, *187*, 217–222.

Maeno, E., Takahashi, N., & Okada, Y. (2006). Dysfunction of regulatory volume increase is a key component of apoptosis. *FEBS Letters*, *580*, 6513–6517.

Maeno, E., Tsubata, T., & Okada, Y. (2012). Apoptotic volume decrease (AVD) is independent of mitochondrial dysfunction and initiator caspase activation. *Cells*, *1*, 1156–1167.

Majno, G., & Joris, I. (1995). Apoptosis, oncosis, and necrosis. An overview of cell death. *The American Journal of Pathology*, *146*, 3–15.

Manabe, K., Shimizu, T., Morishima, S., & Okada, Y. (2004). Regulatory volume increase after secretory volume decrease in colonic epithelial cells under muscarinic stimulation. *Pflügers Archiv*, *448*, 596–604.

Marklund, L., Andersson, B., Behnam-Motlagh, P., Sandstrom, P. E., Henriksson, R., & Grankvist, K. (2004). Cellular potassium ion deprivation enhances apoptosis induced by cisplatin. *Basic & Clinical Pharmacology & Toxicology*, *94*, 245–251.

Marklund, L., Henriksson, R., & Grankvist, K. (2001). Cisplatin-induced apoptosis of mesothelioma cells is affected by potassium ion flux modulator amphotericin B and bumetanide. *International Journal of Cancer*, *93*, 577–583.

Marks, P. A., Richon, V. M., & Rifkind, R. A. (2000). Histone deacetylase inhibitors: Inducers of differentiation or apoptosis of transformed cells. *Journal of the National Cancer Institute*, *92*, 1210–1216.

Marks, P. A., & Rifkind, R. A. (1978). Erythroleukemic differentiation. *Annual Review of Biochemistry*, *47*, 419–448.

Martin, L. J., Al-Abdulla, N. A., Brambrink, A. M., Kirsch, J. R., Sieber, F. E., & Portera-Cailliau, C. (1998). Neurodegeneration in excitotoxicity, global cerebral ischemia, and target deprivation: A perspective on the contributions of apoptosis and necrosis. *Brain Research Bulletin*, *46*, 281–309.

Matesic, D. F., & Lin, R. C. (1994). Microtubule-associated protein 2 as an early indicator of ischemia-induced neurodegeneration in the gerbil forebrain. *Journal of Neurochemistry*, *63*, 1012–1020.

Matsuoka, S., Ehara, T., & Noma, A. (1990). Chloride-sensitive nature of the adrenaline-induced current in guinea-pig cardiac myocytes. *The Journal of Physiology*, *425*, 579–598.

McLarnon, J. G., & Kim, S. U. (1991). Ion channels in cultured adult human Schwann cells. *Glia*, *4*, 534–539.

McManus, M., Fischbarg, J., Sun, A., Hebert, S., & Strange, K. (1993). Laser light-scattering system for studying cell volume regulation and membrane transport processes. *The American Journal of Physiology*, *265*, C562–C570.

McNulty, S., & Fonfria, E. (2005). The role of TRPM channels in cell death. *Pflügers Archiv*, *451*, 235–242.

Meng, X. J., Carruth, M. W., & Weinman, S. A. (1997). Leukotriene D4 activates a chloride conductance in hepatocytes from lipopolysaccharide-treated rats. *The Journal of Clinical Investigation*, *99*, 2915–2922.

Min, X. J., Li, H., Hou, S. C., He, W., Liu, J., Hu, B., et al. (2011). Dysfunction of volume-sensitive chloride channels contributes to cisplatin resistance in human lung adenocarcinoma cells. *Experimental Biology and Medicine (Maywood, N.J.)*, *236*, 483–491.

Minieri, L., Pivonkova, H., Harantova, L., Anderova, M., & Ferroni, S. (2015). Intracellular Na^+ inhibits volume-regulated anion channel in rat cortical astrocytes. *Journal of Neurochemistry*, *132*, 286–300.

Minke, B., & Cook, B. (2002). TRP channel proteins and signal transduction. *Physiological Reviews*, *82*, 429–472.

Mitchell, C. H., Wang, L., & Jacob, T. J. (1997). A large-conductance chloride channel in pigmented ciliary epithelial cells activated by GTPgammaS. *The Journal of Membrane Biology*, *158*, 167–175.

Mongin, A. A., & Orlov, S. N. (2001). Mechanisms of cell volume regulation and possible nature of the cell volume sensor. *Pathophysiology*, *8*, 77–88.

Mori, S., Morishima, S., Takasaki, M., & Okada, Y. (2002). Impaired activity of volume-sensitive anion channel during lactacidosis-induced swelling in neuronally differentiated NG108-15 cells. *Brain Research*, *957*, 1–11.

Moriguchi, I., Hirono, S., Liu, Q., Nakagome, I., & Matsushita, Y. (1992). Simple method of calculating octanol/water partition coefficient. *Chemical & Pharmaceutical Bulletin (Tokyo)*, *40*, 127–130.

Morihata, H., Nakamura, F., Tsutada, T., & Kuno, M. (2000). Potentiation of a voltage-gated proton current in acidosis-induced swelling of rat microglia. *The Journal of Neuroscience*, *20*, 7220–7227.

Nabekura, T., Morishima, S., Cover, T. L., Mori, S., Kannan, H., Komune, S., et al. (2003). Recovery from lactacidosis-induced glial cell swelling with the aid of exogenous anion channels. *Glia*, *41*, 247–259.

Nagasaki, M., Ye, L., Duan, D., Horowitz, B., & Hume, J. R. (2000). Intracellular cyclic AMP inhibits native and recombinant volume-regulated chloride channels from mammalian heart. *The Journal of Physiology*, *523*(Pt. 3), 705–717.

Nakagawa, T., Shimizu, S., Watanabe, T., Yamaguchi, O., Otsu, K., Yamagata, H., et al. (2005). Cyclophilin D-dependent mitochondrial permeability transition regulates some necrotic but not apoptotic cell death. *Nature*, *434*, 652–658.

Nakahari, T., Murakami, M., Sasaki, Y., Kataoka, T., Imai, Y., Shiba, Y., et al. (1991). Dose effects of acetylcholine on the cell volume of rat mandibular salivary acini. *The Japanese Journal of Physiology*, *41*, 153–168.

Nedergaard, M., & Goldman, S. A. (1993). Carrier-mediated transport of lactic acid in cultured neurons and astrocytes. *The American Journal of Physiology*, *265*, R282–R289.

Nedergaard, M., Goldman, S. A., Desai, S., & Pulsinelli, W. A. (1991). Acid-induced death in neurons and glia. *The Journal of Neuroscience*, *11*, 2489–2497.

Nicotera, P., Leist, M., & Ferrando-May, E. (1998). Intracellular ATP, a switch in the decision between apoptosis and necrosis. *Toxicology Letters*, *102–103*, 139–142.

Nieminen, A. L. (2003). Apoptosis and necrosis in health and disease: Role of mitochondria. *International Review of Cytology*, *224*, 29–55.

Nietsch, H. H., Roe, M. W., Fiekers, J. F., Moore, A. L., & Lidofsky, S. D. (2000). Activation of potassium and chloride channels by tumor necrosis factor alpha. Role in liver cell death. *The Journal of Biological Chemistry*, *275*, 20556–20561.

Nilius, B., Eggermont, J., Voets, T., Buyse, G., Manolopoulos, V., & Droogmans, G. (1997). Properties of volume-regulated anion channels in mammalian cells. *Progress in Biophysics and Molecular Biology*, *68*, 69–119.

Ninomiya, H., Otani, H., Lu, K., Uchiyama, T., Kido, M., & Imamura, H. (2002). Complementary role of extracellular ATP and adenosine in ischemic preconditioning in the rat heart. *American Journal of Physiology. Heart and Circulatory Physiology*, *282*, H1810–H1820.

Nukui, M., Shimizu, T., & Okada, Y. (2006). Normotonic cell shrinkage induced by Na^+ deprivation results in apoptotic cell death in human epithelial HeLa cells. *The Journal of Physiological Sciences*, *56*, 335–339.

Numata, T., & Okada, Y. (2008). Proton conductivity through the human TRPM7 channel and its molecular determinants. *The Journal of Biological Chemistry*, *283*, 15097–15103.

Numata, T., Sato, K., Christmann, J., Marx, R., Mori, Y., Okada, Y., et al. (2012). The deltaC splice-variant of TRPM2 is the hypertonicity-induced cation channel in HeLa cells, and the ecto-enzyme CD38 mediates its activation. *The Journal of Physiology*, *590*, 1121–1138.

Numata, T., Sato, K., Okada, Y., & Wehner, F. (2008). Hypertonicity-induced cation channels rescue cells from staurosporine-elicited apoptosis. *Apoptosis*, *13*, 895–903.

Numata, T., Sato-Numata, K., Okada, Y., & Inoue, R. (2018). Cellular mechanism for herbal medicine Junchoto to facilitate intestinal Cl^-/water secretion that involves cAMP-dependent activation of CFTR. *Journal of Natural Medicines*, *72*, 694–705.

Numata, T., Shimizu, T., & Okada, Y. (2007a). Direct mechano-stress sensitivity of TRPM7 channel. *Cellular Physiology and Biochemistry*, *19*, 1–8.

Numata, T., Shimizu, T., & Okada, Y. (2007b). TRPM7 is a stretch- and swelling-activated cation channel involved in volume regulation in human epithelial cells. *American Journal of Physiology. Cell Physiology*, *292*, C460–C467.

Nunez, R., Sancho-Martinez, S. M., Novoa, J. M., & Lopez-Hernandez, F. J. (2010). Apoptotic volume decrease as a geometric determinant for cell dismantling into apoptotic bodies. *Cell Death and Differentiation*, *17*, 1665–1671.

Obayashi, K., Horie, M., Washizuka, T., Nishimoto, T., & Sasayama, S. (1999). On the mechanism of genistein-induced activation of protein kinase A-dependent Cl^- conductance in cardiac myocytes. *Pflügers Archiv*, *438*, 269–277.

O'Connor, E. R., Kimelberg, H. K., Keese, C. R., & Giaever, I. (1993). Electrical resistance method for measuring volume changes in monolayer cultures applied to primary astrocyte cultures. *The American Journal of Physiology*, *264*, C471–C478.

Oike, M., Droogmans, G., & Nilius, B. (1994). The volume-activated chloride current in human endothelial cells depends on intracellular ATP. *Pflügers Archiv*, *427*, 184–186.

Oiki, S., Kubo, M., & Okada, Y. (1994). Mg^{2+} and ATP-dependence of volume-sensitive Cl^- channels in human epithelial cells. *The Japanese Journal of Physiology*, *44*(Suppl. 2), S77–S79.

Okada, Y. (1997). Volume expansion-sensing outward-rectifier Cl^- channel: Fresh start to the molecular identity and volume sensor. *The American Journal of Physiology*, *273*, C755–C789.

Okada, Y. (1998). Cell volume-sensitive chloride channels. *Contributions to Nephrology*, *123*, 21–33.

Okada, Y. (2004). Ion channels and transporters involved in cell volume regulation and sensor mechanisms. *Cell Biochemistry and Biophysics*, *41*, 233–258.

Okada, Y. (2016). Channelling frozen cells to survival after thawing: Opening the door to cryo-physiology. *The Journal of Physiology*, *594*, 1523–1524.

Okada, Y., & Hazama, A. (1989). Volume-regulatory ion channels in epithelial cells. *Physiology*, *4*, 238–242.

Okada, Y., Hazama, A., & Yuan, W. L. (1990). Stretch-induced activation of Ca^{2+}-permeable ion channels is involved in the volume regulation of hypotonically swollen epithelial cells. *Neuroscience Research. Supplement*, *12*, S5–13.

Okada, Y., & Maeno, E. (2001). Apoptosis, cell volume regulation and volume-regulatory chloride channels. *Comparative Biochemistry and Physiology. Part A, Molecular & Integrative Physiology*, *130*, 377–383.

Okada, Y., Maeno, E., & Mori, S. (2005). Essential role of anion channel in induction of apoptotic and necrotic cell death. In J. X.-J. Yuan (Ed.), *Ion channels in the pulmonary vasculature* (pp. 527–544). Boca Raton, FL: Taylor & Francis.

Okada, Y., Maeno, E., Shimizu, T., Dezaki, K., Wang, J., & Morishima, S. (2001). Receptor-mediated control of regulatory volume decrease (RVD) and apoptotic volume decrease (AVD). *The Journal of Physiology*, *532*, 3–16.

Okada, Y., Maeno, E., Shimizu, T., Manabe, K., Mori, S., & Nabekura, T. (2004). Dual roles of plasmalemmal chloride channels in induction of cell death. *Pflügers Archiv*, *448*, 287–295.

Okada, Y., Okada, T., Islam, M. R., & Sabirov, R. Z. (2018). Molecular identities and ATP release activities of two types of volume-regulatory anion channels, VSOR and Maxi-Cl. *Current Topics in Membranes*, *81*, 125–176.

Okada, Y., Okada, T., Sato-Numata, K., Islam, M. R., Ando-Akatsuka, Y., Numata, T., et al. (2019). Cell volume-activated and volume-correlated anion channels in mammalian cells: Their biophysical, molecular, and pharmacological properties. *Pharmacological Reviews*, *71*, 49–88.

Okada, Y., Shimizu, T., Maeno, E., Tanabe, S., Wang, X., & Takahashi, N. (2006). Volume-sensitive chloride channels involved in apoptotic volume decrease and cell death. *The Journal of Membrane Biology*, *209*, 21–29.

Olney, J. W. (1969). Brain lesions, obesity, and other disturbances in mice treated with monosodium glutamate. *Science*, *164*, 719–721.

Olney, J. W., Adamo, N. J., & Ratner, A. (1971). Monosodium glutamate effects. *Science*, *172*, 294.

Olson, J. E., & Li, G. Z. (1997). Increased potassium, chloride, and taurine conductances in astrocytes during hypoosmotic swelling. *Glia*, *20*, 254–261.

Olson, J. E., Sankar, R., Holtzman, D., James, A., & Fleischhacker, D. (1986). Energy-dependent volume regulation in primary cultured cerebral astrocytes. *Journal of Cellular Physiology*, *128*, 209–215.

O'Reilly, N., Xia, Z., Fiander, H., Tauskela, J., & Small, D. L. (2002). Disparity between ionic mediators of volume regulation and apoptosis in N1E 115 mouse neuroblastoma cells. *Brain Research*, *943*, 245–256.

Orlov, S. N., Dam, T. V., Tremblay, J., & Hamet, P. (1996). Apoptosis in vascular smooth muscle cells: Role of cell shrinkage. *Biochemical and Biophysical Research Communications*, *221*, 708–715.

Orlov, S. N., Platonova, A. A., Hamet, P., & Grygorczyk, R. (2013). Cell volume and monovalent ion transporters: Their role in cell death machinery triggering and progression. *American Journal of Physiology. Cell Physiology*, *305*, C361–C372.

Orlov, S. N., Taurin, S., Thorin-Trescases, N., Dulin, N. O., Tremblay, J., & Hamet, P. (2000). Inversion of the intracellular Na^+/K^+ ratio blocks apoptosis in vascular smooth muscle cells by induction of RNA synthesis. *Hypertension*, *35*, 1062–1068.

Oz, M. C., & Sorota, S. (1995). Forskolin stimulates swelling-induced chloride current, not cardiac cystic fibrosis transmembrane-conductance regulator current, in human cardiac myocytes. *Circulation Research*, *76*, 1063–1070.

Paddle, B. M., & Burnstock, G. (1974). Release of ATP from perfused heart during coronary vasodilatation. *Blood Vessels*, *11*, 110–119.

Pangrsic, T., Potokar, M., Haydon, P. G., Zorec, R., & Kreft, M. (2006). Astrocyte swelling leads to membrane unfolding, not membrane insertion. *Journal of Neurochemistry*, *99*, 514–523.

Parkerson, K. A., & Sontheimer, H. (2003). Contribution of chloride channels to volume regulation of cortical astrocytes. *American Journal of Physiology. Cell Physiology*, *284*, C1460–C1467.

Parkerson, K. A., & Sontheimer, H. (2004). Biophysical and pharmacological characterization of hypotonically activated chloride currents in cortical astrocytes. *Glia*, *46*, 419–436.

Parpura, V., Basarsky, T. A., Liu, F., Jeftinija, K., Jeftinija, S., & Haydon, P. G. (1994). Glutamate-mediated astrocyte-neuron signalling. *Nature*, *369*, 744–747.

Parpura, V., Fang, Y., Basarsky, T., Jahn, R., & Haydon, P. G. (1995). Expression of synaptobrevin II, cellubrevin and syntaxin but not SNAP-25 in cultured astrocytes. *FEBS Letters*, *377*, 489–492.

Parpura, V., Liu, F., Brethorst, S., Jeftinija, K., Jeftinija, S., & Haydon, P. G. (1995). Alpha-latrotoxin stimulates glutamate release from cortical astrocytes in cell culture. *FEBS Letters*, *360*, 266–270.

Parsons, D. F., Cole, R. W., & Kimelberg, H. K. (1989). Shape, size, and distribution of cell structures by 3-D graphics reconstruction and stereology. I. The regulatory volume decrease of astroglial cells. *Cell Biophysics*, *14*, 27–42.

Pasantes Morales, H., & Schousboe, A. (1988). Volume regulation in astrocytes: A role for taurine as an osmoeffector. *Journal of Neuroscience Research*, *20*, 503–509.

Pasantes-Morales, H., Alavez, S., Sanchez Olea, R., & Moran, J. (1993). Contribution of organic and inorganic osmolytes to volume regulation in rat brain cells in culture. *Neurochemical Research*, *18*, 445–452.

Pasantes-Morales, H., Chacon, E., Sanchez-Olea, R., & Moran, J. (1994). Volume regulation in cultured neurons: Pivotal role of taurine. *Advances in Experimental Medicine and Biology*, *359*, 317–323.

Pasantes-Morales, H., Maar, T. E., & Moran, J. (1993). Cell volume regulation in cultured cerebellar granule neurons. *Journal of Neuroscience Research*, *34*, 219–224.

Pasantes-Morales, H., Murray, R. A., Lilja, L., & Moran, J. (1994). Regulatory volume decrease in cultured astrocytes. I. Potassium- and chloride-activated permeability. *The American Journal of Physiology*, *266*, C165–C171.

Patel, A. J., Lauritzen, I., Lazdunski, M., & Honore, E. (1998). Disruption of mitochondrial respiration inhibits volume-regulated anion channels and provokes neuronal cell swelling. *The Journal of Neuroscience*, *18*, 3117–3123.

Pedersen, S. F., Hoffmann, E. K., & Novak, I. (2013). Cell volume regulation in epithelial physiology and cancer. *Frontiers in Physiology*, *4*, 233.

Pedersen, S. F., Okada, Y., & Nilius, B. (2016). Biophysics and physiology of the volume-regulated anion channel (VRAC)/volume-sensitive outwardly rectifying anion channel (VSOR). *Pflügers Archiv*, *468*, 371–383.

Peti-Peterdi, J., Morishima, S., Bell, P. D., & Okada, Y. (2002). Two-photon excitation fluorescence imaging of the living juxtaglomerular apparatus. *American Journal of Physiology. Renal Physiology*, *283*, F197–F201.

Pin, J. P., & Bockaert, J. (1989). Two distinct mechanisms, differentially affected by excitatory amino acids, trigger GABA release from fetal mouse striatal neurons in primary culture. *The Journal of Neuroscience*, *9*, 648–656.

Piwnica-Worms, D., Jacob, R., Horres, C. R., & Lieberman, M. (1985). Potassium-chloride cotransport in cultured chick heart cells. *The American Journal of Physiology*, *249*, C337–C344.

Planells-Cases, R., Lutter, D., Guyader, C., Gerhards, N. M., Ullrich, F., Elger, D. A., et al. (2015). Subunit composition of VRAC channels determines substrate specificity and cellular resistance to Pt-based anti-cancer drugs. *The EMBO Journal*, *34*, 2993–3008.

Platonova, A., Koltsova, S. V., Hamet, P., Grygorczyk, R., & Orlov, S. N. (2012). Swelling rather than shrinkage precedes apoptosis in serum-deprived vascular smooth muscle cells. *Apoptosis*, *17*, 429–438.

Platonova, A., Koltsova, S., Maksimov, G. V., Grygorczyk, R., & Orlov, S. N. (2011). The death of ouabain-treated renal epithelial C11-MDCK cells is not mediated by swelling-induced plasma membrane rupture. *The Journal of Membrane Biology*, *241*, 145–154.

Plesnila, N., Haberstok, J., Peters, J., Kolbl, I., Baethmann, A., & Staub, F. (1999). Effect of lactacidosis on cell volume and intracellular pH of astrocytes. *Journal of Neurotrauma*, *16*, 831–841.

Plettenberg, S., Weiss, E. C., Lemor, R., & Wehner, F. (2008). Subunits alpha, beta and gamma of the epithelial Na^+ channel (ENaC) are functionally related to the hypertonicity-induced cation channel (HICC) in rat hepatocytes. *Pflügers Archiv*, *455*, 1089–1095.

Porcelli, A. M., Ghelli, A., Zanna, C., Valente, P., Ferroni, S., & Rugolo, M. (2003). Staurosporine induces apoptotic volume decrease (AVD) in ECV304 cells. *Annals of the New York Academy of Sciences*, *1010*, 342–346.

Porcelli, A. M., Ghelli, A., Zanna, C., Valente, P., Ferroni, S., & Rugolo, M. (2004). Apoptosis induced by staurosporine in ECV304 cells requires cell shrinkage and upregulation of Cl- conductance. *Cell Death and Differentiation*, *11*, 655–662.

Portera-Cailliau, C., Price, D. L., & Martin, L. J. (1997a). Excitotoxic neuronal death in the immature brain is an apoptosis-necrosis morphological continuum. *The Journal of Comparative Neurology*, *378*, 70–87.

Portera-Cailliau, C., Price, D. L., & Martin, L. J. (1997b). Non-NMDA and NMDA receptor-mediated excitotoxic neuronal deaths in adult brain are morphologically distinct: Further evidence for an apoptosis-necrosis continuum. *The Journal of Comparative Neurology*, *378*, 88–104.

Poulsen, K. A., Andersen, E. C., Hansen, C. F., Klausen, T. K., Hougaard, C., Lambert, I. H., et al. (2010). Deregulation of apoptotic volume decrease and ionic movements in multidrug-resistant tumor cells: Role of chloride channels. *American Journal of Physiology. Cell Physiology*, *298*, C14–C25.

Praetorius, H. A., & Leipziger, J. (2009). Released nucleotides amplify the cilium-dependent, flow-induced [Ca2+]i response in MDCK cells. *Acta Physiologica (Oxford, England)*, *197*, 241–251.

Price, M. T., Olney, J. W., Samson, L., & Labruyere, J. (1985). Calcium influx accompanies but does not cause excitotoxin-induced neuronal necrosis in retina. *Brain Research Bulletin*, *14*, 369–376.

Pulsinelli, W. A., Brierley, J. B., & Plum, F. (1982). Temporal profile of neuronal damage in a model of transient forebrain ischemia. *Annals of Neurology*, *11*, 491–498.

Qiu, Z., Dubin, A. E., Mathur, J., Tu, B., Reddy, K., Miraglia, L. J., et al. (2014). SWELL1, a plasma membrane protein, is an essential component of volume-regulated anion channel. *Cell*, *157*, 447–458.

Quinn, R. H., & Pierce, S. K. (1992). The ionic basis of the hypo-osmotic depolarization in neurons from the opisthobranch mollusc *Elysia chlorotica*. *The Journal of Experimental Biology*, *163*, 169–186.

Ransom, B. R., Yamate, C. L., & Connors, B. W. (1985). Activity-dependent shrinkage of extracellular space in rat optic nerve: A developmental study. *The Journal of Neuroscience*, *5*, 532–535.

Rasmusson, R. L., Davis, D. G., & Lieberman, M. (1993). Amino acid loss during volume regulatory decrease in cultured chick heart cells. *The American Journal of Physiology*, *264*, C136–C145.

Raucci, F. J., Jr., Wijesinghe, D. S., Chalfant, C. E., & Baumgarten, C. M. (2010). Exogenous and endogenous ceramides elicit volume-sensitive chloride current in ventricular myocytes. *Cardiovascular Research*, *86*, 55–62.

Ren, Z., & Baumgarten, C. M. (2005). Antagonistic regulation of swelling-activated Cl^- current in rabbit ventricle by Src and EGFR protein tyrosine kinases. *American Journal of Physiology. Heart and Circulatory Physiology*, *288*, H2628–H2636.

Ren, Z., Raucci, F. J., Jr., Browe, D. M., & Baumgarten, C. M. (2008). Regulation of swelling-activated Cl^- current by angiotensin II signalling and NADPH oxidase in rabbit ventricle. *Cardiovascular Research*, *77*, 73–80.

Ringel, F., Chang, R. C., Staub, F., Baethmann, A., & Plesnila, N. (2000). Contribution of anion transporters to the acidosis-induced swelling and intracellular acidification of glial cells. *Journal of Neurochemistry*, *75*, 125–132.

Ringel, F., Plesnila, N., Chang, R. C., Peters, J., Staub, F., & Baethmann, A. (1997). Role of calcium ions in acidosis-induced glial swelling. *Acta Neurochirurgica. Supplement*, *70*, 144–147.

Rojas-Rivera, D., Diaz-Elizondo, J., Parra, V., Salas, D., Contreras, A., Toro, B., et al. (2009). Regulatory volume decrease in cardiomyocytes is modulated by calcium influx and reactive oxygen species. *FEBS Letters*, *583*, 3485–3492.

Rossato, M., Di Virgilio, F., & Foresta, C. (1996). Involvement of osmo-sensitive calcium influx in human sperm activation. *Molecular Human Reproduction*, *2*, 903–909.

Rossi, D. J., Brady, J. D., & Mohr, C. (2007). Astrocyte metabolism and signaling during brain ischemia. *Nature Neuroscience*, *10*, 1377–1386.

Rothman, S. M., Thurston, J. H., & Hauhart, R. E. (1987). Delayed neurotoxicity of excitatory amino acids in vitro. *Neuroscience*, *22*, 471–480.

Roy, G. (1995). Amino acid current through anion channels in cultured human glial cells. *The Journal of Membrane Biology*, *147*, 35–44.

Rozencweig, M., von Hoff, D. D., Slavik, M., & Muggia, F. M. (1977). Cis-diamminedichloroplatinum (II). A new anticancer drug. *Annals of Internal Medicine*, *86*, 803–812.

Rubio, R., Berne, R. M., & Katori, M. (1969). Release of adenosine in reactive hyperemia of the dog heart. *The American Journal of Physiology*, *216*, 56–62.

Rungta, R. L., Choi, H. B., Tyson, J. R., Malik, A., Dissing-Olesen, L., Lin, P. J. C., et al. (2015). The cellular mechanisms of neuronal swelling underlying cytotoxic edema. *Cell*, *161*, 610–621.

Ryder, C., McColl, K., Zhong, F., & Distelhorst, C. W. (2012). Acidosis promotes Bcl-2 family-mediated evasion of apoptosis: Involvement of acid-sensing G protein-coupled receptor Gpr65 signaling to Mek/Erk. *The Journal of Biological Chemistry*, *287*, 27863–27875.

Sabirov, R. Z., Dutta, A. K., & Okada, Y. (2001). Volume-dependent ATP-conductive large-conductance anion channel as a pathway for swelling-induced ATP release. *The Journal of General Physiology*, *118*, 251–266.

Sabirov, R. Z., Merzlyak, P. G., Islam, M. R., Okada, T., & Okada, Y. (2016). The properties, functions, and pathophysiology of maxi-anion channels. *Pflügers Archiv*, *468*, 405–420.

Sabirov, R. Z., Merzlyak, P. G., Okada, T., Islam, M. R., Uramoto, H., Mori, T., et al. (2017). The organic anion transporter SLCO2A1 constitutes the core component of the Maxi-Cl channel. *The EMBO Journal*, *36*, 3309–3324.

Sabirov, R. Z., & Okada, Y. (2004). Wide nanoscopic pore of maxi-anion channel suits its function as an ATP-conductive pathway. *Biophysical Journal*, *87*, 1672–1685.

Sabirov, R. Z., & Okada, Y. (2005). ATP release via anion channels. *Purinergic Signal*, *1*, 311–328.

Sabirov, R. Z., & Okada, Y. (2009). The maxi-anion channel: A classical channel playing novel roles through an unidentified molecular entity. *The Journal of Physiological Sciences*, *59*, 3–21.

Sabirov, R. Z., Prenen, J., Droogmans, G., & Nilius, B. (2000). Extra- and intracellular proton-binding sites of volume-regulated anion channels. *The Journal of Membrane Biology*, *177*, 13–22.

Sakaguchi, M., Matsuura, H., & Ehara, T. (1997). Swelling-induced Cl^- current in guinea-pig atrial myocytes: Inhibition by glibenclamide. *The Journal of Physiology*, *505*(Pt. 1), 41–52.

Sakai, R., Hagiwara, N., Kasanuki, H., & Hosoda, S. (1995). Chloride conductance in human atrial cells. *Journal of Molecular and Cellular Cardiology*, *27*, 2403–2408.

Sanchez-Olea, R., Pena, C., Moran, J., & Pasantes-Morales, H. (1993). Inhibition of volume regulation and efflux of osmoregulatory amino acids by blockers of Cl^- transport in cultured astrocytes. *Neuroscience Letters*, *156*, 141–144.

Saraste, A., Pulkki, K., Kallajoki, M., Henriksen, K., Parvinen, M., & Voipio-Pulkki, L. M. (1997). Apoptosis in human acute myocardial infarction. *Circulation*, *95*, 320–323.

Sarfaraz, D., & Fraser, C. L. (1999). Effects of arginine vasopressin on cell volume regulation in brain astrocyte in culture. *The American Journal of Physiology*, *276*, E596–E601.

Sarkadi, B., & Parker, J. C. (1991). Activation of ion transport pathways by changes in cell volume. *Biochimica et Biophysica Acta*, *1071*, 407–427.

Sato, K., Numata, T., Saito, T., Ueta, Y., & Okada, Y. (2011). V_2 receptor-mediated autocrine role of somatodendritic release of AVP in rat vasopressin neurons under hypo-osmotic conditions. *Science Signaling*, *4*, ra5.

Sato-Numata, K., Numata, T., Inoue, R., & Okada, Y. (2016). Distinct pharmacological and molecular properties of the acid-sensitive outwardly rectifying (ASOR) anion channel from those of the volume-sensitive outwardly rectifying (VSOR) anion channel. *Pflügers Archiv*, *468*, 795–803.

Sato-Numata, K., Numata, T., Inoue, R., Sabirov, R. Z., & Okada, Y. (2017). Distinct contributions of LRRC8A and its paralogs to the VSOR anion channel from those of the ASOR anion channel. *Channels (Austin, Tex)*, *11*, 167–172.

Sato-Numata, K., Numata, T., & Okada, Y. (2014). Temperature sensitivity of acid-sensitive outwardly rectifying (ASOR) anion channels in cortical neurons is involved in hypothermic neuroprotection against acidotoxic necrosis. *Channels (Austin, Tex.)*, *8*, 278–283.

Sato-Numata, K., Numata, T., Okada, T., & Okada, Y. (2013). Acid-sensitive outwardly rectifying (ASOR) anion channels in human epithelial cells are highly sensitive to temperature and independent of ClC-3. *Pflügers Archiv*, *465*, 1535–1543.

Sato-Numata, K., Ueta, Y., & Okada, Y. (2014). AVP neurons possess the volume regulation mechanism under hyperosmotic conditions. *The Journal of Physiological Sciences*, *64*(Suppl. 1), S115.

Scaffidi, P., Misteli, T., & Bianchi, M. E. (2002). Release of chromatin protein HMGB1 by necrotic cells triggers inflammation. *Nature*, *418*, 191–195.

Schlichter, L. C., Grygorczyk, R., Pahapill, P. A., & Grygorczyk, C. (1990). A large, multiple-conductance chloride channel in normal human T lymphocytes. *Pflügers Archiv*, *416*, 413–421.

Schober, A. L., Wilson, C. S., & Mongin, A. A. (2017). Molecular composition and heterogeneity of the LRRC8-containing swelling-activated osmolyte channels in primary rat astrocytes. *The Journal of Physiology*, *595*, 6939–6951.

Schomberg, S. L., Bauer, J., Kintner, D. B., Su, G., Flemmer, A., Forbush, B., et al. (2003). Cross talk between the GABA(A) receptor and the Na-K-Cl cotransporter is mediated by intracellular Cl. *Journal of Neurophysiology*, *89*, 159–167.

Schrantz, N., Blanchard, D. A., Auffredou, M. T., Sharma, S., Leca, G., & Vazquez, A. (1999). Role of caspases and possible involvement of retinoblastoma protein during TGFbeta-mediated apoptosis of human B lymphocytes. *Oncogene*, *18*, 3511–3519.

Schumann, M. A., Gardner, P., & Raffin, T. A. (1993). Recombinant human tumor necrosis factor alpha induces calcium oscillation and calcium-activated chloride current in human neutrophils. The role of calcium/calmodulin-dependent protein kinase. *The Journal of Biological Chemistry*, *268*, 2134–2140.

Schwab, A., Nechyporuk-Zloy, V., Fabian, A., & Stock, C. (2007). Cells move when ions and water flow. *Pflügers Archiv*, *453*, 421–432.

Shen, M., Wang, L., Wang, B., Wang, T., Yang, G., Shen, L., et al. (2014). Activation of volume-sensitive outwardly rectifying chloride channel by ROS contributes to ER stress and cardiac contractile dysfunction: Involvement of CHOP through Wnt. *Cell Death & Disease*, *5* e1528.

Shen, M., Wang, L., Yang, G., Gao, L., Wang, B., Guo, X., et al. (2014). Baicalin protects the cardiomyocytes from ER stress-induced apoptosis: Inhibition of CHOP through induction of endothelial nitric oxide synthase. *PLoS One*, *9* e88389.

Shi, L., Xu, M., Liu, J., Zhang, Z., Bao, Z., Wang, Y., et al. (2009). K(ATP) channels are involved in regulatory volume decrease in rat cardiac myocytes. *Physiological Research*, *58*, 645–652.

Shi, J., Zhao, Y., Wang, K., Shi, X., Wang, Y., Huang, H., et al. (2015). Cleavage of GSDMD by inflammatory caspases determines pyroptotic cell death. *Nature*, *526*, 660–665.

Shimizu, T., Lee, E. L., Ise, T., & Okada, Y. (2008). Volume-sensitive Cl^- channel as a regulator of acquired cisplatin resistance. *Anticancer Research*, *28*, 75–83.

Shimizu, T., Maeno, E., & Okada, Y. (2007). Prerequisite role of persistent cell shrinkage in apoptosis of human epithelial cells. *Sheng Li Xue Bao*, *59*, 512–516.

Shimizu, T., Numata, T., & Okada, Y. (2004). A role of reactive oxygen species in apoptotic activation of volume-sensitive Cl^- channel. *Proceedings of the National Academy of Sciences of the United States of America*, *101*, 6770–6773.

Shuba, L. M., Asai, T., Pelzer, S., & McDonald, T. F. (1996). Activation of cardiac chloride conductance by the tyrosine kinase inhibitor, genistein. *British Journal of Pharmacology*, *119*, 335–345.

Shuba, L. M., & McDonald, T. F. (1997). Synergistic activation of guinea-pig cardiac cystic fibrosis transmembrane conductance regulator by the tyrosine kinase inhibitor genistein and cAMP. *The Journal of Physiology*, *505*(Pt. 1), 23–40.

Shuba, L. M., Ogura, T., & McDonald, T. F. (1996). Kinetic evidence distinguishing volume-sensitive chloride current from other types in guinea-pig ventricular myocytes. *The Journal of Physiology*, *491*(Pt. 1), 69–80.

Shuba, Y. M., Prevarskaya, N., Lemonnier, L., Van Coppenolle, F., Kostyuk, P. G., Mauroy, B., et al. (2000). Volume-regulated chloride conductance in the LNCaP human prostate cancer cell line. *American Journal of Physiology. Cell Physiology*, *279*, C1144–C1154.

Siemkowicz, E., & Hansen, A. J. (1981). Brain extracellular ion composition and EEG activity following 10 minutes ischemia in normo- and hyperglycemic rats. *Stroke*, *12*, 236–240.

Silver, I. A., & Erecinska, M. (1990). Intracellular and extracellular changes of [Ca^{2+}] in hypoxia and ischemia in rat brain in vivo. *The Journal of General Physiology*, *95*, 837–866.

Simoes, F., Ousingsawat, J., Wanitchakool, P., Fonseca, A., Cabrita, I., Benedetto, R., et al. (2018). CFTR supports cell death through ROS-dependent activation of TMEM16F (anoctamin 6). *Pflügers Archiv*, *470*, 305–314.

Small, D. L., Tauskela, J., & Xia, Z. (2002). Role for chloride but not potassium channels in apoptosis in primary rat cortical cultures. *Neuroscience Letters*, *334*, 95–98.

Smets, I., Ameloot, M., Steels, P., & Van Driessche, W. (2002). Loss of cell volume regulation during metabolic inhibition in renal epithelial cells (A6): Role of intracellular pH. *American Journal of Physiology Cell Physiology*, *283*, C535–C544.

Sorensen, B. H., Dam, C. S., Sturup, S., & Lambert, I. H. (2016). Dual role of LRRC8A-containing transporters on cisplatin resistance in human ovarian cancer cells. *Journal of Inorganic Biochemistry*, *160*, 287–295.

Sorensen, B. H., Nielsen, D., Thorsteinsdottir, U. A., Hoffmann, E. K., & Lambert, I. H. (2016). Downregulation of LRRC8A protects human ovarian and alveolar carcinoma cells against cisplatin-induced expression of p53, MDM2, p21Waf1/Cip1, and caspase-9/-3 activation. *American Journal of Physiology. Cell Physiology*, *310*, C857–C873.

Sorensen, B. H., Thorsteinsdottir, U. A., & Lambert, I. H. (2014). Acquired cisplatin resistance in human ovarian A2780 cancer cells correlates with shift in taurine homeostasis and ability to volume regulate. *American Journal of Physiology. Cell Physiology*, *307*, C1071–C1080.

Sorota, S. (1992). Swelling-induced chloride-sensitive current in canine atrial cells revealed by whole-cell patch-clamp method. *Circulation Research*, *70*, 679–687.

Sorota, S. (1994). Pharmacologic properties of the swelling-induced chloride current of dog atrial myocytes. *Journal of Cardiovascular Electrophysiology*, *5*, 1006–1016.

Sorota, S. (1995). Tyrosine protein kinase inhibitors prevent activation of cardiac swelling-induced chloride current. *Pflügers Archiv*, *431*, 178–185.

Sorota, S., & Du, X. Y. (1998). Delayed activation of cardiac swelling-induced chloride current after step changes in cell size. *Journal of Cardiovascular Electrophysiology*, *9*, 825–831.

Souktani, R., Berdeaux, A., Ghaleh, B., Giudicelli, J. F., Guize, L., Le Heuzey, J. Y., et al. (2000). Induction of apoptosis using sphingolipids activates a chloride current in *Xenopus laevis* oocytes. *American Journal of Physiology. Cell Physiology*, *279*, C158–C165.

Souza, M. M., Boyle, R. T., & Lieberman, M. (2000). Different physiological mechanisms control isovolumetric regulation and regulatory volume decrease in chick embryo cardiomyocytes. *Cell Biology International*, *24*, 713–721.

Staub, F., Baethmann, A., Peters, J., Weigt, H., & Kempski, O. (1990). Effects of lactacidosis on glial cell volume and viability. *Journal of Cerebral Blood Flow and Metabolism*, *10*, 866–876.

Staub, F., Mackert, B., Kempski, O., Peters, J., & Baethmann, A. (1993). Swelling and death of neuronal cells by lactic acid. *Journal of the Neurological Sciences*, *119*, 79–84.

Staub, F., Winkler, A., Haberstok, J., Plesnila, N., Peters, J., Chang, R. C., et al. (1996). Swelling, intracellular acidosis, and damage of glial cells. *Acta Neurochirurgica. Supplement*, *66*, 56–62.

Stout, C. E., Costantin, J. L., Naus, C. C., & Charles, A. C. (2002). Intercellular calcium signaling in astrocytes via ATP release through connexin hemichannels. *The Journal of Biological Chemistry*, *277*, 10482–10488.

Strange, K. (1991). Volume regulatory Cl^- loss after Na^+ pump inhibition in CCT principal cells. *The American Journal of Physiology*, *260*, F225–F234.

Strange, K., Emma, F., & Jackson, P. S. (1996). Cellular and molecular physiology of volume-sensitive anion channels. *The American Journal of Physiology*, *270*, C711–C730.

Strange, K., & Jackson, P. S. (1995). Swelling-activated organic osmolyte efflux: A new role for anion channels. *Kidney International*, *48*, 994–1003.

Subramanyam, M., Takahashi, N., Hasegawa, Y., Mohri, T., & Okada, Y. (2010). Inhibition of protein kinase Akt1 by apoptosis signal-regulating kinase-1 (ASK1) is involved in apoptotic inhibition of regulatory volume increase. *The Journal of Biological Chemistry*, *285*, 6109–6117.

Suleymanian, M. A., & Baumgarten, C. M. (1996). Osmotic gradient-induced water permeation across the sarcolemma of rabbit ventricular myocytes. *The Journal of General Physiology*, *107*, 503–514.

Suleymanian, M. A., Clemo, H. F., Cohen, N. M., & Baumgarten, C. M. (1995). Stretch-activated channel blockers modulate cell volume in cardiac ventricular myocytes. *Journal of Molecular and Cellular Cardiology*, *27*, 721–728.

Sun, D. A., Sombati, S., & DeLorenzo, R. J. (2001). Glutamate injury-induced epileptogenesis in hippocampal neurons: An in vitro model of stroke-induced "epilepsy" *Stroke*, *32*, 2344–2350.

Sun, X. P., Supplisson, S., & Mayer, E. (1993). Chloride channels in myocytes from rabbit colon are regulated by a pertussis toxin-sensitive G protein. *The American Journal of Physiology*, *264*, G774–G785.

Szabo, I., Lepple-Wienhues, A., Kaba, K. N., Zoratti, M., Gulbins, E., & Lang, F. (1998). Tyrosine kinase-dependent activation of a chloride channel in CD95-induced apoptosis in T lymphocytes. *Proceedings of the National Academy of Sciences of the United States of America*, *95*, 6169–6174.

Sztriha, L., Joo, F., & Szerdahelyi, P. (1986). Time-course of changes in water, sodium, potassium and calcium contents of various brain regions in rats after systemic kainic acid administration. *Acta Neuropathologica*, *70*, 169–176.

Takahashi, N., Wang, X., Tanabe, S., Uramoto, H., Jishage, K., Uchida, S., et al. (2005). ClC-3-independent sensitivity of apoptosis to Cl^- channel blockers in mouse cardiomyocytes. *Cellular Physiology and Biochemistry*, *15*, 263–270.

Tanabe, S., Wang, X., Takahashi, N., Uramoto, H., & Okada, Y. (2005). HCO_3^--independent rescue from apoptosis by stilbene derivatives in rat cardiomyocytes. *FEBS Letters*, *579*, 517–522.

Taouil, K., Giancola, R., Morel, J. E., & Hannaert, P. (1998). Hypotonically induced calcium increase and regulatory volume decrease in newborn rat cardiomyocytes. *Pflügers Archiv*, *436*, 565–574.

Taouil, K., & Hannaert, P. (1999). Evidence for the involvement of K^+ channels and K^+-Cl^- cotransport in the regulatory volume decrease of newborn rat cardiomyocytes. *Pflügers Archiv*, *439*, 56–66.

Taruno, A. (2018). ATP release channels. *International Journal of Molecular Sciences*, *19* E808.

Tasaki, I., & Byrne, P. M. (1990). Heat production associated with neuromuscular transmission in the toad nerve-muscle preparation. *The Japanese Journal of Physiology*, *40*, 181–187.

Ternovsky, V. I., Okada, Y., & Sabirov, R. Z. (2004). Sizing the pore of the volume-sensitive anion channel by differential polymer partitioning. *FEBS Letters*, *576*, 433–436.

Thornton, J. D., Liu, G. S., Olsson, R. A., & Downey, J. M. (1992). Intravenous pretreatment with A1-selective adenosine analogues protects the heart against infarction. *Circulation*, *85*, 659–665.

Thorsteinsdottir, U. A., Thorsteinsdottir, M., & Lambert, I. H. (2016). Protolichesterinic acid, isolated from the lichen *Cetraria islandica*, reduces LRRC8A expression and volume-sensitive release of organic osmolytes in human lung epithelial cancer cells. *Phytotherapy Research*, *30*, 97–104.

Tilly, B. C., Bezstarosti, K., Boomaars, W. E., Marino, C. R., Lamers, J. M., & de Jonge, H. R. (1996). Expression and regulation of chloride channels in neonatal rat cardiomyocytes. *Molecular and Cellular Biochemistry*, *157*, 129–135.

Todd, R. C., & Lippard, S. J. (2009). Inhibition of transcription by platinum antitumor compounds. *Metallomics*, *1*, 280–291.

Tosteson, D. C., & Hoffman, J. F. (1960). Regulation of cell volume by active cation transport in high and low potassium sheep red cells. *The Journal of General Physiology*, *44*, 169–194.

Toychiev, A. H., Sabirov, R. Z., Takahashi, N., Ando-Akatsuka, Y., Liu, H., Shintani, T., et al. (2009). Activation of maxi-anion channel by protein tyrosine dephosphorylation. *American Journal of Physiology. Cell Physiology*, *297*, C990–1000.

Tranum-Jensen, J., Janse, M. J., Fiolet, W. T., Krieger, W. J., D'Alnoncourt, C. N., & Durrer, D. (1981). Tissue osmolality, cell swelling, and reperfusion in acute regional myocardial ischemia in the isolated porcine heart. *Circulation Research*, *49*, 364–381.

Tseng, G. N. (1992). Cell swelling increases membrane conductance of canine cardiac cells: Evidence for a volume-sensitive Cl channel. *The American Journal of Physiology*, *262*, C1056–C1068.

Tymianski, M. (2011). Emerging mechanisms of disrupted cellular signaling in brain ischemia. *Nature Neuroscience*, *14*, 1369–1373.

Uckermann, O., Vargova, L., Ulbricht, E., Klaus, C., Weick, M., Rillich, K., et al. (2004). Glutamate-evoked alterations of glial and neuronal cell morphology in the guinea pig retina. *The Journal of Neuroscience*, *24*, 10149–10158.

Uramoto, H., Okada, T., & Okada, Y. (2012). Protective role of cardiac CFTR activation upon early reperfusion against myocardial infarction. *Cellular Physiology and Biochemistry*, *30*, 1023–1038.

Uramoto, H., Takahashi, N., Dutta, A. K., Sabirov, R. Z., Ando-Akatsuka, Y., Morishima, S., et al. (2003). Ischemia-induced enhancement of CFTR expression on the plasma membrane in neonatal rat ventricular myocytes. *The Japanese Journal of Physiology*, *53*, 357–365.

Van Belle, H., Goossens, F., & Wynants, J. (1987). Formation and release of purine catabolites during hypoperfusion, anoxia, and ischemia. *The American Journal of Physiology*, *252*, H886–H893.

van Borren, M. M., Verkerk, A. O., Vanharanta, S. K., Baartscheer, A., Coronel, R., & Ravesloot, J. H. (2002). Reduced swelling-activated Cl^- current densities in hypertrophied ventricular myocytes of rabbits with heart failure. *Cardiovascular Research*, *53*, 869–878.

Vandenberg, J. I., Rees, S. A., Wright, A. R., & Powell, T. (1996). Cell swelling and ion transport pathways in cardiac myocytes. *Cardiovascular Research*, *32*, 85–97.

Vandenberg, J. I., Yoshida, A., Kirk, K., & Powell, T. (1994). Swelling-activated and isoprenaline-activated chloride currents in guinea pig cardiac myocytes have distinct electrophysiology and pharmacology. *The Journal of General Physiology*, *104*, 997–1017.

Varela, D., Simon, F., Riveros, A., Jorgensen, F., & Stutzin, A. (2004). NAD(P)H oxidase-derived H_2O_2 signals chloride channel activation in cell volume regulation and cell proliferation. *The Journal of Biological Chemistry*, *279*, 13301–13304.

Vercammen, D., Brouckaert, G., Denecker, G., Van de Craen, M., Declercq, W., Fiers, W., et al. (1998). Dual signaling of the Fas receptor: Initiation of both apoptotic and necrotic cell death pathways. *The Journal of Experimental Medicine*, *188*, 919–930.

Vessey, D. A., Li, L., & Kelley, M. (2011). P2X7 receptor agonists pre- and postcondition the heart against ischemia-reperfusion injury by opening pannexin-1/P2X_7 channels. *American Journal of Physiology. Heart and Circulatory Physiology*, *301*, H881–H887.

Vial, C., Owen, P., Opie, L. H., & Posel, D. (1987). Significance of release of adenosine triphosphate and adenosine induced by hypoxia or adrenaline in perfused rat heart. *Journal of Molecular and Cellular Cardiology*, *19*, 187–197.

Viana, F., de la Pena, E., Pecson, B., Schmidt, R. F., & Belmonte, C. (2001). Swelling-activated calcium signalling in cultured mouse primary sensory neurons. *The European Journal of Neuroscience*, *13*, 722–734.

Vick, J. S., & Askwith, C. C. (2015). ASICs and neuropeptides. *Neuropharmacology*, *94*, 36–41.

Videbaek, J., Christensen, N. J., & Sterndorff, B. (1972). Serial determination of plasma catecholamines in myocardial infarction. *Circulation*, *46*, 846–855.

Voss, F. K., Ullrich, F., Munch, J., Lazarow, K., Lutter, D., Mah, N., et al. (2014). Identification of LRRC8 heteromers as an essential component of the volume-regulated anion channel VRAC. *Science*, *344*, 634–638.

Vu, C. C., Bortner, C. D., & Cidlowski, J. A. (2001). Differential involvement of initiator caspases in apoptotic volume decrease and potassium efflux during Fas- and UV-induced cell death. *The Journal of Biological Chemistry*, *276*, 37602–37611.

Wahl, M., Whalley, E. T., Unterberg, A., Schilling, L., Parsons, A. A., Baethmann, A., et al. (1996). Vasomotor and permeability effects of bradykinin in the cerebral microcirculation. *Immunopharmacology*, *33*, 257–263.

Wang, R., Lu, Y., Gunasekar, S., Zhang, Y., Benson, C. J., Chapleau, M. W., et al. (2017). The volume-regulated anion channel (LRRC8) in nodose neurons is sensitive to acidic pH. *JCI Insight*, *2* e90632.

Wang, Z., Mitsuiye, T., Rees, S. A., & Noma, A. (1997). Regulatory volume decrease of cardiac myocytes induced by beta-adrenergic activation of the Cl^- channel in guinea pig. *The Journal of General Physiology*, *110*, 73–82.

Wang, J., Morishima, S., & Okada, Y. (2003). IK channels are involved in the regulatory volume decrease in human epithelial cells. *American Journal of Physiology. Cell Physiology*, *284*, C77–C84.

Wang, L., Shen, M., Guo, X., Wang, B., Xia, Y., Wang, N., et al. (2017). Volume-sensitive outwardly rectifying chloride channel blockers protect against high glucose-induced apoptosis of cardiomyocytes via autophagy activation. *Scientific Reports*, 7, 44265.

Wang, H. Y., Shimizu, T., Numata, T., & Okada, Y. (2007). Role of acid-sensitive outwardly rectifying anion channels in acidosis-induced cell death in human epithelial cells. *Pflügers Archiv*, *454*, 223–233.

Wang, X., Takahashi, N., Uramoto, H., & Okada, Y. (2005). Chloride channel inhibition prevents ROS-dependent apoptosis induced by ischemia-reperfusion in mouse cardiomyocytes. *Cellular Physiology and Biochemistry*, *16*, 147–154.

Webster, C. M., Hokari, M., McManus, A., Tang, X. N., Ma, H., Kacimi, R., et al. (2013). Microglial P2Y12 deficiency/inhibition protects against brain ischemia. *PLoS One*, *8* e70927.

Wee, S., Peart, J. N., & Headrick, J. P. (2007). P2 purinoceptor-mediated cardioprotection in ischemic-reperfused mouse heart. *The Journal of Pharmacology and Experimental Therapeutics*, *323*, 861–867.

Wehner, F., Bondarava, M., ter Veld, F., Endl, E., Nurnberger, H. R., & Li, T. (2006). Hypertonicity-induced cation channels. *Acta Physiologica (Oxford, England)*, *187*, 21–25.

Wehner, F., Olsen, H., Tinel, H., Kinne-Saffran, E., & Kinne, R. K. (2003). Cell volume regulation: Osmolytes, osmolyte transport, and signal transduction. *Reviews of Physiology, Biochemistry and Pharmacology*, *148*, 1–80.

Wei, W. L., Sun, H. S., Olah, M. E., Sun, X., Czerwinska, E., Czerwinski, W., et al. (2007). TRPM7 channels in hippocampal neurons detect levels of extracellular divalent cations. *Proceedings of the National Academy of Sciences of the United States of America, 104*, 16323–16328.

Wei, L., Xiao, A. Y., Jin, C., Yang, A., Lu, Z. Y., & Yu, S. P. (2004). Effects of chloride and potassium channel blockers on apoptotic cell shrinkage and apoptosis in cortical neurons. *Pflügers Archiv, 448*, 325–334.

Weiss, S. (1988). Excitatory amino acid-evoked release of gamma-[3H]aminobutyric acid from striatal neurons in primary culture. *Journal of Neurochemistry, 51*, 435–441.

Wesselborg, S., & Kabelitz, D. (1993). Activation-driven death of human T cell clones: Time course kinetics of the induction of cell shrinkage, DNA fragmentation, and cell death. *Cellular Immunology, 148*, 234–241.

Whalen, D. A., Jr., Hamilton, D. G., Ganote, C. E., & Jennings, R. B. (1974). Effect of a transient period of ischemia on myocardial cells. I. Effects on cell volume regulation. *The American Journal of Pathology, 74*, 381–397.

Whelan, R. S., Kaplinskiy, V., & Kitsis, R. N. (2010). Cell death in the pathogenesis of heart disease: Mechanisms and significance. *Annual Review of Physiology, 72*, 19–44.

Wilson, C. S., & Mongin, A. A. (2018). Cell volume control in healthy brain and neuropathologies. *Current Topics in Membranes, 81*, 385–455.

Wolf, C. M., Reynolds, J. E., Morana, S. J., & Eastman, A. (1997). The temporal relationship between protein phosphatase, ICE/CED-3 proteases, intracellular acidification, and DNA fragmentation in apoptosis. *Experimental Cell Research, 230*, 22–27.

Wu, K. L., Khan, S., Lakhe-Reddy, S., Wang, L., Jarad, G., Miller, R. T., et al. (2003). Renal tubular epithelial cell apoptosis is associated with caspase cleavage of the NHE1 Na^+/H^+ exchanger. *American Journal of Physiology. Renal Physiology, 284*, F829–F839.

Wyllie, A. H., Kerr, J. F., & Currie, A. R. (1980). Cell death: The significance of apoptosis. *International Review of Cytology, 68*, 251–306.

Xiang, S. Y., Ye, L. L., Duan, L. L., Liu, L. H., Ge, Z. D., Auchampach, J. A., et al. (2011). Characterization of a critical role for CFTR chloride channels in cardioprotection against ischemia/reperfusion injury. *Acta Pharmacologica Sinica, 32*, 824–833.

Xiao, A. Y., Wei, L., Xia, S., Rothman, S., & Yu, S. P. (2002). Ionic mechanism of ouabain-induced concurrent apoptosis and necrosis in individual cultured cortical neurons. *The Journal of Neuroscience, 22*, 1350–1362.

Xiong, Z. G., Zhu, X. M., Chu, X. P., Minami, M., Hey, J., Wei, W. L., et al. (2004). Neuroprotection in ischemia: Blocking calcium-permeable acid-sensing ion channels. *Cell, 118*, 687–698.

Xu, W. X., Kim, S. J., So, I., Kang, T. M., Rhee, J. C., & Kim, K. W. (1997). Volume-sensitive chloride current activated by hyposmotic swelling in antral gastric myocytes of the guinea-pig. *Pflügers Archiv, 435*, 9–19.

Yamamoto, S., & Ehara, T. (2006). Acidic extracellular pH-activated outwardly rectifying chloride current in mammalian cardiac myocytes. *American Journal of Physiology. Heart and Circulatory Physiology, 290*, H1905–H1914.

Yamamoto, S., Ehara, T., & Shioya, T. (2001). Changes in cell volume induced by activation of the cyclic amp-dependent chloride channel in guinea-pig cardiac myocytes. *The Japanese Journal of Physiology, 51*, 31–41.

Yamamoto, S., Ichishima, K., & Ehara, T. (2008). Regulation of volume-regulated outwardly rectifying anion channels by phosphatidylinositol 3,4,5-trisphosphate in mouse ventricular cells. *Biomedical Research, 29*, 307–315.

Yamamoto, S., Ichishima, K., & Ehara, T. (2009). Reduced volume-regulated outwardly rectifying anion channel activity in ventricular myocyte of type 1 diabetic mice. *The Journal of Physiological Sciences, 59*, 87–96.

Yamamoto, S., Ishihara, K., Ehara, T., & Shioya, T. (2004). Cell-volume regulation by swelling-activated chloride current in guinea-pig ventricular myocytes. *The Japanese Journal of Physiology*, *54*, 31–38.

Yamamoto, S., Kita, S., Iyoda, T., Yamada, T., & Iwamoto, T. (2011). New molecular mechanisms for cardiovascular disease: Cardiac hypertrophy and cell-volume regulation. *Journal of Pharmacological Sciences*, *116*, 343–349.

Yamamoto-Mizuma, S., Wang, G. X., Liu, L. L., Schegg, K., Hatton, W. J., Duan, D., et al. (2004). Altered properties of volume-sensitive osmolyte and anion channels (VSOACs) and membrane protein expression in cardiac and smooth muscle myocytes from $Clcn3^{-/-}$ mice. *The Journal of Physiology*, *557*, 439–456.

Yamazaki, J., & Hume, J. R. (1997). Inhibitory effects of glibenclamide on cystic fibrosis transmembrane regulator, swelling-activated, and Ca^{2+}-activated Cl^{-} channels in mammalian cardiac myocytes. *Circulation Research*, *81*, 101–109.

Yang, C., He, L., Chen, G., Ning, Z., & Xia, Z. (2019). LRRC8A potentiates temozolomide sensitivity in glioma cells via activating mitochondria-dependent apoptotic pathway. *Human Cell*, *32*, 41–50.

Yang, H., Huang, L. Y., Zeng, D. Y., Huang, E. W., Liang, S. J., Tang, Y. B., et al. (2012). Decrease of intracellular chloride concentration promotes endothelial cell inflammation by activating nuclear factor-kappaB pathway. *Hypertension*, *60*, 1287–1293.

Yang, X., Zhu, L., Lin, J., Liu, S., Luo, H., Mao, J., et al. (2015). Cisplatin activates volume-sensitive like chloride channels via purinergic receptor pathways in nasopharyngeal carcinoma cells. *The Journal of Membrane Biology*, *248*, 19–29.

Ye, Z. C., Oberheim, N., Kettenmann, H., & Ransom, B. R. (2009). Pharmacological "cross-inhibition" of connexin hemichannels and swelling activated anion channels. *Glia*, *57*, 258–269.

Yermolaieva, O., Leonard, A. S., Schnizler, M. K., Abboud, F. M., & Welsh, M. J. (2004). Extracellular acidosis increases neuronal cell calcium by activating acid-sensing ion channel 1a. *Proceedings of the National Academy of Sciences of the United States of America*, *101*, 6752–6757.

Yitzhaki, S., Shneyvays, V., Jacobson, K. A., & Shainberg, A. (2005). Involvement of uracil nucleotides in protection of cardiomyocytes from hypoxic stress. *Biochemical Pharmacology*, *69*, 1215–1223.

Yu, Q., Guo, Z., Liu, X., Ouyang, Q., He, C., Burnstock, G., et al. (2013). Block of P2X7 receptors could partly reverse the delayed neuronal death in area CA1 of the hippocampus after transient global cerebral ischemia. *Purinergic Signal*, *9*, 663–675.

Yue, Z., Zhang, Y., Xie, J., Jiang, J., & Yue, L. (2013). Transient receptor potential (TRP) channels and cardiac fibrosis. *Current Topics in Medicinal Chemistry*, *13*, 270–282.

Yurinskaya, V. E., Moshkov, A. V., Wibberley, A. V., Lang, F., Model, M. A., & Vereninov, A. A. (2012). Dual response of human leukemia U937 cells to hypertonic shrinkage: Initial regulatory volume increase (RVI) and delayed apoptotic volume decrease (AVD). *Cellular Physiology and Biochemistry*, *30*, 964–973.

Zamaraeva, M. V., Sabirov, R. Z., Maeno, E., Ando-Akatsuka, Y., Bessonova, S. V., & Okada, Y. (2005). Cells die with increased cytosolic ATP during apoptosis: A bioluminescence study with intracellular luciferase. *Cell Death and Differentiation*, *12*, 1390–1397.

Zausinger, S., Lumenta, D. B., Pruneau, D., Schmid-Elsaesser, R., Plesnila, N., & Baethmann, A. (2002). Effects of LF 16-0687 Ms, a bradykinin B_2 receptor antagonist, on brain edema formation and tissue damage in a rat model of temporary focal cerebral ischemia. *Brain Research*, *950*, 268–278.

Zhang, Z., & Bourque, C. W. (2003). Osmometry in osmosensory neurons. *Nature Neuroscience*, *6*, 1021–1022.

Zhang, H., Cao, H. J., Kimelberg, H. K., & Zhou, M. (2011). Volume regulated anion channel currents of rat hippocampal neurons and their contribution to oxygen-and-glucose deprivation induced neuronal death. *PLoS One*, *6* e16803.

Zhang, Y., Zhang, H., Feustel, P. J., & Kimelberg, H. K. (2008). DCPIB, a specific inhibitor of volume regulated anion channels (VRACs), reduces infarct size in MCAo and the release of glutamate in the ischemic cortical penumbra. *Experimental Neurology*, *210*, 514–520.

Zhao, B., Gu, L., Liu, K., Zhang, M., & Liu, H. (2017). Maxi-anion channels play a key role in glutamate-induced ATP release from mouse astrocytes in primary culture. *Neuroreport*, *28*, 380–385.

Zhivotovsky, B., Gahm, A., Ankarcrona, M., Nicotera, P., & Orrenius, S. (1995). Multiple proteases are involved in thymocyte apoptosis. *Experimental Cell Research*, *221*, 404–412.

Zhou, S. S., Hazama, A., & Okada, Y. (1998). Tyrosine kinase-independent extracellular action of genistein on the CFTR Cl^- channel in guinea pig ventricular myocytes and CFTR-transfected mouse fibroblasts. *The Japanese Journal of Physiology*, *48*, 389–396.

CHAPTER NINE

Molecular mechanisms for the regulation of blood pressure by potassium

Adrián Rafael Murillo-de-Ozores[a,b], Gerardo Gamba[a,c,d], Maria Castañeda-Bueno[a,*]

[a]Molecular Physiology Unit, Department of Nephrology and Mineral Metabolism, Instituto Nacional de Ciencias Médicas y Nutrición Salvador Zubirán, Mexico City, Mexico
[b]Facultad de Medicina, Universidad Nacional Autónoma de México, Mexico City, Mexico
[c]Instituto de Investigaciones Biomédicas, Universidad Nacional Autónoma de México, Mexico City, Mexico
[d]Tecnológico de Monterrey, Escuela de Medicina y Ciencias de la Salud, Monterrey, NL, Mexico
*Corresponding author: e-mail addresses: maria.castanedab@incmnsz.mx; mcasta85@yahoo.com.mx

Contents

Abstract

It has been well documented that the amount of potassium in the diet is associated with blood pressure levels in the population: the higher the potassium consumption, the lower the blood pressure and the cardiovascular mortality. In the last few years certain mechanisms for potassium regulation of salt reabsorption in the kidney have been elucidated at the molecular level. In this work we discuss the evidence demonstrating the relationship between potassium intake and blood pressure levels in human populations and in animal models, as well as the experimental data that reveal the effects of potassium on transepithelial Na^+ reabsorption in different nephron segments. We also discuss the physiological relevance of K^+-induced natriuresis, and finally, we focus on the molecular mechanisms by which extracellular potassium modulates the activity of the renal NaCl cotransporter, which is the mechanism that has been best dissected so far.

Current Topics in Membranes, Volume 83
ISSN 1063-5823
https://doi.org/10.1016/bs.ctm.2019.01.004

Abbreviations

ASDN aldosterone-sensitive distal nephron
BK Maxi-K channel
$[Cl^-]_I$ intracellular chloride concentration
CCD cortical collecting duct
CNT connecting tubule
CUL3 Cullin 3
DCT distal convoluted tubule
ENaC epithelial Na^+ channel
FHHt familial hyperkalemic hypertension
$[K^+]_e$ extracellular potassium concentration
KIR inward rectifier K^+ channel
KLHL3 kelch like family member 3
NCC Na^+:Cl^- cotransporter
NKCC Na^+:K^+:$2Cl^-$ cotransporter
OSR1 oxidative stress-responsive 1 protein
ROMK renal outer medullary K^+ channel
SPAK Ste20/SPS1-related proline-alanine rich kinase
TAL thick ascending limb of Henle's loop
WHO World Health Organization
WNK with no lysine (K) kinase

1. Evidences of the impact of dietary potassium intake on blood pressure levels

Hypertension affects >1 billion people around the world and it is one of the leading risk factors for cardiovascular disease and death (Mente et al., 2014). Although it has been thoroughly documented that sodium intake plays a determinant role in the pathogenesis of hypertension, potassium content in the diet might be as important as sodium in the regulation of blood pressure.

Compared to ancient diets, the current typical Western diet has a relatively low potassium content (~54 mmol or 2.1 g/day) and a high sodium content (~214 mmol or 4.9 g/day) (Mente et al., 2014). For example, individuals from the Yanomamo tribe in South America have a diet rich in potassium (~150 mmol or 5.9 g/day) (Oliver, Cohen, & Neel, 1975), with a potassium content well above the one recommended by the World Health Organization (WHO) in order to lower the risk of cardiovascular disease (at least 90 mmol or 3.5 g/day) (WHO, 2012b). In contrast, sodium intake of Yanomamo Indians averages (~1 mmol or 0.023 g/day), while WHO recommends ingesting <86 mmol or 2 g/day (WHO, 2012a). It is

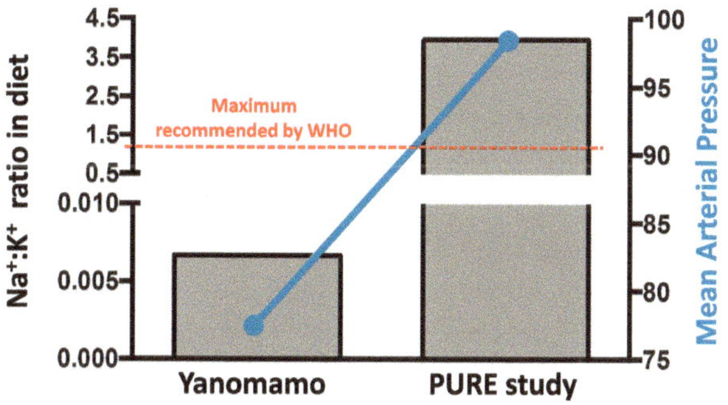

Fig. 1 The ratio of Na+:K$^+$ contained in diets from different cultures and its relationship to blood pressure. Yanomamo Indians have a diet with an exceptionally low Na$^+$:K$^+$ ratio and it is likely that the low mean arterial pressure observed in those individuals is related to this variable. On the other hand, individuals from civilized environments, such as the ones included in the PURE cohort (Mente et al., 2014), have diets with very high Na$^+$:K$^+$ ratios, even higher than the maximum values recommended by the WHO, and it is possible that this is an important factor contributing to the high prevalence of hypertension around the world.

possible that the lower prevalence of hypertension in these isolated communities is the result of the low Na$^+$:K$^+$ ratio in their diet, as it has been shown that this parameter has a stronger correlation with blood pressure than potassium or sodium intakes alone (Dahl, Leitl, & Heine, 1972; Khaw & Barrett-Connor, 1988; Mente et al., 2014) (Fig. 1).

Several basic, clinical and epidemiological studies have highlighted the inverse relationship between potassium consumption and blood pressure levels. In 1928, Addison reported that potassium salts could alleviate elevated blood pressure in hypertensive patients (Addison, 1928), and this finding was confirmed by subsequent reports by different groups (Kempner, 1974; Mcquarrie, Thompson, & Anderson, 1935; Meneely, Ball, & Youmans, 1957; Priddle, 1931). It has been shown that lowering potassium content in the diet increases blood pressure levels in patients with primary hypertension (Krishna & Kapoor, 1991). Additionally, part of the hypotensive properties of the dietary approaches to stop hypertension (DASH) diet has been attributed to the high content of potassium in such diet (Sacks et al., 2001). Interestingly, potassium depletion may also be involved in the pathogenesis of secondary hypertension caused by mineralocorticoid excess such as primary aldosteronism, as suggested by a rat model where potassium supplementation prevented the increase in blood pressure caused by deoxycorticosterone acetate (DOCA)-salt treatment (Fujita & Sato, 1983). This might also be true in humans, as it has been shown that potassium supplementation (~42 mmol or 3.1 gK^+/day) reduces sodium retention in healthy volunteers that were administered fludrocortisone for 10 days (Krishna & Kapoor, 1993).

More recent studies published in the New England Journal of Medicine, where a large and heterogeneous cohort of >100,000 individuals was analyzed as part of the Prospective Urban Rural Epidemiology (PURE) study, showed that potassium excretion (measured as a surrogate for potassium intake) is negatively correlated with risk of major cardiovascular events (O'Donnell et al., 2014) and it is possible that this effect is related to the fact that, in this same cohort, higher potassium excretion is related to lower blood pressure levels, independently to the effect of sodium intake (Mente et al., 2014). A recent meta-analysis also showed a significant effect of potassium supplementation on lowering systolic and diastolic blood pressure, and it has been suggested that potassium could serve as an adjuvant antihypertensive agent for patients with essential hypertension (Poorolajal et al., 2017).

The kidney is one of the major organs responsible for both potassium and blood pressure homeostasis. Ninety percent of the ingested K^+ is excreted in the urine, while the remaining 10% is excreted along with the stool. Accordingly, renal failure is often associated with hyperkalemia (Zacchia, Abategiovanni, Stratigis, & Capasso, 2016). The kidney also plays a determinant role in the long-term regulation of blood pressure, and it has been proposed that defective sodium handling by the kidney is a pathogenic hallmark of hypertension (Guyton, 1991). Supporting this concept, all the monogenic forms of hypertension described so far affect, directly or indirectly, sodium reabsorption by the kidney (Lifton, Gharavi, Geller, & Hughes, 2001). Additionally, normotensive kidney transplant recipients whose donor had a familial history of hypertension have a higher probability to develop hypertension later in life (Guidi et al., 1996; Guidi, Cozzi, Minetti, & Bianchi, 1998), and it has been shown that angiotensin II requires the presence of its receptor specifically in the kidney in order to cause hypertension and cardiac hypertrophy (Crowley et al., 2006). The mechanisms linking altered sodium handling by the kidney to hypertension and other mechanisms explaining the "salt-sensitivity" of hypertension have been thoroughly reviewed elsewhere (Lifton et al., 2001; Salt, 2015, complete issue).

Recent evidence has shown that renal handling of sodium, and therefore blood pressure, is directly affected by potassium intake. In this review, we focus on classic micropuncture and microperfusion studies that shed light on the effect of potassium on renal sodium handling by different nephron segments, and in more recent genetic and molecular studies that have brought attention to the WNK4-SPAK-NCC signaling pathway in the distal convoluted tubule as an important player in the regulation of blood pressure by potassium.

2. Natriuretic and diuretic effects of potassium; nephron segments involved in the natriuretic effect of high potassium intake

The diuretic effect of potassium salts is not new knowledge. In 1935, Keith and Binger wrote: "For two centuries certain potassium salts have been employed as diuretics in clinical medicine" (Keith & Binger, 1935). They go on to cite the works of several physicians who administered potassium salts in large doses that produced diuresis and satisfactory outcomes in patients with edema. In their 1935 work, Keith and Binger analyzed the diuretic effect in normal individuals and edematous patients of potassium chloride, bicarbonate, nitrate, acetate, and citrate, and showed that all potassium salts produced diuresis, with the greater effect produced by potassium nitrate, followed by potassium chloride. A sudden decrease in water excretion was reported after discontinuing administration of the salt (Keith & Binger, 1935). The observed diuresis has been shown to be accompanied by increased rates of sodium and chloride excretion (Kahn & Bohrer, 1967).

In animal models, administration of large amounts of potassium salts has allowed to observe dramatic effects of potassium on diuresis. In these studies, the administered high potassium diets usually contain 5–10 times the amount of K^+ contained in the "control" diets. With a diet containing five times more K^+, urinary volumes more than doubled (~2.5 ×) (Castaneda-Bueno et al., 2014). With diets containing 10 times more K^+, urinary volumes observed were 4–6 times larger than those observed with the control diets (Cheng, Truong, Baum, & Huang, 2012; O'Reilly et al., 2006). The increase in urinary volume was paralleled by an increase in natriuresis (Castaneda-Bueno et al., 2014; Cheng et al., 2012) and a negative Na^+ balance (Jung et al., 2011).

Despite these clear observations, the diuresis/natriuresis induced by high potassium intake has generally been disregarded as one of the main mechanisms linking high potassium consumption with lower blood pressure (Adrogué & Madias, 2007; He et al., 2010). Other mechanisms including effects on vasculature, sympathetic nervous system, etc. have gained more attention. This is striking, given that diuretics are among the most common drugs used for the treatment of hypertension.

In 1970, Vander showed that the direct intra-arterial infusion of KCl in the right kidneys of dogs increased Na^+ excretion by the right kidneys, but not by the contralateral kidney (Vander, 1970). This demonstrated that the natriuretic effect of potassium is due to direct renal actions of potassium.

Experiments performed during the micropuncture and microperfusion era began to delineate the nephron segments in which high K^+ intake exerts this inhibitory action on Na^+ reabsorption (Table 1). Free flow micropuncture experiments showed inhibition of absolute rates of fluid reabsorption by the proximal convoluted tubules in rats perfused intravenously with KCl (Brandis et al., 1972). Microperfusion with 10 mM KCl of peritubular capillaries irrigating the micropunctured proximal convoluted tubules also decreased fluid reabsorption, suggesting that the increase in K^+ concentration in peritubular fluid was responsible for such inhibition. Kirchner et al., however, failed to observe an effect on proximal fluid or Cl^- reabsorption in micropunctured tubules of rats acutely infused with KNO_3 (Kirchner, 1983). In rats on normal or high K^+ diets, Wright et al. observed inhibition of proximal tubule Na^+ reabsorption after acute KCl infusion by free flow micropuncture, although the segment they considered as "proximal tubule" comprised proximal tubule and loop of Henle (Wright et al., 1971).

Several works also showed an inhibitory effect of K^+ on Na^+ and Cl^- reabsorption in the loop of Henle. For example, increasing K^+ concentration in the peritubular or luminal fluid of in vitro microperfused medullary thick ascending limbs of mice or rabbits decreased Na^+ and Cl^- reabsorption (Stokes et al., 1982). In vivo, micropuncture of proximal and distal rat nephrons showed that acute K^+ infusion inhibited reabsorption of Cl^- in the loop of Henle (Kirchner, 1983). In rats maintained on a high K^+ diet for 10–14 days, a decrease in the Na^+ reabsorption rate of the loop of Henle was measured by in vivo microperfusion (Unwin et al., 1994). More recently, Cheng et al. performed in vitro microperfusion and detected decreased levels of Na^+ reabsorption in cortical thick ascending limbs of mice that were maintained on high K^+ diet when compared to tubules of mice maintained on normal diet (Cheng et al., 2012). A decrease in furosemide sensitive Na^+ excretion in mice on high K^+ diet has also been reported (Wang et al., 2017).

The increased diuresis on high K^+ diets is not only secondary to the increase in Na^+ and Cl^- excretion, as experiments show that an increase in free water excretion occurs (Cheng et al., 2012; Kahn & Bohrer, 1967).

Table 1 Nephron segments in which a decrease in Na+, Cl^-, or fluid reabsorption has been observed in response to a K+ challenge.

Authors (year)	Nephron segments studied	Employed methodology	K^+ challenge	Effect observed
Brandis, Keyes, and Windhager (1972)	Proximal convoluted tubule	Free flow micropuncture, rats	Intravenous perfusion of KCl	↓ fluid reabsorption
Brandis et al. (1972)	Proximal convoluted tubule	Free flow micropuncture, rats	Microperfusion with 10 mM KCl of peritubular capillaries irrigating the tubules	↓ fluid reabsorption
Kirchner (1983)	Proximal tubule (superficial nephrons)	Micropuncture, rats	Intravenous acute infusion of KNO3	No change in fluid of Cl^- reabsorption
Kirchner (1983)	Loop of Henle (superficial nephrons)	Micropuncture, rats	Intravenous acute infusion of KNO3	↓ in Cl^- reabsorption
Wright, Strieder, Fowler, and Giebisch (1971)	Proximal tubule and loop of Henle	Free flow micropuncture, rats	Acute infusion of KCl in rats maintained on normal or high K^+ diets	↓ in Na^+ reabsorption
Stokes, Lee, and Williams (1982)	Medullary thick ascending limb of Henle's loop	In vitro microperfusion, mice or rabbits	Increasing K^+ concentration n peritubular or luminal fluid	↓ in Na^+ and Cl^- reabsorption
Unwin, Capasso, and Giebisch (1994)	Loop of Henle (superficial nephrons)	In vivo microperfusion, rats	High K^+ diet for 10–14 days	↓ in Na^+ reabsorption
Cheng et al. (2012)	Cortical thick ascending limbs of Henle's loop	In vitro microperfusion, mice	High K^+ diet for 2 weeks	↓ in Na^+ reabsorption

Continued

Table 1 Nephron segments in which a decrease in Na+, Cl^-, or fluid reabsorption has been observed in response to a K+ challenge.—cont'd

Authors (year)	Nephron segments studied	Employed methodology	K^+ challenge	Effect observed
Wang, Wen, Li, Wang-france, and Sansom (2017)	Thick ascending limb of Henle's loop	Measurement of furosemide-sensitive Na + excretion, mice	High K^+ diet for 4–7 days	↓ in furosemide-sensitive Na^+ reabsorption
Vallon, Schroth, Lang, Kuhl, and Uchida (2009)	Distal convoluted tubule	Measurement of NCC phosphorylation (Western blot), mice	Low or high K^+ diets for 7 days	↓ in NCC phosphorylation with increased dietary K^+
Frindt and Palmer (2010)	Distal convoluted tubule	Measurement of NCC abundance in apical membrane (Western blot), rats	Low or high K^+ diets for 6–8 days	↓ in NCC surface expression with increased dietary K^+
Castaneda-Bueno et al. (2014)	Distal convoluted tubule	Measurement of NCC phosphorylation (Western blot), mice	High KCl diets for 4 days High K-citrate diet for 4 days	↓ in NCC phosphor. With high KCl diet ↑ in NCC phosphor. With high K-citrate diet
Sorensen et al. (2013)	Distal convoluted tubule	Measurement of NCC phosphorylation (Western blot), mice	Oral gavage with 2% K^+ (acute effect)	↓ in NCC phosphorylation (15 min after gavage)

Rengarajan et al. (2014)	Distal convoluted tubule	Measurement of NCC phosphorylation (Western blot), rats	Intravenous KCl infusion	↓ in NCC phosphorylation
Terker, Zhang, Erspamer, et al. (2015); Terker, Zhang, McCormick, et al. (2015)	Distal convoluted tubule	Measurement of NCC phosphorylation (Western blot), mice	Amiloride admin. For 5–7 days (to increase plasma K^+)	↓ in NCC expression and phosphorylation
Terker et al. (2016)	Distal convoluted tubule	Measurement of NCC phosphorylation (Western blot), mice	Inducible, nephron specific MR-knockout mice (that develop hyperkalemia when treated with doxycycline)	↓ in NCC expression and phosphorylation
Jung et al. (2011)	Various segments	Measurement of transporter expression (Western blot)	High K^+ diet for 1 week and 3 weeks	↓ in NHE3, NBC1, and NCC expression
Yang, Xu, et al. (2018)	Various segments	Measurement of transporter expression and phosphoryl. (Western blot)	High K^+ diet for 1 week	↓ in NHE3 and NCC expression ↓ in NKCC2 and NCC phosphoryl.

MR, mineralocorticoid receptor.

In this regard, the decrease in NaCl reabsorption by the thick ascending limb may contribute to a decrease in the urine concentrating ability of the kidney and explain, at least in part, this observation.

Recent evidence shows that high K^+ intakes and increases in plasma K^+ concentration also inhibit Na^+ and Cl^- reabsorption in the distal convoluted tubule (DCT). Phosphorylation of the thiazide sensitive Na-Cl cotransporter NCC, which is the major apical transporter involved in Na^+ and Cl^- reabsorption in this nephron segment, is drastically decreased by high K^+ intake or elevations in plasma K^+ (Sorensen et al., 2013; Vallon et al., 2009). It has been well demonstrated that phosphorylation of NCC is a good indicator of cotransporter activity, and thus, decreased phosphorylation levels of NCC are indicative of low Na^+ and Cl^- reabsorption levels in the DCT.

Finally, in a recent mathematical model of the rat kidney by Weinstein (2017), the effect on Na^+ reabsorption in different nephron segments of increasing peritubular K^+ concentration from 5 to 5.5 mM was analyzed. It was predicted that a 2.3-fold increase in natriuresis would result. The author concluded that, although Na^+ reabsorption in the proximal convoluted tubule, ascending loop of Henle, and DCT were predicted to be inhibited, the main contributor to natriuresis was be the inhibition of proximal convoluted tubule reabsorption. The reason for this, according to this model, is that the loop of Henle and the DCT respond to increases in Na^+ delivery with increased Na^+ reabsorption.

3. Role of the kidney in potassium homeostasis

Given the steep K^+ concentration gradient that is maintained between the intracellular and extracellular compartments (120 mM intracellular vs 3.5–5.5 mM extracellular), which is crucial for a variety of cell functions, the majority of K^+ contained in the body is found within the cells, and only a minor part is found in the extracellular fluid, i.e., approximately 63 mmol in a young adult male of 70 kg with and extracellular fluid volume of ∼14 L (14 L × 4.5 mmol/L). More K^+ may be present in a meal than that contained in the extracellular fluid. For instance, the average daily K^+ intake observed in the population of the PURE study was 2.1 g (54 mmol) (Mente et al., 2014). Thus, the dietary intake of K^+ could significantly modify the K^+ concentration of the extracellular fluid if it were not for the existence of mechanisms that buffer these increases by promoting the internalization of K^+ to the cells. Simultaneously, mechanisms to increase renal K^+ excretion are activated in order to match K^+ excretion to K^+ intake and achieve K^+ balance.

No matter how much K^+ is ingested in the diet, the kidneys filter approximately 720 mmol of K^+ per day (180 L × 4 mmol/L), of which 90% are reabsorbed in the proximal tubule and thick ascending limb of Henle's loop (TAL) (Boulpaep & Boron, 2016). Therefore, the distal nephron, in particular the aldosterone-sensitive distal nephron (ASDN), which comprises the late portion of the DCT, the connecting tubule (CNT), and the cortical collecting duct (CCD), is the segment in charge of adjusting urinary excretion to dietary intake in order to maintain K^+ balance. When dietary intake is normal to high, K^+ is secreted in the distal nephron. Secreted K^+ can reach values of 20%–180% of the filtered K^+ and urinary excretion levels can reach to amounts of 10%–150% of the filtered K^+ (because some of the secreted K^+ is reabsorbed in the collecting duct). In contrast, when dietary intake is low, net K^+ reabsorption is observed in the distal nephron, and K^+ excretion is reduced to as much as 2% of the filtered amount.

Within the ASDN, the late portion of the DCT and the CNT are the segments which appear to play a major role in the regulation of K^+ balance (Meneton, Loffing, & Warnock, 2004). Compared to the collecting duct, the late DCT and CNT show a several-fold higher Na^+:K^+-ATPase activity per millimeter of tubule length in different mammalian species (Katz, Doucet, & Morel, 1979). In mice, apical accumulation of the K^+ channel ROMK, which participates in K^+ secretion, is only observed in the late DCT and CNT when mice are fed a moderately high K^+ diet, while apical ROMK accumulation is observed in the CCD only when K^+ intake levels are very high (Loffing-Cueni et al., 2003). The activity of the epithelial Na^+ channel ENaC (which contributes to establishing the driving force for K^+ secretion as explained below) is also several times larger in the late DCT/CNT than in the CCD of rats treated with aldosterone or fed a high K^+ diet (Frindt & Palmer, 2004). Moreover, genetic mouse models deficient in ENaC activity in the CCD do not display a salt-losing or K^+ retaining phenotype, even when placed on low Na^+ diet or high K^+ diet, respectively (Rubera et al., 2003), whereas mice lacking ENaC activity in the entire ASDN (Boscardin et al., 2017, 2018; Perrier et al., 2016) or lacking ENaC activity specifically in the CNT (Christensen et al., 2010) do present a Na^+ wasting phenotype when on low Na^+ diet which is accompanied with K^+ retention. Thus, it has been proposed that the collecting duct makes an important contribution to K^+ secretion and for achieving K^+ balance only when K^+ intake levels are very high. With the low K^+ intake levels observed in modern societies' diets it is thus likely that the majority of aldosterone-regulated K+ secretion takes place in the late DCT and CNT (Meneton et al., 2004).

The most relevant molecular players involved in the process of K^+ secretion in the distal nephron are the epithelial Na^+ channel (ENaC) and the K^+ channels renal outer medullary K^+ channel (ROMK), and the Maxi-K channel (BK).

ENaC expression in the nephron is found in the apical membrane of the ASDN, specifically in principal cells of the CNT and CCD, and in similar cells that are present in the late portion of the DCT (Loffing et al., 2000). Na^+ entry to the cell via ENaC promotes the depolarization of the apical membrane, thus generating a lumen-negative transepithelial voltage and contributing to the driving force for K^+ secretion. ENaC's role in K^+ secretion is evidenced by the existence of genetic diseases such as pseudohypoaldosteronism type I, which presents with hyperkalemia and is caused by inactivating mutations in any of the three subunits forming the channel (Chang et al., 1996) (Strautnieks, Thompson, Gardiner, & Chung, 1996), or Liddle syndrome, which features hypokalemia and is caused by mutations in β- or γ-ENaC subunits that prevent channel retrieval from the plasma membrane and thus lead to channel overactivation (Hansson et al., 1995; Schild et al., 1996; Shimkets et al., 1994). ENaC is one of the principal targets for regulation by aldosterone that mainly affects the channel's retrieval from the plasma membrane mediated by Nedd4-2 (Debonneville et al., 2002). Another important target for regulation by aldosterone is the Na^+:K^+-ATPase (Féraille et al., 1993). Coordinated upregulation of the Na^+:K^+-ATPase at the basolateral membrane and ENaC at the apical membrane raises the Na^+ reabsorption capacity of the ASDN, which stimulates K^+ secretion.

Potassium secretion across the apical membrane of the ASDN can occur via two mayor pathways: ROMK and BK channels. ROMK expression in the nephron begins in the TAL and continues all the way to the CD (Wade et al., 2011). However, net K^+ secretion mediated by ROMK is only observed in the segments comprising the ASDN, segments in which ROMK is coexpressed with ENaC in principal cells. A high K^+ intake promotes an increase in the apical conductance of K^+ and in the activity of ROMK (Palmer, 1999). Given that ROMK channels have a high basal open probability that virtually makes them constitutively active, regulation of ROMK-mediated K^+ secretion is achieved through the regulation of translocation of channels from subapical compartments to the plasma membrane (Frindt & Palmer, 2010; Wade et al., 2011). Modulation of the channel's synthesis rate has not been observed (Frindt, Zhou, Sackin, & Palmer, 1998). Although certain controversy exists, several lines of evidence suggest that the regulation of ROMK's translocation to the apical membrane by high K^+ intake is independent of aldosterone (Huang et al., 2004; Wei, Bloom, Lin, Gu, & Wang, 2001).

BK channels are activated by membrane depolarization, elevation of intracellular calcium concentration, hypoosmotic stress, or membrane stretch (Carrisoza-Gaytan, Carattino, Kleyman, & Satlin, 2016). Unlike ROMK, BK channels are normally inactive and only become activated under conditions of high tubular flow rate, such as that observed under high K^+ intake (Woda, Bragin, Kleyman, & Satlin, 2001). Flow-mediated activation is associated with an increase in intracellular calcium (Carrisoza-Gaytan et al., 2016) (Welling, 2013). BK channels in the kidney are expressed in the apical membrane of the ASDN, both in principal cells and intercalated cells (all types); however, expression is low in principal cells and much more robust in intercalated cells (Carrisoza-Gaytan et al., 2016; Welling, 2013). Despite this high expression, it is still under debate whether BK channels expressed in intercalated cells can mediate flow induced K^+ secretion (FIKS). Indeed principal cells present robust expression of the Na^+:K^+-ATPase in the basolateral membrane that mediates entry of K^+ ions into the cell that can be extruded to the tubule lumen via BK. However, very low amounts of Na^+:K^+-ATPase are present in intercalated cells (Pácha, Frindt, Sackin, & Palmer, 1991). It has been proposed that basolateral K^+ entry to these cells may be mediated by Na^+:K^+:$2Cl^-$ cotransporter type 1 (NKCC1) activity. Evidence favoring this concept is that BK-mediated FIKS is dependent on a basolateral bumetanide sensitive- Cl^- dependent transport pathway (Liu et al., 2011). However, it is unclear how Na^+ entering the cells via NKCC1 can be recycled back to the interstitial fluid, which would be important to maintain NKCC1 activity. The importance of BK channels in K^+ secretion is demonstrated by the fact that its activity can compensate ROMK deficiency in ROMK knockout mice or patients with Bartter syndrome type II (caused by inactivating mutations in ROMK) (Simon, Karet, et al., 1996; Wagner et al., 2008). These mice and patients develop mild hypokalemia which is due to the impaired reabsorption capacity of the TAL which leads to increased distal Na^+ and flow delivery that activates K^+ secretion via BK channels. Finally, high dietary K^+ stimulates BK transcription and channel translocation to the plasma membrane, a response that is independent of aldosterone (Welling, 2013). However, aldosterone indirectly affects BK-mediated K^+ secretion through the stimulation of ENaC (Cornelius et al., 2015).

The natriuretic and diuretic effects of high K^+ diets described in the previous section, which appear to be due to an inhibition of Na^+ reabsorption pathways in the proximal tubule, TAL, and DCT, are important to maximize renal K^+ excretion (Fig. 2). It is generally argued that this is because: (1) the increase in Na^+ delivery to the ASDN allows the electrogenic Na^+-K^+ exchange through ENaC and ROMK/BK channels; (2) the increased fluid delivery to the ASDN slows down the rise in luminal K^+ concentration which is produced as a consequence of K^+ secretion; (3) increased flow stimulates the flow-activated BK channels. Nevertheless, recent data

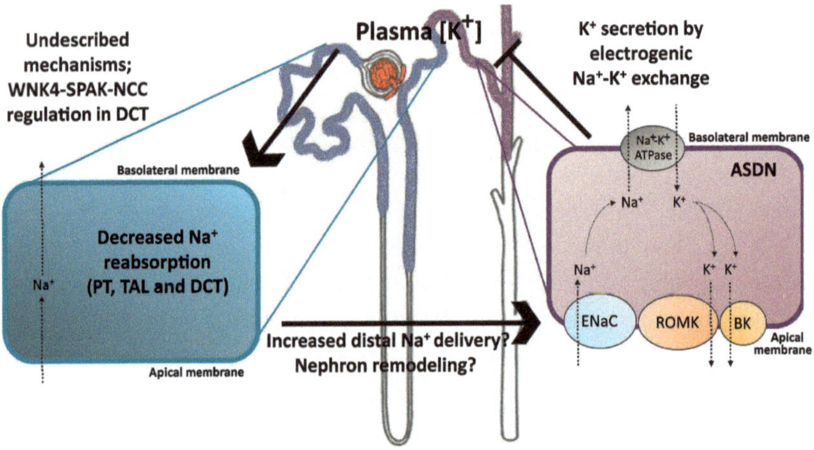

Fig. 2 Physiological relevance of K^+-induced natriuresis. Experimental evidence suggests that an increase in K^+ intake, and the consequent rise in plasma K^+ concentration, induce natriuresis due to a decrease in Na^+ reabsorption in the proximal tubule (PT), thick ascending limb of Henle's loop (TAL), and distal convoluted tubule (DCT). The resulting increase in Na^+ delivery and flow to the aldosterone-sensitive distal nephron (ASDN) stimulate the K^+ secretory apparatus. However, it is still under debate if high Na^+ delivery and flow directly stimulate K^+ secretion by stimulating ENaC mediated Na^+ reabsorption that generates a lumen negative transepithelial voltage increasing the driving force for K^+ secretion, and by activating flow induced K^+ secretion through BK channels. Recent evidence suggests that acute inhibition of proximal Na^+ reabsorption by itself does not stimulate K^+ secretion (Grimm, Coleman, Delpire, & Welling, 2017; Hunter, Craigie, Homer, Mullins, & Bailey, 2014), but that in a more chronic setting (3 days in mouse), remodeling of the distal nephron occurs with hypertrophy of ASDN, promoting K^+ secretion.

questions the idea that the increase in Na^+ delivery acutely increases the capacity of the ASDN to secrete K^+. This would mean that not only the higher availability of Na^+ in the tubule lumen favoring ENaC-mediated Na^+ reabsorption may be sufficient to promote Na^+ reabsorption. For example, early in vivo microperfusion studies suggested that luminal Na^+ concentration plays no role in modulating K^+ secretion (Good & Wright, 1979). Moreover, it has been shown recently that acute inhibition of NCC with hydrochlorothiazide elicited immediate natriuresis that was sustained for a 2-h period after diuretic administration, but no kaliuresis was observed (Hunter et al., 2014). Finally, in transgenic mice expressing a constitutively active version of Ste20/SPS1-realted proline-alanine rich kinase (SPAK; CA-SPAK mice) in which NCC is thus also constitutively activated, hyperkalemia is observed. However, thiazide administration elicits an immediate natriuretic effect, but a delayed kaliuretic effect, and thiazide normalizes plasma K^+ levels only after 3 days of thiazide administration (Grimm et al., 2017). These mice show remodeling of the distal tubule with hypertrophy of the DCT and hypotrophy of the CNT. Thiazide administration for 3 days normalizes these structural changes, correlating with the correction of plasma K^+ levels. Interestingly, nephron remodeling in response to altered Na^+ and K^+ intake in rabbits was reported decades ago (Kaissling & Le Hir, 1982), and although further investigation is needed to address the molecular mechanisms at play in this process, it is becoming clear that distal Na^+ delivery (or inhibition of proximal Na^+ reabsorption) may affect K^+ secretion in a more complex way than previously thought.

4. Molecular mechanisms implicated in the natriuretic and diuretic effects of potassium

Although, as discussed above, it has been shown that sodium reabsorption by different nephron segments is inhibited by high K^+ intake, most of the molecular mechanisms responsible for these phenomena remain elusive. Some of them, however, are just recently beginning to be understood. For example, decreased protein levels of the sodium-hydrogen exchanger NHE3 and the sodium-bicarbonate cotransporter NBC-1 could explain the effect on proximal tubule and possibly on the loop of Henle (Jung et al., 2011). Negative modulation of the Na^+:K^+:$2Cl^-$ cotransporter NKCC2 phosphorylation could also explain the inhibitory effect of high potassium upon sodium reabsorption in TAL (Yang, Xu, et al., 2018).

4.1 Regulation of the Na-Cl cotransporter NCC by potassium

While it has been suggested by a mathematical model that the proximal nephron is determinant for increased natriuresis by high K^+ intake (Weinstein, 2017), experimental evidence has shown that the distal nephron plays a key role in the relationship between K^+ and blood pressure. For example, it has been shown that the hypertensive effect of a high sodium/low potassium diet, such as the typical western diet, could be explained by increased salt reabsorption by NCC in the DCT in a mouse model. Wild-type mice fed with such diet displayed a significant increase in blood pressure, while NCC knockout mice did not (Terker, Zhang, McCormick, et al., 2015).

So far the best understood molecular mechanism explaining the inverse relationship between K^+ intake and blood pressure levels involves NCC regulation in the DCT. This understanding began with the description of monogenic diseases in which NCC activity is dysregulated, causing alterations in potassium and blood pressure homeostasis.

Gitelman syndrome (OMIM #263800) is caused by loss of function mutations in the gene *SLC12A3*, that codifies NCC, and it is characterized by hypokalemia, low blood pressure, metabolic alkalosis, hypocalciuria, and hypomagnesemia (Simon, Nelson-Williams, et al., 1996). Accordingly, this disease is reminiscent of thiazide overdose, and it manifests that NCC plays a determinant role not only in blood pressure, but also in K^+ regulation.

On the other hand, familial hyperkalemic hypertension (FHHt) (OMIM #614491-2 and #614495-6), also called Gordon syndrome or pseudohypoaldosteronism type II (PHAII), is associated with hyperkalemia, hypertension, metabolic acidosis and hypercalciuria (Wilson et al., 2001). Low doses of thiazide diuretics are able to correct all physiological abnormalities of FHHt (Mayan, Vered, Mouallem, Tzadok-witkon, & Pauzner, 2002), which suggests that NCC overactivation is involved in this disease. Nevertheless, no gain of function mutations have been described in *SLC12A3*. In 2001 Richard Lifton's group reported that mutations in the genes *PRKWNK1* and *PRKWNK4* caused FHHt (Wilson et al., 2001). These genes codify two members of the with no lysine (K) (WNK) kinase family that contains four members in total in human. These serine-threonine kinases are characterized by the atypical positioning of their catalytic lysine and they have been described as important regulators of transepithelial ion transport. In particular, WNK4 and WNK1 have been shown to be key regulators of NCC (Alessi et al., 2014) (Hadchouel, Ellison, & Gamba, 2016).

Gain of function mutations in WNK4 cause familial hyperkalemic hypertension (FHHt) due to NCC overactivation (Lalioti et al., 2006; Wilson et al., 2001; Yang et al., 2007). FHHt mutations in WNK4 cause an amino acid change in a region of the protein denominated "acidic domain," responsible for the interaction with kelch like family member 3 (KLHL3), an adaptor protein and receptor for the KLHL3-CUL3-RING ubiquitin ligase complex. Mutated WNK4 cannot interact with this complex and its proteasomal degradation is compromised, leading to increased WNK4 levels in the DCT (Ohta et al., 2013; Shibata, Zhang, Puthumana, Stone, & Lifton, 2013; Wakabayashi et al., 2013). Mutations in the genes coding KLHL3 and Cullin 3 (CUL3) are also a cause for FHHt (Boyden et al., 2012; Louis-Dit-Picard et al., 2012).

NCC activity is positively regulated by phosphorylation at several sites in its cytoplasmic N-terminal domain (Pacheco-Alvarez et al., 2006) by the kinases SPAK and OSR1 (oxidative stress-responsive 1 protein), which in turn are phosphorylated and activated by WNK kinases (Richardson et al., 2008; Vitari, Deak, Morrice, & Alessi, 2005). Although WNK1 and WNK3 are also expressed in the kidney, it has been suggested that WNK4 is the main WNK kinase responsible for NCC regulation in vivo (Castaneda-Bueno et al., 2012; Oi et al., 2012; Susa et al., 2017). Accordingly, WNK4 knockout mice have a phenotype resembling Gitelman syndrome (Castaneda-Bueno et al., 2012; Takahashi et al., 2014;Yang, Xie, et al., 2018).

Kinase activity of WNK kinases has been shown to be negatively modulated by chloride. X-ray crystallography showed that the chloride anion binds close to the active site, within the kinase domain of WNK1. Binding stabilizes an inactive conformation of the kinase. The decrease in Cl^- concentration ($[Cl^-]$) promotes Cl^- dissociation, and thus WNK1 activation and autophosphorylation in the activation loop (Piala et al., 2014). Later, it was shown that WNK4 has an even higher sensitivity to chloride (Bazua-Valenti et al., 2015) and this modulation could be important for WNK4 activity in vivo, as mathematical modeling of the DCT predicts that intracellular chloride concentration ($[Cl^-]_i$) fluctuates between ~12 and ~20 mM, in function of extracellular potassium (Terker, Zhang, McCormick, et al., 2015).

To our knowledge, the first report of the inverse relationship between NCC activity and plasma $[K^+]$ was published in 2009 by Uchida's group. In this work, mice were fed with high (5%) K^+ diets, which prompted a reciprocal inverse modulation of NCC phosphorylation at Thr53, Thr58 and Ser71 (Vallon et al., 2009). Later, several groups published similar findings performing dietary manipulations in mouse models (Castaneda-Bueno et al., 2014; Frindt & Palmer, 2010; Lubbe et al., 2013; Terker, Zhang, Erspamer, et al., 2015) and even in humans (Terker, Zhang, McCormick, et al., 2015). Sorensen and collaborators used gastric gavage of 2% K^+ (15 μL/g) in order to study the rapid response of NCC to acute increases in plasma [K^+] (Sorensen et al., 2013). K^+ loading promoted a marked decrease in NCC phosphorylation, accompanied by increased natriuresis, which has significantly blunted in NCC knockout mice. These results suggest that the natriuretic effect of potassium salts depend, at least in part, on decreased salt reabsorption in the DCT.

In 2015, Terker and collaborators proposed that extracellular potassium concentration ($[K^+]_e$) has a direct effect on DCT's $[Cl^-]_i$ due to effects on cell voltage (Fig. 3). They proposed that low $[K^+]_e$ promotes apical membrane hyperpolarization which increases the driving force for chloride efflux from the cell, leading to WNK activation (Terker, Zhang, McCormick, et al., 2015). This hypothesis has been supported by experimental evidence in in vitro and ex vivo models (Penton et al., 2016; Terker, Zhang, Erspamer, et al., 2015). Therefore, the current view is that NCC activity is increased in response to decreases in extracellular [K^+] due to higher WNK4 activity.

In the DCT, the basolateral conductance of K^+ is given by KIR4.1/KIR5.1 heterotetramer potassium channels, characterized by ~40pS K^+ currents. These channels are responsible for the recycling of K^+ that entered through the Na^+-K^+ ATPase (Cuevas et al., 2017; Su & Wang, 2016). KIR4.1 has been called the 'potassium sensor' of the DCT, as *Kcnj10* (which codifies KIR4.1) knockout mice lose NCC modulation by potassium (Cuevas et al., 2017). In fact, loss of function of KIR4.1 causes EAST/SeSAME syndrome, a complex disorder with a renal phenotype similar to Gitelman syndrome (Bockenhauer et al., 2009; Scholl et al., 2009).

In contrast, *Kcnj16* (codifying KIR5.1) knockout mice have increased sensibility to thiazide diuretics, a surrogate for NCC activity in vivo (Paulais et al., 2011). This observation is likely explained because KIR4.1 homomultimers are insensitive to intracellular pH, while KIR4.1/5.1 channels heteromers are sensitive to physiological levels of intracellular pH. Whether modulation of intracellular pH of the DCT is physiologically relevant to NCC modulation remains elusive. However, it has been shown that chronic K^+ depletion causes increased activity of KIR4.1/5.1

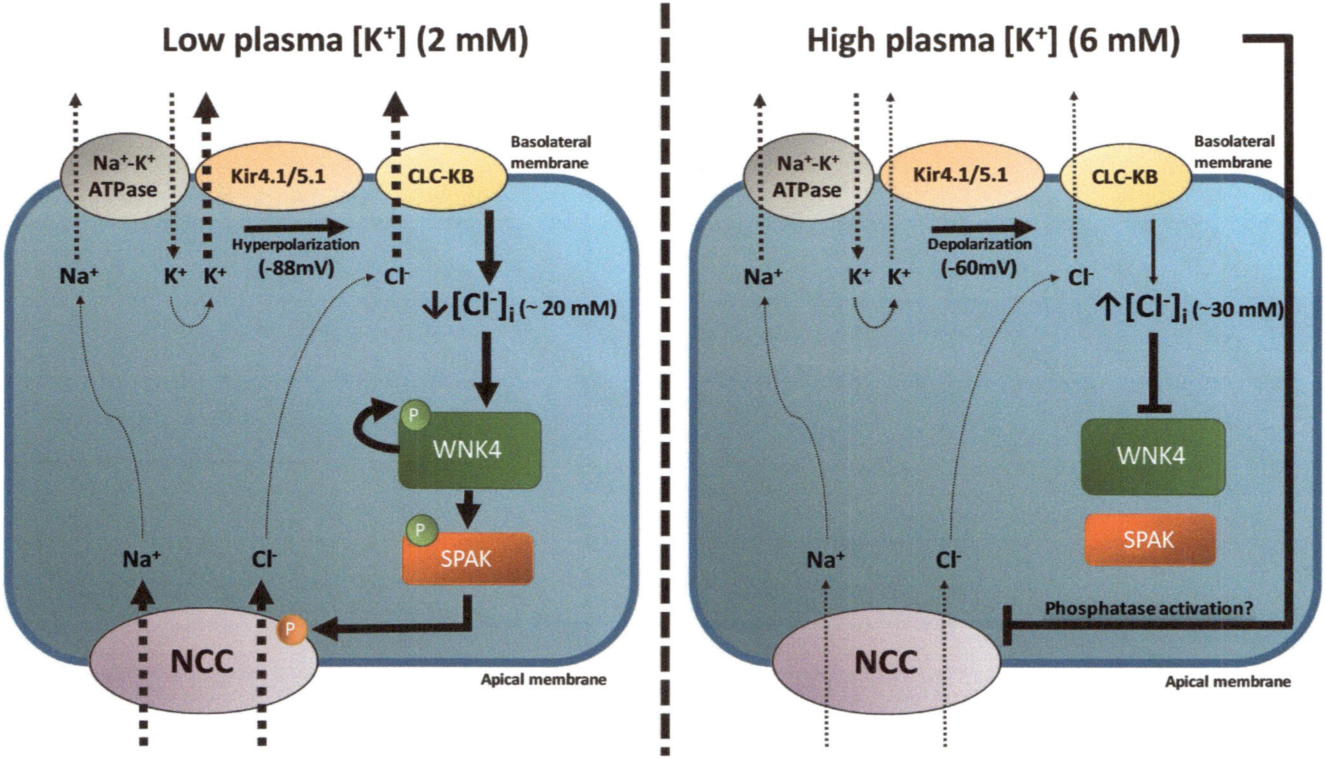

Fig. 3 Molecular mechanisms by which plasma K^+ levels regulate NCC activity in the DCT. During hypokalemic states ($[K^+] = 2$ mM), K^+ extrusion through basolateral Kir4.1/5.1 channels increases, which promotes hyperpolarization (−88 mV) of distal convoluted tubule (DCT) cells that leads to increased chloride efflux through CLC-KB. The decrease in intracellular chloride concentration ($[Cl^-]_i \sim 20$ mM) causes phosphorylation and activation of WNK4 kinase, which promotes SPAK and, therefore, NCC activation. On the other hand, high plasma K^+ ($[K^+] = 6$ mM) depolarizes DCT cells (−60 mV), which prevents chloride efflux. High $[Cl^-]_i$ ($\sim$30 mM) renders WNK4 inactive, decreasing SPAK and NCC phosphorylation. Additional mechanisms promote specific NCC dephosphorylation, probably due to activation of protein phosphatases. (Values stated were calculated through mathematical modeling by Terker, Zhang, McCormick, et al. (2015).)

heterotetramers, promoting increased potassium currents, and hyperpolarization of the distal convoluted tubule membrane (Wang et al., 2018).

Basolateral chloride efflux from the DCT cells occurs mainly through the ClC-KB (ClC-K2 in mouse) chloride channel (Hennings et al., 2017), whose translocation to the basolateral membrane and stability depend on the protein Barttin. Nomura and collaborators showed that hypomorphic mice with a mutated Barttin are able to thrive, but display a Gitelman-like phenotype, with the inability to upregulate NCC by further K^+ depletion (Nomura et al., 2018). This was the first evidence that chloride efflux was responsible for the activation of NCC by hypokalemia in vivo.

4.2 Additional mechanisms for potassium-mediated regulation of NCC

Inhibition of NCC by high extracellular potassium seems to involve additional mechanisms independent of $[Cl^-]_i$. For example, in an ex vivo model of mouse kidney slices (Penton et al., 2016) observed that NCC dephosphorylation by high extracellular K^+ occurs even when WNK-SPAK activity is clamped on a low chloride medium. In this work, the mechanism responsible for NCC dephosphorylation remained elusive, as protein phosphatase 1 (PP1)/protein phosphatase 2 (PP2) and calcineurin inhibitors could not prevent dephosphorylation. In contrast, Shoda et al. showed that in mice, rapid NCC dephosphorylation in response to acute K^+ loading was prevented by inhibition of the phosphatase calcineurin with tacrolimus (Shoda et al., 2017). Conflicting data between these works may be due to the use of different models. For example, the results obtained in the in vivo approach may be influenced by systemic factors additional to the direct effects of K^+ on DCT. Further investigation will be helpful to understand the intricacies of the NCC dephosphorylation process and its relationship to potassium homeostasis.

A recent work from our group showed that binding of Cl^- to the kinase domain of WNK4 is not the sole regulatory mechanism of kinase activity. We described that at least two phosphorylation sites, one located within the N-terminal domain (S64) and another one within the C-terminal domain (S1196), modulate WNK4 activity measured as WNK4-mediated SPAK phosphorylation and T-loop autophosphorylation (Castañeda-Bueno et al., 2017). These sites are target for phosphorylation by Protein kinase C (PKC) and Protein kinase A (PKA) and we initially showed that its phosphorylation is induced in response to stimulation of cells with angiotensin II. More recently, we have observed that hypokalemia in mouse is also an important stimulus that promotes phosphorylation of these sites (Murillo-de-Ozores, Grajeda-Medina, Gamba, & Castañeda-Bueno, 2018). In an in vitro cell model, we observed that intracellular chloride depletion, secondary to cell incubation in a low K^+ medium, is responsible for such increase in the phosphorylation levels of these sites. Thus, we have proposed that DCT intracellular Cl^- depletion secondary to decreases in extracellular $[K^+]$, may not only promote WNK4 activation due to Cl^- dissociation, but also through the phosphorylation of these sites.

Finally, it has been observed in several mouse models that hypokalemia promotes an increase in WNK4 expression levels (Ishizawa et al., 2016; Terker, Zhang, McCormick, et al., 2015). It has been recently shown that the molecular mechanism behind this effect involves a decrease in KLHL3-CUL3-RING-mediated degradation of WNK4. Hypokalemia promotes KLHL3 phosphorylation at Ser433, which is a site located within the substrate recognition domain of KLHL3. Thus, phosphorylation prevents substrate binding and subsequent ubiquitination and degradation (Ishizawa et al., 2016).

5. Summary and conclusions

For a long time, it has been known that potassium salts have diuretic and natriuretic effects that ameliorate hypertension and edematous states. Evidence suggests that increased potassium intake decreases sodium reabsorption by the proximal tubule, TAL, and DCT, even though the molecular mechanisms at play are mostly unknown at the moment. However, several groups across the world have shown that the WNK4-SPAK-NCC pathway in the DCT plays a pivotal role in the physiological relationship between potassium and blood pressure regulation. This pathway functions as a molecular switch that determines salt reabsorption in the DCT in response to plasma $[K^+]$. Turning off the switch by increased potassium intake allows appropriate K^+ secretion by the ASDN, while K^+ depletion turns on NCC to prevent electrogenic Na^+-K^+ exchange, and indirectly promotes an increase in blood pressure.

Further work will be required to assess the quantitative contribution of different nephron segments to the phenomenon of K^+-induced natriuresis. In the meanwhile, it is evident that eating more avocados and bananas is good for your health.

Acknowledgments

A.R.M.-O. is a graduate student from the "Programa de Doctorado en Ciencias Biomédicas, Universidad Nacional Autónoma de México (UNAM)" and received a fellowship 606808 from CONACYT.

This work was supported by National Institutes of Health Grant DK51496 (to G.G.) and Conacyt Grants No. 23 (to G.G.) and 257726 (to M.C.-B.).

References

Addison, W. L. T. (1928). The use of sodium chloride, potassium chloride, sodium bromide, and potassium bromide in cases of arterial hypertension which are amenable to potassium chloride. *Canadian Medical Association Journal*, *18*(3), 281–285.

Adrogué, H. J., & Madias, N. E. (2007). Sodium and potassium in the pathogenesis of hypertension. *New England Journal of Medicine*, *356*(19), 1966–1978.

Alessi, D. R., Zhang, J., Khanna, A., Hochdorfer, T., Shang, Y., & Kahle, K. T. (2014). The WNK-SPAK/OSR1 pathway: Master regulator of cation-chloride cotransporters. *Science Signaling*, 7(334), re3.

Bazua-Valenti, S., Chavez-Canales, M., Rojas-Vega, L., Gonzalez-Rodriguez, X., Vazquez, N., Rodriguez-Gama, A., et al. (2015). The effect of WNK4 on the Na+-Cl^- cotransporter Is modulated by intracellular chloride. *Journal of the American Society of Nephrology*, *26*(8), 1781–1786.

Bockenhauer, D., Feather, S., Stanescu, H. C., Bandulik, S., Zdebik, A. A., Reichold, M., et al. (2009). Epilepsy, ataxia, sensorineural deafness, tubulopathy, and KCNJ10 mutations. *New England Journal of Medicine*, *360*(19), 1960.

Boscardin, E., Perrier, R., Sergi, C., Maillard, M., Loffing, J., Loffing-Cueni, D., et al. (2017). Severe hyperkalemia is rescued by low-potassium diet in renal βENaC-deficient mice. *Pflügers Archiv - European Journal of Physiology*, *469*(10), 1387–1399.

Boscardin, E., Perrier, R., Sergi, C., Maillard, M. P., Loffing, J., Loffing-Cueni, D., et al. (2018). Plasma potassium determines NCC abundance in adult kidney-specific γ ENaC knockout. *Journal of the American Society of Nephrology*, *29*(3), 977–990.

Boulpaep, E. L., & Boron, W. F. (2016). *Medical physiology* (3rd Ed.). Elsevier.

Boyden, L. M., Choi, M., Choate, K. A., Nelson-Williams, C. J., Farhi, A., Toka, H. R., et al. (2012). Mutations in kelch-like 3 and cullin 3 cause hypertension and electrolyte abnormalities. *Nature*, *482*(7383), 98–102.

Brandis, M., Keyes, J., & Windhager, E. (1972). Potassium-induced inhibition of proximal tubular fluid reabsorption in rats. *The American Journal of Physiology*, *222*(2), 421–427.

Carrisoza-Gaytan, R., Carattino, M. D., Kleyman, T. R., & Satlin, L. M. (2016). An unexpected journey: Conceptual evolution of mechanoregulated potassium transport in the distal nephron. *American Journal of Physiology. Cell Physiology*, *310*, C243–C259.

Castañeda-Bueno, M., Arroyo, J. P., Zhang, J., Puthumana, J., Yarborough, O., Shibata, S., et al. (2017). Phosphorylation by PKC and PKA regulate the kinase activity and downstream signaling of WNK4. *Proceedings of the National Academy of Sciences*, *114*(5), E879–E886. https://doi.org/10.1073/pnas.1620315114.

Castaneda-Bueno, M., Cervantes-Perez, L. G., Rojas-Vega, L., Arroyo-Garza, I., Vazquez, N., Moreno, E., et al. (2014). Modulation of NCC activity by low and high K+ intake: Insights into the signaling pathways involved. *American Journal of Physiology Renal Physiology*, *306*(12), F1507–F1519.

Castaneda-Bueno, M., Cervantes-Perez, L. G., Vazquez, N., Uribe, N., Kantesaria, S., Morla, L., et al. (2012). Activation of the renal Na+: Cl- cotransporter by angiotensin II is a WNK4-dependent process. *Proceedings of the National Academy of Sciences*, *109*(20), 7929–7934.

Chang, S. S., Grunder, S., Hanukoglu, A., Rösler, A., Mathew, P. M., Hanukoglu, I., et al. (1996). Mutations in subunits of the epithelial sodium channel cause salt wasting with hyperkalaemic acidosis, pseudohypoaldosteronism type 1. *Nature Genetics*, *12*(3), 248–253.

Cheng, C.-J., Truong, T., Baum, M., & Huang, C.-L. (2012). Kidney-specific WNK1 inhibits sodium reabsorption in the cortical thick ascending limb. *American Journal of Physiology. Renal Physiology*, *303*(5), F667–F673.

Christensen, B. M., Perrier, R., Wang, Q., Zuber, A. M., Maillard, M., Mordasini, D., et al. (2010). Sodium and potassium balance depends on αENaC expression in connecting tubule. *Journal of the American Society of Nephrology*, *21*(11), 1942–1951.

Cornelius, R. J., Wen, D., Li, H., Yuan, Y., Wang-France, J., Warner, P. C., et al. (2015). Low Na, high K diet and the role of aldosterone in BK-mediated K excretion. *PLoS One*, *10*(1), 1–16.

Crowley, S. D., Gurley, S. B., Herrera, M. J., Ruiz, P., Griffiths, R., Kumar, A. P., et al. (2006). Angiotensin II causes hypertension and cardiac hypertrophy through its receptors in the kidney. *Proceedings of the National Academy of Sciences*, *103*(47), 17985–17990.

Cuevas, C. A., Su, X.-T., Wang, M.-X., Terker, A. S., Lin, D.-H., McCormick, J. A., et al. (2017). Potassium sensing by renal distal tubules requires kir4.1. *Journal of the American Society of Nephrology*, *28*(6), 1814–1825.

Dahl, L. K., Leitl, G., & Heine, M. (1972). Influence of dietary potassium and sodium/potassium molar ratios on the development of salt hypertension. *The Journal of Experimental Medicine*, *136*, 318–330.

Debonneville, C., Flores, S. Y., Kamynina, E., Plant, P. J., Tauxe, C., Thomas, M. A., et al. (2002). Phosphorylation of Nedd4-2 by Sgk1 regulates epithelial Na+ channel cell surface expression. *EMBO Journal*, *20*(24), 7052–7059.

Féraille, E., Vogt, B., Rousselot, M., Barlet-Bas, C., Cheval, L., Doucet, A., et al. (1993). Mechanism of enhanced Na-K-ATPase activity in cortical collecting duct from rats with nephrotic syndrome. *Journal of Clinical Investigation*, *91*(4), 1295–1300.

Frindt, G., & Palmer, L. G. (2004). Na channels in the rat connecting tubule. *American Journal of Physiology. Renal Physiology*, *286*, F669–F674.

Frindt, G., & Palmer, L. G. (2010). Effects of dietary K on cell-surface expression of renal ion channels and transporters. *American Journal of Physiology. Renal Physiology*, *299*(4), F890–F897. https://doi.org/10.1152/ajprenal.00323.2010.

Frindt, G., Zhou, H., Sackin, H., & Palmer, L. G. (1998). Dissociation of K channel density and ROMK mRNA in rat cortical collecting tubule during K adaptation. *American Journal of Physiology. Renal Physiology*, *274*(3, Pt 2), F525–F531.

Fujita, T., & Sato, Y. (1983). Natriuretic and antihypertensive effects of potassium in DOCA-salt hypertensive rats. *Kidney International*, *24*(6), 731–739.

Good, D. W., & Wright, F. S. (1979). Luminal influences on potassium secretion: Sodium concentration and fluid flow rate. *American Journal of Physiology. Renal Physiology*, *236*(2), F192–F205.

Grimm, P. R., Coleman, R., Delpire, E., & Welling, P. A. (2017). Constitutively active SPAK causes hyperkalemia by activating NCC and remodeling distal tubules. *Journal of the American Society of Nephrology*, *28*(9), 2597–2606.

Guidi, E., Cozzi, M. G., Minetti, E., & Bianchi, G. (1998). Donor and recipient family histories of hypertension influence renal impairment and blood pressure during acute rejections. *Journal of the American Society of Nephrology*, *9*(11), 2102–2107.

Guidi, E., Menghetti, D., Milani, S., Montagnino, G., Palazzi, P., & Bianchi, G. (1996). Hypertension may be transplanted with the kidney in humans: A long-term historical prospective follow-up of recipients grafted with kidneys coming from donors with or without hypertension in their families. *Journal of the American Society of Nephrology: JASN*, *7*(8), 1131–1138.

Guyton, A. C. (1991). Blood pressure control—special role of the kidneys and body fluids. *Science*, *252*(5014), 1813–1816.

Hadchouel, J., Ellison, D. H., & Gamba, G. (2016). Regulation of renal electrolyte transport by WNK and SPAK-OSR1 kinases. *Annual Review of Physiology*, *78*(1), 367–389.

Hansson, J. H., Schild, L., Lu, Y., Wilson, T. A., Gautschi, I., Shimkets, R., et al. (1995). A de novo missense mutation of the beta subunit of the epithelial sodium channel causes hypertension and Liddle syndrome, identifying a proline-rich segment critical for regulation of channel activity. *Proceedings of the National Academy of Sciences*, *92*(25), 11495–11499.

He, F. J., Marciniak, M., Carney, C., Markandu, N. D., Anand, V., Fraser, W. D., et al. (2010). Effects of potassium chloride and potassium bicarbonate on endothelial function, cardiovascular risk factors, and bone turnover in mild hypertensives. *Hypertension*, *55*(3), 681–688.

Hennings, J. C., Andrini, O., Picard, N., Paulais, M., Huebner, A. K., Cayuqueo, I. K. L., et al. (2017). The ClC-K2 chloride channel is critical for salt handling in the distal nephron. *Journal of the American Society of Nephrology*, *28*(1), 209–217.

Huang, D. Y., Wulff, P., Völkl, H., Loffing, J., Richter, K., Kuhl, D., et al. (2004). Impaired regulation of renal K+ elimination in the SGK1-Knockout mouse. *Journal of the American Society of Nephrology*, *15*(4), 885–891.

Hunter, R. W., Craigie, E., Homer, N. Z. M., Mullins, J. J., & Bailey, M. A. (2014). Acute inhibition of NCC does not activate distal electrogenic Na+ reabsorption or kaliuresis. *American Journal of Physiology. Renal Physiology*, *306*(4), F457–F467.

Ishizawa, K., Xu, N., Loffing, J., Lifton, R. P., Fujita, T., Uchida, S., et al. (2016). Potassium depletion stimulates Na-Cl cotransporter via phosphorylation and inactivation of the ubiquitin ligase Kelch-like 3. *Biochemical and Biophysical Research Communications*, *480*(4), 745–751.

Jung, J. Y., Kim, S., Lee, J. W., Jung, E. S., Heo, N. J., Son, M.-J., et al. (2011). Effects of potassium on expression of renal sodium transporters in salt-sensitive hypertensive rats induced by uninephrectomy. *American Journal of Physiology. Renal Physiology*, *300*(6), F1422–F1430.

Kahn, M., & Bohrer, N. K. (1967). Effect of potassium-induced diuresis on renal concentration and dilution. *American Journal of Physiology*, *212*(6), 1365–1375.

Kaissling, B., & Le Hir, M. (1982). Distal tubular segments of the rabbit kidney after adaptation to altered Na- and K-intake I. Structural changes. *Cell and Tissue Research*, *224*(3), 469–492.

Katz, A. I., Doucet, A., & Morel, F. (1979). Na-K-ATPase activity along the rabbit, rat, and mouse nephron. *American Journal of Physiology*, *237*(2), F114–F120.

Keith, N. M., & Binger, M. W. (1935). Diuretic action of potassium salts. *Journal of the American Medical Association*, *105*(20), 1584–1591.

Kempner, W. (1974). Treatment of hypertensive vascular disease with rice diet. *Archives of Internal Medicine*, *133*(5), 758–790.

Khaw, K. T., & Barrett-Connor, E. (1988). The association between blood pressure, age, and dietary sodium and potassium: A population study. *Circulation*, 77(1), 53–61. https://doi.org/10.1161/01.CIR.77.1.53.

Kirchner, K. A. (1983). Effect of acute potassium infusion on loop segment chloride reabsorption in the rat. *American Journal of Physiology. Renal Physiology*, *244*(6), F599–F605.

Krishna, G. G., & Kapoor, S. C. (1991). Potassium depletion exacerbates essential hypertension. *Annals of Internal Medicine*, *115*(2), 77–83.

Krishna, G. G., & Kapoor, S. C. (1993). Potassium supplementation ameliorates mineralocorticoid-induced sodium retention. *Kidney International*, *43*(5), 1097–1103.

Lalioti, M. D., Zhang, J., Volkman, H. M., Kahle, K. T., Hoffmann, K. E., Toka, H. R., et al. (2006). Wnk4 controls blood pressure and potassium homeostasis via regulation of mass and activity of the distal convoluted tubule. *Nature Genetics, 38*(10), 1124–1132.

Lifton, R. P., Gharavi, A. G., Geller, D. S., & Hughes, H. (2001). Molecular mechanisms of human hypertension. *Cell, 104*, 545–556.

Liu, W., Schreck, C., Coleman, R. a., Wade, J. B., Hernandez, Y., Zavilowitz, B., et al. (2011). Role of NKCC in BK channel-mediated net K+ secretion in the CCD. *American Journal of Physiology. Renal Physiology, 301*(5), F1088–F1097.

Loffing, J., Pietri, L., Aregger, F., Bloch-Faure, M., Ziegler, U., Meneton, P., et al. (2000). Differential subcellular localization of ENaC subunits in mouse kidney in response to high- and low-Na diets. *American Journal of Physiology. Renal Physiology, 279*(2), F252–F258.

Loffing-Cueni, D., Schlaepfer, M., Bloch-Faure, M., Hummler, E., Rossier, B., Kaissling, B., et al. (2003). Apical translocation of ROMK and ENaC in mouse distal nephron in response to high K+ diet (Abstract). *Nephrology, Dialysis, Transplantation, 18*, 552.

Louis-Dit-Picard, H., Barc, J., Trujillano, D., Miserey-Lenkei, S., Bouatia-Naji, N., Pylypenko, O., et al. (2012). KLHL3 mutations cause familial hyperkalemic hypertension by impairing ion transport in the distal nephron. *Nature Genetics, 44*(4), 456–460.

Lubbe, N., Moes, A., Rosenbaek, L., Schoep, S., Meima, M., Danser, A., et al. (2013). K+-induced natriuresis is preserved during Na+ depletion and accompanied by inhibition of the Na+-Cl- cotransporter. *American Journal of Physiology. Renal Physiology, 305*(8), F1177–F1188.

Mayan, H., Vered, I., Mouallem, M., Tzadok-witkon, M., & Pauzner, R. (2002). Pseudo-hypoaldosteronism Type II: Marked sensitivity to Thiazides, hypercalciuria, normomagnesemia, and low bone mineral density. *The Journal of Clinical Endocrinology and Metabolism, 87*(7), 3248–3254.

Mcquarrie, I., Thompson, W. H., & Anderson, J. A. (1935). Effects of excessive ingestion of sodium and potassium salts on carbohydrate metabolism and blood pressure in diabetic children. *The Journal of Nutrition, 11*(1), 77–101.

Meneely, G. R., Ball, C. O., & Youmans, J. B. (1957). Chronic sodium chloride toxicity: The protective effect of added potassium chloride. *Annals of Internal Medicine, 47*(2), 263–273.

Meneton, P., Loffing, J., & Warnock, D. G. (2004). Sodium and potassium handling by the aldosterone-sensitive distal nephron: The pivotal role of the distal and connecting tubule. *American Journal of Physiology. Renal Physiology, 287*(4), F593–F601.

Mente, A., O'Donnell, M. J., Rangarajan, S., McQueen, M. J., Poirier, P., Wielgosz, A., et al. (2014). Association of urinary sodium and potassium excretion with blood pressure. *New England Journal of Medicine, 371*(7), 601–611.

Murillo-de-Ozores, A. R., Grajeda-Medina, L. I., Gamba, G., & Castañeda-Bueno, M. (2018). Intracellular chloride depletion promotes WNK4-RRXS phosphorylation by a PKC/PKA dependent mechanism. *Journal of the American Society of Nephrology, 29*, 1005 (meeting abstract).

Nomura, N., Shoda, W., Wang, Y., Mandai, S., Furusho, T., Takahashi, D., et al. (2018). Role of ClC-K and barttin in low-potassium induced sodium-chloride cotransporter activation and hypertension in mouse kidney. *Bioscience Reports, 38*(1).

O'Donnell, M., Mente, A., Rangarajan, S., McQueen, M. J., Wang, X., Liu, L., et al. (2014). Urinary sodium and potassium excretion, mortality, and cardiovascular events. *New England Journal of Medicine, 371*(7), 612–623.

Ohta, A., Schumacher, F., Mehellou, Y., Johnson, C., Knebel, A., Macartney, T. J., et al. (2013). The CUL3-KLHL3 E3 ligase complex mutated in Gordon's hypertension syndrome interacts with and ubiquitylates WNK isoforms: Disease-causing mutations in KLHL3 and WNK4 disrupt interaction. *Biochemical Journal*, *451*, 111–122.

Oi, K., Sohara, E., Rai, T., Misawa, M., Chiga, M., Alessi, D. R., et al. (2012). A minor role of WNK3 in regulating phosphorylation of renal NKCC2 and NCC co-transporters in vivo. *Biology Open*, *1*(2), 120–127.

Oliver, W. J., Cohen, E. L., & Neel, J. V. (1975). Blood pressure, sodium intake, and sodium related hormones in the Yanomamo Indians, a "no-salt" culture. *Circulation*, *52*(1), 146–151.

O'Reilly, M., Marshall, E., Macgillivray, T., Mittal, M., Xue, W., Kenyon, C. J., et al. (2006). Dietary electrolyte-driven responses in the renal WNK kinase pathway in vivo. *Journal of the American Society of Nephrology*, *17*(9), 2402–2413.

Pácha, J., Frindt, G., Sackin, H., & Palmer, L. G. (1991). Apical maxi K channels in intercalated cells of CCT. *The American Journal of Physiology*, *261*(30), F696–F705.

Pacheco-Alvarez, D., San Cristóbal, P., Meade, P., Moreno, E., Vazquez, N., Muñoz, E., et al. (2006). The Na+:Cl- cotransporter is activated and phosphorylated at the amino-terminal domain upon intracellular chloride depletion. *Journal of Biological Chemistry*, *281*(39), 28755–28763.

Palmer, L. G. (1999). Potassium secretion and the regulation of distal nephron K channels. *American Journal of Physiology. Renal Physiology*, *277*(46), F821–F833.

Paulais, M., Bloch-Faure, M., Picard, N., Jacques, T., Ramakrishnan, S. K., Keck, M., et al. (2011). Renal phenotype in mice lacking the Kir5.1 (Kcnj16) K+ channel subunit contrasts with that observed in SeSAME/EAST syndrome. *Proceedings of the National Academy of Sciences*, *108*(25), 10361–10366.

Penton, D., Czogalla, J., Wengi, A., Himmerkus, N., Loffing-Cueni, D., Carrel, M., et al. (2016). Extracellular K + rapidly controls NaCl cotransporter phosphorylation in the native distal convoluted tubule by Cl^--dependent and independent mechanisms. *Journal of Physiology*, *594*(21), 6319–6331.

Perrier, R., Boscardin, E., Malsure, S., Sergi, C., Maillard, M. P., Loffing, J., et al. (2016). Severe salt-losing syndrome and hyperkalemia induced by adult nephron-specific knock-out of the epithelial sodium channel-subunit. *Journal of the American Society of Nephrology*, *27*(8), 2309–2318.

Piala, A. T., Moon, T. M., Akella, R., He, H., Cobb, M. H., & Goldsmith, E. J. (2014). Chloride sensing by WNK1 involves inhibition of autophosphorylation. *Science Signaling*, 7(324), ra41.

Poorolajal, J., Zeraati, F., Soltanian, A. R., Sheikh, V., Hooshmand, E., & Maleki, A. (2017). Oral potassium supplementation for management of essential hypertension: A meta-analysis of randomized controlled trials. *PLoS One*, *12*(4), 1–16.

Priddle, W. W. (1931). Observations on the management of hypertension. *Canadian Medical Association Journal*, *25*(1), 5–8.

Rengarajan, S., Lee, D. H., Oh, Y. T., Delpire, E., Youn, J. H., & McDonough, A. A. (2014). Increasing plasma [K+] by intravenous potassium infusion reduces NCC phosphorylation and drives kaliuresis and natriuresis. *American Journal of Physiology. Renal Physiology*, *306*(9), F1059–F1068.

Richardson, C., Rafiqi, F. H., Karlsson, H. K. R., Moleleki, N., Vandewalle, A., Campbell, D. G., et al. (2008). Activation of the thiazide-sensitive Na+-Cl- cotransporter by the WNK-regulated kinases SPAK and OSR1. *Journal of Cell Science*, *121*(5), 675–684.

Rubera, I., Loffing, J., Palmer, L. G., Frindt, G., Fowler-jaeger, N., Sauter, D., et al. (2003). Collecting duct-specific gene inactivation of aENaC in the mouse kidney does not impair sodium and potassium balance. *Journal of Clinical Investigation*, *112*(4), 554–565. https://doi.org/10.1172/JCI200316956.Introduction.

Sacks, F. M., Svetkey, L. P., Vollmer, W. M., Appel, L. J., Bray, G. A., & Harsha, D. DASH-Sodium Collaborative Research Group. (2001). Effects on blood pressure of reduced dietary sodium and the dietary approaches to stop hypertension (DASH) diet. *New England Journal of Medicine, 344*(1), 3–10.

Salt. (2015). *Pflügers Archiv - European Journal of Physiology, 467*(3). complete issue.

Schild, L., Lu, Y., Gautschi, I., Schneeberger, E., Lifton, R. P., & Rossier, B. C. (1996). Identification of a PY motif in the epithelial Na channel subunits as a target sequence for mutations causing channel activation found in Liddle syndrome. *The EMBO Journal, 15*(10), 2381–2387.

Scholl, U. I., Choi, M., Liu, T., Ramaekers, V. T., Hausler, M. G., Grimmer, J., et al. (2009). Seizures, sensorineural deafness, ataxia, mental retardation, and electrolyte imbalance (SeSAME syndrome) caused by mutations in KCNJ10. *Proceedings of the National Academy of Sciences, 106*(14), 5842–5847.

Shibata, S., Zhang, J., Puthumana, J., Stone, K. L., & Lifton, R. P. (2013). Kelch-like 3 and Cullin 3 regulate electrolyte homeostasis via ubiquitination and degradation of WNK4. *Proceedings of the National Academy of Sciences, 110*(19), 7838–7843.

Shimkets, R. A., Warnock, D. G., Bositis, C. M., Nelson-Williams, C., Hansson, J. H., Schambelan, M., et al. (1994). Liddle's syndrome: Heritable human hypertension caused by mutations in the beta subunit of the epithelial sodium channel. *Cell, 79*(3), 407–414.

Shoda, W., Nomura, N., Ando, F., Mori, Y., Mori, T., Sohara, E., et al. (2017). Calcineurin inhibitors block sodium-chloride cotransporter dephosphorylation in response to high potassium intake. *Kidney International, 91*(2), 402–411.

Simon, D. B., Karet, F. E., RodriguezSoriano, J., Hamdan, J. H., DiPietro, A., Trachtman, H., et al. (1996). Genetic heterogeneity of Bartter's syndrome revealed by mutations in the K+ channel, ROMK. *Nature Genetics, 14*(2), 152–156. https://doi.org/10.1038/ng1096-152.

Simon, D. B., Nelson-Williams, C., Bia, M. J., Ellison, D., Karet, F. E., Molina, A. M., et al. (1996). Gitelman's variant of Bartter's syndrome, inherited hypokalaemic alkalosis, is caused by mutations in the thiazide-sensitive Na-Cl cotransporter. *Nature Genetics, 12*(1), 24–30.

Sorensen, M. V., Grossmann, S., Roesinger, M., Gresko, N., Todkar, A. P., Barmettler, G., et al. (2013). Rapid dephosphorylation of the renal sodium chloride cotransporter in response to oral potassium intake in mice. *Kidney International, 83*(5), 811–824.

Stokes, J. B., Lee, I., & Williams, A. (1982). Consequences of potassium recycling in the renal medulla. *Journal of Clinical Investigation, 70*(2), 219–229.

Strautnieks, S. S., Thompson, R. J., Gardiner, R. M., & Chung, E. (1996). A novel splice-site mutation in the y subunit of the epitheliall sodium channel gene in three pseudohypoaldosteronism type 1 families. *Nature Genetics, 13*(2), 248–250.

Su, X.-T., & Wang, W.-H. (2016). The expression, regulation, and function of Kir4.1 (*Kcnj10*) in the mammalian kidney. *American Journal of Physiology. Renal Physiology, 311*(1), F12–F15.

Susa, K., Sohara, E., Takahashi, D., Okado, T., Rai, T., & Uchida, S. (2017). WNK4 is indispensable for the pathogenesis of pseudohypoaldosteronism type II caused by mutant KLHL3. *Biochemical and Biophysical Research Communications, 491*(3), 727–732.

Takahashi, D., Mori, T., Nomura, N., Khan, M. Z. H., Araki, Y., Zeniya, M., et al. (2014). WNK4 is the major WNK positively regulating NCC in the mouse kidney. *Bioscience Reports, 34*(3), 195–206. https://doi.org/10.1042/BSR20140047.

Terker, A. S., Yarbrough, B., Ferdaus, M. Z., Lazelle, R. A., Erspamer, K. J., Meermeier, N. P., et al. (2016). Direct and indirect mineralocorticoid effects determine distal salt transport. *Journal of the American Society of Nephrology, 27*(8), 2436–2445. https://doi.org/10.1681/ASN.2015070815.

Terker, A. S., Zhang, C., Erspamer, K. J., Gamba, G., Yang, C.-L., & Ellison, D. H. (2015). Unique chloride-sensing properties of WNK4 permit the distal nephron to modulate potassium homeostasis. *Kidney International, 89*(1), 1–8. https://doi.org/10.1038/ki.2015.289.

Terker, A. S., Zhang, C., McCormick, J. A., Lazelle, R. A., Zhang, C., Meermeier, N. P., et al. (2015). Potassium modulates electrolyte balance and blood pressure through effects on distal cell voltage and chloride. *Cell Metabolism, 21*(1), 39–50.

Unwin, R., Capasso, G., & Giebisch, G. (1994). Potassium and sodium transport along the loop of Henle: Effects of altered dietary potassium intake. *Kidney International, 46*(4), 1092–1099.

Vallon, V., Schroth, J., Lang, F., Kuhl, D., & Uchida, S. (2009). Expression and phosphorylation of the Na+-Cl- cotransporter NCC in vivo is regulated by dietary salt, potassium, and SGK1. *American Journal of Physiology. Renal Physiology, 297*(3), F704–F712.

Vander, A. J. (1970). Direct effects of potassium on renin secretion and renal function. *The American Journal of Physiology, 219*(2), 455–459.

Vitari, A. C., Deak, M., Morrice, N. A., & Alessi, D. R. (2005). The WNK1 and WNK4 protein kinases that are mutated in Gordon's hypertension syndrome phosphorylate and activate SPAK and OSR1 protein kinases. *Biochemical Journal, 391*(1), 17–24.

Wade, J. B., Fang, L., Coleman, R. A., Liu, J., Grimm, P. R., Wang, T., et al. (2011). Differential regulation of ROMK (Kir1.1) in distal nephron segments by dietary potassium. *American Journal of Physiology. Renal Physiology, 300*(6), F1385–F1393.

Wagner, C. A., Loffing-Cueni, D., Yan, Q., Schulz, N., Fakitsas, P., Carrel, M., et al. (2008). Mouse model of type II Bartter's syndrome. II. Altered expression of renal sodium- and water-transporting proteins. *American Journal of Physiology. Renal Physiology, 294*(6), F1373–F1380.

Wakabayashi, M., Mori, T., Isobe, K., Sohara, E., Susa, K., Araki, Y., et al. (2013). Impaired KLHL3-mediated ubiquitination of WNK4 causes human hypertension. *Cell Reports, 3*(3), 858–868.

Wang, M. X., Cuevas, C. A., Su, X. T., Wu, P., Gao, Z. X., Lin, D. H., et al. (2018). Potassium intake modulates the thiazide-sensitive sodium-chloride cotransporter (NCC) activity via the Kir4.1 potassium channel. *Kidney International, 93*(4), 893–902. https://doi.org/10.1016/j.kint.2017.10.023.

Wang, B., Wen, D., Li, H., Wang-france, J., & Sansom, S. C. (2017). Net K+ secretion in the thick ascending limb of mice on a low-Na, high-K diet. *Kidney International, 92*(4), 864–875.

Wei, Y., Bloom, P., Lin, D., Gu, R., & Wang, W. H. (2001). Effect of dietary K intake on apical small-conductance K channel in CCD: Role of protein tyrosine kinase. *American Journal of Physiology. Renal Physiology, 281*(2), F206–F212.

Weinstein, A. M. (2017). A mathematical model of the rat nephron: K+-induced natriuresis. *American Journal of Physiology. Renal Physiology, 312*, F925–F950.

Welling, P. A. (2013). Regulation of renal potassium secretion: Molecular mechanisms. *Seminars in Nephrology, 33*(3), 215–228.

WHO. (2012a). *Guideline: Sodium intake for adults and children.* Geneva: World Health Organization (WHO).

WHO. (2012b). *Guideline: Potassium intake for adults and children.* Geneva: World Health Organization (WHO). https://doi.org/9789241549028.

Wilson, F. H., Disse-Nicodème, S., Choate, K. A., Ishikawa, K., Nelson-Williams, C., Desitter, I., et al. (2001). Human hypertension caused by mutations in WNK kinases. *Science, 293*(5532), 1107–1112.

Woda, C. B., Bragin, A., Kleyman, T. R., & Satlin, L. M. (2001). Flow-dependent K(+) secretion in the cortical collecting duct is mediated by a maxi-K channel. *American Journal of Physiology. Renal Physiology, 280*(5), F786–F793.

Wright, F. S., Strieder, N., Fowler, N. B., & Giebisch, G. (1971). Potassium secretion by distal tubule after potassium adaptation. *The American Journal of Physiology*, *221*(2), 437–448.

Yang, S. S., Morimoto, T., Rai, T., Chiga, M., Sohara, E., Ohno, M., et al. (2007). Molecular pathogenesis of pseudohypoaldosteronism type II: Generation and analysis of a Wnk4D561A/+ knockin mouse model. *Cell Metabolism*, *5*(5), 331–344.

Yang, Y., Xie, J., Yang, S., Lin, S., Huang, C., Drive, H., et al. (2018). Differential roles of WNK4 in regulation of NCC in vivo. *American Journal of Physiology. Renal Physiology*, *314*(5), F999–F1007.

Yang, L., Xu, S., Guo, X., Uchida, S., Weinstein, A. M., Wang, T., et al. (2018). Regulation of renal Na transporters in response to dietary K. *American Journal of Physiology. Renal Physiology*, *315*(4), F1032–F1041.

Zacchia, M., Abategiovanni, M. L., Stratigis, S., & Capasso, G. (2016). Potassium: From physiology to clinical implications. *Kidney Diseases (Basel, Switzerland)*, *2*(2), 72–79.

CHAPTER TEN

Hormonal regulation of Na^+-K^+-ATPase from the evolutionary perspective

Sergej Pirkmajer[a], Alexander V. Chibalin[b,c,*]
[a]Institute of Pathophysiology, Faculty of Medicine, University of Ljubljana, Ljubljana, Slovenia
[b]Department of Molecular Medicine and Surgery, Integrative Physiology, Karolinska Institutet, Stockholm, Sweden
[c]National Research Tomsk State University, Tomsk, Russia
*Corresponding author: e-mail address: alexander.chibalin@ki.se

Contents

Abstract

Na^+-K^+-ATPase, an α/β heterodimer, is an ancient enzyme that maintains Na^+ and K^+ gradients, thus preserving cellular ion homeostasis. In multicellular organisms, this basic housekeeping function is integrated to fulfill the needs of specialized organs and preserve whole-body homeostasis. In vertebrates, Na^+-K^+-ATPase is essential for many fundamental physiological processes, such as nerve conduction, muscle contraction, nutrient absorption, and urine excretion. During vertebrate evolution, three key developments contributed to diversification and integration of Na^+-K^+-ATPase functions. Generation of novel α- and β-subunits led to formation of multiple Na^+-K^+-ATPase isoenyzmes with distinct functional characteristics. Development of a complex endocrine system enabled efficient coordination of diverse Na^+-K^+-ATPase functions. Emergence of FXYDs, small transmembrane proteins that regulate Na^+-K^+-ATPase, opened new ways to modulate

Current Topics in Membranes, Volume 83
ISSN 1063-5823
https://doi.org/10.1016/bs.ctm.2019.01.009

its function. FXYDs are a vertebrate innovation and an important site of hormonal action, suggesting they played an especially prominent role in evolving interaction between Na^+-K^+-ATPase and the endocrine system in vertebrates.

Abbreviations

MYA million years ago
NCX Na^+/Ca^{2+} exchanger
PKA protein kinase A
PKC protein kinase C
SERCA sarco-endoplasmic reticulum Ca^{2+}-ATPase

1. Introduction

Na^+-K^+-ATPase (sodium pump), originally described by Skou (1957), is a heterodimeric (α/β) (Craig & Kyte, 1980; Kyte, 1971; Morth et al., 2007) transmembrane ion pump that hydrolyzes ATP to transport three Na^+ ions from the cell and two K^+ ions into the cell (Post & Jolly, 1957; Sen & Post, 1964). The catalytic α-subunit is a 100–112 kDa type IIC P-ATPase, which exists in four isoforms (α1–4) in mammals (Blanco & Mercer, 1998). The non-catalytic β-subunit, a 35–60 kDa glycoprotein, exists in three isoforms (β1–3) in mammals and four isoforms (β1–4) in reptiles, fishes, and birds (Blanco & Mercer, 1998; Geering, 2008; Pestov et al., 2007; Pestov, Zhao, Basrur, & Modyanov, 2011) and is essential for assembly, maturation, and function of α/β heterodimers (Ackermann & Geering, 1990; Geering, 2008; McDonough, Geering, & Farley, 1990). While the α/β-heterodimer represents a fully functional core of Na^+-K^+-ATPase, it is often associated with a small transmembrane protein from the FXYD family, thus forming a heterotrimeric complex (α/β/FXYD) (Geering, 2006, 2008; Morth et al., 2007; Sweadner & Rael, 2000). FXYDs are therefore sometimes referred to as the third (auxiliary) subunit or the γ-subunit of Na^+-K^+-ATPase (Rossier, Baker, & Studer, 2015).

In animals, the intracellular fluid has low Na^+ and high K^+ and concentrations, while the extracellular fluid has high Na^+ and low K^+ concentrations. Due to such transmembrane distribution of ions, animal cells passively take up Na^+ and lose K^+ via various ion channels and carriers. Without actively opposing passive Na^+ and K^+ currents, cells would not be able to maintain ion and water homeostasis, membrane potentials, and Na^+-coupled transport of various substances. Na^+-K^+-ATPase, which represents the only mechanism that opposes Na^+ uptake and K^+ loss in animal cells (Rossier et al., 2015), is therefore essential for survival of animal cells.

Maintenance of intracellular Na^+ and K^+ concentrations is the most basic function of Na^+-K^+-ATPase in all animal cells. In multicellular organisms, this function is integrated in functions of various tissues and organs. Among these, excitable tissues, such as the nervous and muscle tissue, and tissues specialized for ion and water transport, such as the intestinal and kidney epithelia, are especially prominent. Indeed, Na^+-K^+-ATPase is vital for various fundamental physiological processes, including nerve conduction, muscle contraction, nutrient absorption in the gut, as well as urine excretion in the kidney. Involvement of Na^+-K^+-ATPase in so many diverse physiological functions implies existence of distinct mechanisms that enable organ-specific regulation of Na^+-K^+-ATPase. Clearly, under homeostatic conditions, Na^+-K^+-ATPase has to meet the demands of each individual cell, entire organs, as well as organism as a whole. At the level of whole organism, as much as 19–28% of total ATP synthesized in oxidative processes is used for Na^+-K^+-ATPase (Rolfe & Brown, 1997), underlining the workload put on this enzyme/pump.

Homeostatic demands of cells and organs are linked to Na^+-K^+-ATPase via various regulatory mechanisms. When and how vertebrates acquired regulatory mechanisms that enable integration of diverse cellular, tissue, and whole-body functions of Na^+-K^+-ATPase is unclear. However, three evolutionary developments were likely especially important: appearance of novel Na^+-K^+-ATPase isoforms, appearance of auxiliary regulatory proteins from the FXYD family, and evolution of a complex endocrine system. In continuation we first provide a short overview of mechanisms underlying Na^+-K^+-ATPase regulation and then focus on their integrative and evolutionary aspects.

2. Tissue perspective: Local and systemic regulation of Na^+-K^+-ATPase

From the tissue perspective regulatory mechanisms can be divided into local and systemic. Local mechanisms arise from the tissue itself. For instance, contractions of skeletal muscle lead to marked perturbation in transmembrane gradients of Na^+ and K^+ ions (Clausen, 2003; Pirkmajer & Chibalin, 2016). Due to increased intracellular concentration of Na^+ and extracellular concentration of K^+ activity of Na^+-K^+-ATPase in contracting skeletal muscle is stimulated directly without involvement of systemic factors (Clausen, 2003; Clausen, Everts, & Kjeldsen, 1987; DiFranco, Hakimjavadi, Lingrel, & Heiny, 2015). Activation of Na^+-K^+-ATPase driven by local factors helps to match the demands for ion transport in actively contracting muscle, but does not alter activity of Na^+-K^+-ATPase in resting muscles. Aside from ion concentrations, hypoxia, cellular energy status, purines, oxidative stress, pH, and formation of NO can also contribute to local regulation of Na^+-K^+-ATPase (Clausen, 2003; Pirkmajer & Chibalin, 2016).

Local regulation is, by definition, spatially limited and therefore rather specific in its effects. However, unlike local factors, systemic factors can potentially affect the entire pool of Na^+-K^+-ATPase at the same time. Under physiological conditions, hormones are major factors involved in systemic regulation of Na^+-K^+-ATPase. Specificity of this regulation is in part conferred by differential tissue responsiveness to hormones. This is achieved by differential expression of hormonal receptors and/or enzymes involved in hormone (in)activation. For instance, cortisol can bind to glucocorticoid and mineralocorticoid receptors. However, mineralocorticoid (aldosterone) responsive tissue express 11β-hydroxysteroid dehydrogenase 2, which converts glucocorticoid cortisol (corticosterone in rodents) to inactive cortisone (11-dehydrocorticosterone in rodents), thus protecting these tissues from effects of glucocorticoids (Chapman, Holmes, & Seckl, 2013). In addition to differences in hormone responsiveness, mechanisms that contribute to organ-specific regulation of Na^+-K^+-ATPase include differential expression of Na^+-K^+-ATPase isoenzymes and FXYDs.

3. Molecular perspective: Mechanisms of regulation of Na^+-K^+-ATPase

From the molecular perspective, there are three levels of regulation of Na^+-K^+-ATPase: regulation of total cellular Na^+-K^+-ATPase content, regulation of subcellular distribution of Na^+-K^+-ATPase, and regulation of intrinsic activity of Na^+-K^+-ATPase. Hormones contribute to regulation of Na^+-K^+-ATPase at all three levels.

Total content of Na^+-K^+-ATPase in the cell depends on gene expression and degradation of Na^+-K^+-ATPase subunits. Regulation of total Na^+-K^+-ATPase content enables cells to chronically adapt to altered demands in ion transport. Such situations typically include altered intake and/or loss of ions and water or altered chronic level of physical activity. Challenges to homeostasis due to these alterations can be profound, such as migration of fish from sea to fresh water. Mechanisms underlying adaptations in Na^+-K^+-ATPase expression involve hormonal regulation as well as local factors, such as those triggered by muscle contractions. Among hormones, the most prominent regulators of Na^+-K^+-ATPase expression in mammals are thyroxine and aldosterone, which both increase its total cellular content (Clausen, 2003; Everts, Dorup, Flyvbjerg, Marshall, & Jorgensen, 1990; Feraille & Dizin, 2016; Lin & Akera, 1978). Insulin treatment in diabetic rats increases Na^+-K^+-ATPase content in skeletal muscle, suggesting a role for insulin in regulation of its expression (Schmidt Hasselbalch, Farrell, Vestergaard, & Kjeldsen, 1994). Pharmacological treatment with glucocorticoids also increases Na^+-K^+-ATPase in organs like skeletal muscle, but endogenous glucocorticoids are likely not a major regulator of Na^+-K^+-ATPase under physiological conditions (Clausen, 2003; Pirkmajer & Chibalin, 2016). In addition to various corticosteroids (Rossier et al., 2015), growth hormone, insulin-like growth factor 1 (IGF-1), and prolactin regulate Na^+-K^+-ATPase expression in fishes (Rossier et al., 2015; Tipsmark & Madsen, 2009). Effects of these hormones are complex and context-dependent, including the presence or absence of other hormones (hormonal milieu) and environmental conditions (e.g., water salinity). Detailed description of these effects is therefore beyond the scope of this review.

The second level of regulation of Na^+-K^+-ATPase involves regulation of its subcellular distribution between the plasma membrane and the intracellular compartment (Benziane & Chibalin, 2008; Chibalin, Katz, Berggren, & Bertorello, 1997; Hundal et al., 1992; Marette et al., 1993; Pirkmajer & Chibalin, 2016). Indeed, Na^+-K^+-ATPase is localized both in the plasma membrane as well as in the intracellular compartment (Hundal et al., 1994). To perform ion transport Na^+-K^+-ATPase has to be present in the plasma membrane. Total content of Na^+-K^+-ATPase therefore determines potential maximal capacity of the cell to perform transport of Na^+ and K^+, while membrane abundance determine actual capacity for ion transport. By stimulating exocytosis and/or suppressing endocytosis of Na^+-K^+-ATPase the cell can rapidly increase its abundance in the plasma membrane. Recruitment of the intracellular reserve pool of Na^+-K^+-ATPase to the plasma membrane, enables the cell to acutely increase its capacity to perform transmembrane transport of Na^+ and K^+. Mechanisms that increase the translocation of Na^+-K^+-ATPase to the plasma membrane include insulin and skeletal muscle contraction (Al-Khalili et al., 2004; Al-Khalili, Yu, & Chibalin, 2003; Benziane & Chibalin, 2008; Hundal et al., 1992; Pirkmajer & Chibalin, 2016). Conversely, dopamine reduces membrane abundance of Na^+-K^+-ATPase by stimulating endocytosis (Chibalin et al., 1999). Together, regulation of total Na^+-K^+-ATPase content and its subcellular distribution represent an important, but a relatively crude, mechanism of Na^+ and K^+ transport. Indeed, if all Na^+-K^+-ATPases were maximally active at the same time, this would lead to a lethal decrease in extracellular concentration of K^+ in seconds (Clausen, 2010).

The third level of regulation is regulation of intrinsic activity of Na^+-K^+-ATPase in the plasma membrane. Intrinsic activity is regulated by concentrations of substrates (Na^+, K^+, and ATP), covalent modification of Na^+-K^+-ATPase subunits, and interaction with FXYD proteins. Na^+ and K^+ ions stimulate enzymatic activity of isolated Na^+-K^+-ATPase (Skou, 1957). Elevations in intracellular Na^+ or extracellular K^+ concentrations also increase Na^+-K^+-ATPase activity in organs, such as skeletal muscle, thus helping to oppose the disturbance in ion concentrations (Clausen, 2003; Clausen et al., 1987; DiFranco et al., 2015). Alterations in ion concentrations may also arise due to hormonal stimulation. For instance, thyroxine increases Na^+ flux and/or K^+ efflux in target cells, thus indirectly increasing intrinsic activity and content of Na^+-K^+-ATPase (Clausen, 2003; Mullur, Liu, & Brent, 2014; Silva, 2006). While activity of Na^+-K^+-ATPase depends on ATP (Blanco & Mercer, 1998; Skou, 1957), changes in its concentrations are likely not a major regulatory mechanism under physiological conditions. Indeed, excepting pathophysiological states, ATP concentrations are relatively stable even when ion transport in cells is markedly activated (Heineman & Balaban, 1990; Hochachka & McClelland, 1997).

The α-subunit of Na^+-K^+-ATPase is a direct target for protein kinase A (PKA) (Bertorello, Aperia, Walaas, Nairn, & Greengard, 1991; Chibalin, Vasilets, Hennekes, Pralong, & Geering, 1992; Yeh, Ling, English, & Cantley, 1983), protein kinase C (PKC) (Bertorello et al., 1991; Chibalin et al., 1992), extracellular signal-regulated kinase 1/2 (ERK1/2) (Al-Khalili et al., 2004), and cGMP-activated protein kinase (PKG) (Fotis, Tatjanenko, & Vasilets, 1999). Furthermore, phosphorylation of the $\alpha 1$-subunit is regulated by AMP-activated protein kinase (AMPK) indirectly (Benziane et al., 2012). The α-subunit has Ser, Thr, and Tyr residues that can be phosphorylated in living cells (Al-Khalili et al., 2004; Chibalin et al., 2001, 1998, 1999; Feraille et al., 1999). Depending on the site of phosphorylation, intrinsic activity of Na^+-K^+-ATPase is stimulated, suppressed, or remains unaltered. Regulation of Na^+-K^+-ATPase activity by phosphorylation might be important under some conditions, such as hormonal stimulation or contraction of muscle fibers. Nevertheless, phosphorylation is likely more important as regulator of translocation of Na^+-K^+-ATPase between the plasma membrane and the intracellular compartment (Benziane & Chibalin, 2008). Indeed, phosphorylation that alters subcellular distribution of Na^+-K^+-ATPase is not necessarily associated with changes in its intrinsic activity (Al-Khalili et al., 2004; Chibalin et al., 1999), highlighting that these two processes are regulated separately.

Glutathionylation is another mode of covalent modification of Na^+-K^+-ATPase. It involves formation of a reversible covalent bond between glutathione and thiol residues on the α- and β-subunits and results in inactivation of Na^+-K^+-ATPase (Figtree et al., 2009; Juel, 2014; Juel, Hostrup, & Bangsbo, 2015; Liu et al., 2012; Petrushanko et al., 2012). In some tissues, such as skeletal muscle, a significant fraction (up to ~10–20%) of the α- and the β-subunits may be glutathionylated (Juel, 2014; Juel et al., 2015), indicating that part of the Na^+-K^+-ATPase pool is inactive and not available for activation. Glutathionylation of Na^+-K^+-ATPase is increased by oxidative stress (Figtree et al., 2009; Petrushanko et al., 2012) and by adrenergic signaling via the β_2-adrenoceptors (Juel et al., 2015), which provides another link between hormones and regulation of Na^+ and K^+ transport.

In addition to regulation by substrates and covalent modification, intrinsic activity of Na^+-K^+-ATPase is importantly regulated by protein-protein interactions with FXYDs, which modulate its kinetic properties, such as maximal ion transport rate (I_{max}), maximal enzymatic (ATPase) rate (V_{max}), and affinity for Na^+ ($K_{1/2,Na}$) and K^+ ($K_{1/2,K}$) (Cornelius & Mahmmoud, 2003; Geering, 2006, 2008) (see Table 1). These modulatory effects and expression of FXYDs are regulated by hormones (see Tables 2 and 3), adding another layer of complexity to hormonal regulation of Na^+-K^+-ATPase. The role of FXYDs in regulation of Na^+-K^+-ATPase and their evolutionary history are described in sections below.

4. Diversification of Na^+-K^+-ATPase in the vertebrate lineage: Evolutionary history and functional consequences

All vertebrate α- and β-subunits, which form Na^+-K^+-ATPase and a closely related H^+-K^+-ATPase (aka proton pump), probably developed from one α- and one β-subunit in deuterostomes (Saez, Lozano, & Zaldivar-Riveron, 2009; Studer, Person, Robinson-Rechavi, & Rossier, 2011). Mammals share with the fishes the α1–3, but not the α4-subunit of Na^+-K^+-ATPase. The α4-subunit appeared in therian mammals (Saez et al., 2009). Na^+-K^+-ATPase β-subunit probably had a similar evolutionary

Table 1 Effect of FXYD proteins on kinetic parameters of Na^+-K^--ATPase.

	Effects of FXYD protein on Na^+-K^+-ATPase					
	$K_{1/2,Na}$	$K_{1/2,K}$	V_{max}, I_{max}	Effect of phosphorylation	Comments	References
FXYD1	↑↑	↔ or ↑	↔ or ↓	Phosphorylation reverses effects of unphosphorylated FXYD1 on $K_{1/2,Na}$ (and V_{max} (I_{max}) in some models).	FXYD1 associates with α1/β1, α2/β1, α3/β1, α1/β2, α2/β2, α3/β2 (*Xenopus* oocytes). The association is more efficient with α1/β than α2/β (heart, skeletal muscle). Effects of FXYD1 on α1/β1 are more pronounced than on α2/β1.	Bibert, Roy, Schaer, Horisberger, and Geering (2008), Bossuyt et al. (2009), Crambert, Fuzesi, Garty, Karlish, and Geering (2002), Despa et al. (2005), Han, Bossuyt, Martin, Despa, and Bers (2010), and Mishra et al. (2015)
FXYD2	↑	↔ or ↑	↔ or ↓	Lacks phosphorylation sites.	Effects depend on splice variants of FXYD2.	Arystarkhova, Donnet, Asinovski, and Sweadner (2002), Arystarkhova, Wetzel, Asinovski, and Sweadner (1999), Beguin et al. (2001), and Wetzel, Pascoa, and Arystarkhova (2004)
FXYD-3/4	↔	↔	↓	PKC induces dissociation from the α-subunit and increases V_{max}.		Mahmmoud et al. (2000) and Mahmmoud et al. (2003)

FXYD3	↑	↑	↔	Unknown.	Associates with H^+-K^+-ATPase (*Xenopus* oocytes).	Crambert, Li, Claeys, and Geering (2005)
FXYD4	↓	↑	↔	Unknown.		Beguin et al. (2001)
FXYD5	↔	↔	↑	Unknown.	Associates with α1/β1 (*Xenopus* oocytes).	Lubarski, Karlish, and Garty (2007) and Lubarski, Pihakaski-Maunsbach, Karlish, Maunsbach, and Garty (2005)
FXYD6	↑↑ (α1/β1) ↓ (α1/β2)	↑ (α1/β1) ↓ (α1/β2)	↔	Unknown.	Associates with α1/β1 and α1/β2 (*Xenopus* oocytes). Associates with H^+-K^+-ATPase (*Xenopus* oocytes).	Delprat et al. (2007)
FXYD6f	ND	ND	ND	Unknown	Expression of FXYD6f (aka FXYD8) decreases activity of Na^+-K^+-ATPase in *Xenopus* oocytes.	Wang, Yang, Lin, Hwang, and Lee (2017)
FXYD7	↔	↑	↔ or ↓	Unknown.	Associates with α/β1 (*Xenopus* oocytes). Associates with α1/β1 (brain).	Beguin et al. (2002) and Crambert, Li, Swee, and Geering (2004)
					Does not associate with α/β2. Does not affect $K_{1/2,K}$ of α3/β1.	

Association between an FXYD protein and Na^+,K^+-ATPase modulates its affinity for Na^+ ($K_{1/2,Na}$), K^+ ($K_{1/2,K}$), and/or its maximal enzymatic rate (V_{max}) or ion transport rate (I_{max}). All FXYDs except FXYD2 have putative phosphorylation sites (Mahmmoud, Vorum, & Cornelius, 2000; Pirkmajer et al., 2017; Sweadner & Rael, 2000; Tipsmark, 2008), but their functional significance is unknown. *ND*, not determined; *PKC*, protein kinase C.

Table 2 Hormonal regulation of FXYD1 phosphorylation.

Hormone or its analog	Effect on phosphorylation of FXYD	References
Pituitary hormones		
AVP (ADH)	↑ Phosphorylation of the "16-kDa protein" (possibly FXYD1) in cultured rat aortic smooth muscle cells. Effect was blocked by the V1-receptor antagonist PMT-VP	Boulanger-Saunier and Stoclet (1987)
dDAVP	↓ Phosphorylation of FXYD1 (at Ser^{63} and Ser^{68}/Ser^{69}) in the inner kidney medulla 4 h after injection of the V2-receptor agonist dDAVP in mice	Arystarkhova, Bouley, Liu, and Sweadner (2017)
Pancreatic hormones		
Insulin	↑ Phosphorylation of the "15-kDa protein" during incubation of sarcolemma-enriched membrane preparation (from rat hind limb skeletal muscles) in the presence of insulin	Walaas et al. (1977)
	↑ Phosphorylation of the "15-kDa protein" in the rat diaphragm muscle	Walaas, Horn, and Walaas (1991)
	↑ Phosphorylation of FXYD1 (Ser^{63} and Ser^{68}) in the rat diaphragm muscle	Walaas, Czernik, Olstad, Sletten, and Walaas (1994)
	↑ Phosphorylation of FXYD1 in isolated adipocytes (rat).	Walaas, Horn, and Walaas (1999)
	↑ Phosphorylation of FXYD1 (Ser^{63} and Ser^{68}) in rat, mouse, and human skeletal muscle	Benziane et al. (2011)
Adrenal hormones		
Epinephrine (adrenaline)	↑ Phosphorylation of FXYD1 (Ser^{68}) in the rat diaphragm muscle	Walaas et al. (1994)
Isoproterenol (isoprenaline)	↑ Phosphorylation of the sarcolemmal "15-kDa protein" (possibly FXYD1) in isolated ventricles (guinea pig)	Presti, Jones, and Lindemann (1985)
	↑ Phosphorylation of the sarcolemmal "15-kDa protein" (possibly FXYD1) in isolated ventricles (rat)	Lindemann (1986)
	↑ Phosphorylation of the sarcolemmal "16-kDa protein" (possibly FXYD1) in plasma membrane-enriched fraction from aortic myocytes (rat)	Boulanger-Saunier, Kattenburg, and Stoclet, (1985)

Table 2 Hormonal regulation of FXYD1 phosphorylation.—cont'd

Hormone or its analog	Effect on phosphorylation of FXYD	References
	↑ Phosphorylation of FXYD1 (Ser^{68}) in isolated cardiac myocytes (rat) ↔ Phosphorylation of FXYD1 (Ser^{63}) in isolated cardiac myocytes (rat)	Despa et al. (2005)
	↑ Phosphorylation of FXYD1 (Ser^{63}, Ser^{68}, but not Thr^{69}) in cultured ventricular myocytes (rat)	Fuller et al. (2009)
	↑ Phosphorylation of FXYD1 (Ser^{68}) in isolated ventricular myocytes (mouse)	Wang et al. (2014)
Terbutaline	Terbutaline tended to increase phosphorylation of FXYD1 (Ser^{68}) in non-fatigued muscle vastus lateralis muscle in humans.	Hostrup et al. (2014)
Phenylephrine	↑ Phosphorylation of the sarcolemmal "15-kDa protein" (possibly FXYD1) in isolated ventricles (rat)	Lindemann (1986)
	↑ Phosphorylation of FXYD1 (Ser^{63}, Ser^{68}, and Thr^{69}) in cultured ventricular myocytes (rat)	Fuller et al. (2009)

Endogenous hormones and synthetic compounds that modulate their action are classified according to the endocrine gland from which the endogenous hormone originates. FXYD1 was first described as the "15-kDa protein" or "16-kDa protein" (Boulanger-Saunier & Stoclet, 1987; Palmer, Scott, & Jones, 1991; Walaas et al., 1994, 1991, 1977). Isoproterenol (isoprenaline) and terbutaline are agonists of β-adrenoceptors, while phenylephrine is agonist of α-adrenoceptors. Abbreviations: AVP, vasopressin (arginine vasopressin); also known as ADH (antidiuretic hormone), dDAVP, desmopressin (deamino-Cys^1,D-Arg^8-vasopressin); an agonist of the vasopressin V2 receptor, PMT-VP, 1-(2-mercapto-2, 2-cyclopentamethylene propionic acid)-2-(*O*-methyl)tyrosine-8-arginine-vasopressin, an antagonist of the vasopressin V1 receptor.

history (Studer et al., 2011). However, while there are four β-subunits (β1–4) in fishes and birds, there are only three β-subunits in placental mammals (β1–3). The fourth β-subunit, which is referred to as the βm-subunit, lost its ancestral function in placental mammals and functions as a regulator of transcription (Korneenko et al., 2016; Pestov et al., 2007, 2011; Zhao, Pestov, Korneenko, Shakhparonov, & Modyanov, 2004). Diversification of ion transporters is a common theme in the vertebrate evolution. Indeed, the Na^+/Ca^{2+} exchanger (NCX) had a similar history. Tunicates, such as *Ciona*, have one NCX, while four NCX isoforms (NCX1–4) are present

Table 3 Hormonal regulation of FXYD expression.

Procedure or compound	Effect on FXYD mRNA and/or protein expression	References
Pituitary hormones		
Hypophysectomy	↓ (or ↔) FXYD3/4-1 (FXYD11) mRNA in gills of FW *Oreochromis mossambicus*[a] ↑ (or ↔) FXYD3/4-1 (FXYD11) mRNA in gills of SW *O. mossambicus*[a] ↔ FXYD3/4-2 (FXYD9) mRNA in gills of FW and SW *O. mossambicus*	Tipsmark et al. (2011)
GH	↑ FXYD5 mRNA in BaF/3 cell line	Conway-Campbell et al. (2007)
	↑ FXYD3/4-1 (FXYD11) mRNA and protein in gills of FW (but not SW) *Salmo salar* ↔ FXYD3/4-2 (FXYD9) mRNA in gills of FW and SW *Salmo salar*	Tipsmark, Mahmmoud, Borski, and Madsen (2010)
	GH tended to decrease FXYD3/4-2 (FXYD9) mRNA in gills of hypophysectomized FW *O. mossambicus*	Tipsmark et al. (2011)
PRL	↔ FXYD3/4-1 (FXYD11) mRNA, but PRL tended to decrease its protein level in gills of SW *S. salar* ↔ FXYD3/4-2 (FXYD9) mRNA in gills of SW *S. salar*	Tipsmark et al. (2010)
	↑ FXYD3/4-1 (FXYD11) mRNA in gills of hypophysectomized FW *O. mossambicus* ↔ FXYD3/4-2 (FXYD9) mRNA in gills of hypophysectomized FW *O. mossambicus*	Tipsmark et al. (2011)
Adrenal hormones		
Adrenalectomy	↓ FXYD4 mRNA in the distal colon, ↔ FXYD4 mRNA in the kidney (rat, 7 days after adrenalectomy)	Wald, Popovtzer, and Garty, (1997)
	↓ FXYD4 mRNA in the distal colon, ↔ FXYD4 mRNA in the kidney (rat)	Capurro et al. (1997)
	↓ FXYD4 mRNA in the distal colon, ↔ FXYD4 mRNA in the kidney (rat, 24–48 h after adrenalectomy)	Brennan and Fuller (1999)

Table 3 Hormonal regulation of FXYD expression.—cont'd

Procedure or compound	Effect on FXYD mRNA and/or protein expression	References
Dexamethasone	↑ FXYD4 mRNA in colon (rat, up to 66 h)	Attali, Latter, Rachamim, and Garty (1995)
	↑ FXYD4 mRNA in the colon, ↔ FXYD4 mRNA in the kidney (rat, 3 and/or 5 days) ↔ FXYD4 mRNA in cultured rat kidney cells (2 days)	Wald, Goldstein, Asher, Yagil, and Garty (1996)
	↑FXYD4 mRNA in the distal colon, ↔FXYD4 mRNA in the kidney (rat, 2 days)	Capurro et al. (1997)
	↑ FXYD4 mRNA in the distal colon (adrenalectomized rat, 2–4 h) ↑ FXYD4 mRNA in the proximal and transverse colon (rat, 4 h) ↔ FXYD4 mRNA in the kidney (adrenalectomized rat, 4 h) ↔ FXYD2 mRNA in the kidney and the colon (intact and adrenalectomized rat, 4 h)	Brennan and Fuller (1999)
RU28362	↑ FXYD4 mRNA in colon (rat, 2 h)	Brennan and Fuller (1999)
Cortisol	↔ FXYD4 mRNA in cultured rat kidney cells (2 days)	Wald et al. (1996)
	↑ FXYD3/4-1 (FXYD11) mRNA and protein in gills of FW (but not SW) *S. salar* ↑ FXYD3/4-1 (FXYD11) mRNA and protein in organ culture of gills of FW and SW *S. salar* ↔ FXYD3/4-2 (FXYD9) mRNA in gills of FW and SW *S. salar*	Tipsmark et al. (2010)
	↑ FXYD3/4-1 (FXYD11) mRNA in gills of hypophysectomized FW *O. mossambicus* ↓ FXYD3/4-1 (FXYD11) mRNA in gills of hypophysectomized SW *O. mossambicus* Tended to increase FXYD3/4–2 (FXYD9) mRNA in gills of hypophysectomized SW *O. mossambicus*	Tipsmark et al. (2011)

Continued

Table 3 Hormonal regulation of FXYD expression.—cont'd

Procedure or compound	Effect on FXYD mRNA and/or protein expression	References
Aldosterone	↑ FXYD4 mRNA in colon (rat, up to 66 h)	Attali et al. (1995)
	↑ FXYD4 mRNA in the distal colon (rat, from 3 h to 5 days)	Wald et al. (1996)
	↑ FXYD4 mRNA in the distal colon, ↔FXYD4 mRNA in the kidney (adrenalectomized rat on maintenance dose of corticosterone, 7 days)	Wald et al. (1997)
	↑ FXYD4 mRNA in the distal colon, ↔FXYD4 mRNA in the kidney (rat, 7 days)	Capurro et al. (1997)
	↑ FXYD4 mRNA in the distal colon (adrenalectomized rat, 1–4 h) ↔ FXYD2 mRNA in the colon and the kidney (adrenalectomized rat, 2 or 4 h)	Brennan and Fuller (1999)
Carbenoxolone	↑ FXYD4 mRNA in colon (rat, 4 h)	Brennan and Fuller (1999)

[a]Results depend on experimental conditions; for exact experimental protocol, see Tipsmark et al. (2011). The table provides an overview of endocrine glands and hormones (as well as their synthetic analogs or modulators) that regulate expression of FXYDs. Endogenous hormones and synthetic compounds that modulate their action are classified according to the endocrine gland from which the endogenous hormone originates. Carbenoxolone: inhibitor of 11β-hydroxysteroid dehydrogenase, RU28362: agonist of glucocorticoid receptor. FW: freshwater, SW: seawater.

in teleosts (Fig. 1). NCX4 was later lost, so that only NCX1–3 are present in birds and mammals (On, Marshall, Chen, Moyes, & Tibbits, 2008). The major mechanism for diversification of Na^+-K^+-ATPase, H^+-K^+-ATPase, and NCX was probably two rounds of whole-genome duplication at the base of vertebrate lineage, followed by local duplications and/or loss of genes.

Existence of multiple α- and β-subunits enables formation of Na^+-K^+-ATPase isoenzymes (α/β-heterodimers) with distinct kinetic properties (Blanco & Mercer, 1998), regulatory mechanisms, and tissue-specific expression patterns (Blanco & Mercer, 1998). For instance, in mammals, the α1/β-heterodimers predominate in most tissues, including the kidney (Shyjan & Levenson, 1989), α2/β in skeletal muscle (Hansen, 2001; He et al., 2001; Lytton, Lin, & Guidotti, 1985; Orlowski & Lingrel, 1988; Sweadner, McGrail, & Khaw, 1992), α3/β in the nervous system (Orlowski & Lingrel, 1988; Shyjan & Levenson, 1989), and α4/β

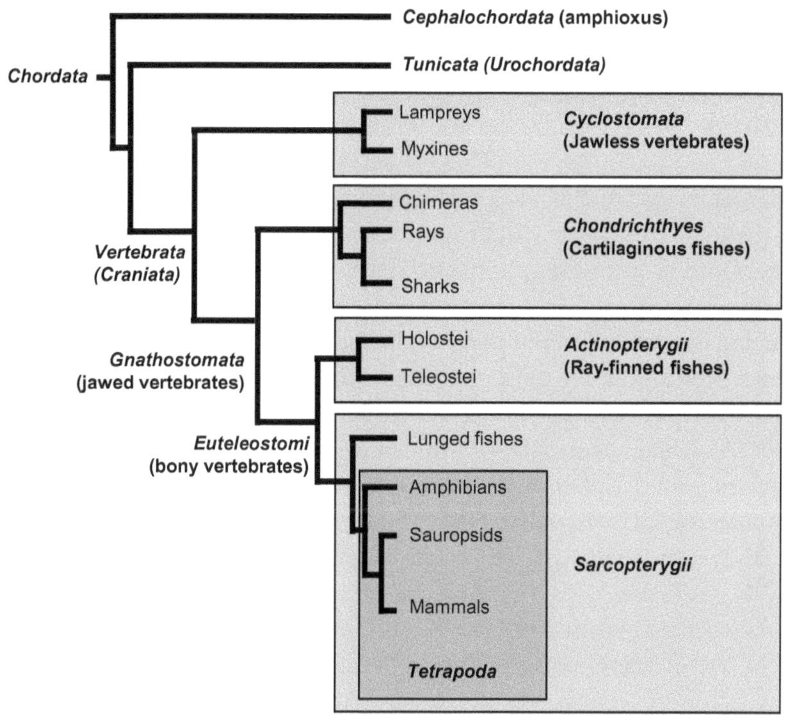

Fig. 1 The phylogenetic tree of chordates. The phylum *Chordata* has three subphyla: *Cephalochordata*, *Tunicata (Urochordata)*, and *Vertebrata*. The most basal group of verebrates are *Cyclostomata* (jawless vertebrates), commonly referred to *Agnatha*, which include lampreys and myxines (hagfishes), according to the currently prevailing model of chordate evolution. The *Gnathostomata* include vertebrates with cartilaginous skeleton, *Chondrichthyes* (cartilaginous fishes), and vertebrates with bony skeleton, *Euteleostomi*. The bony vertebrates comprise *Actinopterygii* (ray-finned fishes), which represent 99% of extant fish species, and *Sarcopterygii*, which include all terrestrial vertebrates (*Tetrapoda*), such as Amphibians, Sauropsids (reptiles and birds), and Mammals.

in the testis (Shamraj & Lingrel, 1994). Expression of multiple isoenzymes has important functional consequences since they display different characteristics at all three levels of Na^+-K^+-ATPase regulation. For instance, excepting thyroxine and aldosterone, which induce expression of the α- and the β-subunits across many different types of cells (Clausen, 2003; Li & Langhans, 2015; Rossier et al., 2015), hormones frequently have distinct effects on expression of Na^+-K^+-ATPase. For instance, treatment of 3T3-L1 cells with insulin increases expression of the α2-subunit, does not alter expression of the α1-subunit, and decreases expression of the β1-subunit (Russo & Sweadner, 1993). On the other hand, antidiuretic hormone (vasopressin) increases expression of the α1-subunit, but does not affect the β1-subunit in the collecting ducts of the kidney (Blot-Chabaud et al., 2001).

Na^+-K^+-ATPase subunits also differ in phosphorylation sites that are targeted by protein kinases. These differences underlie divergent effects of protein kinases on Na^+-K^+-ATPase function across tissues and species. For instance, treatment with dopamine activates PKC in rat proximal tubular cells and induces phosphorylation of Ser^{18} of the α1-subunit. Increased phosphorylation of Ser^{18} triggers endocytosis of Na^+-K^+-ATPase, thus reducing its membrane abundance (Chibalin et al., 1998, 1999). In cultured human skeletal muscle cells, insulin treatment increases membrane abundance of the α1- and α2-subunits and activation of PKC is required for this effect (Al-Khalili et al., 2004). Human α1- and α2-subunits, as well as the rat α2-subunit, lack Ser^{18} (Feschenko & Sweadner, 1995) and are not good substrates for PKC (Al-Khalili et al., 2004), which provides one explanation for divergent effects of PKC activation on subcellular distribution of Na^+-K^+-ATPase.

As regards regulation of Na^+-K^+-ATPase, differences in kinetic properties of α/β-heterodimers are also important since they affect responses to local factors. For instance, the α2/β-heterodimers, have lower affinities for extracellular K^+ ($K_{1/2,K}$=1.3–4.8 mM) than α1/β-heterodimers ($K_{1/2,K}$=0.92–1.9 mM), although estimated values differ between experimental models (Blanco & Mercer, 1998; Crambert et al., 2000; DiFranco et al., 2015). This indicates that the α2/β-heterodimers, especially the α2/β2-heterodimers, which most consistently display low affinity to K^+ (Blanco & Mercer, 1998; Crambert et al., 2000; Larsen, Stoica, & MacAulay, 2016), are regulated by physiological fluctuations in extracellular K^+ concentrations to a much greater degree than the α1/β-heterodimers (DiFranco et al., 2015).

K^+-stimulated activation of the α2/β-heterodimers is therefore probably important during muscle contractions, when extracellular K^+ concentrations in T-tubules rise markedly (DiFranco et al., 2015). Similarly, in the central nervous system, α2/β2-heterodimers, which are preferentially expressed in astrocytes (Blanco & Mercer, 1998; Cameron, Klein, Shyjan, Rakic, & Levenson, 1994), likely have an important role in normalization of extracellular K^+ concentrations, which are increased during brain activity (Larsen et al., 2016). Indeed, in both cases low affinity to K^+ would enable the α2/β2-heterodimers to be further activated, when other α/β-heterodimers are already saturated (DiFranco et al., 2015;

Larsen et al., 2016). On the other hand, the α3/β1-heterodimers, which are predominantly expressed in neurons, have the lowest affinity to Na^+ among the Na^+-K^+-ATPase α/β-heterodimers (Blanco & Mercer, 1998; Crambert et al., 2000) and are therefore especially well-suited to effectively respond to increase in intracellular Na^+ concentrations (Larsen et al., 2016), which are sharply increased during neuronal excitation (Baeza-Lehnert et al., 2018).

Clearly, existence of different Na^+-K^+-ATPase isoenzymes, which arose via gene duplications during vertebrate evolution, contributes to a multi-layered tissue-specific regulation of total cellular content, subcellular distribution, and intrinsic activity of Na^+-K^+-ATPase. An additional layer of complexity is added by FXYDs, which associate with Na^+-K^+-ATPase and modulate its function.

5. Origins and diversification of FXYDs in vertebrates

FXYDs (FXYD1–8) (Tables 1 and 4) are small transmembrane regulators of Na^+-K^+-ATPase that are characterized by the eponymous motif Phe-Xaa-Tyr-Asp (FXYD) (Sweadner & Rael, 2000). FXYDs are apparently an invention that appeared early during the vertebrate evolution. Indeed, so far *Petromyzon marinus* (sea lamprey) is the most basal vertebrate (Fig. 1), in which expression of an FXYD gene was determined, while no homologs seem to exist in non-vertebrates (Pirkmajer et al., 2017). In contrast, other small transmembrane regulators of ion transport, such as regulators of sarco-endoplasmic reticulum Ca^{2+}-ATPase (SERCA) are present in vertebrates (phospholamban, Simmerman, Collins, Theibert, Wegener, & Jones, 1986; Tada, Ohmori, Yamada, & Abe, 1979; sarcolipin, Odermatt et al., 1998, 1997; myoregulin, Anderson et al., 2015; and dwarf reading frame (DWORF), Nelson et al., 2016) and non-vertebrates (sarcolamban; Magny et al., 2013).

Seven FXYDs (FXYD1–7) were originally described in mammals (Sweadner & Rael, 2000). At least nine FXYDs were found in fishes (FXYD2 and FXYD5–12) (Cornelius, Mahmmoud, Meischke, & Cramb, 2005; Mahmmoud et al., 2000; Tipsmark, 2008). Based on this classification some FXYDs would be mammal-specific (FXYD1, FXYD3, FXYD4), while others would be fish-specific (FXYD8–12). However, due to their relatively small size (~60–100 amino acid residues, except FXYD5, which has ~178 amino acid residues) and low sequence homology outside the

Table 4 Comparison of the revised and old FXYD classifications in fishes and primates.

Revised classification	Other names	Cartilaginous fishes	Ray-finned fishes	Primates	Comments, including reasons for reclassification
FXYD1	PLM	–	–	+	
FXYD2	γ-subunit	+	+	+	
FXYD2f	FXYD12	+	+	–	Named FXYD2f to indicate that this is a fish paralog of FXYD2
FXYD3/4	PLMS	+	–	–	Named FXYD3/4 to indicate that this is an ancestral form of the FXYD3-FXYD4 subfamily
FXYD3/4-1	FXYD11	–	+	–	Named FXYD3/4-1 to indicate that this is one of the two FXYD3/4 paralogs in ray-finned fishes
FXYD3/4-2	FXYD9	–	+	–	Named FXYD3/4-2 to indicate that this is one of the two FXYD3/4 paralogs in ray-finned fishes
FXYD3	Mat-8	–	–	+	
FXYD4	CHIF	–	–	+	
FXYD5	Dysadherin, RIC	+	+	+	
FXYD6	Phosphohippolin	+	+	+	
FXYD6f	FXYD8	–	+	–	Named to FXYD6f to indicate that this is a paralog of FXYD6 in ray-finned fishes
FXYD6p		–	–	+	FXYD6 paralog in primates.
FXYD7		+	+	+	
FXYD8		–	–	+	FXYD6 paralog in primates, pseudogenized in the *Homo-Pan* lineage

The presence or absence of a particular FXYD in cartilaginous fishes, ray-finned fishes, and primates is indicated by "+" or "–," respectively. Suffix "f" and "p" denotes fish and primate, respectively. Abbreviations: CHIF, corticosteroid hormone-inducible factor or channel-inducing factor; Mat-8, Mammary tumor (protein of) 8 kDa; PLM, phospholemman; PLMS, PLM-like protein from shark; RIC, related to ion channel.

Table is based on the data from Pirkmajer, S., Kirchner, H., Lundell, L.S., Zelenin, P.V., Zierath, J.R., Makarova, K.S., et al. (2017). Early vertebrate origin and diversification of small transmembrane regulators of cellular ion transport. *The Journal of Physiology*. *595*, 4611–4630.

signature motif it was challenging to establish phylogenetic relationships among FXYDs across different vertebrate species. If a newly discovered fish FXYD was not an obvious ortholog of a mammalian FXYD, it was given the next number (e.g., FXYD9, FXYD10, FXYD11) (Cornelius et al., 2005; Mahmmoud et al., 2003; Tipsmark, 2008) with or without prefixes (e.g., OdFXYD9, OlFXYD9; Od = *Oryzias dancena*; Ol = *Oryzias latipes*) (Yang et al., 2013) and suffixes (e.g., FXYD9dr; dr = *Danio rerio*) (Sweadner & Rael, 2000) indicating the species, in which it was discovered. Sometimes a new FXYD was given a species indicator without a number (e.g., pFXYD, p = pufferfish) (Wang, Lin, Hwang, & Lee, 2008).

Usage of common names, which do not reflect phylogenetic relationships, was useful since more precise annotation was not possible. However, on the other hand, nomenclature that is not consistent across vertebrate species created an important barrier to comparative studies investigating FXYD physiology in vertebrates. Indeed, while experimental findings in lower vertebrates can provide valuable information for understanding human physiology, pathophysiology, and pharmacology (Asnani & Peterson, 2014; Cai et al., 2013; Donovan et al., 2000; MacRae & Peterson, 2015), it is difficult to relate the findings regarding FXYD function in fishes to human and other mammals without having a clear understanding of phylogenetic relationships.

To resolve this issue, a recent study used a combination of sequence-based phylogenetic analysis and conservation of local chromosome context (Pirkmajer et al., 2017). According to this study, FXYDs in extant vertebrates can be classified into 13 gene subfamilies, which do not always correspond to the established FXYD classification. A new, revised, classification of FXYDs was therefore proposed, which takes into account current understanding of evolutionary history of FXYDs and thus provides a consistent, phylogeny-based, nomenclature for FXYDs across all vertebrate species (Table 4). The purpose of the revision was to create consistent names that non-equivocally indicate homologous FXYDs across all vertebrates with minimum changes to the established nomenclature. The names of mammalian FXYDs remain the same as they had been originally defined (Sweadner & Rael, 2000). The new nomenclature therefore mainly affects the names of fish FXYDs. A comparison of the old and revised classifications for fish and primate FXYDs is presented in Table 4.

In addition to providing sufficient data for phylogeny-based classification of FXYDs, evolutionary scenario of the FXYD family has also been revised. Briefly, based on the currently available data, FXYDs probably appeared ca.

600–800 million years ago (MYA) in the *Chordata* lineage, leading to the *Craniata* common ancestor, which had at least one FXYD gene (Fig. 1). Extant *Cyclostomata* (jawless vertebrates), such as lampries, have at least one FXYD, which might represent an early version of FXYD5 (proto-FXYD5). The *Gnathostomata* (jawed vertebrates) ancestor (ca. 525 MYA) probably had at least six FXYDs, including FXYD5, FXYD6, FXYD7, FXYD3/4 (aka FXYD10 or PLMS) (Cornelius et al., 2005; Mahmmoud et al., 2003, 2000), FXYD2 and FXYD2f (aka FXYD12). These FXYDs are present also in extant *Chondrichthyes* (cartilaginous fishes). FXYD3/4 is an ancestor of fish-specific and tetrapod-specific FXYDs (Pirkmajer et al., 2017). On the one hand, fish-specific duplications in *Actinopterygii* produced FXYD3/4-1 (aka FXYD11) (Tipsmark, 2008) and FXYD3/4-2 (aka FXYD9) (Pirkmajer et al., 2017; Tipsmark, 2008). On the other hand, FXYD3/4 duplicated into FXYD3 and FXYD4 in the *Tetrapoda* ancestor clade (ca. 360–455 MYA). FXYD1 also emerged within this timeframe (Pirkmajer et al., 2017). FXYD8 was produced via duplication of FXYD6, but was pseudogenized in the *Pan-Homo* lineage (Pirkmajer et al., 2017). In summary, the new evolutionary scenario and the revised classification resolve the issue of phylogenetic relationship between the fish FXYDs (FXYD9–12) and the mammalian FXYDs.

It is interesting to note that diversification of FXYDs paralleled the diversification of vertebrate SERCA regulators phospholamban, sarcolipin, and myoregulin (Pirkmajer et al., 2017), diversification of the α- and β-subunits of Na^+-K^+-ATPase and H^+-K^+-ATPase (Saez et al., 2009; Studer et al., 2011), as well as appearance of new NCX isoforms (On et al., 2008). Notably, ion transporters were not only markedly diversified, but also new functional interactions were established. Indeed, FXYDs regulate Na^+-K^+-ATPase (Geering, 2008), NCX1 (Zhang et al., 2006, 2003), and potentially H^+-K^+-ATPase (Crambert et al., 2005; Delprat et al., 2007).

6. FXYDs are small transmembrane regulators of ion transport

Although sequences of FXYDs are variable outside the signature motif, their function appears to be conserved across the vertebrate lineage. Indeed, all mammalian FXYDs regulate Na^+-K^+-ATPase (Geering, 2008) (Table 1). Association of Na^+-K^+-ATPase isoenzymes with FXYDs produces distinct changes in their kinetic properties, such as apparent affinity for Na^+ ($K_{1/2,Na}$) and K^+ ($K_{1/2,K}$), maximal enzymatic rate (V_{max}), or maximal ion transport rate (I_{max}) (Table 1). Taken together with tissue-specific distribution of FXYDs as well as Na^+-K^+-ATPase α/β-heterodimers, this means that multiple, functionally distinct, heterotrimers are possible. Such diversity likely contributes to functional specialization of ion transport in different cell types (Beguin et al., 2001). For instance, association between FXYD4, which is expressed in collecting ducts in the kidney, and Na^+-K^+-ATPase increases its apparent affinity for Na^+ (Beguin et al., 2001). Aldosterone stimulates Na^+ reabsorption and K^+ excretion in the collecting ducts. Expression of Na^+-K^+-ATPase/FXYD4 complexes in the collecting ducts would therefore promote Na^+ reabsorption under aldosterone-stimulated conditions.

The role of fish FXYDs has been less well characterized; nevertheless, data suggest that they also regulate Na^+-K^+-ATPase. Shark FXYD3/4 (aka FXYD10 or PLMS) associates with the α-subunit of Na^+-K^+-ATPase and modulates its activity (Mahmmoud et al., 2003, 2000). As assessed by immunohistochemistry, immunoprecipitation, and/or proximity-ligation assay, Na^+-K^+-ATPase also co-localizes and interacts with FXYD2f (FXYD12) in *Oryzias dancena* and *Oryzias latipes* (Yang et al., 2018, 2013; Yang, Kang, Hsu, Lin, & Lee, 2016), FXYD3/4-1 (FXYD11) in *Salmo salar* (Tipsmark et al., 2010), *Danio rerio* (Saito et al., 2010) and *Oryzias dancena* (Chang et al., 2016), as well as FXYD3/4-2 (pFXYD, FXYD9) and FXYD6f (FXYD8) in *Tetraodon nigroviridis* (Wang et al., 2008, 2017). Furthermore, expression of FXYD6f reduces Na^+-K^+-ATPase activity in *Xenopus* oocytes (Wang et al., 2017). It is therefore likely that all FXYDs regulate Na^+-K^+-ATPase.

While FXYDs play an important role in regulation of Na^+-K^+-ATPase and are often regarded as the third Na^+-K^+-ATPase subunit, it has to be stressed that they also have other physiological roles. For instance, FXYD1 regulates the NCX1 (Zhang et al., 2003) and the L-type Ca^{2+} channels (Zhang et al., 2015) in cardiomyocytes, while FXYD3 and FXYD5 affect paracellular permeability of intestinal epithelium and the kidney tubules, respectively (Lubarski, Asher, & Garty, 2011; Okuda et al., 2010). Interestingly, FXYD3 (Crambert et al., 2005) and FXYD6 (Delprat et al., 2007) are able to associate with the H^+-K^+-ATPase in *Xenopus* oocytes. The physiological significance of this association is uncertain, since FXYD3 and H^+-K^+-ATPase are not expressed in the same cell types in the stomach (Crambert et al., 2005). Nevertheless, salinity-dependent expression of a putative H^+-K^+-ATPase ortholog in gills of *Dasyatis sabina*

(Choe, Verlander, Wingo, & Evans, 2004) indirectly suggests FXYDs might play a role in its regulation in gills.

Furthermore, a recent study showed that FXYD1 regulates translocation of aquaporin 2-containing vesicles in the kidney and is therefore important for water retention stimulated by the antidiuretic hormone (vasopressin) (Arystarkhova et al., 2017). Clearly, FXYDs should be regarded as small transmembrane regulators of ion and water transport and not just regulators of Na^+-K^+-ATPase. Interestingly, their involvement in regulation of transmembrane transport may go even beyond transport of ions and water. Indeed, FXYD1 regulates translocation of glucose transporter GLUT4 in adipocytes (Walaas et al., 1999), indicating it may have a more general regulatory role in vesicular transport.

7. FXYDs provide a link between hormonal stimuli and Na^+-K^+-ATPase

FXYDs provide a link between hormones and Na^+-K^+-ATPase at two levels. First, hormones acutely modulate interaction between Na^+-K^+-ATPase and FXYDs. Second, hormones regulate expression of FXYDs. Hormonal regulation of interaction with Na^+-K^+-ATPase is characterized by FXYD1 (phospholemman) (Crambert et al., 2002; Despa et al., 2005; Palmer et al., 1991). Among FXYDs, FXYD1 is the most important regulator of Na^+-K^+-ATPase in skeletal muscle and the heart. FXYD1 has several phosphorylation sites in the intracellular C-terminal region (Ser^{63}, Ser^{68}, and Thr^{69}/Ser^{68}) (Fuller et al., 2009; Palmer et al., 1991; Walaas et al., 1994). Association with unphosphorylated FXYD1 suppresses activity of Na^+-K^+-ATPase by decreasing affinity for intracellular Na^+ (Crambert et al., 2002) (Table 1). Phosphorylation of FXYD1 at Ser^{63} and Ser^{68} increases affinity for Na^+, which disinhibits Na^+-K^+-ATPase and increases its activity. Phosphorylation of FXYD1 may also increase maximal ATPase rate (V_{max}) and/or transport rate (I_{max}), although this effect was not observed consistently (Bibert et al., 2008; Crambert et al., 2002; Despa et al., 2005; Han et al., 2010) (see Table 1). Phosphorylation of Ser^{63} is catalyzed primarily by PKC (and under some conditions also by PKA), phosphorylation of Ser^{68} by PKC and PKA, and phosphorylation of Thr^{69} by PKC (Despa et al., 2005; Fuller et al., 2009; Walaas et al., 1994). Since PKC and/or PKA are activated by various hormones, including insulin and catecholamines, FXYD1 provides a direct link between hormonal stimulation and acute activation of Na^+-K^+-ATPase in skeletal muscle and heart (Pavlovic, Fuller, & Shattock, 2013; Pirkmajer & Chibalin, 2016; Walaas et al., 1994).

Interestingly, regulation of Na^+-K^+-ATPase via phosphorylation of FXYD1 closely resembles regulation of SERCA by phospholamban and other SERCA regulators. For instance, treatment with catecholamines leads to phosphorylation of phospholamban, which in turn disinhibits SERCA and increases its activity (Shaikh, Sahoo, & Periasamy, 2016). FXYD1 and phospholamban even have closely similar phosphorylation motifs: [–RSSIRRLSTR–] and [–RSAIRASSTI-] (Palmer et al., 1991), respectively, although they do not share common evolutionary origins (Pirkmajer et al., 2017). Extensive analyses of phosphorylation sites in vertebrate FXYDs revealed that all FXYD clades except the FXYD2 clade contain putative phosphorylation sites in their C-terminals (Crambert et al., 2004; Mahmmoud et al., 2000; Pirkmajer et al., 2017; Sweadner & Rael, 2000; Tipsmark, 2008; Yamaguchi, Yamaguchi, Tai, Sugimoto, & Tokuda, 2001). However, aside from FXYD1, only FXYD3/4 (aka FXYD10 or PLMS) was shown to be regulated by phosphorylation (Cornelius et al., 2005; Mahmmoud et al., 2003, 2000). For other FXYDs the role of phosphorylation was not tested or was challenging to ascertain. For instance, putative phosphorylation sites in the C-terminal part of FXYD7 were tested by exchanging three phosphorylatable Ser residues with Ala residues or phosphomimetic Asp residues (Crambert et al., 2004). However, these mutants did not display any functional differences compared with the wild type FXYD7 (Crambert et al., 2004). Clearly, phosphorylation of one or two Ser residues might produce a different effect, but this remains to be determined.

Hormonal regulation of gene expression has been demonstrated for FXYD4, which was previously named corticosteroid hormone-inducible factor (CHIF) (Table 4) (Beguin et al., 2001; Wald et al., 1996). FXYD4 is expressed primarily in aldosterone target tissues, such as the kidney collecting tubules and the distal colon (Aizman et al., 2002; Attali et al., 1995). In the distal colon, its expression is induced by aldosterone, by inhibition of 11β-hydroxysteroid dehydrogenase, an enzyme that inactivates cortisol (and corticosterone), by carbenoxolone, as well as by glucocorticoids, such as dexamethasone (Attali et al., 1995; Brennan & Fuller, 1999) (Table 3). Surprisingly, its expression in the kidney is not responsive to corticosteroids (Table 3). The reason for this tissue-specific effect is unclear.

8. Diversity of the FXYD family and complexity of the endocrine system

The most distant FXYD ortholog identified is proto-FXYD5 in *P. marinus* (Pirkmajer et al., 2017). There is no evidence for the existence of FXYDs outside the vertebrate lineage, including other chordates, such as *Cephalochordata* and *Tunicata* (Fig. 1). Yet Na^+-K^+-ATPase and NCX, which are regulated by FXYDs in vertebrates, are present in these non-vertebrate chordates. Lancelet (amphioxus), a cephalochordate (Table 1), also has muscle, gills, rudimentary nervous system and subendostylar vessel, which can be considered a myocardium homolog (Holland & Chen, 2001). In vertebrates FXYDs are thought to regulate ion transport in all of these (or related) tissues. So, an interesting hypothetical question is what did vertebrates gain by acquisition of FXYDs? There are at least two, non-mutually exclusive, possibilities.

One possibility is that FXYDs enable more complex regulation of ion transport at the level of individual cell, which promoted development of ever more complex tissue functions during evolution. It is especially tempting to speculate that FXYDs were important for the development of a complex nervous system that characterizes vertebrates. FXYD6 (Kadowaki et al., 2004; Yamaguchi et al., 2001) and FXYD7 (Beguin et al., 2002; Crambert et al., 2003), which are prominently and selectively (especially FXYD7) expressed in the brain, probably emerged no later than in the ancestor of *Gnathostomata* (Pirkmajer et al., 2017), indicating they were important for vertebrate evolution. FXYD1, which is also expressed in the brain (Feschenko et al., 2003), appeared only in tetrapods (Table 4). Notably, variants of FXYD1 and FXYD6 genes have been linked to schizophrenia (Chaumette et al., 2018; Choudhury et al., 2007), which demonstrates that FXYDs play a role in higher brain function. It could also be speculated that FXYDs enabled refinement of mechanisms involved in ion and water transport. This notion is again supported by human studies which linked allelic variant of FXYD2 to renal loss of magnesium (Meij et al., 2000).

Another possibility is that FXYDs act as adaptors through which hormones regulate ion, water, and (possibly) nutrient transport. As described above, FXYDs in modern animals are often downstream effectors of various hormones. Acute stimulation of Na^+-K^+-ATPase activity in the heart by catecholamines and insulin is mediated by phosphorylation of FXYD1 (Pavlovic et al., 2013; Pirkmajer & Chibalin, 2016). Similarly, hormones that regulate activation of PKC may be involved in regulation of Na^+-K^+-ATPase via FXYD3/4 (FXYD10 or PLMS) in the shark rectal gland (Cornelius et al., 2005; Mahmmoud et al., 2003, 2000). FXYD4 (Attali et al., 1995; Capurro et al., 1997; Goldschmidt et al., 2004) and its fish paralog FXYD3/4-1 (FXYD11) (Tang, Lai, & Lee, 2012; Tipsmark et al., 2010, 2011) are regulated by corticosteroids (Table 3). In addition, FXYD1 was shown to play a role in antidiuretic hormone (vasopressin)-stimulated translocation of aquaporin 2 in the kidney (Arystarkhova et al., 2017) as well as insulin-stimulated GLUT4 translocation in adipocytes (Walaas et al., 1999).

If the primary role of FXYDs is maintenance of ion homeostasis of individual cells, existence of ancestral FXYDs in lancelet could be expected. However, if the primary role of FXYDs is to regulate ion transport in response to various hormonal signals, the absence of FXYDs in lancelet in not so surprising. Indeed, lancelet has only a very rudimentary endocrine system. For instance, lancelet lacks distinct corticosteroid receptors and cannot synthesize cortisol or its precursors, which are important regulators of NKA and FXYDs in vertebrates (Baker, Funder, & Kattoula, 2013; Baker & Katsu, 2017; Holland et al., 2008; Rossier et al., 2015). Lancelet also lacks growth hormone and prolactin (Holland et al., 2008), which regulate Na^+-K^+-ATPase and FXYDs in fishes (Tipsmark et al., 2011; Tipsmark & Madsen, 2009). In contrast, *Cyclostomata* already have the hypothalamo-pituitary axis and produce insulin as well as corticosteroids (Conlon et al., 1995; Rossier et al., 2015; Sower, 2018). In the vertebrate lineage FXYDs may have developed in parallel with the appearance of more sophisticated and complex endocrine systems.

One example of potential parallelism between diversification of FXYD functions and development of the endocrine system, is the evolution of the mineralocorticoid system. Mineralocorticoids regulate expression of FXYDs in fishes and *Tetrapoda* (Table 3). Regulation of FXYDs by mineralocorticoids is thought to play an important in regulation of electrolyte and water homeostasis. In most fishes cortisol or its precursors or analogs are the main mineralocorticoids (Rossier et al., 2015). Cortisol induces expression of FXYD3/4-1 (aka FXYD11) and, possibly at least under some conditions, FXYD3/4-2 (aka FXYD9), which both descended from FXYD3/4 (aka FXYD10), in various teleost species (Tang et al., 2012; Tipsmark et al., 2011, 2010). According to a recent analysis of evolutionary history of FXYDs (Pirkmajer et al., 2017) the most distant FXYD3/4 ortholog is present in cartilaginous fishes, which are also the first vertebrates with a separate mineralocorticoid and glucocorticoid receptor (Baker et al., 2013; Baker & Katsu, 2017; Rossier et al., 2015).

In contrast, lampreys apparently lack FXYD3/4 and have only a single corticosteroid receptor (Baker et al., 2013; Rossier et al., 2015). Aldosterone is absent from the ray-finned fishes and appeared only after the divergence of lobe-finned fishes (Baker et al., 2013), from which *Tetrapoda* developed. FXYD3/4 was duplicated into FXYD3 and FXYD4 in the *Tetrapoda* ancestor (Pirkmajer et al., 2017). This suggests that the appearance of aldosterone-responsive FXYD4 closely followed the appearance of aldosterone. Taken together, available evidence suggests that diversification of FXYDs paralleled evolution of the aldosterone system.

9. Coordinated control of metabolic and ion homeostasis: Do FXYDs provide a link?

Given the extent of energy usage by Na^+-K^+-ATPase, regulation of Na^+ and K^+ transport and energy metabolism must be closely coordinated. Indeed, some types of cells, notably kidney tubular cells and neurons, may spend at least 50% of ATP to power Na^+-K^+-ATPase (Dienel, 2019; Rolfe & Brown, 1997). Challenges to osmotic homeostasis, therefore represent also challenges to energy homeostasis. Not surprisingly, many hormones exert effects on ion transport as well as metabolism. Notably, insulin, a prototypical metabolic hormone, markedly affects K^+ transport, an effect that was first demonstrated just a few years after its discovery (Briggs, Koechig, Doisy, & Weber, 1924). This raises the question whether FXYDs may be involved not only in regulation of ion and water transport but also in regulation of metabolism. Indirect evidence suggests that this is the case. First, FXYD1 is involved in insulin-stimulated GLUT4 translocation (Walaas et al., 1999). Second, high-fat diet in sedentary rats increases FXYD1 expression, while concomitantly reducing Na^+-K^+-ATPase content and impairing insulin-induced translocation of Na^+-K^+-ATPase α2-subunits (Galuska et al., 2009). This suggests that unhealthy diet and low physical activity lead not only to metabolic dysregulation but also to dysregulation of Na^+-K^+-ATPase and FXYD1 function. Notably, glucose intolerant men have reduced intracellular Na^+-to-K^+ ratio in skeletal muscle (Landin, Lindgarde, Saltin, & Wilhelmsen, 1989), which demonstrates that insulin resistance suppresses activity of Na^+-K^+-ATPase in skeletal muscle, thus providing another link between regulation of metabolism and ion transport. Taken together, we may speculate that FXYD1 and, possibly, other FXYDs provide a link in integrative regulation of ion homeostasis and metabolism.

10. Conclusions and perspectives

When and how vertebrates acquired regulatory mechanisms enabling integration of diverse cellular, tissue, and whole-body functions of Na^+-K^+-ATPase is unclear. Generation of novel α- and β-subunits must have been important, but existence of multiple Na^+-K^+-ATPase isoenzymes is insufficient to explain how their functions are integrated across diverse tissues and organs, indicating vertebrates acquired other important innovations in regulation of Na^+-K^+-ATPase. Development of a highly complex endocrine system and emergence of FXYDs were most likely two innovations that were important for development of integrative regulatory mechanisms. It is therefore unfortunate, that there has been relatively little progress in understanding of FXYD functions during the last 15 years. Indeed, with the exception of FXYD1, FXYD2, FXYD5, and FXYD10 (PLMS) only few functional studies have been conducted, thus leaving many key questions unanswered.

For instance, the functional role of putative phosphorylation sites, which are present in all FXYDs except those from the FXYD2 clade, should be established. Tissue-specific roles of FXYDs should also be investigated in more detail. For instance, very little is known about FXYD6 and FXYD7, which are expressed primarily in the brain and might be important for higher brain functions. It would also be important to determine the function of the newly discovered FXYD in *P. marinus* (proto-FXYD5) and evolution of FXYD functions in the vertebrate lineage.

Finally, the role of FXYD1 and, possibly, other FXYDs in hormone-regulated vesicular traffic, including GLUT4 and aquaporin 2 should be further examined. It would be especially important to establish if involvement of FXYD1 in regulation of vesicular traffic is specific for GLUT4 and aquaporin 2 or it plays a more general role in hormone-induced translocation of transporters and channels.

Acknowledgments

A.V.C. is supported by funding from the Swedish Research Council, Novo Nordisk Research Foundation, and the Strategic Research Programme in Diabetes at Karolinska Institutet. S.P. is supported by funding from the Slovenian Research Agency (grants P3-0043, J3-9263, J7-7138, and J7-8276). A.V.C. and S.P. are also supported by the bilateral grant for research cooperation between Republic of Slovenia and Russian Federation funded by the Slovenian Research Agency (BI-RU/19-20-039).

References

Ackermann, U., & Geering, K. (1990). Mutual dependence of Na, K-ATPase alpha- and beta-subunits for correct posttranslational processing and intracellular transport. *FEBS Letters*, *269*, 105–108.

Aizman, R., Asher, C., Fuzesi, M., Latter, H., Lonai, P., Karlish, S. J., et al. (2002). Generation and phenotypic analysis of CHIF knockout mice. *American Journal of Physiology. Renal Physiology*, *283*, F569–F577.

Al-Khalili, L., Kotova, O., Tsuchida, H., Ehren, I., Feraille, E., Krook, A., et al. (2004). ERK1/2 mediates insulin stimulation of Na(+),K(+)-ATPase by phosphorylation of the alpha-subunit in human skeletal muscle cells. *The Journal of Biological Chemistry*, *279*, 25211–25218.

Al-Khalili, L., Yu, M., & Chibalin, A. V. (2003). Na+,K+-ATPase trafficking in skeletal muscle: insulin stimulates translocation of both alpha 1- and alpha 2-subunit isoforms. *FEBS Letters*, *536*, 198–202.

Anderson, D. M., Anderson, K. M., Chang, C. L., Makarewich, C. A., Nelson, B. R., McAnally, J. R., et al. (2015). A micropeptide encoded by a putative long noncoding RNA regulates muscle performance. *Cell*, *160*, 595–606.

Arystarkhova, E., Bouley, R., Liu, Y. B., & Sweadner, K. J. (2017). Impaired AQP2 trafficking in Fxyd1 knockout mice: A role for FXYD1 in regulated vesicular transport. *PLoS One*, *12*(11), e0188006. https://doi.org/10.1371/journal.pone.018800.

Arystarkhova, E., Donnet, C., Asinovski, N. K., & Sweadner, K. J. (2002). Differential regulation of renal Na, K-ATPase by splice variants of the gamma subunit. *The Journal of Biological Chemistry*, *277*, 10162–10172.

Arystarkhova, E., Wetzel, R. K., Asinovski, N. K., & Sweadner, K. J. (1999). The gamma subunit modulates Na(+) and K(+) affinity of the renal Na, K-ATPase. *The Journal of Biological Chemistry*, *274*, 33183–33185.

Asnani, A., & Peterson, R. T. (2014). The zebrafish as a tool to identify novel therapies for human cardiovascular disease. *Disease Models & Mechanisms*, 7, 763–767.

Attali, B., Latter, H., Rachamim, N., & Garty, H. (1995). A corticosteroid-induced gene expressing an "IsK-like" K+ channel activity in Xenopus oocytes. *Proceedings of the National Academy of Sciences of the United States of America*, *92*, 6092–6096.

Baeza-Lehnert, F., Saab, A. S., Gutierrez, R., Larenas, V., Diaz, E., Horn, M., et al. (2018). Non-canonical control of neuronal energy status by the Na(+) pump. *Cell Metabolism*. https://doi.org/10.1016/j.cmet.2018.11.005 [Epub ahead of print].

Baker, M. E., Funder, J. W., & Kattoula, S. R. (2013). Evolution of hormone selectivity in glucocorticoid and mineralocorticoid receptors. *The Journal of Steroid Biochemistry and Molecular Biology*, *137*, 57–70.

Baker, M. E., & Katsu, Y. (2017). 30 Years of the mineralocorticoid receptor: Evolution of the mineralocorticoid receptor: Sequence, structure and function. *Journal of Endocrinology*, *234*, T1–T16.

Beguin, P., Crambert, G., Guennoun, S., Garty, H., Horisberger, J. D., & Geering, K. (2001). CHIF, a member of the FXYD protein family, is a regulator of Na, K-ATPase distinct from the gamma-subunit. *The EMBO Journal*, *20*, 3993–4002.

Beguin, P., Crambert, G., Monnet-Tschudi, F., Uldry, M., Horisberger, J. D., Garty, H., et al. (2002). FXYD7 is a brain-specific regulator of Na, K-ATPase alpha 1-beta isozymes. *The EMBO Journal*, *21*, 3264–3273.

Benziane, B., Bjornholm, M., Pirkmajer, S., Austin, R. L., Kotova, O., Viollet, B., et al. (2012). Activation of AMP-activated protein kinase stimulates Na+,K+-ATPase activity in skeletal muscle cells. *The Journal of Biological Chemistry*, *287*, 23451–23463.

Benziane, B., & Chibalin, A. V. (2008). Frontiers: Skeletal muscle sodium pump regulation: a translocation paradigm. *American Journal of Physiology. Endocrinology and Metabolism*, *295*, E553–E558.

Benziane, B., Widegren, U., Pirkmajer, S., Henriksson, J., Stepto, N. K., & Chibalin, A. V. (2011). Effect of exercise and training on phospholemman phosphorylation in human skeletal muscle. *American Journal of Physiology. Endocrinology and Metabolism, 301*, E456–E466.

Bertorello, A. M., Aperia, A., Walaas, S. I., Nairn, A. C., & Greengard, P. (1991). Phosphorylation of the catalytic subunit of Na+,K(+)-ATPase inhibits the activity of the enzyme. *Proceedings of the National Academy of Sciences of the United States of America, 88*, 11359–11362.

Bibert, S., Roy, S., Schaer, D., Horisberger, J. D., & Geering, K. (2008). Phosphorylation of phospholemman (FXYD1) by protein kinases a and C modulates distinct Na, K-ATPase isozymes. *The Journal of Biological Chemistry, 283*, 476–486.

Blanco, G., & Mercer, R. W. (1998). Isozymes of the Na-K-ATPase: Heterogeneity in structure, diversity in function. *The American Journal of Physiology, 275*, F633–F650.

Blot-Chabaud, M., Djelidi, S., Courtois-Coutry, N., Fay, M., Cluzeaud, F., Hummler, E., et al. (2001). Coordinate control of Na, K-atpase mRNA expression by aldosterone, vasopressin and cell sodium delivery in the cortical collecting duct. *Cellular and Molecular Biology (Noisy-le-Grand, France), 47*, 247–253.

Bossuyt, J., Despa, S., Han, F., Hou, Z., Robia, S. L., Lingrel, J. B., et al. (2009). Isoform specificity of the Na/K-ATPase association and regulation by phospholemman. *The Journal of Biological Chemistry, 284*, 26749–26757.

Boulanger-Saunier, C., Kattenburg, D. M., & Stoclet, J. C. (1985). Cyclic AMP-dependent phosphorylation of a 16 kDa protein in a plasma membrane-enriched fraction of rat aortic myocytes. *FEBS Letters, 193*, 283–288.

Boulanger-Saunier, C., & Stoclet, J. C. (1987). A 16 kDa protein substrate for protein kinase C and its phosphorylation upon stimulation of vasopressin receptors in rat aortic myocytes. *Biochemical and Biophysical Research Communications, 143*, 517–524.

Brennan, F. E., & Fuller, P. J. (1999). Acute regulation by corticosteroids of channel-inducing factor gene messenger ribonucleic acid in the distal colon. *Endocrinology, 140*, 1213–1218.

Briggs, A. P., Koechig, I., Doisy, E. A., & Weber, C. J. (1924). Some changes in the composition of blood due to the injection of insulin. *The Journal of Biological Chemistry, 58*, 721–730.

Cai, S. Y., Lionarons, D. A., Hagey, L., Soroka, C. J., Mennone, A., & Boyer, J. L. (2013). Adult Sea lamprey tolerates biliary atresia by altering bile salt composition and renal excretion. *Hepatology, 57*, 2418–2426.

Cameron, R., Klein, L., Shyjan, A. W., Rakic, P., & Levenson, R. (1994). Neurons and astroglia express distinct subsets of Na, K-ATPase alpha and beta subunits. *Brain Research. Molecular Brain Research, 21*, 333–343.

Capurro, C., Coutry, N., Bonvalet, J. P., Escoubet, B., Garty, H., & Farman, N. (1997). Specific expression and regulation of CHIF in kidney and colon. *Annals of the New York Academy of Sciences, 834*, 562–564.

Chang, C. H., Yang, W. K., Lin, C. H., Kang, C. K., Tang, C. H., & Lee, T. H. (2016). FXYD11 mediated modulation of Na(+)/K(+)-ATPase activity in gills of the brackish medaka (Oryzias dancena) when transferred to hypoosmotic or hyperosmotic environments. *Comparative Biochemistry and Physiology. Part A, Molecular & Integrative Physiology, 194*, 19–26.

Chapman, K., Holmes, M., & Seckl, J. (2013). 11beta-hydroxysteroid dehydrogenases: Intracellular gate-keepers of tissue glucocorticoid action. *Physiological Reviews, 93*, 1139–1206.

Chaumette, B., Ferrafiat, V., Ambalavanan, A., Goldenberg, A., Dionne-Laporte, A., Spiegelman, D., et al. (2018). Missense variants in ATP1A3 and FXYD gene family are associated with childhood-onset schizophrenia. *Molecular Psychiatry*. https://doi.org/10.1038/s41380-018-0103-8 [Epub ahead of print].

Chibalin, A. V., Katz, A. I., Berggren, P. O., & Bertorello, A. M. (1997). Receptor-mediated inhibition of renal Na(+)-K(+)-ATPase is associated with endocytosis of its alpha- and beta-subunits. *The American Journal of Physiology, 273*, C1458–C1465.

Chibalin, A. V., Kovalenko, M. V., Ryder, J. W., Feraille, E., Wallberg-Henriksson, H., & Zierath, J. R. (2001). Insulin- and glucose-induced phosphorylation of the Na(+),K(+)-adenosine triphosphatase alpha-subunits in rat skeletal muscle. *Endocrinology, 142*, 3474–3482.

Chibalin, A. V., Ogimoto, G., Pedemonte, C. H., Pressley, T. A., Katz, A. I., Feraille, E., et al. (1999). Dopamine-induced endocytosis of Na+,K+-ATPase is initiated by phosphorylation of Ser-18 in the rat alpha subunit and Is responsible for the decreased activity in epithelial cells. *The Journal of Biological Chemistry, 274*, 1920–1927.

Chibalin, A. V., Pedemonte, C. H., Katz, A. I., Feraille, E., Berggren, P. O., & Bertorello, A. M. (1998). Phosphorylation of the catalyic alpha-subunit constitutes a triggering signal for Na+,K+-ATPase endocytosis. *The Journal of Biological Chemistry, 273*, 8814–8819.

Chibalin, A. V., Vasilets, L. A., Hennekes, H., Pralong, D., & Geering, K. (1992). Phosphorylation of Na, K-ATPase alpha-subunits in microsomes and in homogenates of Xenopus oocytes resulting from the stimulation of protein kinase a and protein kinase C. *The Journal of Biological Chemistry, 267*, 22378–22384.

Choe, K. P., Verlander, J. W., Wingo, C. S., & Evans, D. H. (2004). A putative H+-K+-ATPase in the Atlantic stingray, Dasyatis sabina: Primary sequence and expression in gills. *American Journal of Physiology. Regulatory, Integrative and Comparative Physiology, 287*, R981–R991.

Choudhury, K., McQuillin, A., Puri, V., Pimm, J., Datta, S., Thirumalai, S., et al. (2007). A genetic association study of chromosome 11q22-24 in two different samples implicates the FXYD6 gene, encoding phosphohippolin, in susceptibility to schizophrenia. *American Journal of Human Genetics, 80*, 664–672.

Clausen, T. (2003). Na+-K+ pump regulation and skeletal muscle contractility. *Physiological Reviews, 83*, 1269–1324.

Clausen, T. (2010). Hormonal and pharmacological modification of plasma potassium homeostasis. *Fundamental & Clinical Pharmacology, 24*, 595–605.

Clausen, T., Everts, M. E., & Kjeldsen, K. (1987). Quantification of the maximum capacity for active sodium-potassium transport in rat skeletal muscle. *The Journal of Physiology, 388*, 163–181.

Conlon, J. M., Bondareva, V., Rusakov, Y., Plisetskaya, E. M., Mynarcik, D. C., & Whittaker, J. (1995). Characterization of insulin, glucagon, and somatostatin from the river lamprey, Lampetra fluviatilis. *General and Comparative Endocrinology, 100*, 96–105.

Conway-Campbell, B. L., Wooh, J. W., Brooks, A. J., Gordon, D., Brown, R. J., Lichanska, A. M., et al. (2007). Nuclear targeting of the growth hormone receptor results in dysregulation of cell proliferation and tumorigenesis. *Proceedings of the National Academy of Sciences of the United States of America, 104*, 13331–13336.

Cornelius, F., & Mahmmoud, Y. A. (2003). Functional modulation of the sodium pump: The regulatory proteins "fixit". *News in Physiological Sciences, 18*, 119–124.

Cornelius, F., Mahmmoud, Y. A., Meischke, L., & Cramb, G. (2005). Functional significance of the shark Na, K-ATPase N-terminal domain. Is the structurally variable N-terminus involved in tissue-specific regulation by FXYD proteins? *Biochemistry, 44*, 13051–13062.

Craig, W. S., & Kyte, J. (1980). Stoichiometry and molecular weight of the minimum asymmetric unit of canine renal sodium and potassium ion-activated adenosine triphosphatase. *The Journal of Biological Chemistry, 255*, 6262–6269.

Crambert, G., Beguin, P., Uldry, M., Monnet-Tschudi, F., Horisberger, J. D., Garty, H., et al. (2003). FXYD7, the first brain- and isoform-specific regulator of Na, K-ATPase: Biosynthesis and function of its posttranslational modifications. *Annals of the New York Academy of Sciences*, *986*, 444–448.

Crambert, G., Fuzesi, M., Garty, H., Karlish, S., & Geering, K. (2002). Phospholemman (FXYD1) associates with Na, K-ATPase and regulates its transport properties. *Proceedings of the National Academy of Sciences of the United States of America*, *99*, 11476–11481.

Crambert, G., Hasler, U., Beggah, A. T., Yu, C., Modyanov, N. N., Horisberger, J. D., et al. (2000). Transport and pharmacological properties of nine different human Na, K-ATPase isozymes. *The Journal of Biological Chemistry*, *275*, 1976–1986.

Crambert, G., Li, C., Claeys, D., & Geering, K. (2005). FXYD3 (Mat-8), a new regulator of Na, K-ATPase. *Molecular Biology of the Cell*, *16*, 2363–2371.

Crambert, G., Li, C., Swee, L. K., & Geering, K. (2004). FXYD7, mapping of functional sites involved in endoplasmic reticulum export, association with and regulation of Na, K-ATPase. *The Journal of Biological Chemistry*, *279*, 30888–30895.

Delprat, B., Schaer, D., Roy, S., Wang, J., Puel, J. L., & Geering, K. (2007). FXYD6 is a novel regulator of Na, K-ATPase expressed in the inner ear. *The Journal of Biological Chemistry*, *282*, 7450–7456.

Despa, S., Bossuyt, J., Han, F., Ginsburg, K. S., Jia, L. G., Kutchai, H., et al. (2005). Phospholemman-phosphorylation mediates the beta-adrenergic effects on Na/K pump function in cardiac myocytes. *Circulation Research*, *97*, 252–259.

Dienel, G. A. (2019). Brain glucose metabolism: Integration of energetics with function. *Physiological Reviews*, *99*, 949–1045.

DiFranco, M., Hakimjavadi, H., Lingrel, J. B., & Heiny, J. A. (2015). Na, K-ATPase alpha2 activity in mammalian skeletal muscle T-tubules is acutely stimulated by extracellular K. *The Journal of General Physiology*, *146*, 281–294.

Donovan, A., Brownlie, A., Zhou, Y., Shepard, J., Pratt, S. J., Moynihan, J., et al. (2000). Positional cloning of zebrafish ferroportin1 identifies a conserved vertebrate iron exporter. *Nature*, *403*, 776–781.

Everts, M. E., Dorup, I., Flyvbjerg, A., Marshall, S. M., & Jorgensen, K. D. (1990). Na(+)-K+ pump in rat muscle: Effects of hypophysectomy, growth hormone, and thyroid hormone. *The American Journal of Physiology*, *259*, E278–E283.

Feraille, E., Carranza, M. L., Gonin, S., Beguin, P., Pedemonte, C., Rousselot, M., et al. (1999). Insulin-induced stimulation of Na+,K(+)-ATPase activity in kidney proximal tubule cells depends on phosphorylation of the alpha-subunit at Tyr-10. *Molecular Biology of the Cell*, *10*, 2847–2859.

Feraille, E., & Dizin, E. (2016). Coordinated control of ENaC and Na+,K+-ATPase in renal collecting duct. *Journal of the American Society of Nephrology*, *27*, 2554–2563.

Feschenko, M. S., Donnet, C., Wetzel, R. K., Asinovski, N. K., Jones, L. R., & Sweadner, K. J. (2003). Phospholemman, a single-span membrane protein, is an accessory protein of Na, K-ATPase in cerebellum and choroid plexus. *The Journal of Neuroscience*, *23*, 2161–2169.

Feschenko, M. S., & Sweadner, K. J. (1995). Structural basis for species-specific differences in the phosphorylation of Na, K-ATPase by protein kinase C. *The Journal of Biological Chemistry*, *270*, 14072–14077.

Figtree, G. A., Liu, C. C., Bibert, S., Hamilton, E. J., Garcia, A., White, C. N., et al. (2009). Reversible oxidative modification: A key mechanism of Na+-K+ pump regulation. *Circulation Research*, *105*, 185–193.

Fotis, H., Tatjanenko, L. V., & Vasilets, L. A. (1999). Phosphorylation of the alpha-subunits of the Na+/K+-ATPase from mammalian kidneys and Xenopus oocytes by cGMP-dependent protein kinase results in stimulation of ATPase activity. *European Journal of Biochemistry*, *260*, 904–910.

Fuller, W., Howie, J., McLatchie, L. M., Weber, R. J., Hastie, C. J., Burness, K., et al. (2009). FXYD1 phosphorylation in vitro and in adult rat cardiac myocytes: Threonine 69 is a novel substrate for protein kinase C. *American Journal of Physiology. Cell Physiology*, *296*, C1346–C1355.

Galuska, D., Kotova, O., Barres, R., Chibalina, D., Benziane, B., & Chibalin, A. V. (2009). Altered expression and insulin-induced trafficking of Na+-K+-ATPase in rat skeletal muscle: Effects of high-fat diet and exercise. *American Journal of Physiology. Endocrinology and Metabolism*, *297*, E38–E49.

Geering, K. (2006). FXYD proteins: New regulators of Na-K-ATPase. *American Journal of Physiology. Renal Physiology*, *290*, F241–F250.

Geering, K. (2008). Functional roles of Na, K-ATPase subunits. *Current Opinion in Nephrology and Hypertension*, *17*, 526–532.

Goldschmidt, I., Grahammer, F., Warth, R., Schulz-Baldes, A., Garty, H., Greger, R., et al. (2004). Kidney and colon electrolyte transport in CHIF knockout mice. *Cellular Physiology and Biochemistry*, *14*, 113–120.

Han, F., Bossuyt, J., Martin, J. L., Despa, S., & Bers, D. M. (2010). Role of phospholemman phosphorylation sites in mediating kinase-dependent regulation of the Na+-K+-ATPase. *American Journal of Physiology. Cell Physiology*, *299*, C1363–C1369.

Hansen, O. (2001). The alpha1 isoform of Na+,K+-ATPase in rat soleus and extensor digitorum longus. *Acta Physiologica Scandinavica*, *173*, 335–341.

He, S., Shelly, D. A., Moseley, A. E., James, P. F., James, J. H., Paul, R. J., et al. (2001). The alpha(1)- and alpha(2)-isoforms of Na-K-ATPase play different roles in skeletal muscle contractility. *American Journal of Physiology. Regulatory, Integrative and Comparative Physiology*, *281*, R917–R925.

Heineman, F. W., & Balaban, R. S. (1990). Phosphorus-31 nuclear magnetic resonance analysis of transient changes of canine myocardial metabolism in vivo. *The Journal of Clinical Investigation*, *85*, 843–852.

Hochachka, P. W., & McClelland, G. B. (1997). Cellular metabolic homeostasis during large-scale change in ATP turnover rates in muscles. *The Journal of Experimental Biology*, *200*, 381–386.

Holland, L. Z., Albalat, R., Azumi, K., Benito-Gutierrez, E., Blow, M. J., Bronner-Fraser,- M., et al. (2008). The amphioxus genome illuminates vertebrate origins and cephalochordate biology. *Genome Research*, *18*, 1100–1111.

Holland, N. D., & Chen, J. (2001). Origin and early evolution of the vertebrates: New insights from advances in molecular biology, anatomy, and palaeontology. *BioEssays*, *23*, 142–151.

Hostrup, M., Kalsen, A., Ortenblad, N., Juel, C., Morch, K., Rzeppa, S., et al. (2014). Beta2-adrenergic stimulation enhances Ca2+ release and contractile properties of skeletal muscles, and counteracts exercise-induced reductions in Na+-K+-ATPase Vmax in trained men. *The Journal of Physiology*, *592*, 5445–5459.

Hundal, H. S., Marette, A., Mitsumoto, Y., Ramlal, T., Blostein, R., & Klip, A. (1992). Insulin induces translocation of the alpha 2 and beta 1 subunits of the Na+/K(+)-ATPase from intracellular compartments to the plasma membrane in mammalian skeletal muscle. *The Journal of Biological Chemistry*, *267*, 5040–5043.

Hundal, H. S., Maxwell, D. L., Ahmed, A., Darakhshan, F., Mitsumoto, Y., & Klip, A. (1994). Subcellular distribution and immunocytochemical localization of Na, K-ATPase subunit isoforms in human skeletal muscle. *Molecular Membrane Biology*, *11*, 255–262.

Juel, C. (2014). Oxidative stress (glutathionylation) and Na, K-ATPase activity in rat skeletal muscle. *PLoS One*, *9*, e110514.

Juel, C., Hostrup, M., & Bangsbo, J. (2015). The effect of exercise and beta2-adrenergic stimulation on glutathionylation and function of the Na, K-ATPase in human skeletal muscle. *Physiological Reports*, *3*(8), e12515. https://doi.org/10.14814/phy2.12515.

Kadowaki, K., Sugimoto, K., Yamaguchi, F., Song, T., Watanabe, Y., Singh, K., et al. (2004). Phosphohippolin expression in the rat central nervous system. *Brain Research. Molecular Brain Research, 125*, 105–112.
Korneenko, T. V., Pestov, N. B., Ahmad, N., Okkelman, I. A., Dmitriev, R. I., Shakhparonov, M. I., et al. (2016). Evolutionary diversification of the BetaM interactome acquired through co-option of the ATP1B4 gene in placental mammals. *Scientific Reports, 6*, 22395.
Kyte, J. (1971). Purification of the sodium- and potassium-dependent adenosine triphosphatase from canine renal medulla. *The Journal of Biological Chemistry, 246*, 4157–4165.
Landin, K., Lindgarde, F., Saltin, B., & Wilhelmsen, L. (1989). Increased skeletal muscle Na/K-ratio in obese men, but not in women, with glucose intolerance. *Journal of Internal Medicine, 225*, 89–94.
Larsen, B. R., Stoica, A., & MacAulay, N. (2016). Managing brain extracellular K(+) during neuronal activity: The physiological role of the Na(+)/K(+)-ATPase subunit isoforms. *Frontiers in Physiology, 7*, 141.
Li, Z., & Langhans, S. A. (2015). Transcriptional regulators of Na, K-ATPase subunits. *Frontiers in Cell and Development Biology, 3*, 66.
Lin, M. H., & Akera, T. (1978). Increased (Na+,K+)-ATPase concentrations in various tissues of rats caused by thyroid hormone treatment. *The Journal of Biological Chemistry, 253*, 723–726.
Lindemann, J. P. (1986). Alpha-adrenergic stimulation of sarcolemmal protein phosphorylation and slow responses in intact myocardium. *The Journal of Biological Chemistry, 261*, 4860–4867.
Liu, C. C., Garcia, A., Mahmmoud, Y. A., Hamilton, E. J., Galougahi, K. K., Fry, N. A., et al. (2012). Susceptibility of beta1 Na+-K+ pump subunit to glutathionylation and oxidative inhibition depends on conformational state of pump. *The Journal of Biological Chemistry, 287*, 12353–12364.
Lubarski, I., Asher, C., & Garty, H. (2011). FXYD5 (dysadherin) regulates the paracellular permeability in cultured kidney collecting duct cells. *American Journal of Physiology. Renal Physiology, 301*, F1270–F1280.
Lubarski, I., Karlish, S. J., & Garty, H. (2007). Structural and functional interactions between FXYD5 and the Na+-K+-ATPase. *American Journal of Physiology. Renal Physiology, 293*, F1818–F1826.
Lubarski, I., Pihakaski-Maunsbach, K., Karlish, S. J., Maunsbach, A. B., & Garty, H. (2005). Interaction with the Na,K-ATPase and tissue distribution of FXYD5 (related to ion channel). *The Journal of Biological Chemistry, 280*, 37717–37724.
Lytton, J., Lin, J. C., & Guidotti, G. (1985). Identification of two molecular forms of (Na+,K+)-ATPase in rat adipocytes. Relation to insulin stimulation of the enzyme. *The Journal of Biological Chemistry, 260*, 1177–1184.
MacRae, C. A., & Peterson, R. T. (2015). Zebrafish as tools for drug discovery. *Nature Reviews. Drug Discovery, 14*, 721–731.
Magny, E. G., Pueyo, J. I., Pearl, F. M., Cespedes, M. A., Niven, J. E., Bishop, S. A., et al. (2013). Conserved regulation of cardiac calcium uptake by peptides encoded in small open reading frames. *Science, 341*, 1116–1120.
Mahmmoud, Y. A., Cramb, G., Maunsbach, A. B., Cutler, C. P., Meischke, L., & Cornelius, F. (2003). Regulation of Na,K-ATPase by PLMS, the phospholemman-like protein from shark: Molecular cloning, sequence, expression, cellular distribution, and functional effects of PLMS. *The Journal of Biological Chemistry, 278*, 37427–37438.
Mahmmoud, Y. A., Vorum, H., & Cornelius, F. (2000). Identification of a phospholemman-like protein from shark rectal glands. Evidence for indirect regulation of Na, K-ATPase by protein kinase c via a novel member of the FXYDY family. *The Journal of Biological Chemistry, 275*, 35969–35977.

Marette, A., Krischer, J., Lavoie, L., Ackerley, C., Carpentier, J. L., & Klip, A. (1993). Insulin increases the Na(+)-K(+)-ATPase alpha 2-subunit in the surface of rat skeletal muscle: Morphological evidence. *The American Journal of Physiology*, *265*, C1716–C1722.

McDonough, A. A., Geering, K., & Farley, R. A. (1990). The sodium pump needs its beta subunit. *The FASEB Journal*, *4*, 1598–1605.

Meij, I. C., Koenderink, J. B., van Bokhoven, H., Assink, K. F., Groenestege, W. T., de Pont, J. J., et al. (2000). Dominant isolated renal magnesium loss is caused by misrouting of the Na(+),K(+)-ATPase gamma-subunit. *Nature Genetics*, *26*, 265–266.

Mishra, N. K., Habeck, M., Kirchner, C., Haviv, H., Peleg, Y., Eisenstein, M., et al. (2015). Molecular mechanisms and kinetic effects of FXYD1 and phosphomimetic mutants on purified human Na, K-ATPase. *The Journal of Biological Chemistry*, *290*, 28746–28759.

Morth, J. P., Pedersen, B. P., Toustrup-Jensen, M. S., Sorensen, T. L., Petersen, J., Andersen, J. P., et al. (2007). Crystal structure of the sodium-potassium pump. *Nature*, *450*, 1043–1049.

Mullur, R., Liu, Y. Y., & Brent, G. A. (2014). Thyroid hormone regulation of metabolism. *Physiological Reviews*, *94*, 355–382.

Nelson, B. R., Makarewich, C. A., Anderson, D. M., Winders, B. R., Troupes, C. D., Wu, F., et al. (2016). A peptide encoded by a transcript annotated as long noncoding RNA enhances SERCA activity in muscle. *Science*, *351*, 271–275.

Odermatt, A., Becker, S., Khanna, V. K., Kurzydlowski, K., Leisner, E., Pette, D., et al. (1998). Sarcolipin regulates the activity of SERCA1, the fast-twitch skeletal muscle sarcoplasmic reticulum Ca2+-ATPase. *The Journal of Biological Chemistry*, *273*, 12360–12369.

Odermatt, A., Taschner, P. E., Scherer, S. W., Beatty, B., Khanna, V. K., Cornblath, D. R., et al. (1997). Characterization of the gene encoding human sarcolipin (SLN), a proteolipid associated with SERCA1: Absence of structural mutations in five patients with Brody disease. *Genomics*, *45*, 541–553.

Okuda, J., Hayashi, N., Okamoto, M., Sawada, S., Minagawa, S., Yano, Y., et al. (2010). Translocation of *Pseudomonas aeruginosa* from the intestinal tract is mediated by the binding of ExoS to an Na,K-ATPase regulator, FXYD3. *Infection and Immunity*, *78*, 4511–4522.

On, C., Marshall, C. R., Chen, N., Moyes, C. D., & Tibbits, G. F. (2008). Gene structure evolution of the Na+-Ca2+ exchanger (NCX) family. *BMC Evolutionary Biology*, *8*, 127.

Orlowski, J., & Lingrel, J. B. (1988). Tissue-specific and developmental regulation of rat Na, K-ATPase catalytic alpha isoform and beta subunit mRNAs. *The Journal of Biological Chemistry*, *263*, 10436–10442.

Palmer, C. J., Scott, B. T., & Jones, L. R. (1991). Purification and complete sequence determination of the major plasma membrane substrate for cAMP-dependent protein kinase and protein kinase C in myocardium. *The Journal of Biological Chemistry*, *266*, 11126–11130.

Pavlovic, D., Fuller, W., & Shattock, M. J. (2013). Novel regulation of cardiac Na pump via phospholemman. *Journal of Molecular and Cellular Cardiology*, *61*, 83–93.

Pestov, N. B., Ahmad, N., Korneenko, T. V., Zhao, H., Radkov, R., Schaer, D., et al. (2007). Evolution of Na,K-ATPase beta m-subunit into a coregulator of transcription in placental mammals. *Proceedings of the National Academy of Sciences of the United States of America*, *104*, 11215–11220.

Pestov, N. B., Zhao, H., Basrur, V., & Modyanov, N. N. (2011). Isolation and characterization of BetaM protein encoded by ATP1B4–a unique member of the Na, K-ATPase beta-subunit gene family. *Biochemical and Biophysical Research Communications*, *412*, 543–548.

Petrushanko, I. Y., Yakushev, S., Mitkevich, V. A., Kamanina, Y. V., Ziganshin, R. H., Meng, X., et al. (2012). S-glutathionylation of the Na, K-ATPase catalytic alpha subunit is a determinant of the enzyme redox sensitivity. *The Journal of Biological Chemistry, 287*, 32195–32205.

Pirkmajer, S., & Chibalin, A. V. (2016). Na,K-ATPase regulation in skeletal muscle. *American Journal of Physiology. Endocrinology and Metabolism, 311*, E1–E31.

Pirkmajer, S., Kirchner, H., Lundell, L. S., Zelenin, P. V., Zierath, J. R., Makarova, K. S., et al. (2017). Early vertebrate origin and diversification of small transmembrane regulators of cellular ion transport. *The Journal of Physiology, 595*, 4611–4630.

Post, R. L., & Jolly, P. C. (1957). The linkage of sodium, potassium, and ammonium active transport across the human erythrocyte membrane. *Biochimica et Biophysica Acta, 25*, 118–128.

Presti, C. F., Jones, L. R., & Lindemann, J. P. (1985). Isoproterenol-induced phosphorylation of a 15-kilodalton sarcolemmal protein in intact myocardium. *The Journal of Biological Chemistry, 260*, 3860–3867.

Rolfe, D. F., & Brown, G. C. (1997). Cellular energy utilization and molecular origin of standard metabolic rate in mammals. *Physiological Reviews, 77*, 731–758.

Rossier, B. C., Baker, M. E., & Studer, R. A. (2015). Epithelial sodium transport and its control by aldosterone: The story of our internal environment revisited. *Physiological Reviews, 95*, 297–340.

Russo, J. J., & Sweadner, K. J. (1993). Na(+)-K(+)-ATPase subunit isoform pattern modification by mitogenic insulin concentration in 3T3-L1 preadipocytes. *The American Journal of Physiology, 264*, C311–C316.

Saez, A. G., Lozano, E., & Zaldivar-Riveron, A. (2009). Evolutionary history of Na, K-ATPases and their osmoregulatory role. *Genetica, 136*, 479–490.

Saito, K., Nakamura, N., Ito, Y., Hoshijima, K., Esaki, M., Zhao, B., et al. (2010). Identification of zebrafish Fxyd11a protein that is highly expressed in ion-transporting epithelium of the gill and skin and its possible role in ion homeostasis. *Frontiers in Physiology, 1*, 129.

Schmidt, T. A., Hasselbalch, S., Farrell, P. A., Vestergaard, H., & Kjeldsen, K. (1994). Human and rodent muscle Na(+)-K(+)-ATPase in diabetes related to insulin, starvation, and training. *Journal of Applied Physiology (1985), 76*, 2140–2146.

Sen, A. K., & Post, R. L. (1964). Stoichiometry and localization of adenosine triphosphate-dependent sodium and potassium transport in the erythrocyte. *The Journal of Biological Chemistry, 239*, 345–352.

Shaikh, S. A., Sahoo, S. K., & Periasamy, M. (2016). Phospholamban and sarcolipin: Are they functionally redundant or distinct regulators of the Sarco(Endo)Plasmic reticulum calcium ATPase? *Journal of Molecular and Cellular Cardiology, 91*, 81–91.

Shamraj, O. I., & Lingrel, J. B. (1994). A putative fourth Na+,K(+)-ATPase alpha-subunit gene is expressed in testis. *Proceedings of the National Academy of Sciences of the United States of America, 91*, 12952–12956.

Shyjan, A. W., & Levenson, R. (1989). Antisera specific for the alpha 1, alpha 2, alpha 3, and beta subunits of the Na,K-ATPase: Differential expression of alpha and beta subunits in rat tissue membranes. *Biochemistry, 28*, 4531–4535.

Silva, J. E. (2006). Thermogenic mechanisms and their hormonal regulation. *Physiological Reviews, 86*, 435–464.

Simmerman, H. K., Collins, J. H., Theibert, J. L., Wegener, A. D., & Jones, L. R. (1986). Sequence analysis of phospholamban. Identification of phosphorylation sites and two major structural domains. *The Journal of Biological Chemistry, 261*, 13333–13341.

Skou, J. C. (1957). The influence of some cations on an adenosine triphosphatase from peripheral nerves. *Biochimica et Biophysica Acta, 23*, 394–401.

Sower, S. A. (2018). Landmark discoveries in elucidating the origins of the hypothalamic-pituitary system from the perspective of a basal vertebrate, sea lamprey. *General and Comparative Endocrinology*, *264*, 3–15.
Studer, R. A., Person, E., Robinson-Rechavi, M., & Rossier, B. C. (2011). Evolution of the epithelial sodium channel and the sodium pump as limiting factors of aldosterone action on sodium transport. *Physiological Genomics*, *43*, 844–854.
Sweadner, K. J., McGrail, K. M., & Khaw, B. A. (1992). Discoordinate regulation of isoforms of Na,K-ATPase and myosin heavy chain in the hypothyroid postnatal rat heart and skeletal muscle. *The Journal of Biological Chemistry*, *267*, 769–773.
Sweadner, K. J., & Rael, E. (2000). The FXYD gene family of small ion transport regulators or channels: cDNA sequence, protein signature sequence, and expression. *Genomics*, *68*, 41–56.
Tada, M., Ohmori, F., Yamada, M., & Abe, H. (1979). Mechanism of the stimulation of Ca2+−dependent ATPase of cardiac sarcoplasmic reticulum by adenosine 3′:5′-monophosphate-dependent protein kinase. Role of the 22,000-dalton protein. *The Journal of Biological Chemistry*, *254*, 319–326.
Tang, C. H., Lai, D. Y., & Lee, T. H. (2012). Effects of salinity acclimation on Na(+)/K(+)-ATPase responses and FXYD11 expression in the gills and kidneys of the Japanese eel (Anguilla japonica). *Comparative Biochemistry and Physiology. Part A, Molecular & Integrative Physiology*, *163*, 302–310.
Tipsmark, C. K. (2008). Identification of FXYD protein genes in a teleost: Tissue-specific expression and response to salinity change. *American Journal of Physiology. Regulatory, Integrative and Comparative Physiology*, *294*, R1367–R1378.
Tipsmark, C. K., Breves, J. P., Seale, A. P., Lerner, D. T., Hirano, T., & Grau, E. G. (2011). Switching of Na+, K+-ATPase isoforms by salinity and prolactin in the gill of a cichlid fish. *The Journal of Endocrinology*, *209*, 237–244.
Tipsmark, C. K., & Madsen, S. S. (2009). Distinct hormonal regulation of Na(+),K(+)-atpase genes in the gill of Atlantic salmon (Salmo salar L.). *The Journal of Endocrinology*, *203*, 301–310.
Tipsmark, C. K., Mahmmoud, Y. A., Borski, R. J., & Madsen, S. S. (2010). FXYD-11 associates with Na+-K+-ATPase in the gill of Atlantic salmon: Regulation and localization in relation to changed ion-regulatory status. *American Journal of Physiology. Regulatory, Integrative and Comparative Physiology*, *299*, R1212–R1223.
Walaas, S. I., Czernik, A. J., Olstad, O. K., Sletten, K., & Walaas, O. (1994). Protein kinase C and cyclic AMP-dependent protein kinase phosphorylate phospholemman, an insulin and adrenaline-regulated membrane phosphoprotein, at specific sites in the carboxy terminal domain. *The Biochemical Journal*, *304*(Pt 2), 635–640.
Walaas, O., Horn, R. S., & Walaas, S. I. (1991). Insulin and phorbol ester stimulate phosphorylation of a 15,000 Dalton membrane protein in rat diaphragm in a similar manner. *Biochimica et Biophysica Acta*, *1094*, 92–102.
Walaas, O., Horn, R. S., & Walaas, S. I. (1999). Inhibition of insulin-stimulated phosphorylation of the intracellular domain of phospholemman decreases insulin-dependent GLUT4 translocation in streptolysin-O-permeabilized adipocytes. *The Biochemical Journal*, *343*(Pt 1), 151–157.
Walaas, O., Walaas, E., Lystad, E., Alertsen, A. R., Horn, R. S., & Fossum, S. (1977). A stimulatory effect of insulin on phosphorylation of a peptide in sarcolemma-enriched membrane preparation from rat skeletal muscle. *FEBS Letters*, *80*, 417–422.
Wald, H., Goldstein, O., Asher, C., Yagil, Y., & Garty, H. (1996). Aldosterone induction and epithelial distribution of CHIF. *The American Journal of Physiology*, *271*, F322–F329.
Wald, H., Popovtzer, M. M., & Garty, H. (1997). Differential regulation of CHIF mRNA by potassium intake and aldosterone. *The American Journal of Physiology*, *272*, F617–F623.

Wang, P. J., Lin, C. H., Hwang, H. H., & Lee, T. H. (2008). Branchial FXYD protein expression in response to salinity change and its interaction with Na+/K+-ATPase of the euryhaline teleost Tetraodon nigroviridis. *The Journal of Experimental Biology, 211*, 3750–3758.

Wang, J., Song, J., Gao, E., Zhang, X. Q., Gu, T., Yu, D., et al. (2014). Induced overexpression of phospholemman S68E mutant improves cardiac contractility and mortality after ischemia-reperfusion. *American Journal of Physiology. Heart and Circulatory Physiology, 306*, H1066–H1077.

Wang, P. J., Yang, W. K., Lin, C. H., Hwang, H. H., & Lee, T. H. (2017). FXYD8, a novel regulator of renal Na(+)/K(+)-ATPase in the Euryhaline teleost, Tetraodon nigroviridis. *Frontiers in Physiology, 8*, 576.

Wetzel, R. K., Pascoa, J. L., & Arystarkhova, E. (2004). Stress-induced expression of the gamma subunit (FXYD2) modulates Na,K-ATPase activity and cell growth. *Journal of Biological Chemistry, 279*, 41750–41757.

Yamaguchi, F., Yamaguchi, K., Tai, Y., Sugimoto, K., & Tokuda, M. (2001). Molecular cloning and characterization of a novel phospholemman-like protein from rat hippocampus. *Brain Research. Molecular Brain Research, 86*, 189–192.

Yang, W. K., Hsu, A. D., Kang, C. K., Lai, I. P., Liao, P. S., & Lee, T. H. (2018). Intestinal FXYD12 and sodium-potassium ATPase: A comparative study on two euryhaline medakas in response to salinity changes. *PLoS One, 13*, e0201252.

Yang, W. K., Kang, C. K., Chang, C. H., Hsu, A. D., Lee, T. H., & Hwang, P. P. (2013). Expression profiles of branchial FXYD proteins in the brackish medaka Oryzias dancena: A potential saltwater fish model for studies of osmoregulation. *PLoS One, 8*, e55470.

Yang, W. K., Kang, C. K., Hsu, A. D., Lin, C. H., & Lee, T. H. (2016). Different modulatory mechanisms of renal FXYD12 for Na(+)-K(+)-ATPase between two closely related Medakas upon salinity challenge. *International Journal of Biological Sciences, 12*, 730–745.

Yeh, L. A., Ling, L., English, L., & Cantley, L. (1983). Phosphorylation of the (Na,K)-ATPase by a plasma membrane-bound protein kinase in friend erythroleukemia cells. *The Journal of Biological Chemistry, 258*, 6567–6574.

Zhang, X. Q., Ahlers, B. A., Tucker, A. L., Song, J., Wang, J., Moorman, J. R., et al. (2006). Phospholemman inhibition of the cardiac Na+/Ca2+ exchanger. Role of phosphorylation. *J Biol Chem, 281*, 7784–7792.

Zhang, X. Q., Qureshi, A., Song, J., Carl, L. L., Tian, Q., Stahl, R. C., et al. (2003). Phospholemman modulates Na+/Ca2+ exchange in adult rat cardiac myocytes. *American Journal of Physiology. Heart and Circulatory Physiology, 284*, H225–H233.

Zhang, X. Q., Wang, J., Song, J., Rabinowitz, J., Chen, X., Houser, S. R., et al. (2015). Regulation of L-type calcium channel by phospholemman in cardiac myocytes. *Journal of Molecular and Cellular Cardiology, 84*, 104–111.

Zhao, H., Pestov, N. B., Korneenko, T. V., Shakhparonov, M. I., & Modyanov, N. N. (2004). Accumulation of beta (m), a structural member of X,K-ATPase beta-subunit family, in nuclear envelopes of perinatal myocytes. *American Journal of Physiology Cell Physiology, 286*, C757–C767.

CHAPTER ELEVEN

Ion channels and transporters in diabetic kidney disease

Denisha Spires[a,†], Anna D. Manis[a,†], Alexander Staruschenko[a,b,*]
[a]Department of Physiology, Medical College of Wisconsin, Milwaukee, WI, United States
[b]Clement J. Zablocki VA Medical Center, Milwaukee, WI, United States
*Corresponding author: e-mail address: staruschenko@mcw.edu

Contents

Abstract

Type 1 and 2 diabetes mellitus are major medical epidemics affecting millions of patients worldwide. Diabetes mellitus is the leading cause of diabetic kidney disease (DKD), which is the most common cause of end-stage renal disease (ESRD). DKD is associated with significant changes in renal hemodynamics and electrolyte transport. Alterations in renal ion transport triggered by pathophysiological conditions in diabetes can

† These authors contributed equally to this work.

Current Topics in Membranes, Volume 83
ISSN 1063-5823
https://doi.org/10.1016/bs.ctm.2019.01.001

exacerbate hypertension, accelerate renal injury, and are integral to the development of DKD. Renal ion transporters and electrolyte homeostasis play a fundamental role in functional changes and injury to the kidney during DKD. With the large number of ion transporters involved in DKD, understanding the roles of individual transporters as well as the complex cascades through which they interact is essential in the development of effective treatments for patients suffering from this disease. This chapter aims to gather current knowledge of the major renal ion transporters with altered expression and activity under diabetic conditions, and provide a comprehensive overview of their interactions and collective functions in DKD.

Abbreviations

AGEs	advanced glycation end-products
BK	large-conductance Ca^{2+}-activated K^+channel
Cav2.2	N-type voltage-gated calcium channel
CD	collecting duct
CKD	chronic kidney disease
DCT	distal convoluted tubule
DKD	diabetic kidney disease
DM	diabetes mellitus
DN	diabetic nephropathy
DPP-4	dipeptidyl peptidase-4
ENaC	epithelial Na^+ channel
ESRD	end-stage renal disease
ET-1	endothelin 1
FSGS	focal segmental glomerulosclerosis
GFB	glomerular filtration barrier
GFR	glomerular filtration rate
GLP-1	glucagon-like peptide-1
GLUT	glucose transporter
IP_3	inositol 1,4,5-trisphosphate
K_{ATP}	ATP-sensitive K^+ channels
mTOR	mechanistic target of rapamycin
Na^+/K^+ ATPase	sodium-potassium ATPase
NaT-DC	sodium-transporting carboxylic acid decarboxylase
NCC	Na^+-Cl^- cotransporter
NHE	Na^+/H^+ exchanger
NKCC	Na^+-K^+-$2Cl^-$ cotransporter
NO	nitric oxide
PT	proximal tubule
RAAS	renin-angiotensin-aldosterone system
ROS	reactive oxygen species
SGK1	serum and glucocorticoid-regulated kinase
SGLT1	sodium-glucose transporter 1
SGLT2	sodium-glucose transporter 2
SLC41A2	solute carrier family 41 member 2
STZ	streptozotocin

T1DM	type 1 diabetes mellitus
T2DM	type 2 diabetes mellitus
TAL	thick ascending limb
TGF	tubuloglomerular feedback
TGF-β	transforming growth factor beta
TRP	transient receptor potential channel
TRPC	transient receptor potential canonical channel
TRPM	transient receptor potential melastatin channel

1. Introduction

Diabetic kidney disease (DKD) is the primary contributor to the development of end-stage renal disease (ESRD) and is associated with the onset of cardiovascular disease and stroke (Hagg et al., 2013; Maqbool, Cooper, & Jandeleit-Dahm, 2018; Umanath & Lewis, 2018). DKD is a subtype of chronic kidney disease (CKD) in which hyperglycemia, in combination with other manifestations of diabetes mellitus (DM), leads to the development of severe renal complications. DKD is a steadily growing epidemic, with approximately 660,000 Americans diagnosed annually and Medicare expenditures in excess of $31 billion. In 2014, it was estimated that nearly 29 million Americans suffered from diabetes and an additional 86 million from prediabetes. 44% of newly reported ESRD cases resulted from diabetes for 2014. The Center for Disease Control and Prevention predicted that diabetes could affect nearly 1 in 3 U.S. adults by 2050 based on current trends. Worldwide, the number of diabetic patients is expected to increase to approximately 350 million by the year 2035, with more than 40% of these patients developing CKD (Gheith, Farouk, Nampoory, Halim, & Al-Otaibi, 2016; Pavkov, Collins, Coresh, & Nelson, 2018). With the increasing urgency of the health risk posed by diabetes-related renal complications, the development of effective therapies and strategies for prevention is paramount. However, the pathogenesis of DKD has not been fully elucidated. Renal ion transporters are central to the intricate pathophysiological mechanisms of DKD and its progression. This chapter will review research concerning the contribution of key renal ion transporters to the progression of DKD and assess their utility as candidate therapeutic targets.

2. Diabetic kidney disease

In DKD, also known as diabetic nephropathy (DN), hyperglycemia overwhelms the kidney's functionality, resulting in a breakdown of the glomerular filtration barrier (GFB) and overall dysfunction of the kidney. Hyperglycemia can lead to a variety of pathological cascades that affect ion transport in the kidney. The defining characteristics of DKD include a greater than 50% decline in glomerular filtration rate (GFR) over the course of the disease, microalbuminuria resulting from progressive GFB deterioration, and histological evidence of renal injury (interstitial fibrosis and glomerulosclerosis) (Brosius et al., 2009; Schena & Gesualdo, 2005; Vallon & Komers, 2011). The substantial reduction in GFR associated with DKD is not observed during the initial stages of the disease. Patients initially exhibit hyperfiltration, an elevation in GFR, which gradually declines as DKD progresses over the course of 5–10 years. The initial hyperfiltration stage is an attempt by the kidneys to compensate for the apparent decrease in sodium delivery resulting from hyperglycemia. Once the tubuloglomerular feedback (TGF) system's ability to compensate for increased sodium and glucose reabsorption reaches saturation, kidney function, and GFR decline (Brosius et al., 2009; Schena & Gesualdo, 2005; Tuttle, 2017; Vallon & Komers, 2011). The second hallmark of DKD, microalbuminuria, is caused by the breakdown of the GFB during disease progression. The GFB is comprised of podocytes and their foot processes (the slit diaphragm), the glomerular basement membrane, and endothelial cells. These components normally function together with a network of proteins to filter the contents of the glomerular capillaries and prevent the passage of substances larger than approximately 69kDa (the molecular weight of albumin). As renal function declines and glomerular hypertrophy ensues, the GFB deteriorates. As a result, albumin leaks from the capillaries and is excreted in the urine (Jefferson, Shankland, & Pichler, 2008). Renal histologic changes typical of DKD include cellular and tissue injury, mesangial matrix expansion, nodular glomerular lesions, arteriolar hyalinosis, thickening of glomerular basement membranes, and renal interstitial fibrosis.

Hypertension is a frequent comorbidity in diabetic patients and is implicated in the progression of DKD (Fig. 1). Hypertension is typically twice as common in patients with diabetes compared to the general population. DM interacts synergistically with hypertension to promote kidney injury (Staruschenko, 2017). Therefore, the effects of both hyperglycemia and hypertension are required for kidney injury to occur in DKD (Wang et al., 2017). The development of hypertension in diabetes results from over-activation of the renin-angiotensin-aldosterone system (RAAS), upregulation of endothelin 1 (ET-1; a vasoconstrictor secreted by endothelial cells), overproduction of reactive oxygen species (ROS), downregulation

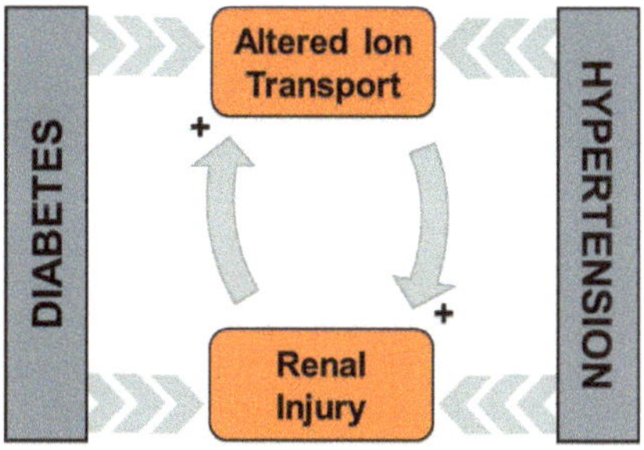

Fig. 1 The positive feedback relationship between diabetes, hypertension, altered ion transport, and renal injury. This represents a simplified explanation for a very complex relationship that defines diabetic kidney disease (DKD).

of nitric oxide (NO; a vasodilator), as well as other aberrant signaling (Arora & Singh, 2013; Kohan, Rossi, Inscho, & Pollock, 2011; Patney, Chaudhary, & Whaley-Connell, 2018; Patney, Whaley-Connell, & Bakris, 2015).

In DKD, hyperglycemia leads to the activation of numerous pathways including RAAS, inflammatory cytokines, and oxidative stress cascades (Fig. 2), which ultimately result in the renal impairment characteristic of this disease. The RAAS is particularly important in the progression of DKD, given that the primary treatment for DKD is anti-RAAS drugs, such as angiotensin converting enzyme (ACE) inhibitors and angiotensin receptor blockers (ARBs) (Anders, Huber, Isermann, & Schiffer, 2018; Umanath & Lewis, 2018). In response to hyperglycemic conditions, renal cells begin to secrete angiotensin II (Ang II). Intrarenal Ang II has been found to be substantially elevated when compared to circulating Ang II in DKD patients. This redistribution of Ang II, which likely has a considerable impact on renal ion transport, has been shown to have a causative influence on multiple distinctive features of DKD, including podocyte injury and apoptosis (Leehey, Singh, Alavi, & Singh, 2000; Vallon & Komers, 2011).

This introduction to DKD represents only a glimpse of the complexity involved in the development and progression of this disease. The variation in renal ion transporter expression and activity and their intersecting pathways, signaling cascades, and feedback loops both respond and contribute to pathophysiological states during the progression of diabetes to DKD. These pathological states, including hyperglycemia, hypertension, and dysfunctional insulin signaling substantially dysregulate renal ion transport and electrolyte homeostasis, accelerating renal injury characteristic of DKD. The intricacy of DKD makes it particularly challenging to fully understand all details involved in its progression.

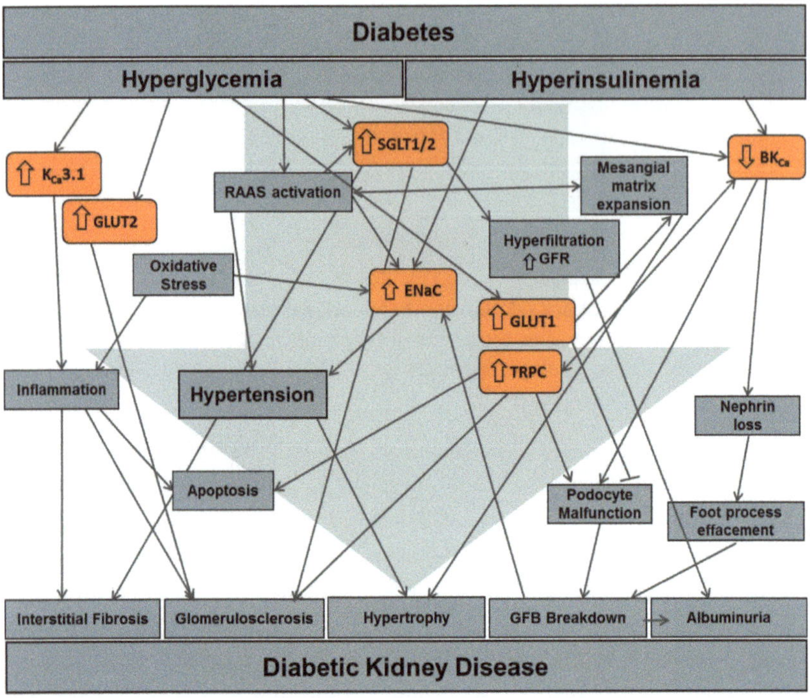

Fig. 2 Schematic for the pathogenesis of diabetic kidney disease (DKD). Hyperglycemia and hyperinsulinemia induced by diabetes lead to changes in the direct and indirect activation and expression of the different channels and transporters located along the nephron. The multitude of these channel alterations results in interstitial fibrosis, glomerulosclerosis, hypertrophy, breakdown of the glomerular filtration barrier (GFB), and albuminuria which ultimately leads to DKD. Abbreviations: $K_{Ca}3.1$, calcium activated potassium channel 3.1; GLUT1 and 2, glucose transporters 1 and 2; SGLT1 and 2, sodium glucose cotransporter 1 and 2; BK_{Ca}, large-conductance Ca^{2+}-activated K^+ channel; ENaC, epithelial Na^+ channel; TRPC, transient receptor potential canonical channel.

3. Glucose transporters

3.1 Sodium-glucose cotransporters

The kidneys play a major role in glucose regulation and are responsible for reabsorbing 99% of plasma glucose. Glucose reabsorption in the kidney occurs via sodium-glucose transporter 2 (SGLT2) in the early proximal tubule (PT) and to a lesser extent via sodium-glucose transporter 1 (SGLT1) in the late PT. Several SGLT1, SGLT2, or dual SGLT1/SGLT2 inhibitors have been shown to lower blood glucose by preventing glucose reabsorption at the PT, and act as effective antiglycemic drugs that may have utility in the treatment or prevention of DKD (Fig. 3) (Rieg & Vallon, 2018). These inhibitors are well described and have been marketed for treatment of type 2 diabetes mellitus (T2DM).

Members of the SGLT family of glucose transporters are involved in the reuptake of glucose across the apical cell membrane. Their structure is composed of 14 transmembrane helices (Deng & Yan, 2016). There are six members of the SGLT family (SGLT1–6), but only SGLT1 and SGLT2 are well-characterized and have been shown to be highly expressed in the kidney (Harada & Inagaki, 2012; Poulsen, Fenton, & Rieg, 2015). Although SGLT1 primarily functions within the small intestine, it does contribute to the maintenance of normal glucose balance in the kidney (Tahrani, Barnett, & Bailey, 2013); whereas SGLT2 is predominantly responsible for glucose uptake within the kidney. SGLT2 is located in the brush border of segments 1 and 2 of the PT and is responsible for approximately 90% of the glucose reabsorption in this part of the nephron. The remaining 10% is reabsorbed via SGLT1 in late segments of the PT (Hediger & Rhoads, 1994; Poulsen et al., 2015). The driving force for these cotransporters is the active movement of sodium via the sodium-potassium ATPase (Na^+/K^+ ATPase) causing reuptake of glucose by the cell. In a study of these two cotransporters, it was found that only one sodium ion is required by SGLT2 for the reabsorption of one glucose molecule through the transporter. In contrast, SGLT1 requires two sodium ions to be moved for each glucose molecule absorbed ($K_m \sim 0.4$ mM). Normally, SGLT2 ($K_m \leq 6$ mM) works at 50% capacity, only becoming fully saturated at a glucose level greater than or equal to 35 mM (Ferrannini & Solini, 2012; Ghezzi, Loo, & Wright, 2018; Harada & Inagaki, 2012; Hummel et al., 2011; Szablewski, 2017). Together with Na^+/K^+ ATPase SGLT1, and SGLT2 make up the first stage of glucose transport and prevent excessive loss of glucose in the urine. Powell and colleagues confirmed the important role of the SGLTs in glucose transport showing that in a double knockout mouse model, the absence of both SGLT1 and SGLT2 resulted in the excretion of the entire filtered load of glucose (Ghezzi et al., 2018; Powell et al., 2013). Given these transporters' substantial impact on overall glucose homeostasis, their implication in the development of DM and its progression to DKD is unsurprising.

3.2 SGLT1 and SGLT2 in DKD

In treating DM, controlling blood glucose levels is imperative to prevent disease progression. The tubules reabsorb glucose at a maximal rate of

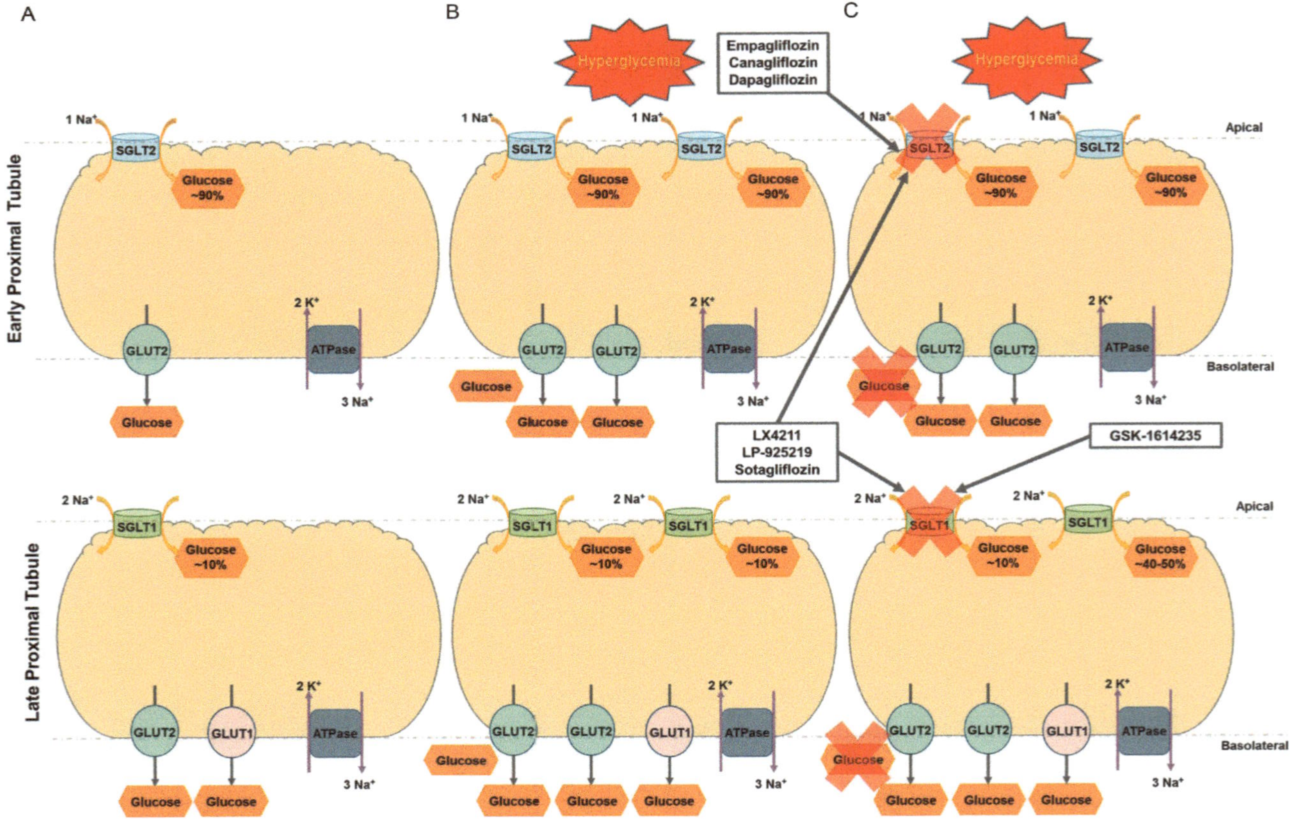

Fig. 3 Glucose transport within the proximal tubule of the kidneys. (A) Healthy proximal tubule and normal function of sodium glucose cotransporters (SGLT1 and SGLT2) and their glucose transporter (GLUT1 and GLUT2) counterparts. (B) The altered function of SGLT and GLUT transporters under DKD conditions. (C) Examples of the inhibitors specific for SGLT2, SGLT1, and dual inhibitors for both transporters and their general effects under DKD conditions.

approximately 375 mg/min, after which their capacity for transport is saturated. In diabetes, plasma glucose increases to hyperglycemic levels, bombarding the TGF system, and the macula densa with overwhelming quantities of glucose. The overloading of this system causes hyperfiltration, which as previously mentioned, ultimately leads to glucose excretion in the urine and DKD (Abdul-Ghani, Norton, & DeFronzo, 2015; Brenner, 1983; Farber, Berger, & Earle, 1951; Ruggenenti et al., 2012; Vallon et al., 2011). SGLT transporters, especially SGLT2, have been targeted by multiple therapeutics in attempt to regulate hyperglycemia in diabetes. The goal of SGLT inhibitors is to increase the urinary output of glucose, thereby decreasing circulating glucose content and minimizing damage (Spatola, Finazzi, Angelini, Dauriz, & Badalamenti, 2018). Vallon et al. found that administering an SGLT2 inhibitor (empagliflozin) to the Akita mouse, a model of type 1 diabetes mellitus (T1DM), prevented hyperfiltration and reduced kidney hypertrophy and albumin excretion in the early stages of DKD. In the T2DN rat, a model of type 2 diabetic nephropathy (DN), the progression of DKD was slowed using a SGLT2 inhibitor which led to an appreciable reduction in GFR, glomerulosclerosis, tubulointerstitial fibrosis, and proteinuria. Similar results were seen in a mouse model of type 2 DKD with SGLT2 inhibition causing a reduction in mesangial expansion and expression of inflammatory markers (Gembardt et al., 2014; Koepsell, 2017; Kojima, Williams, Takahashi, Miyata, & Roman, 2013). Interestingly, insulin receptor deletion significantly reduced SGLT2 expression and increased urinary glucose excretion and urine flow (Nizar, Shepard, Vo, & Bhalla, 2018).

Unlike T1DM where insulin treatments are greatly beneficial and contribute to an improved quality of life, insulin treatment in T2DM is not as effective. These patients are, to some extent, insulin resistant. For T2DM patients, research has been focused on therapies targeting glucose transporters, specifically SGLT2 because the bulk of glucose reuptake occurs through this transporter (Defronzo, 2009; Tahrani et al., 2013). Additionally, this cotransporter is primarily located in the kidney which minimizes off target effects from globally inhibiting SGLT2 (Koepsell, 2017). Patients with T2DM have similar renal protein expression of SGLT1 and SGLT2 compared to normoglycemic patients, although Norton and colleagues found that SGLT1 and SGLT2 mRNA expression levels in T2DM are higher than in normoglycemic patients. SGLT2 inhibitors reduce glucose reabsorption in the PT, ultimately resulting in improved kidney function (Sano, Takei, Shiraishi, & Suzuki, 2016; Zou, Zhou, & Xu, 2017). SGLT2 inhibitors also result in increased sodium delivery to the macula densa, which activates the

TGF system to constrict the afferent arteriole and decrease GFR (Cherney et al., 2014; Vallon, Blantz, & Thomson, 2003; Zou et al., 2017). In the late stages of T2DM and DKD, the SGLT2 inhibitor canagliflozin was found to improve glycemic control and reduce albuminuria. Kohan et al. found that dapagliflozin, another SGLT2 inhibitor, reduced the body weight and blood pressure of T2DM patients without the change in glycemic control that Yale et al. observed (Kohan, Fioretto, Tang, & List, 2014; Yale et al., 2013; Zou et al., 2017). Similar results were found by other groups showing multiple benefits of SGLT2 inhibitors in diabetes and DKD, including improvements to renal oxygenation, natriuresis, and oxidative stress (Dekkers, Gansevoort, & Heerspink, 2018; Tanaka et al., 2018; Wang et al., 2017). SGLT2 expression and/or activity has been found to be upregulated in both type 1 and 2 diabetes, with an increase in the maximum glucose transport of approximately 20%. Administration of these inhibitors decreases this maximum capacity by roughly 30%–50% (DeFronzo et al., 2013; Gallo, Wright, & Vallon, 2015). In addition, a study by Ferrannini et al. found that empagliflozin improved pancreatic beta cell function and insulin sensitivity in T2DM patients (Ferrannini et al., 2014). Inhibitors of SGLT2 also reduced urate concentrations in the blood, which may contribute to the protective effects of these inhibitors in the progression of DKD (Ficociello et al., 2010; Haring et al., 2014; Jabbour, Hardy, Sugg, & Parikh, 2014; Koepsell, 2017; Kovacs et al., 2014; Rosenstock et al., 2014; Wilding et al., 2012). Fig. 3 summarizes the effects of hyperglycemia on these transporters and the effects of their various inhibitors in DKD.

Although SGLT2 is the primary target for treatment of T2DM, it has been proposed that SGLT1 expression increases as a result of SGLT2 inhibition. As SGLT2 becomes overloaded and is inhibited, saturating glucose transport in the early PT, there is an increased need for SGLT1 contribution to handle the increase in glucose delivery to the later segment of the PT. Also, it has been hypothesized that SGLT2 inhibitors are less effective in T2DM patients with renal impairments, and SGLT1 inhibitors have been found to improve glycemic control in these cases (Gorboulev et al., 2012; Liu, Lee, & DeFronzo, 2012; Spatola et al., 2018; Yale et al., 2013). SGLT1 inhibitors also improve glucose homeostasis by exerting substantial effects on the gastrointestinal system, increasing the release of glucagon-like peptide-1 (GLP-1) and reducing the absorption of glucose in the gut (Song, Onishi, Koepsell, & Vallon, 2016).

Although broadly beneficial for the treatment of diabetes, there are also complications associated with SGLT inhibitors (Ferrannini & Ryden, 2018; Lupsa & Inzucchi, 2018). Adverse effects of SGLT2 blockers include genital mycotic and urinary tract infections. Euglycemic diabetic ketoacidosis has been found in some cases, likely due to increases in glucagon secretion and stimulation of lipolysis and ketogenesis. Furthermore, SGLT2 inhibitor monotherapy seems to be ineffective at maintaining long-term control of hyperglycemia in some cases of DKD (Hershon, 2016; Zou et al., 2017). Combined inhibitors of SGLT1 and SGLT2, such as sotagliflozin, have been shown to effectively reduce glucose and insulin levels in the plasma of T2DM patients (Powell et al., 2013; Zambrowicz et al., 2012). Recent studies also revealed beneficial effects of sotagliflozin in combination with insulin treatment in patients with T1DM (Garg et al., 2017). SGLT2 inhibitors have also shown to be effective in combination with RAAS blockers by reducing cardiovascular events, albuminuria, hyperfiltration, and blood pressure in DKD (Bautista et al., 2004; Kojima et al., 2015, 2013; Zou et al., 2017). Combining SGLT2 inhibitors with other targets of the glycemic control pathway, specifically dipeptidyl peptidase-4 (DPP-4) inhibitors and glucagon-like peptide-1 (GLP-1) receptor agonists, has also shown to be beneficial in treating DKD. GLP-1 receptor agonists can potentially counteract increases in glucagon secretion caused by SGLT2 inhibitors; whereas DPP-4 inhibitors combined with SGLT2 inhibitors showed reduced hypoglycemia, albuminuria, hyperglycemia, and blood pressure in DKD (DeFronzo et al., 2015; Scheen & Delanaye, 2018; Secrest, Udell, & Filion, 2017).

3.3 Glucose transporters

In addition to the sodium-glucose cotransporters, glucose transporters (GLUTs) are also vital for proper glucose homeostasis, and therefore have implications in the development of DM and DKD. GLUTs along with SGLTs are members of the major facilitator superfamily (Pao, Paulsen, & Saier, 1998; Thorens & Mueckler, 2010). GLUTs facilitate the energy independent movement of glucose down its electrochemical gradient. They are expressed in every cell in the body and are essential for energy metabolism. There are 17 GLUT proteins in the *SLC2* family, which are further divided into three classes based on structure. Class I contains GLUT1–4, class II contains GLUT5, 7, 9, and 11, and class III contains GLUT6, 8, 10, 12 as well as the H^+/myo-inositol transporter. Expression patterns, regulation, and properties of GLUTs are tissue specific. During different disease stages, GLUT expression levels tend to vary as well (Szablewski, 2017). This review will focus on specific GLUTs with prominent effects in the kidney.

3.4 GLUTs in DKD

GLUTs 1, 2, 4, 5, 8–10, and 12 all function in different segments of the kidney to facilitate glucose transport (Chin et al., 1997; Heilig, Brosius, & Cunningham, 2006; Heilig et al., 1995; Mather & Pollock, 2011). GLUT1 has been found to be upregulated in the renal cortex in diabetes as well as in glomerular hypertension. Similar to SGLT1 in the PT, GLUT1 is a low capacity glucose transporter in the glomerulus with a high affinity for glucose (Heilig, Brosius, & Henry, 1997). Wang et al. showed that when GLUT1 is overexpressed in glomerular mesangial cells of the C57BL/6 mouse (a relatively DKD-resistant strain), the glomerulus develops damage like that of the glomerulosclerosis typically seen in DKD. More interestingly, these mice were not hyperglycemic or hypertensive during overexpression, insinuating that GLUT1 plays a potential role in the development of glomerulosclerosis in DKD (Wang et al., 2010). GLUT1 is also expressed in the podocytes of the glomerulus, where it is widely localized to both the apical and basolateral membrane, within vesicles in the cytoplasm and plasma membrane of the foot processes. GLUT1 plays an important role in the proper function of podocytes in the GFB, which impacts overall kidney function (Brosius, Briggs, Marcus, Barac-Nieto, & Charron, 1992; Brosius & Heilig, 2005; Coward et al., 2005; Jefferson et al., 2008; Wasik & Lehtonen, 2018). GLUT1 is activated in models of streptozotocin (STZ) induced T1DM as well as in mouse models of T2DM (Chen, Heilig, Brosius, & Heilig, 2003; D'Agord Schaan et al., 2001). Although GLUT1 overexpression in mesangial cells leads to the further development of DKD, its overexpression in podocytes appears to be protective in the progression of DKD (Wang et al., 2010; Wasik & Lehtonen, 2018; Zhang et al., 2010). Several studies also implied that genetic variations in GLUT1 results in a genetic predisposition for DKD (Brosius & Heilig, 2005; Grzeszczak et al., 2001; Hodgkinson, Millward, & Demaine, 2001; Hsu et al., 2011; Liu, Guan, Chen, & Li, 1999; Ng et al., 2002; Tarnow, Grarup, Hansen, Parving, & Pedersen, 2001; Vaulont & Kahn, 1994).

In the later stages of glucose reuptake by the kidney, GLUT2 is the primary transporter responsible for the basolateral movement of glucose on the brush border of the PT (Ghezzi et al., 2018; Mather & Pollock, 2011). In the diabetic kidney, it has been found that there is an increase in GLUT2 expression in the PT. For example, Chin et al. found that GLUT2 mRNA expression in the PT was increased in a T1DM animal model (Chin et al., 1997). Kamran et al. found that GLUT2 is overexpressed in both STZ-treated Sprague-Dawley rats and diabetic Zucker rats (Kamran, Peterson, & Dominguez, 1997). Marks et al. found similar results in the PT of STZ-treated rats (Marks, Carvou, Debnam, Srai, & Unwin, 2003). GLUT1 and 2 work within the PT together with SGLT1 and 2 (Fig. 3). However, SGLT1 and 2 work more closely through GLUT2 (high capacity and low glucose affinity) to handle the bulk of glucose transport in the PT (Ghezzi et al., 2018; Mather & Pollock, 2011). In contrast to SGLTs, finding specific inhibitors of GLUT2 has proven challenging considering the close homology between members of the GLUT family (Ghezzi et al., 2018; Yan, 2015).

In addition to GLUT1 and 2, GLUTs 4, 5, 8, 9, and 10 have also been detected in the kidney. In podocytes, GLUT4 is located on both the apical and basolateral membranes and within intracellular vesicles in the podocyte foot processes. GLUT4 is also expressed in mesangial cells, whereas GLUT8 is only expressed in podocytes (Brosius et al., 1992; Brosius & Heilig, 2005; Heilig et al., 2006; Mather & Pollock, 2011). GLUT4 and 8 are considered to be insulin responsive transporters, another key function linking them to DKD. It was found that GLUT8 mRNA and protein levels are regulated by plasma glucose in both normal conditions and in cases of diabetes, including insulin resistant forms of diabetes. GLUT4 relocation to the plasma membrane has been found to be induced by the insulin-stimulated increase in phosphoinositide 3-kinase/protein kinase B (PI3K/AKT) pathway (Brosius & Heilig, 2005; Marcus et al., 1994; Schiffer et al., 2005; Wasik & Lehtonen, 2018). In a model of STZ-induced T1DM, GLUT4 expression in glomeruli was reduced (Marcus et al., 1994). Additionally, Coward et al. found that GLUT4 redistributes to the basal membrane in podocytes in a T2DM model. This suggests that despite a marked reduction in GLUT4 expression, functionality may not be affected during DKD progression (Coward et al., 2005). Additional studies have found that GLUT1 and 4 are responsive to insulin in the podocyte specifically, being stimulated in some fashion by insulin in these cells. Insulin causes activation of GLUT4 and its translocation from the perinuclear and cytosolic vesicular structures to the plasma membrane of the podocyte. Additional studies of GLUT4 deficient mice with DKD demonstrate that reduced GLUT4 activity protects podocytes from DKD by reducing mechanistic target of rapamycin (mTOR) activity (Coward et al., 2005; Guzman et al., 2014; Wasik & Lehtonen, 2018).

In addition to GLUT expression in glomeruli and the PT, Linden et al. found that GLUT12 is located within the distal tubule and collecting duct (CD) of the nephron. GLUT12 protein was found to be predominantly located in the cytoplasm and apical membranes of these segments. Using STZ-treated *Ren-2* (renin gene) transgenic rats as a model of DKD, they saw an increase in both GLUT12 and GLUT1, indicating that both GLUT transporters have some involvement in the progression of this disease (Linden et al., 2006; Mather & Pollock, 2011). Expression levels for GLUT3 were also reported, but protein localization has not been determined (Heilig et al., 1995). GLUT5, which has been proposed to be fructose specific, has shown increased mRNA expression in the PT during chronic T1DM induced by STZ. Thus, GLUTs, together with SGLTs, play critical roles in glucose transport and contribute toward DKD progression.

4. Sodium transporters

Sodium absorption in the kidney is precisely regulated and controlled by numerous physiological mechanisms. Under pathological conditions such as DM, sodium transport is significantly altered. Eriguchi et al. recently performed a sodium transporter profile immunoblot analysis in wild-type mice where T1DM was induced by injection of STZ. After 6 months of STZ injections, the authors found no significant changes in total Na^+/H^+ exchanger (NHE) and sodium-potassium-2 chloride cotransporter (NKCC) expression. However, NKCC2, NKCC isoform, phosphorylation was significantly increased in diabetic mice compared with nondiabetic controls. Similarly, analysis of DCT transporters revealed increased expression in both total and phosphorylated sodium-chloride cotransporter (NCC), and in subunits of the epithelial sodium channel (ENaC). Diabetic animals exhibited elevations in total α- and β-ENaC as well as cleaved forms of α- and γ-ENaC (Eriguchi et al., 2018). These data demonstrate that STZ-mediated hypoinsulinemia and hyperglycemia cause upregulation of most major sodium transporters. This section will delineate additional details about specific sodium transporters, and their potential contributions to DKD.

4.1 Epithelial sodium channel

Expressed primarily in principal cells of the distal nephron, ENaC plays a central role in maintaining salt and water homeostasis, regulating extracellular fluid volume, controlling blood pressure, and overall renal function (Hanukoglu & Hanukoglu, 2016; Kleyman, Kashlan, & Hughey, 2018; Pavlov & Staruschenko, 2017; Staruschenko, 2012). Diabetes and DKD have been associated with increased ENaC activity and expression, which may reflect or contribute to the pathophysiology of DKD. Studies in humans and animals present multiple mechanisms by which the diabetic state can elicit changes in ENaC, which interfere with renal blood pressure control, exacerbate hypertension, and thereby contribute to the progression of DKD.

ENaC subunits are located on the apical membrane of principal cells in the aldosterone-sensitive distal nephron where they are tightly controlled by various hormones and mediate fine-tuning of sodium absorption in the kidney (Staruschenko, 2012). We and others have shown that insulin augments ENaC expression and activity (Gonzalez-Rodriguez, Gaeggeler, & Rossier, 2007; Ilatovskaya, Levchenko, Brands, Pavlov, & Staruschenko, 2015; Mansley et al., 2016; Pao, 2016; Pavlov et al., 2013; Tiwari, Nordquist, Halagappa, & Ecelbarger, 2007). As an example, single-channel analysis in freshly isolated, split-open tubules demonstrated that ENaC activity was acutely activated by insulin. Moreover, insulin receptor knockout mice have significantly lower activity compared to their wild-type littermates (Pavlov et al., 2013). Interestingly, high fat-fed mice had no increase in ENaC activity (Nizar et al., 2016). Recent studies by Irsik, Blazer-Yost, Staruschenko, and Brands (2017) and Irsik and Brands (2018) have utilized a sophisticated insulin-clamping technique in rats, which allowed them to test the role of daily variations in insulin on sodium excretion. They found that rats whose insulin was clamped to prevent increases in response to carbohydrate showed elevated sodium excretion over the first 4 h post carbohydrate administration (Irsik & Brands, 2018).

One proposed mechanism suggests that ENaC involvement with DKD is inexorably linked to the serum and glucocorticoid-regulated kinase (SGK1) protein. SGK1 is stimulated by insulin, which causes more ENaC to be translocated to the membrane (through the Nedd4–2 signaling pathway) increasing sodium reabsorption from the tubule. This may lead to excess renal sodium retention, hypertension, and ultimately renal damage associated with DKD (McCormick, Bhalla, Pao, & Pearce, 2005; Pearce et al., 2015). In vitro studies found that both ENaC and SGK1 are upregulated by high levels of extracellular glucose (Hills, Bland, Bennett, Ronco, & Squires, 2006). It has been well established that the over-activity of ENaC can result in hypertension, and increased ENaC expression has been identified in animal models of both type 1 and type 2 diabetes (Chang et al., 2007). In a rat model of STZ-induced T1DM, increased glucose was correlated with upregulation of all three ENaC subunits, attributed to elevations in aldosterone and vasopressin (Song, Knepper, Verbalis, & Ecelbarger, 2003). Another proposed mechanism of ENaC increases in DKD involves the serine protease, plasmin (Kleyman et al., 2018; Ray et al., 2018). Urinary plasmin has been found to be elevated in human subjects with DKD as well as in the puromycin aminonucleoside rat model of nephrotic syndrome. Dysfunction of the GFB in DKD causes plasmin to be filtered to the tubules where it activates ENaC and increases sodium reabsorption (Svenningsen, Skott, & Jensen, 2012). In a study of patients with T2DM, microalbuminuria, a hallmark of GFB breakdown, is associated with increased aberrant filtration of plasmin. This surge of filtered plasmin was shown to be sufficient to increase the open probability for ENaC and was proposed as a possible mechanism contributing to hypertension in diabetes (Buhl et al., 2014). Clinical studies have also found that amiloride, an ENaC blocker, may be protective in DKD as it significantly increased sodium excretion, and reduced blood pressure, albuminuria, and plasmin in urine of diabetic patients (Andersen et al., 2015). Recently, a pilot randomized cross-over study comparing the effects of daily administration of either oral amiloride or the NCC inhibitor, hydrochlorothiazide (HCTZ), to patients with type 2 diabetes and proteinuria revealed similar effects with both drugs resulting in reduced systolic blood pressure (Unruh et al., 2017).

It is widely accepted that oxidative stress plays a central role in diabetes-induced renal injury. Prolonged hyperglycemia causes excess glucose to contact and react with proteins and lipids resulting in advanced glycation end-products (AGEs), which are known to cause multiple complications in diabetic patients and are implicated in DKD. The role of AGEs in DKD may be especially important to understand as they are capable of having substantial effects, including oxidative stress, that persist long term even after blood glucose control is regained in the patient (Singh, Bali, Singh, & Jaggi, 2014). AGEs have been shown to be upregulated in diabetic subjects with hypertension, with an especially pronounced elevation in the distal nephron where ENaC is highly expressed (Schleicher, Wagner, & Nerlich, 1997). When applied to cultured tubular epithelial cells in concentrations comparable to what occurs in diabetes, AGEs increased ENaC mRNA and protein and stimulated ENaC activity by inhibiting catalase and increasing intracellular ROS production (Wang et al., 2015). The effect on ENaC activity persisted for more than 72h after removal of AGEs. This sustained ENaC elevation may be key to understanding why DKD often continues to progress despite adequate glucose control and provide key insights necessary for the development of more effective treatments. From these studies, it is evident that diabetes creates pathophysiological conditions that affect ENaC via multiple pathways, causing a sustained increase in activity or expression, ultimately resulting in blood pressure elevation (Fig. 4). As hypertension is one of the most important risk factors in the progression from diabetes to DKD, ENaC is a critical mechanistic and potential therapeutic target in DKD research.

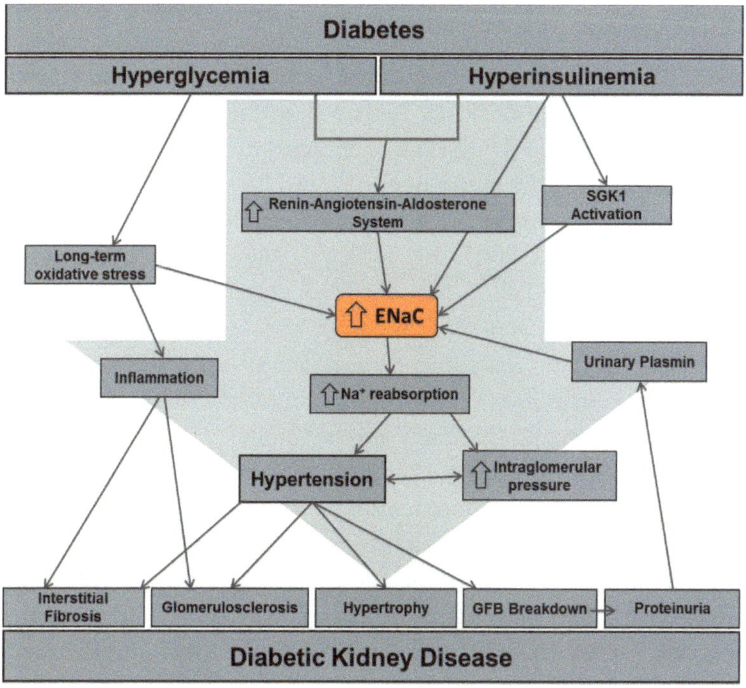

Fig. 4 Schematic for epithelial Na^+ channel (ENaC) induced tubular renal injury in DKD. Hyperglycemia and hyperinsulinemia in diabetes cause over-activation of the renin-angiotensin-aldosterone system (RAAS), long-term oxidative stress, and serum and glucocorticoid-regulated kinase (SGK) 1 activation, increase ENaC activation and/or expression. This increase in ENaC leads to various factors shown in the above pathway, such as hypertension that contributes to the development of the major characteristics of DKD.

4.2 Sodium hydrogen exchanger

Sodium-hydrogen exchangers (NHE) directly and indirectly contribute to the maintenance of blood volume and whole body acid-base homeostasis. The inward movement of sodium down its electrochemical gradient supplies the energy for the active transport of hydrogen against its gradient. In the human genome there are nine isoforms of NHE belonging to one of three subfamilies: cation-proton antiporters (CPA1 and CPA2) and sodium-transporting carboxylic acid decarboxylase (NaT-DC). NHE1–4 and 6–9 are members of the CPA1 family and are found in various parts of the kidney (Bobulescu, Di Sole, & Moe, 2005; Bobulescu & Moe, 2009; Brett, Donowitz, & Rao, 2005; Chang, Lin, Keith Studley, Tran, & Saier, 2004; Orlowski & Grinstein, 2004). NHE isoforms 1–4 and 8 are located on either the apical or basolateral membranes of renal epithelial cells, whereas NHE6, 7, and 9 are found only on membranes of organelles. NHE3 and NHE8 are located on the apical membrane of the PT; NHE2 and NHE3 are on the apical membrane of the thick ascending limb (TAL) of the loop of henle; NHE2 is on the apical membrane of the distal convoluted tubule (DCT) and connecting tubule; NHE4 is on the basolateral membrane of the entire nephron; and NHE1 is located on the basolateral membrane everywhere except the macula densa and intercalated cells (Chambrey et al., 2001, 1998; Goyal, Mentone, & Aronson, 2005; Nakamura, Tanaka, Teko, Mitsui, & Kanazawa, 2005; Orlowski & Grinstein, 2004; Xu, Chen, & Ghishan, 2005).

Given the location and the general function of NHE exchangers, it is expected that certain NHE isoforms are involved in the progression of DKD. STZ-induced T1DM in a mouse model with a loss-of-function mutation in NHE1 (*swe/swe* mouse created on the C57BL/6 mouse background) resulted in the development of DKD characteristics. *Swe/swe* mice without STZ-induced T1DM exhibited renal tubular epithelial cell atrophy, and STZ treatment resulted in the additional development of albuminuria and increased tubulointerstital pathology (Khan, Wu, Sedor, Abu Jawdeh, & Schelling, 2006; Wu et al., 2003). Numerous studies have found that NHE3 in particular is implicated in DKD development. A link between SGLT2 and NHE3 has been proposed to contribute to the renoprotective effects of SGLT2 inhibitors and to potentially contribute to the reduction in sodium reabsorption during treatment with SGLT2 inhibitors. However, contradictory explanations have been reported, suggesting that SGLT2 inhibitors increase rather than reduce sodium absorption and that the nephro-protective effects of these inhibitors is not dependent on the TGF system (Wakisaka, 2016; Wakisaka, Nagao, & Yoshinari, 2016; Wright, Loo, & Hirayama, 2011; Zeni, Norden, Cancarini, & Unwin, 2017). The connection between NHE3 and SGLT2 has even been hypothesized to have benefits in treating some of the characteristics of DKD, including the alteration of renal hemodynamics (Pessoa, Campos, Carraro-Lacroix, Girardi, & Malnic, 2014; Tonneijck et al., 2017). The "tubular theory" for hyperfiltration (a hallmark of DKD) suggests that the relationship between the glomerulus, and the tubule is the key to explaining diabetes-induced renal dysfunction and abnormalities, and proposes that hyperfiltration is caused by increased sodium reabsorption combined with tubular hypertrophy and upregulation of SGLTs and NHE3. The theory suggests that the combination of these factors inhibit TGF (Tonneijck et al., 2017; Tuttle, 2017; Zeni et al., 2017). It has also been discovered that the GLP-1 receptor agonist, liraglutide, reduces GFR and albumin excretion in patients with T2DM by GLP-1-mediated inhibition of NHE3 and DPP-4 assembly in the PT brush border. Inhibiting this relationship causes a reduction in sodium reabsorption and GFR by activating the TGF system (Muskiet, Smits, Morsink, & Diamant, 2014; Tonneijck et al., 2017).

4.3 Sodium-potassium-2 chloride and sodium-chloride transporters

The furosemide-sensitive cotransporter NKCC2 and the thiazide-sensitive cotransporter NCC play important roles in renal salt handling and extracellular volume regulation in the TAL and DCT, respectively. Similar to other sodium transporters, expression of both total and active forms of NKCC2 and NCC is increased under diabetic conditions, which has been reported in several rodent models (Cipriani et al., 2012; Eriguchi et al., 2018; Riazi, Khan, Tiwari, Hu, & Ecelbarger, 2006; Riazi, Maric, & Ecelbarger, 2006). Metformin, an antidiabetic drug that is widely used to treat patients with DM, was shown to increase urinary sodium excretion by reducing phosphorylation of NCC. Interestingly, the activity of other renal sodium transporters, such as NKCC2, ENaC, and NHE3 did not show significant changes during metformin treatment (Hashimoto et al., 2018). Similar to this finding, our data also revealed that metformin-treated Dahl salt-sensitive rats fed a high salt diet had no difference in the activity of ENaC (Pavlov et al., 2017).

Hyperinsulinism is associated with increased expression of NCC along with Na^+/K^+ ATPase and ENaC (Bickel, Verbalis, Knepper, & Ecelbarger, 2001). In vitro studies of insulin effects revealed that insulin induces activation and phosphorylation of NCC, which could contribute to sodium imbalance and the progression of DKD in hyperinsulinemic states (Chavez-Canales et al., 2013). Therefore, there is some evidence demonstrating the potential contribution of sodium cotransporters in the TAL and DCT (especially for NCC). However, additional studies are warranted.

5. Potassium channels

Although renal potassium channels have not been definitively identified as causal in the development of DKD, it is likely that some of these channels are involved in the disease progression. The kidney is responsible for maintaining whole body potassium homeostasis, which is essential for the proper control of blood glucose, as insulin is secreted from pancreatic beta cells in response to a potassium induced depolarization (Ekmekcioglu, Elmadfa, Meyer, & Moeslinger, 2016). Various clinical studies indicate that insufficient serum potassium or dietary potassium intake is associated with the onset of T2DM. Low potassium diets and hypokalemia contribute to impaired insulin secretion and glucose intolerance (Rowe, Tobin, Rosa, & Andres, 1980; Sagild, Andersen, & Andreasen, 1961). It has also been shown that the treatment of hypertension with thiazide diuretics, which commonly cause potassium depletion as a side effect, has been associated with increased risk of new-onset diabetes (Zillich, Garg, Basu, Bakris, & Carter, 2006). In addition to diabetes onset, potassium may also play a role in the progression of diabetes to more severe cardiovascular and renal impairment. High potassium diets have been established to reduce the risk of development of cardiovascular disease in healthy patients. However, it has been less clear whether increasing dietary potassium intake in patients with diabetes would have similar protective results. A study by Smyth et al. examined how potassium intake may correlate with renal outcomes in nearly 30,000 patients with established diabetes or other vascular disease. This study found that higher potassium was associated with decreased risk for all renal outcomes in these patients. Interestingly, they found that only potassium, not sodium, was predictive for renal outcome (Smyth et al., 2014). A similar study involving over 600 Japanese patients with T2DM also showed that high urinary excretion of potassium (indicative of higher potassium intake), but not sodium, was associated with better cardiorenal outcomes (Araki et al., 2015). These studies together indicate that the improper maintenance of potassium homeostasis, and possible dysfunction of renal potassium channels, may be involved in the development of renal impairments in diabetes. For example, the voltage-gated potassium channel gene subfamily, KCNQ1, which localizes to the brush border of the PT, has been proposed as a marker for DKD. Genetic variants in this gene have significant association with susceptibility for DKD and microalbuminuria in multiple studies involving East Asian and European populations (Lim et al., 2012; Ohshige et al., 2010; Unoki et al., 2008).

The calcium-activated potassium channel 3.1 ($K_{Ca}3.1$) has been identified as a potential target in DKD. This voltage-independent potassium channel is expressed in multiple cell types implicated in tubulointerstitial fibrosis including renal PT epithelial cells, fibroblasts, inflammatory cells, and endothelial cells (Huang, Pollock, & Chen, 2014b). *In vitro* and *in vivo* studies in diabetic mouse models have shown that $K_{Ca}3.1$ is activated by high glucose and produces a proinflammatory response that contributes to renal damage in DKD. Moreover, blocking $K_{Ca}3.1$ suppresses the proinflammatory cytokine chemokine ligand 20 (CCL20), which prevents macrophage accumulation and improves renal fibrosis in diabetic mice, making it a potential therapeutic target for the treatment of DKD (Huang, Pollock, & Chen, 2014a; Huang et al., 2013). One study found that renal injury in DKD may be exacerbated by insufficient autophagy in proximal tubular cells, and that blocking $K_{Ca}3.1$ restores normal autophagy, which may prevent some degree of renal injury in DKD (Huang et al., 2016). $K_{Ca}3.1$ blockers have been studied in clinical trials to treat sickle cell disease and although results showed it may be ineffective for this purpose, these drugs were safe and well-tolerated by patients in the trial (Wulff & Castle, 2010). This supports the claim that $K_{Ca}3.1$ blockers may be beneficial in the treatment of DKD.

As previously discussed, the malfunction of podocytes and loss of nephrin in the GFB is a hallmark of DKD. Studies show that both insulin and exposure to high glucose affect the activity and expression of the large-conductance calcium-activated potassium (BK) channel present in podocytes. Treatment of mouse podocytes with high glucose caused a marked reduction of BK channel current and a decrease in surface expression of BK channels, as well as nephrin, which likely interacts with the channel. However, insulin acts to stimulate BK channel activity and expression, which is blocked by the presence of high glucose. BK channels have been shown to interact with transient receptor potential canonical (TRPC) channels and their large conductance may provide the driving force facilitating movement of calcium into the podocyte through TRPC channels. BK channels in the podocytes are responsive to insulin and glucose and interact with TRPC channels and nephrin, both of which are implicated in DKD. Thus it is feasible that the malfunction of these channels is involved in the progression of DKD (Kim & Dryer, 2011).

In addition to tubular potassium channels, there are renal vascular potassium channels which also contribute to DKD (Salomonsson, Brasen, & Sorensen, 2017). As an example, the involvement of ATP-sensitive potassium channels (K_{ATP} channels) in renal afferent arteriolar dilation was reported during STZ-induced T1DM (Ikenaga, Bast, Fallet, & Carmines, 2000). Similarly, we have shown recently that Dahl salt-sensitive rats with STZ-induced diabetes had an increased vasodilator response to the K_{ATP} channel activator, pinacidil (Miller et al., 2018). In addition to increased K_{ATP} channel activity in diabetes, it was also shown that other renal vascular potassium channels, including BK channels and members of the inward rectifier (K_{ir}) family, contribute to afferent arteriolar dilation in diabetic animal models (Brindeiro, Fallet, Lane, & Carmines, 2008; Carmines & Fujiwara, 2002).

6. Calcium and magnesium channels

6.1 Transient receptor potential channels

The transient receptor potential (TRP) superfamily of cation channels vary in permeability, selectivity, and mode of activation. TRP channels play a pivotal role in the influx of calcium, magnesium, and other ions across the plasma membrane and contribute to a diversity of functions, including their physiological and pathophysiological roles in the kidney (Abramowitz & Birnbaumer, 2009; Marko, Mannaa, Haschler, Kramer, & Gollasch, 2017; Tomilin, Mamenko, Zaika, & Pochynyuk, 2016). The TRP family is subdivided into the following groups based on function and sequence: melastatin-related TRP (TRPM), ankryin transmembrane TRP (TRPA), vanilloid receptor-related TRP (TRPV), mucolipin TRP (TRPML), polycystin TRP (TRPP), and canonical TRP (TRPC). Each group has a varied number of members and is expressed in all cell membranes, excluding mitochondrial membranes, throughout tissue types including the brain, lungs, smooth muscle, and kidneys. Abnormal activity and mutations in these channels have been linked to an assortment of kidney disorders such as nephrotic syndromes, glomerular diseases, polycystic kidney disease (PKD), and DKD (Abramowitz & Birnbaumer, 2009; Harris & Torres, 2009; Nilius & Owsianik, 2011; Tomilin et al., 2016; Woudenberg-Vrenken, Bindels, & Hoenderop, 2009). More recently the TRPC subfamily has become a target for research into possible therapeutic treatments for DKD.

The TRPC subfamily is composed of seven channels that are non-selectively permeable to calcium and sodium (Abramowitz & Birnbaumer, 2009; Vazquez, Wedel, Aziz, Trebak, & Putney, 2004; Woudenberg-Vrenken et al., 2009). TRPC 1–7 are all activated through phospholipase C (PLC) coupled receptors; however, they differ in mode of operation. Some are store-operated, whereas others operate via receptors (Dietrich, Mederos y Schnitzler, Kalwa, Storch, & Gudermann, 2005). Store-operated calcium entry (SOC) occurs when inositol 1,4,5-trisphosphate (IP_3) or another signal causes the release of intracellular calcium stores from the endoplasmic reticulum (ER), reducing the calcium concentration in the ER. This calcium decrease leads to the activation of SOC channels (Dietrich et al., 2005). Receptor-operated calcium entry (ROC) occurs when an agonist binds to and activates the PLC-coupled receptor, which is located on the cell membrane separate from the actual channel (Dietrich et al., 2005). Activation of both SOC and ROC TRPC channels leads to increased intracellular calcium levels (Abramowitz & Birnbaumer, 2009). In the kidney, TRPC channels are present in renal tubules and the glomerulus where their malfunction, overexpression, or mutation is linked to certain renal diseases. Of these channels, only TRPC6 has been genetically linked to a renal disease (Reiser et al., 2005; Winn et al., 2005).

6.2 TRPC6 in DKD

Numerous gain-of-function mutations in *Trpc6* gene have been identified to ultimately lead to the development of focal segmental glomerulosclerosis (FSGS) (Heeringa et al., 2009; Reiser et al., 2005; Winn et al., 2005). A recent analysis of human disease-causing *Trpc6* mutations also revealed a loss-of-function mutation in Trpc6 as an additional cause of hereditary FSGS (Riehle et al., 2016), which demonstrates that not only activation, but also inhibition of TRPC6 activity might lead to FSGS. With its location on the membrane of the podocyte, TRPC6 participates in unison with other integral players of the slit diaphragm such as podocin and nephrin (Dryer &

Reiser, 2010; Ilatovskaya & Staruschenko, 2015; Reiser et al., 2005). TRPC6 normally remains dormant in the cell membrane until activated by a stimulus (Fig. 5) (Ilatovskaya & Staruschenko, 2015). Ang II is increased during the progression of DKD and has been found to activate TRPC6

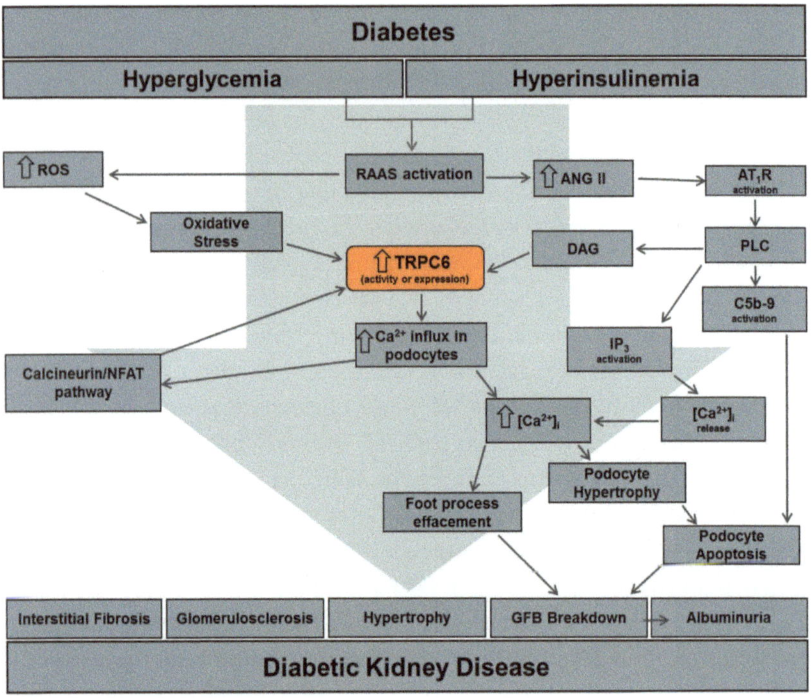

Fig. 5 Schematic for TRPC6-induced glomerular renal injury in the progression of DKD. Activation of the renin-angiotensin-aldosterone system (RAAS) cause an increase in angiotensin II (Ang II) that acts through the angiotensin II receptor type 1 (AT_1R). This receptor activates phospholipase C (PLC) on the podocyte cell membrane. PLC activates three additional targets/pathways in the podocyte: diacylglycerol (DAG) that cause calcium influx through the transient receptor potential canonical 6 (TRPC6) channel which in turn increases intracellular calcium($[Ca^{2+}]_i$); inositol triphosphate (IP_3) activates calcium release from the intracellular stores; and the membrane attack complex of complement C (C5b-9), a transmembrane channel involved in some immune responses, induces podocyte apoptosis. Increases in RAAS also lead to an increase in reactive oxygen species (ROS) that begets oxidative stress activating the calcineurin/nuclear factor of activated T cells (NFAT) pathway that increases the transcription of *Trpc6*. The culmination of the increase in $[Ca^{2+}]_i$ from over-activation of TRPC6 leads to podocyte hypertrophy, apoptosis, and foot process effacement ending in breakdown of the glomerular filtration barrier (GFB). The ultimate result of this pathway is the development of the DKD characteristic, albuminuria.

channels (Anderson, Roshanravan, Khine, & Dryer, 2014; Ilatovskaya et al., 2018; Ilatovskaya, Levchenko, Lowing, et al., 2015; Ilatovskaya et al., 2014; Ilatovskaya, Palygin, Levchenko, Endres, & Staruschenko, 2017; Nijenhuis et al., 2011; Sonneveld et al., 2014). In addition, hyperglycemia together with Ang II leads to overexpression of the TRPC6 channel in the podocyte and the subsequent drastic increase in intracellular calcium flowing through TRPC6. Hyperglycemia alone is insufficient to cause this response (Sonneveld et al., 2014). Ang II influences TRPC6 mRNA and protein levels by increasing the expression of the channel in the podocyte. The Ang II-mediated activation of the TRPC6 channel leads to engorgement of the podocyte with calcium, which causes podocyte cell death and breakdown of the GFB. This inevitably leads to albuminuria which is a hallmark of DKD (Fig. 5) (Adebiyi, Soni, John, & Yang, 2014; Eckel et al., 2011; Evans, Lee, & Ragolia, 2009; Ilatovskaya, Levchenko, Lowing, et al., 2015; Ilatovskaya et al., 2014, 2017; Nijenhuis et al., 2011; Qin Zou & Guo, 2015; Reiser et al., 2005; Sonneveld et al., 2014; Zhang, Ding, Fan, & Liu, 2009). Our recent studies using a TRPC6 knockout on the Dahl salt-sensitive rat background ($SS^{Trpc6-/-}$) indicate that TRPC6 channel inhibition may have at least partial renoprotective effects in the context of type 1 DN (Spires et al., 2018). Further studies revealed the contribution of Nox4-mediated oxidative stress in the regulation of TRPC6 in DKD (Ilatovskaya et al., 2018). In agreement with our studies, Kim et al. reported that the genetic inactivation of TRPC6 in Sprague-Dawley rats confers this protection in a model of severe nephrosis (Kim, Yazdizadeh Shotorbani, & Dryer, 2018). Interestingly, Wang et al. reported that TRPC6 knockout Akita mice exhibit prominent mesangial expansion in the diabetic group, which suggests enhanced susceptibility of glomerular cell types to the adverse effects of the TRPC6 knockout with regard to hyperglycemia (Staruschenko, 2019; Wang, Chang, Buckley, & Spurney, 2019).

6.3 Voltage-gated calcium channels

Although less thoroughly studied than TRPC channels, voltage-gated calcium channels have also been shown to play a role in DKD. In the clinic, calcium channel blockers are commonly prescribed in conjunction with RAAS inhibitors to achieve better control over blood pressure, which is required for hypertensive patients with renal complications. These drugs are promising candidates in the treatment of DKD as they have been shown to be effective antihypertensive medications that are safe and well-tolerated with few reported side effects. The antihypertensive effect of calcium channel blockers is mainly attributed to the blockade of L-type calcium channels, but N- and T-type channels may play a role in renal protective action due to their effects on glomerular capillary pressure, renal fibrotic process, sympathetic nerve activity, and aldosterone synthesis (Sugano, Hayashi, Hosoya, & Yokoo, 2013). Blockers of both the N- and L-type calcium channels have been tested in patients with diabetes to determine whether they impacted renal function. Multiple clinical studies have found that N- and L-type calcium channel blockers are renoprotective and improve proteinuria/microalbuminuria in diabetic patients. However, there is debate as to whether a specific L-type blocker, amlodipine, or a general N-/L-type blocker, cilnidipine, is more protective in the progression of DKD. Ando et al. compared the results of administering the antagonist of both the N- and L-type channels, cilnidipine, and the specific L-type calcium channel blocker, amlodipine, to patients with hypertension and T2DM with microalbuminuria over a 12-month duration and did not find a substantial difference in renal outcomes (Ando et al., 2013). In contrast, a different study in hypertensive diabetic patients found significant differences in patient outcomes between the two drugs. Masuda et al. found that cilnidipine treatment resulted in significantly lower insulin resistance, higher estimated GFR, lower urinary albumin, and lower urinary creatinine. These parameters indicate better renal function and less renal damage, suggesting that cilnidipine rather than amlodipine, better preserves kidney function and is a more appropriate candidate for slowing the progression of DKD (Masuda et al., 2011). Both calcium channel blockers are already safely being prescribed to diabetic patients with hypertension who are at the highest risk for developing DKD; however, more clinical research is needed to determine which drug is most renoprotective. Basic science research on these channels indicates that inhibiting the N-type channel is especially important for the protection against kidney injury and the preservation of function. Thus, it has been predicted that antagonizing the N-type channel would produce a more balanced dilation of the afferent and efferent arterioles, more effectively reducing glomerular pressure. The N-type voltage-gated calcium channel, $Ca_v2.2$, is expressed in podocytes in the glomerulus as well as in the DCT. A global deletion of this channel in the *db/db* diabetic mouse ameliorates many of the renal manifestations of diabetes that are known to contribute to DKD. This knockout mouse had a significant reduction in hyperfiltration, renal injury, blood pressure, and proteinuria. Similarly, applying pharmaceutical antagonists to $Ca_v2.2$ resulted in renoprotection. Blocking this channel in cultured podocytes caused a reduction in transforming growth factor beta (TGF-β) mediated nephrin loss. These studies suggest that treatment with specific $Ca_v2.2$ channel blockers may slow the progression of DKD by protecting podocytes and reducing glomerular injury (Ohno et al., 2016).

6.4 Magnesium homeostasis and TRPM6 and TRPM7 channels

Renal magnesium handling is not classically associated with DKD, as magnesium has long been considered to be independent of endocrine control, often termed an "orphan ion" (Yee, 2018). However, recent studies suggest that magnesium should not be overlooked in the context of DKD. Next to potassium, magnesium is the second most abundant intracellular cation and the kidney is responsible for maintaining its homeostasis. Like potassium, magnesium is associated with increased longevity and cardiovascular health, and is also an important anti-inflammatory molecule. Several studies have found an association between serum magnesium and diabetes. Earlier studies have suggested a link between hypomagnesemia and hyperglycemia, as well as an association between hypomagnesemia and the complications of diabetes (White & Campbell, 1993). A meta-analysis including over 500,000 patients found an inverse correlation between serum magnesium levels and risk for cardiovascular disease (Qu et al., 2013). Furthermore, this protection has been shown to extend to patients with CKD, who are at heightened risk for cardiovascular events (Kanbay, Goldsmith, Uyar, Turgut, & Covic, 2010), and hypertension where magnesium supplementation has been shown to be effective in reducing blood pressure. Additionally, patients with diabetes have lower serum magnesium, on average, than healthy controls. Diabetic patients with microalbuminuria, indicative of progression toward DKD, have significantly lower magnesium than those without (Corsonello et al., 2000). Poor glycemic control in these patients likely causes enhanced urinary magnesium loss, which is associated with a more rapid progression from diabetes to DKD.

There is considerable evidence suggesting that magnesium deficiency is a significant risk factor for the development of insulin resistance and T2DM (Barbagallo & Dominguez, 2007). A recent population-based cohort study by Kieboom et al. revealed that low serum magnesium levels are associated with an increased risk of prediabetes. Furthermore, common variants in magnesium-regulating genes, including the magnesium transporters, SLC41A2 (Solute Carrier Family 41 Member 2) and TRPM6, modify diabetes risk through altering serum magnesium levels (Kieboom et al., 2017).

Transient receptor potential melastatin 6 and 7 (TRPM6 and TRPM7) channels play a central role in magnesium homeostasis, which is critical for maintaining glucose and insulin metabolism. Several loss-of-function mutations in *Trpm6* have been identified among patients with an autosomal-recessive form of hypomagnesemia with secondary hypocalcemia (Lainez et al., 2014; Schlingmann et al., 2002; Walder et al., 2002). Various factors and hormones, including epidermal growth factor, pH, and insulin, contribute to the expression and function of this important channel (de Baaij, Hoenderop, & Bindels, 2015). There is also genetic evidence establishing the potential contribution of TRPM6 channels to the development of diabetes. For instance, Song et al. reported that two common non-synonymous SNPs in *Trpm6* might confer susceptibility to T2DM in women with low magnesium intake (Song et al., 2009). Another study revealed that SNPs in *Trpm6* have been associated with gestational diabetes. Loss of insulin-induced activation of TRPM6 channels results in impaired glucose tolerance during pregnancy (Nair et al., 2012). Insulin binding causes the activation of two cascades resulting in more TRPM6 channels as well as NCC transporters to be inserted in the apical membrane. Hyperinsulinemia and insulin resistance may cause an uncoupling of these cascades, in which NCC increases concomitantly with insulin activity while TRPM6 becomes unresponsive to insulin activity, both effectively contributing to hypertension. A recent genome-wide meta-analysis of magnesium homeostasis and metabolic phenotypes identified two loci associated with urinary magnesium near *Trpm6* (Corre et al., 2018). TRPM6 activity may also be inhibited by oxidative stress, further reducing magnesium uptake in diabetic patients. Diabetic rat models have also shown altered expression levels of TRPM6. Hyperfiltration and increased urinary flow rates in DKD patients are inversely correlated with magnesium reabsorption in the TAL and DCT and may also affect osmotic diuresis and passive reabsorption in the PT (Gommers, Hoenderop, Bindels, & de Baaij, 2016). In addition to TRPM6 and TRPM7 channels, there are other magnesium carriers in the kidney which might contribute to magnesium handling under normal conditions as well as during DKD. Therefore, current studies provide evidence supporting the critical contribution of magnesium channels in diabetes and the potential beneficial role of magnesium supplementation in diabetic patients. However, additional clinical and fundamental research is needed to identify specific mechanisms contributing to magnesium deficiency in diabetes. Despite gaps in research on the complicated role of renal magnesium handling in DKD, it is certain that magnesium imbalance contributes to or is affected by multiple pathways integral to the progression of DKD such as blood pressure control, oxidative stress, inflammation, and hyperfiltration.

7. Conclusions and future perspectives

DKD as a result of DM is a medical pandemic affecting many patients today. Due to the critical role of renal ion transport in kidney damage caused by DKD, these transporters and their pathways represent promising targets for therapeutics. Understanding the roles of the numerous transporters in the disease progression of DKD requires knowledge of both their individual functions as well as their collective interactions, cascades, and mechanisms of regulation. This review provides only a brief description of the involvement of some of the most influential transporters in DKD. More studies are certainly necessary to further validate and evaluate the roles of these and other ion transporters in the pathogenesis of DKD.

Acknowledgments

Work in the authors' laboratories is supported by NHLBI grants R35 HL135749, P01 HL116264, and T32 HL007852, American Heart Association grants 16EIA26720006 and 18PRE34030127, and Department of Veteran Affairs I01 BX004024. We would also like to thank Nnamdi Uche for his literary proofing assistance.

References

Abdul-Ghani, M. A., Norton, L., & DeFronzo, R. A. (2015). Renal sodium-glucose cotransporter inhibition in the management of type 2 diabetes mellitus. *American Journal of Physiology. Renal Physiology, 309*(11), F889–F900. https://doi.org/10.1152/ajprenal.00267.2015.

Abramowitz, J., & Birnbaumer, L. (2009). Physiology and pathophysiology of canonical transient receptor potential channels. *The FASEB Journal, 23*(2), 297–328. https://doi.org/10.1096/fj.08-119495.

Adebiyi, A., Soni, H., John, T. A., & Yang, F. (2014). Lipid rafts are required for signal transduction by angiotensin II receptor type 1 in neonatal glomerular mesangial cells. *Experimental Cell Research, 324*(1), 92–104. https://doi.org/10.1016/j.yexcr.2014.03.011.

Anders, H.-J., Huber, T. B., Isermann, B., & Schiffer, M. (2018). CKD in diabetes: Diabetic kidney disease versus nondiabetic kidney disease. *Nature Reviews Nephrology, 14*(6), 361–377. https://doi.org/10.1038/s41581-018-0001-y.

Andersen, H., Friis, U. G., Hansen, P. B., Svenningsen, P., Henriksen, J. E., & Jensen, B. L. (2015). Diabetic nephropathy is associated with increased urine excretion of proteases plasmin, prostasin and urokinase and activation of amiloride-sensitive current in collecting duct cells. *Nephrology, Dialysis, Transplantation, 30*(5), 781–789. https://doi.org/10.1093/ndt/gfu402.

Anderson, M., Roshanravan, H., Khine, J., & Dryer, S. E. (2014). Angiotensin II activation of TRPC6 channels in rat podocytes requires generation of reactive oxygen species. *Journal of Cellular Physiology*, *229*(4), 434–442. https://doi.org/10.1002/jcp.24461.

Ando, K., Ueshima, K., Tanaka, S., Kosugi, S., Sato, T., Matsuoka, H., et al. (2013). Comparison of the antialbuminuric effects of L-/N-type and L-type calcium channel blockers in hypertensive patients with diabetes and microalbuminuria: The study of assessment for kidney function by urinary microalbumin in randomized (SAKURA) trial. *International Journal of Medical Sciences*, *10*(9), 1209–1216. https://doi.org/10.7150/ijms.5508.

Araki, S., Haneda, M., Koya, D., Kondo, K., Tanaka, S., Arima, H., et al. (2015). Urinary potassium excretion and renal and cardiovascular complications in patients with type 2 diabetes and normal renal function. *Clinical Journal of the American Society of Nephrology*, *10*(12), 2152–2158. https://doi.org/10.2215/cjn.00980115.

Arora, M. K., & Singh, U. K. (2013). Molecular mechanisms in the pathogenesis of diabetic nephropathy: An update. *Vascular Pharmacology*, *58*(4), 259–271. https://doi.org/10.1016/j.vph.2013.01.001.

Barbagallo, M., & Dominguez, L. J. (2007). Magnesium metabolism in type 2 diabetes mellitus, metabolic syndrome and insulin resistance. *Archives of Biochemistry and Biophysics*, *458*(1), 40–47. https://doi.org/10.1016/j.abb.2006.05.007.

Bautista, R., Manning, R., Martinez, F., Avila-Casado Mdel, C., Soto, V., Medina, A., et al. (2004). Angiotensin II-dependent increased expression of Na+-glucose cotransporter in hypertension. *American Journal of Physiology. Renal Physiology*, *286*(1), F127–F133. https://doi.org/10.1152/ajprenal.00113.2003.

Bickel, C. A., Verbalis, J. G., Knepper, M. A., & Ecelbarger, C. A. (2001). Increased renal Na-K-ATPase, NCC, and beta-ENaC abundance in obese Zucker rats. *American Journal of Physiology. Renal Physiology*, *281*(4), F639–F648. https://doi.org/10.1152/ajprenal.2001.281.4.F639.

Bobulescu, I. A., Di Sole, F., & Moe, O. W. (2005). Na+/H+ exchangers: Physiology and link to hypertension and organ ischemia. *Current Opinion in Nephrology and Hypertension*, *14*(5), 485–494.

Bobulescu, I. A., & Moe, O. W. (2009). Luminal Na(+)/H (+) exchange in the proximal tubule. *Pflügers Archiv*, *458*(1), 5–21. https://doi.org/10.1007/s00424-008-0595-1.

Brenner, B. M. (1983). Hemodynamically mediated glomerular injury and the progressive nature of kidney disease. *Kidney International*, *23*(4), 647–655.

Brett, C. L., Donowitz, M., & Rao, R. (2005). Evolutionary origins of eukaryotic sodium/proton exchangers. *American Journal of Physiology. Cell Physiology*, *288*(2), C223–C239. https://doi.org/10.1152/ajpcell.00360.2004.

Brindeiro, C. M. T., Fallet, R. W., Lane, P. H., & Carmines, P. K. (2008). Potassium channel contributions to afferent arteriolar tone in normal and diabetic rat kidney. *American Journal of Physiology. Renal Physiology*, *295*(1), F171–F178. https://doi.org/10.1152/ajprenal.00563.2007.

Brosius, F. C., 3rd, Alpers, C. E., Bottinger, E. P., Breyer, M. D., Coffman, T. M., Gurley, S. B., et al. (2009). Mouse models of diabetic nephropathy. *Journal of the American Society of Nephrology*, *20*(12), 2503–2512. https://doi.org/10.1681/asn.2009070721.

Brosius, F. C., 3rd, Briggs, J. P., Marcus, R. G., Barac-Nieto, M., & Charron, M. J. (1992). Insulin-responsive glucose transporter expression in renal microvessels and glomeruli. *Kidney International*, *42*(5), 1086–1092.

Brosius, F. C., & Heilig, C. W. (2005). Glucose transporters in diabetic nephropathy. *Pediatric Nephrology*, *20*(4), 447–451. https://doi.org/10.1007/s00467-004-1748-x.

Buhl, K. B., Oxlund, C. S., Friis, U. G., Svenningsen, P., Bistrup, C., Jacobsen, I. A., et al. (2014). Plasmin in urine from patients with type 2 diabetes and treatment-resistant hypertension activates ENaC in vitro. *Journal of Hypertension*, *32*(8), 1672–1677. discussion 1677. https://doi.org/10.1097/HJH.0000000000000216.

Carmines, P. K., & Fujiwara, K. (2002). Altered electromechanical coupling in the renal microvasculature during the early stage of diabetes mellitus. *Clinical and Experimental Pharmacology & Physiology*, *29*(1–2), 143–148.

Chambrey, R., St John, P. L., Eladari, D., Quentin, F., Warnock, D. G., Abrahamson, D. R., et al. (2001). Localization and functional characterization of Na+/H+ exchanger isoform NHE4 in rat thick ascending limbs. *American Journal of Physiology. Renal Physiology*, *281*(4), F707–F717. https://doi.org/10.1152/ajprenal.2001.281.4.F707.

Chambrey, R., Warnock, D. G., Podevin, R. A., Bruneval, P., Mandet, C., Belair, M. F., et al. (1998). Immunolocalization of the Na+/H+ exchanger isoform NHE2 in rat kidney. *The American Journal of Physiology*, *275*(3 Pt 2), F379–F386.

Chang, A. B., Lin, R., Keith Studley, W., Tran, C. V., & Saier, M. H., Jr. (2004). Phylogeny as a guide to structure and function of membrane transport proteins. *Molecular Membrane Biology*, *21*(3), 171–181. https://doi.org/10.1080/09687680410001720830.

Chang, C. T., Wu, M. S., Tian, Y. C., Chen, K. H., Yu, C. C., Liao, C. H., et al. (2007). Enhancement of epithelial sodium channel expression in renal cortical collecting ducts cells by advanced glycation end products. *Nephrology, Dialysis, Transplantation*, *22*(3), 722–731. https://doi.org/10.1093/ndt/gfl668.

Chavez-Canales, M., Arroyo, J. P., Ko, B., Vazquez, N., Bautista, R., Castaneda-Bueno, M., et al. (2013). Insulin increases the functional activity of the renal NaCl cotransporter. *Journal of Hypertension*, *31*(2), 303–311. https://doi.org/10.1097/HJH.0b013e32835bbb83.

Chen, S. L., Heilig, K. O., Brosius, F. C., & Heilig, C. W. (2003). Diabetes increases glomerular GLUT1, and antisense-GLUT1 protects against diabetic glomerulosclerosis. In 36th Annual meeting of the American Society of Nephrology, *Journal of the American Society of Nephrology*, p. 46A.

Cherney, D. Z., Perkins, B. A., Soleymanlou, N., Maione, M., Lai, V., Lee, A., et al. (2014). Renal hemodynamic effect of sodium-glucose cotransporter 2 inhibition in patients with type 1 diabetes mellitus. *Circulation*, *129*(5), 587–597. https://doi.org/10.1161/circulationaha.113.005081.

Chin, E., Zamah, A. M., Landau, D., Gronbcek, H., Flyvbjerg, A., LeRoith, D., et al. (1997). Changes in facilitative glucose transporter messenger ribonucleic acid levels in the diabetic rat kidney. *Endocrinology*, *138*(3), 1267–1275. https://doi.org/10.1210/endo.138.3.5015.

Cipriani, P., Kim, S. L., Klein, J. D., Sim, J. H., von Bergen, T. N., & Blount, M. A. (2012). The role of nitric oxide in the dysregulation of the urine concentration mechanism in diabetes mellitus. *Frontiers in Physiology*, *3*, 176. https://doi.org/10.3389/fphys.2012.00176.

Corre, T., Arjona, F. J., Hayward, C., Youhanna, S., de Baaij, J. H. F., Belge, H., et al. (2018). Genome-wide meta-analysis unravels interactions between magnesium homeostasis and metabolic phenotypes. *Journal of the American Society of Nephrology*, *29*(1), 335–348. https://doi.org/10.1681/asn.2017030267.

Corsonello, A., Ientile, R., Buemi, M., Cucinotta, D., Mauro, V. N., Macaione, S., et al. (2000). Serum ionized magnesium levels in type 2 diabetic patients with microalbuminuria or clinical proteinuria. *American Journal of Nephrology*, *20*(3), 187–192. https://doi.org/10.1159/000013582.

Coward, R. J., Welsh, G. I., Yang, J., Tasman, C., Lennon, R., Koziell, A., et al. (2005). The human glomerular podocyte is a novel target for insulin action. *Diabetes*, *54*(11), 3095–3102.

D'Agord Schaan, B., Lacchini, S., Bertoluci, M. C., Irigoyen, M. C., Machado, U. F., & Schmid, H. (2001). Increased renal GLUT1 abundance and urinary TGF-beta 1 in streptozotocin-induced diabetic rats: Implications for the development of nephropathy complicating diabetes. *Hormone and Metabolic Research*, *33*(11), 664–669. https://doi.org/10.1055/s-2001-18683.

de Baaij, J. H. F., Hoenderop, J. G. J., & Bindels, R. J. M. (2015). Magnesium in man: Implications for health and disease. *Physiological Reviews*, *95*, 1–46.

Defronzo, R. A. (2009). Banting lecture. From the triumvirate to the ominous octet: A new paradigm for the treatment of type 2 diabetes mellitus. *Diabetes*, *58*(4), 773–795. https://doi.org/10.2337/db09-9028.

DeFronzo, R. A., Hompesch, M., Kasichayanula, S., Liu, X., Hong, Y., Pfister, M., et al. (2013). Characterization of renal glucose reabsorption in response to dapagliflozin in healthy subjects and subjects with type 2 diabetes. *Diabetes Care*, *36*(10), 3169–3176. https://doi.org/10.2337/dc13-0387.

DeFronzo, R. A., Lewin, A., Patel, S., Liu, D., Kaste, R., Woerle, H. J., et al. (2015). Combination of empagliflozin and linagliptin as second-line therapy in subjects with type 2 diabetes inadequately controlled on metformin. *Diabetes Care*, *38*(3), 384–393. https://doi.org/10.2337/dc14-2364.

Dekkers, C. C. J., Gansevoort, R. T., & Heerspink, H. J. L. (2018). New diabetes therapies and diabetic kidney disease progression: The role of SGLT-2 inhibitors. *Current Diabetes Reports*, *18*(5), 27. https://doi.org/10.1007/s11892-018-0992-6.

Deng, D., & Yan, N. (2016). GLUT, SGLT, and SWEET: Structural and mechanistic investigations of the glucose transporters. *Protein Science*, *25*(3), 546–558. https://doi.org/10.1002/pro.2858.

Dietrich, A., Mederos y Schnitzler, M., Kalwa, H., Storch, U., & Gudermann, T. (2005). Functional characterization and physiological relevance of the TRPC3/6/7 subfamily of cation channels. *Naunyn-Schmiedeberg's Archives of Pharmacology*, *371*(4), 257–265. https://doi.org/10.1007/s00210-005-1052-8.

Dryer, S. E., & Reiser, J. (2010). TRPC6 channels and their binding partners in podocytes: Role in glomerular filtration and pathophysiology. *American Journal of Physiology Renal Physiology*, *299*(4), F689–F701. https://doi.org/10.1152/ajprenal.00298.2010.

Eckel, J., Lavin, P. J., Finch, E. A., Mukerji, N., Burch, J., Gbadegesin, R., et al. (2011). TRPC6 enhances angiotensin II-induced albuminuria. *Journal of the American Society of Nephrology*, *22*(3), 526–535. https://doi.org/10.1681/asn.2010050522.

Ekmekcioglu, C., Elmadfa, I., Meyer, A. L., & Moeslinger, T. (2016). The role of dietary potassium in hypertension and diabetes. *Journal of Physiology and Biochemistry*, *72*(1), 93–106. https://doi.org/10.1007/s13105-015-0449-1.

Eriguchi, M., Bernstein, E. A., Veiras, L. C., Khan, Z., Cao, D. Y., Fuchs, S., et al. (2018). The absence of the ACE N-domain decreases renal inflammation and facilitates sodium excretion during diabetic kidney disease. *Journal of the American Society of Nephrology*, *29*(10), 2546–2561. https://doi.org/10.1681/ASN.2018030323.

Evans, J. F., Lee, J. H., & Ragolia, L. (2009). Ang-II-induced Ca(2+) influx is mediated by the 1/4/5 subgroup of the transient receptor potential proteins in cultured aortic smooth muscle cells from diabetic Goto-Kakizaki rats. *Molecular and Cellular Endocrinology*, *302*(1), 49–57. https://doi.org/10.1016/j.mce.2008.12.004.

Farber, S. J., Berger, E. Y., & Earle, D. P. (1951). Effect of diabetes and insulin of the maximum capacity of the renal tubules to reabsorb glucose. *The Journal of Clinical Investigation*, *30*(2), 125–129. https://doi.org/10.1172/jci102424.

Ferrannini, E., Muscelli, E., Frascerra, S., Baldi, S., Mari, A., Heise, T., et al. (2014). Metabolic response to sodium-glucose cotransporter 2 inhibition in type 2 diabetic patients. *The Journal of Clinical Investigation*, *124*(2), 499–508. https://doi.org/10.1172/jci72227.

Ferrannini, G., & Ryden, L. (2018). Sodium-glucose transporter 2 inhibition and cardiovascular events in patients with diabetes: Information from clinical trials and observational real-world data. *Clinical Science (London, England)*, *132*(18), 2003–2012. https://doi.org/10.1042/CS20171374.

Ferrannini, E., & Solini, A. (2012). SGLT2 inhibition in diabetes mellitus: Rationale and clinical prospects. *Nature Reviews. Endocrinology*, *8*(8), 495–502. https://doi.org/10.1038/nrendo.2011.243.

Ficociello, L. H., Rosolowsky, E. T., Niewczas, M. A., Maselli, N. J., Weinberg, J. M., Aschengrau, A., et al. (2010). High-normal serum uric acid increases risk of early progressive renal function loss in type 1 diabetes: Results of a 6-year follow-up. *Diabetes Care*, *33*(6), 1337–1343. https://doi.org/10.2337/dc10-0227.

Gallo, L. A., Wright, E. M., & Vallon, V. (2015). Probing SGLT2 as a therapeutic target for diabetes: Basic physiology and consequences. *Diabetes & Vascular Disease Research*, *12*(2), 78–89. https://doi.org/10.1177/1479164114561992.

Garg, S. K., Henry, R. R., Banks, P., Buse, J. B., Davies, M. J., Fulcher, G. R., et al. (2017). Effects of Sotagliflozin added to insulin in patients with type 1 diabetes. *The New England Journal of Medicine*, *377*(24), 2337–2348. https://doi.org/10.1056/NEJMoa1708337.

Gembardt, F., Bartaun, C., Jarzebska, N., Mayoux, E., Todorov, V. T., Hohenstein, B., et al. (2014). The SGLT2 inhibitor empagliflozin ameliorates early features of diabetic nephropathy in BTBR Ob/Ob type 2 diabetic mice with and without hypertension. *American Journal of Physiology. Renal Physiology*, *307*(3), F317–F325. https://doi.org/10.1152/ajprenal.00145.2014.

Gheith, O., Farouk, N., Nampoory, N., Halim, M. A., & Al-Otaibi, T. (2016). Diabetic kidney disease: World wide difference of prevalence and risk factors. *Journal of Nephropharmacol*, *5*(1), 49–56.

Ghezzi, C., Loo, D. D. F., & Wright, E. M. (2018). Physiology of renal glucose handling via SGLT1, SGLT2 and GLUT2. *Diabetologia*, *61*(10), 2087–2097. https://doi.org/10.1007/s00125-018-4656-5.

Gommers, L. M., Hoenderop, J. G., Bindels, R. J., & de Baaij, J. H. (2016). Hypomagnesemia in type 2 diabetes: A vicious circle? *Diabetes*, *65*(1), 3–13. https://doi.org/10.2337/db15-1028.

Gonzalez-Rodriguez, E., Gaeggeler, H. P., & Rossier, B. C. (2007). IGF-1 vs insulin: Respective roles in modulating sodium transport via the PI-3 kinase/Sgk1 pathway in a cortical collecting duct cell line. *Kidney International*, *71*(2), 116–125.

Gorboulev, V., Schurmann, A., Vallon, V., Kipp, H., Jaschke, A., Klessen, D., et al. (2012). Na(+)-D-glucose cotransporter SGLT1 is pivotal for intestinal glucose absorption and glucose-dependent incretin secretion. *Diabetes*, *61*(1), 187–196. https://doi.org/10.2337/db11-1029.

Goyal, S., Mentone, S., & Aronson, P. S. (2005). Immunolocalization of NHE8 in rat kidney. *American Journal of Physiology. Renal Physiology*, *288*(3), F530–F538. https://doi.org/10.1152/ajprenal.00229.2004.

Grzeszczak, W., Moczulski, D. K., Zychma, M., Zukowska-Szczechowska, E., Trautsolt, W., & Szydlowska, I. (2001). Role of GLUT1 gene in susceptibility to diabetic nephropathy in type 2 diabetes. *Kidney International*, *59*(2), 631–636. https://doi.org/10.1046/j.1523-1755.2001.059002631.x.

Guzman, J., Jauregui, A. N., Merscher-Gomez, S., Maiguel, D., Muresan, C., Mitrofanova, A., et al. (2014). Podocyte-specific GLUT4-deficient mice have fewer and larger podocytes and are protected from diabetic nephropathy. *Diabetes*, *63*(2), 701–714. https://doi.org/10.2337/db13-0752.

Hagg, S., Thorn, L. M., Putaala, J., Liebkind, R., Harjutsalo, V., Forsblom, C. M., et al. (2013). Incidence of stroke according to presence of diabetic nephropathy and severe diabetic retinopathy in patients with type 1 diabetes. *Diabetes Care*, *36*(12), 4140–4146. https://doi.org/10.2337/dc13-0669.

Hanukoglu, I., & Hanukoglu, A. (2016). Epithelial sodium channel (ENaC) family: Phylogeny, structure-function, tissue distribution, and associated inherited diseases. *Gene*, *579*(2), 95–132. https://doi.org/10.1016/j.gene.2015.12.061.

Harada, N., & Inagaki, N. (2012). Role of sodium-glucose transporters in glucose uptake of the intestine and kidney. *Journal of Diabetes Investigation*, *3*(4), 352–353. https://doi.org/10.1111/j.2040-1124.2012.00227.x.

Haring, H. U., Merker, L., Seewaldt-Becker, E., Weimer, M., Meinicke, T., Broedl, U. C., et al. (2014). Empagliflozin as add-on to metformin in patients with type 2 diabetes: A 24-week, randomized, double-blind, placebo-controlled trial. *Diabetes Care*, *37*(6), 1650–1659. https://doi.org/10.2337/dc13-2105.

Harris, P. C., & Torres, V. E. (2009). Polycystic kidney disease. *Annual Review of Medicine*, *60*, 321–337. https://doi.org/10.1146/annurev.med.60.101707.125712.

Hashimoto, H., Nomura, N., Shoda, W., Isobe, K., Kikuchi, H., Yamamoto, K., et al. (2018). Metformin increases urinary sodium excretion by reducing phosphorylation of the sodium-chloride cotransporter. *Metabolism*, *85*, 23–31. https://doi.org/10.1016/j.metabol.2018.02.009.

Hediger, M. A., & Rhoads, D. B. (1994). Molecular physiology of sodium-glucose cotransporters. *Physiological Reviews*, *74*(4), 993–1026. https://doi.org/10.1152/physrev.1994.74.4.993.

Heeringa, S. F., Moller, C. C., Du, J., Yue, L., Hinkes, B., Chernin, G., et al. (2009). A novel TRPC6 mutation that causes childhood FSGS. *PLoS One*, *4*(11), e7771.

Heilig, C. W., Brosius, F. C., 3rd, & Cunningham, C. (2006). Role for GLUT1 in diabetic glomerulosclerosis. *Expert Reviews in Molecular Medicine*, *8*(4), 1–18. https://doi.org/10.1017/s1462399406010490.

Heilig, C. W., Brosius, F. C., 3rd, & Henry, D. N. (1997). Glucose transporters of the glomerulus and the implications for diabetic nephropathy. *Kidney International. Supplement*, *60*, S91–S99.

Heilig, C., Zaloga, C., Lee, M., Zhao, X., Riser, B., Brosius, F., et al. (1995). Immunogold localization of high-affinity glucose transporter isoforms in normal rat kidney. *Laboratory Investigation*, *73*(5), 674–684.

Hershon, K. S. (2016). Options for empagliflozin in combination therapy in type 2 diabetes mellitus. *International Journal of General Medicine*, *9*, 155–172. https://doi.org/10.2147/ijgm.s100288.

Hills, C. E., Bland, R., Bennett, J., Ronco, P. M., & Squires, P. E. (2006). High glucose up-regulates ENaC and SGK1 expression in HCD-cells. *Cellular Physiology and Biochemistry*, *18*(6), 337–346.

Hodgkinson, A. D., Millward, B. A., & Demaine, A. G. (2001). Polymorphisms of the glucose transporter (GLUT1) gene are associated with diabetic nephropathy. *Kidney International*, *59*(3), 985–989. https://doi.org/10.1046/j.1523-1755.2001.059003985.x.

Hsu, C. C., Kao, W. L., Steffes, M. W., Gambir, T., Brancati, F. L., Heilig, C. W., et al. (2011). Genetic variation of glucose transporter-1 (GLUT1) and albuminuria in 10,278 European Americans and African Americans: A case-control study in the atherosclerosis risk in communities (ARIC) study. *BMC Medical Genetics*, *12*, 16. https://doi.org/10.1186/1471-2350-12-16.

Huang, C., Lin, M. Z., Cheng, D., Braet, F., Pollock, C. A., & Chen, X. M. (2016). KCa3.1 mediates dysfunction of tubular autophagy in diabetic kidneys via PI3k/Akt/mTOR signaling pathways. *Scientific Reports*, *6*, 23884. https://doi.org/10.1038/srep23884.

Huang, C., Pollock, C. A., & Chen, X. M. (2014a). High glucose induces CCL20 in proximal tubular cells via activation of the KCa3.1 channel. *PLoS One*, *9*(4), e95173. https://doi.org/10.1371/journal.pone.0095173.

Huang, C., Pollock, C. A., & Chen, X. M. (2014b). Role of the potassium channel KCa3.1 in diabetic nephropathy. *Clinical Science (London, England)*, *127*(7), 423–433. https://doi.org/10.1042/cs20140075.

Huang, C., Shen, S., Ma, Q., Chen, J., Gill, A., Pollock, C. A., et al. (2013). Blockade of KCa3.1 ameliorates renal fibrosis through the TGF-beta1/Smad pathway in diabetic mice. *Diabetes*, *62*(8), 2923–2934. https://doi.org/10.2337/db13-0135.

Hummel, C. S., Lu, C., Loo, D. D., Hirayama, B. A., Voss, A. A., & Wright, E. M. (2011). Glucose transport by human renal Na+/D-glucose cotransporters SGLT1 and SGLT2. *American Journal of Physiology. Cell Physiology*, *300*(1), C14–C21. https://doi.org/10.1152/ajpcell.00388.2010.

Ikenaga, H., Bast, J. P., Fallet, R. W., & Carmines, P. K. (2000). Exaggerated impact of ATP-sensitive K(+) channels on afferent arteriolar diameter in diabetes mellitus. *Journal of the American Society of Nephrology*, *11*(7), 1199–1207.

Ilatovskaya, D. V., Blass, G., Palygin, O., Levchenko, V., Pavlov, T. S., Grzybowski, M. N., et al. (2018). A NOX4/TRPC6 pathway in podocyte calcium regulation and renal damage in diabetic kidney disease. *Journal of American Society of Nephrology*, *29*(7), 1917–1927. https://doi.org/10.1681/ASN.2018030280.

Ilatovskaya, D. V., Levchenko, V., Brands, M. W., Pavlov, T. S., & Staruschenko, A. (2015). Cross-talk between insulin and IGF-1 receptors in the cortical collecting duct principal cells: Implication for ENaC-mediated Na+ reabsorption. *American Journal of Physiology. Renal Physiology*, *308*(7), F713–F719. https://doi.org/10.1152/ajprenal.00081.2014.

Ilatovskaya, D. V., Levchenko, V., Lowing, A., Shuyskiy, L. S., Palygin, O., & Staruschenko, A. (2015). Podocyte injury in diabetic nephropathy: Implications of angiotensin II-dependent activation of TRPC channels. *Scientific Reports*, *5*, 17637. https://doi.org/10.1038/srep17637.

Ilatovskaya, D. V., Palygin, O., Chubinskiy-Nadezhdin, V., Negulyaev, Y. A., Ma, R., Birnbaumer, L., et al. (2014). Angiotensin II has acute effects on TRPC6 channels in podocytes of freshly isolated glomeruli. *Kidney International*, *86*(3), 506–514. https://doi.org/10.1038/ki.2014.71.

Ilatovskaya, D. V., Palygin, O., Levchenko, V., Endres, B. T., & Staruschenko, A. (2017). The role of angiotensin II in glomerular volume dynamics and podocyte calcium handling. *Scientific Reports*, 7(1), 299. https://doi.org/10.1038/s41598-017-00406-2.

Ilatovskaya, D. V., & Staruschenko, A. (2015). TRPC6 channel as an emerging determinant of the podocyte injury susceptibility in kidney diseases. *American Journal of Physiology. Renal Physiology*, *309*(5), F393–F397. https://doi.org/10.1152/ajprenal.00186.2015.

Irsik, D. L., Blazer-Yost, B. L., Staruschenko, A., & Brands, M. W. (2017). The normal increase in insulin after a meal may be required to prevent postprandial renal sodium and volume losses. *American Journal of Physiology. Regulatory, Integrative and Comparative Physiology*, *312*(6), R965–R972. https://doi.org/10.1152/ajpregu.00354.2016.

Irsik, D. L., & Brands, M. W. (2018). Physiologic hyperinsulinemia caused by acute hyperglycemia minimizes renal sodium loss by direct action on the kidney. *American Journal of Physiology. Regulatory, Integrative and Comparative Physiology*, *315*(3), R547–R552. https://doi.org/10.1152/ajpregu.00016.2018.

Jabbour, S. A., Hardy, E., Sugg, J., & Parikh, S. (2014). Dapagliflozin is effective as add-on therapy to sitagliptin with or without metformin: A 24-week, multicenter, randomized, double-blind, placebo-controlled study. *Diabetes Care*, *37*(3), 740–750. https://doi.org/10.2337/dc13-0467.

Jefferson, J. A., Shankland, S. J., & Pichler, R. H. (2008). Proteinuria in diabetic kidney disease: A mechanistic viewpoint. *Kidney International*, *74*(1), 22–36. https://doi.org/10.1038/ki.2008.128.

Kamran, M., Peterson, R. G., & Dominguez, J. H. (1997). Overexpression of GLUT2 gene in renal proximal tubules of diabetic Zucker rats. *Journal of the American Society of Nephrology*, *8*(6), 943–948.

Kanbay, M., Goldsmith, D., Uyar, M. E., Turgut, F., & Covic, A. (2010). Magnesium in chronic kidney disease: Challenges and opportunities. *Blood Purification, 29*(3), 280–292. https://doi.org/10.1159/000276665.

Khan, S., Wu, K. L., Sedor, J. R., Abu Jawdeh, B. G., & Schelling, J. R. (2006). The NHE1 Na+/H+ exchanger regulates cell survival by activating and targeting ezrin to specific plasma membrane domains. *Cellular and Molecular Biology (Noisy-le-Grand, France), 52*(8), 115–121.

Kieboom, B. C. T., Ligthart, S., Dehghan, A., Kurstjens, S., de Baaij, J. H. F., Franco, O. H., et al. (2017). Serum magnesium and the risk of prediabetes: A population-based cohort study. *Diabetologia, 60*(5), 843–853. https://doi.org/10.1007/s00125-017-4224-4.

Kim, E. Y., & Dryer, S. E. (2011). Effects of insulin and high glucose on mobilization of slo1 BKCa channels in podocytes. *Journal of Cellular Physiology, 226*(9), 2307–2315. https://doi.org/10.1002/jcp.22567.

Kim, E. Y., Yazdizadeh Shotorbani, P., & Dryer, S. E. (2018). Trpc6 inactivation confers protection in a model of severe nephrosis in rats. *Journal of Molecular Medicine (Berlin, Germany), 96*(7), 631–644. https://doi.org/10.1007/s00109-018-1648-3.

Kleyman, T. R., Kashlan, O. B., & Hughey, R. P. (2018). Epithelial Na(+) channel regulation by extracellular and intracellular factors. *Annual Review of Physiology, 80*, 263–281. https://doi.org/10.1146/annurev-physiol-021317-121143.

Koepsell, H. (2017). The Na(+)-D-glucose cotransporters SGLT1 and SGLT2 are targets for the treatment of diabetes and cancer. *Pharmacology & Therapeutics, 170*, 148–165. https://doi.org/10.1016/j.pharmthera.2016.10.017.

Kohan, D. E., Fioretto, P., Tang, W., & List, J. F. (2014). Long-term study of patients with type 2 diabetes and moderate renal impairment shows that dapagliflozin reduces weight and blood pressure but does not improve glycemic control. *Kidney International, 85*(4), 962–971. https://doi.org/10.1038/ki.2013.356.

Kohan, D. E., Rossi, N. F., Inscho, E. W., & Pollock, D. M. (2011). Regulation of blood pressure and salt homeostasis by endothelin. *Physiological Reviews, 91*(1), 1–77.

Kojima, N., Williams, J. M., Slaughter, T. N., Kato, S., Takahashi, T., Miyata, N., et al. (2015). Renoprotective effects of combined SGLT2 and ACE inhibitor therapy in diabetic Dahl S rats. *Physiological Reports, 3*(7), pii: e12436. https://doi.org/10.14814/phy2.12436.

Kojima, N., Williams, J. M., Takahashi, T., Miyata, N., & Roman, R. J. (2013). Effects of a new SGLT2 inhibitor, luseogliflozin, on diabetic nephropathy in T2DN rats. *The Journal of Pharmacology and Experimental Therapeutics, 345*(3), 464–472. https://doi.org/10.1124/jpet.113.203869.

Kovacs, C. S., Seshiah, V., Swallow, R., Jones, R., Rattunde, H., Woerle, H. J., et al. (2014). Empagliflozin improves glycaemic and weight control as add-on therapy to pioglitazone or pioglitazone plus metformin in patients with type 2 diabetes: A 24-week, randomized, placebo-controlled trial. *Diabetes, Obesity & Metabolism, 16*(2), 147–158. https://doi.org/10.1111/dom.12188.

Lainez, S., Schlingmann, K. P., van der Wijst, J., Dworniczak, B., van Zeeland, F., Konrad, M., et al. (2014). New TRPM6 missense mutations linked to hypomagnesemia with secondary hypocalcemia. *European Journal of Human Genetics, 22*(4), 497–504. https://doi.org/10.1038/ejhg.2013.178.

Leehey, D. J., Singh, A. K., Alavi, N., & Singh, R. (2000). Role of angiotensin II in diabetic nephropathy. *Kidney International. Supplement, 77*, S93–S98.

Lim, X. L., Nurbaya, S., Salim, A., Tai, E. S., Maeda, S., Nakamura, Y., et al. (2012). KCNQ1 SNPS and susceptibility to diabetic nephropathy in East Asians with type 2 diabetes. *Diabetologia, 55*(9), 2402–2406. https://doi.org/10.1007/s00125-012-2602-5.

Linden, K. C., DeHaan, C. L., Zhang, Y., Glowacka, S., Cox, A. J., Kelly, D. J., et al. (2006). Renal expression and localization of the facilitative glucose transporters GLUT1 and

GLUT12 in animal models of hypertension and diabetic nephropathy. *American Journal of Physiology. Renal Physiology*, *290*(1), F205–F213. https://doi.org/10.1152/ajprenal.00237.2004.

Liu, Z. H., Guan, T. J., Chen, Z. H., & Li, L. S. (1999). Glucose transporter (GLUT1) allele (XbaI-) associated with nephropathy in non-insulin-dependent diabetes mellitus. *Kidney International*, *55*(5), 1843–1848. https://doi.org/10.1046/j.1523-1755.1999.00449.x.

Liu, J. J., Lee, T., & DeFronzo, R. A. (2012). Why do SGLT2 inhibitors inhibit only 30–50% of renal glucose reabsorption in humans? *Diabetes*, *61*(9), 2199–2204. https://doi.org/10.2337/db12-0052.

Lupsa, B. C., & Inzucchi, S. E. (2018). Use of SGLT2 inhibitors in type 2 diabetes: Weighing the risks and benefits. *Diabetologia*, *61*(10), 2118–2125. https://doi.org/10.1007/s00125-018-4663-6.

Mansley, M. K., Watt, G. B., Francis, S. L., Walker, D. J., Land, S. C., Bailey, M. A., et al. (2016). Dexamethasone and insulin activate serum and glucocorticoid-inducible kinase 1 (SGK1) via different molecular mechanisms in cortical collecting duct cells. *Physiological Reports*, *4*(10), pii: e12792. https://doi.org/10.14814/phy2.12792.

Maqbool, M., Cooper, M. E., & Jandeleit-Dahm, K. A. M. (2018). Cardiovascular disease and diabetic kidney disease. *Seminars in Nephrology*, *38*(3), 217–232. https://doi.org/10.1016/j.semnephrol.2018.02.003.

Marcus, R. G., England, R., Nguyen, K., Charron, M. J., Briggs, J. P., & Brosius, F. C., 3rd. (1994). Altered renal expression of the insulin-responsive glucose transporter GLUT4 in experimental diabetes mellitus. *The American Journal of Physiology*, *267*(5 Pt 2), F816–F824. https://doi.org/10.1152/ajprenal.1994.267.5.F816.

Marko, L., Mannaa, M., Haschler, T. N., Kramer, S., & Gollasch, M. (2017). Renoprotection: Focus on TRPV1, TRPV4, TRPC6 and TRPM2. *Acta Physiologica (Oxford, England)*, *219*(3), 589–612. https://doi.org/10.1111/apha.12828.

Marks, J., Carvou, N. J., Debnam, E. S., Srai, S. K., & Unwin, R. J. (2003). Diabetes increases facilitative glucose uptake and GLUT2 expression at the rat proximal tubule brush border membrane. *The Journal of Physiology*, *553*(Pt. 1), 137–145. https://doi.org/10.1113/jphysiol.2003.046268.

Masuda, T., Ogura, M. N., Moriya, T., Takahira, N., Matsumoto, T., Kutsuna, T., et al. (2011). Beneficial effects of L- and N-type calcium channel blocker on glucose and lipid metabolism and renal function in patients with hypertension and type II diabetes mellitus. *Cardiovascular Therapeutics*, *29*(1), 46–53. https://doi.org/10.1111/j.1755-5922.2009.00126.x.

Mather, A., & Pollock, C. (2011). Glucose handling by the kidney. *Kidney International Supplement*, *79*(120), S1–S6. https://doi.org/10.1038/ki.2010.509.

McCormick, J. A., Bhalla, V., Pao, A. C., & Pearce, D. (2005). SGK1: A rapid aldosterone-induced regulator of renal sodium reabsorption. *Physiology*, *20*, 134–139.

Miller, B. S., Blumenthal, S. R., Shalygin, A., Wright, K. D., Staruschenko, A., Imig, J. D., et al. (2018). Inactivation of p66Shc decreases afferent arteriolar KATP Channel activity and decreases renal damage in diabetic Dahl SS rats. *Diabetes*, *67*(11), 2206–2212. https://doi.org/10.2337/db18-0308.

Muskiet, M. H., Smits, M. M., Morsink, L. M., & Diamant, M. (2014). The gut-renal axis: Do incretin-based agents confer renoprotection in diabetes? *Nature Reviews. Nephrology*, *10*(2), 88–103. https://doi.org/10.1038/nrneph.2013.272.

Nair, A. V., Hocher, B., Verkaart, S., van Zeeland, F., Pfab, T., Slowinski, T., et al. (2012). Loss of insulin-induced activation of TRPM6 magnesium channels results in impaired glucose tolerance during pregnancy. *Proceedings of the National Academy of Sciences*, *109*(28), 11324–11329. https://doi.org/10.1073/pnas.1113811109.

Nakamura, N., Tanaka, S., Teko, Y., Mitsui, K., & Kanazawa, H. (2005). Four Na+/H+ exchanger isoforms are distributed to Golgi and post-Golgi compartments and are

involved in organelle pH regulation. *The Journal of Biological Chemistry, 280*(2), 1561–1572. https://doi.org/10.1074/jbc.M410041200.
Ng, D. P., Canani, L., Araki, S., Smiles, A., Moczulski, D., Warram, J. H., et al. (2002). Minor effect of GLUT1 polymorphisms on susceptibility to diabetic nephropathy in type 1 diabetes. *Diabetes, 51*(7), 2264–2269.
Nijenhuis, T., Sloan, A. J., Hoenderop, J. G., Flesche, J., van Goor, H., Kistler, A. D., et al. (2011). Angiotensin II contributes to podocyte injury by increasing TRPC6 expression via an NFAT-mediated positive feedback signaling pathway. *The American Journal of Pathology, 179*(4), 1719–1732. https://doi.org/10.1016/j.ajpath.2011.06.033.
Nilius, B., & Owsianik, G. (2011). The transient receptor potential family of ion channels. *Genome Biology, 12*(3), 218. https://doi.org/10.1186/gb-2011-12-3-218.
Nizar, J. M., Dong, W., McClellan, R. B., Labarca, M., Zhou, Y., Wong, J., et al. (2016). Na +-sensitive elevation in blood pressure is ENaC independent in diet-induced obesity and insulin resistance. *American Journal of Physiology. Renal Physiology, 310*(9), F812–F820. https://doi.org/10.1152/ajprenal.00265.2015.
Nizar, J. M., Shepard, B. D., Vo, V. T., & Bhalla, V. (2018). Renal tubule insulin receptor modestly promotes elevated blood pressure and markedly stimulates glucose reabsorption. *JCI Insight, 3*(16), pii:95107. https://doi.org/10.1172/jci.insight.95107.
Ohno, S., Yokoi, H., Mori, K., Kasahara, M., Kuwahara, K., Fujikura, J., et al. (2016). Ablation of the N-type calcium channel ameliorates diabetic nephropathy with improved glycemic control and reduced blood pressure. *Scientific Reports, 6*, 27192. https://doi.org/10.1038/srep27192.
Ohshige, T., Tanaka, Y., Araki, S., Babazono, T., Toyoda, M., Umezono, T., et al. (2010). A single nucleotide polymorphism in KCNQ1 is associated with susceptibility to diabetic nephropathy in Japanese subjects with type 2 diabetes. *Diabetes Care, 33*(4), 842–846. https://doi.org/10.2337/dc09-1933.
Orlowski, J., & Grinstein, S. (2004). Diversity of the mammalian sodium/proton exchanger SLC9 gene family. *Pflügers Archiv, 447*(5), 549–565. https://doi.org/10.1007/s00424-003-1110-3.
Pao, A. C. (2016). There and back again: Insulin, ENaC, and the cortical collecting duct. *Physiological Reports, 4*(10), pii:e12809. https://doi.org/10.14814/phy2.12809.
Pao, S. S., Paulsen, I. T., & Saier, M. H., Jr. (1998). Major facilitator superfamily. *Microbiology and Molecular Biology Reviews, 62*(1), 1–34.
Patney, V., Chaudhary, K., & Whaley-Connell, A. (2018). Treatment of diabetic kidney disease with hypertension control and renin angiotensin system inhibition. *Advances in Chronic Kidney Disease, 25*(2), 158–165. https://doi.org/10.1053/j.ackd.2017.11.002.
Patney, V., Whaley-Connell, A., & Bakris, G. (2015). Hypertension management in diabetic kidney disease. *Diabetes Spectrum: A Publication of the American Diabetes Association, 28*(3), 175–180. https://doi.org/10.2337/diaspect.28.3.175.
Pavkov, M. E., Collins, A. J., Coresh, J., & Nelson, R. G. (2018). In C. Cowie, S. Casagrande, A. Menke, M. Cissell, M. Eberhardt, J. Meigs, E. Gregg, W. Knowler, E. Barrett-Connor, D. Becker, F. Brancati, E. Boyko, W. Herman, B. Howard, K. Narayan, M. Rewers, & E. JE Fradkin (Eds.), *Kidney disease in diabetes*. National Institutes of Health. Vol. 22.
Pavlov, T. S., Ilatovskaya, D. V., Levchenko, V., Li, L., Ecelbarger, C. M., & Staruschenko, A. (2013). Regulation of ENaC in mice lacking renal insulin receptors in the collecting duct. *The FASEB Journal, 27*(7), 2723–2732. https://doi.org/10.1096/fj.12-223792.
Pavlov, T. S., Levchenko, V., Ilatovskaya, D. V., Li, H., Palygin, O., Pastor-Soler, N. M., et al. (2017). Lack of effects of metformin and AICAR chronic infusion on the development of hypertension in Dahl salt-sensitive rats. *Frontiers in Physiology, 8*(227), 227. https://doi.org/10.3389/fphys.2017.00227.

Pavlov, T. S., & Staruschenko, A. (2017). Involvement of ENaC in the development of salt-sensitive hypertension. *American Journal of Physiology. Renal Physiology*, *313*(2), F135–F140. https://doi.org/10.1152/ajprenal.00427.2016.

Pearce, D., Soundararajan, R., Trimpert, C., Kashlan, O. B., Deen, P. M. T., & Kohan, D. E. (2015). Collecting duct principal cell transport processes and their regulation. *Clinical Journal of the American Society of Nephrology*, *10*(1), 135–146. https://doi.org/10.2215/cjn.05760513.

Pessoa, T. D., Campos, L. C., Carraro-Lacroix, L., Girardi, A. C., & Malnic, G. (2014). Functional role of glucose metabolism, osmotic stress, and sodium-glucose cotransporter isoform-mediated transport on Na+/H+ exchanger isoform 3 activity in the renal proximal tubule. *Journal of the American Society of Nephrology*, *25*(9), 2028–2039. https://doi.org/10.1681/asn.2013060588.

Poulsen, S. B., Fenton, R. A., & Rieg, T. (2015). Sodium-glucose cotransport. *Current Opinion in Nephrology and Hypertension*, *24*(5), 463–469. https://doi.org/10.1097/mnh.0000000000000152.

Powell, D. R., DaCosta, C. M., Gay, J., Ding, Z. M., Smith, M., Greer, J., et al. (2013). Improved glycemic control in mice lacking Sglt1 and Sglt2. *American Journal of Physiology. Endocrinology and Metabolism*, *304*(2), E117–E130. https://doi.org/10.1152/ajpendo.00439.2012.

Powell, D. R., Smith, M., Greer, J., Harris, A., Zhao, S., DaCosta, C., et al. (2013). LX4211 increases serum glucagon-like peptide 1 and peptide YY levels by reducing sodium/glucose cotransporter 1 (SGLT1)-mediated absorption of intestinal glucose. *The Journal of Pharmacology and Experimental Therapeutics*, *345*(2), 250–259. https://doi.org/10.1124/jpet.113.203364.

Qin Zou, C. Z., & Guo, Y. (2015). TRPC3 and TRPC6 contribute to the pathogenesis of hypertension. *American Journal of Molecular Biology*, *5*(4), 124–133. https://doi.org/10.4236/ajmb.2015.54011.

Qu, X., Jin, F., Hao, Y., Li, H., Tang, T., Wang, H., et al. (2013). Magnesium and the risk of cardiovascular events: A meta-analysis of prospective cohort studies. *PLoS One*, *8*(3). e57720https://doi.org/10.1371/journal.pone.0057720.

Ray, E. C., Miller, R. G., Demko, J. E., Costacou, T., Kinlough, C. L., Demko, C. L., et al. (2018). Urinary plasmin(ogen) as a prognostic factor for hypertension. *Kidney International Reports*, *3*(6), 1434–1442. https://doi.org/10.1016/j.ekir.2018.06.007.

Reiser, J., Polu, K. R., Moller, C. C., Kenlan, P., Altintas, M. M., Wei, C., et al. (2005). TRPC6 is a glomerular slit diaphragm-associated channel required for normal renal function. *Nature Genetics*, *37*(7), 739–744. https://doi.org/10.1038/ng1592.

Riazi, S., Khan, O., Tiwari, S., Hu, X., & Ecelbarger, C. A. (2006). Rosiglitazone regulates ENaC and Na-K-2Cl cotransporter (NKCC2) abundance in the obese Zucker rat. *American Journal of Nephrology*, *26*(3), 245–257. https://doi.org/10.1159/000093783.

Riazi, S., Maric, C., & Ecelbarger, C. A. (2006). 17-beta estradiol attenuates streptozotocin-induced diabetes and regulates the expression of renal sodium transporters. *Kidney International*, *69*(3), 471–480. https://doi.org/10.1038/sj.ki.5000140.

Rieg, T., & Vallon, V. (2018). Development of SGLT1 and SGLT2 inhibitors. *Diabetologia*, *61*(10), 2079–2086. https://doi.org/10.1007/s00125-018-4654-7.

Riehle, M., Buscher, A. K., Gohlke, B. O., Kassmann, M., Kolatsi-Joannou, M., Brasen, J. H., et al. (2016). TRPC6 G757D loss-of-function mutation associates with FSGS. *Journal of the American Society of Nephrology*, *27*(9), 2771–2783. https://doi.org/10.1681/ASN.2015030318.

Rosenstock, J., Jelaska, A., Frappin, G., Salsali, A., Kim, G., Woerle, H. J., et al. (2014). Improved glucose control with weight loss, lower insulin doses, and no increased hypoglycemia with empagliflozin added to titrated multiple daily injections of insulin in obese inadequately controlled type 2 diabetes. *Diabetes Care, 37*(7), 1815–1823. https://doi.org/10.2337/dc13-3055.

Rowe, J. W., Tobin, J. D., Rosa, R. M., & Andres, R. (1980). Effect of experimental potassium deficiency on glucose and insulin metabolism. *Metabolism, 29*(6), 498–502.

Ruggenenti, P., Porrini, E. L., Gaspari, F., Motterlini, N., Cannata, A., Carrara, F., et al. (2012). Glomerular hyperfiltration and renal disease progression in type 2 diabetes. *Diabetes Care, 35*(10), 2061–2068. https://doi.org/10.2337/dc11-2189.

Sagild, U., Andersen, V., & Andreasen, P. B. (1961). Glucose tolerance and insulin responsiveness in experimental potassium depletion. *Acta Medica Scandinavica, 169*, 243–251.

Salomonsson, M., Brasen, J. C., & Sorensen, C. M. (2017). Role of renal vascular potassium channels in physiology and pathophysiology. *Acta Physiologica, 221*(1), 14–31. https://doi.org/10.1111/apha.12882.

Sano, M., Takei, M., Shiraishi, Y., & Suzuki, Y. (2016). Increased hematocrit during sodium-glucose cotransporter 2 inhibitor therapy indicates recovery of tubulointerstitial function in diabetic kidneys. *Journal of Clinical Medical Research, 8*(12), 844–847. https://doi.org/10.14740/jocmr2760w.

Scheen, A. J., & Delanaye, P. (2018). Renal outcomes with dipeptidyl peptidase-4 inhibitors. *Diabetes & Metabolism, 44*(2), 101–111. https://doi.org/10.1016/j.diabet.2017.07.011.

Schena, F. P., & Gesualdo, L. (2005). Pathogenetic mechanisms of diabetic nephropathy. *Journal of the American Society of Nephrology, 16*(Suppl. 1), S30–S33.

Schiffer, M., Susztak, K., Ranalletta, M., Raff, A. C., Bottinger, E. P., & Charron, M. J. (2005). Localization of the GLUT8 glucose transporter in murine kidney and regulation in vivo in nondiabetic and diabetic conditions. *American Journal of Physiology. Renal Physiology, 289*(1), F186–F193. https://doi.org/10.1152/ajprenal.00234.2004.

Schleicher, E. D., Wagner, E., & Nerlich, A. G. (1997). Increased accumulation of the glycoxidation product N(epsilon)-(carboxymethyl)lysine in human tissues in diabetes and aging. *The Journal of Clinical Investigation, 99*(3), 457–468. https://doi.org/10.1172/JCI119180.

Schlingmann, K. P., Weber, S., Peters, M., Niemann, N. L., Vitzthum, H., Klingel, K., et al. (2002). Hypomagnesemia with secondary hypocalcemia is caused by mutations in TRPM6, a new member of the TRPM gene family. *Nature Genetics, 31*(2), 166–170.

Secrest, M. H., Udell, J. A., & Filion, K. B. (2017). The cardiovascular safety trials of DPP-4 inhibitors, GLP-1 agonists, and SGLT2 inhibitors. *Trends in Cardiovascular Medicine, 27*(3), 194–202. https://doi.org/10.1016/j.tcm.2017.01.009.

Singh, V. P., Bali, A., Singh, N., & Jaggi, A. S. (2014). Advanced glycation end products and diabetic complications. *Korean Journal of Physiology and Pharmacology, 18*(1), 1–14. https://doi.org/10.4196/kjpp.2014.18.1.1.

Smyth, A., Dunkler, D., Gao, P., Teo, K. K., Yusuf, S., O'Donnell, M. J., et al. (2014). The relationship between estimated sodium and potassium excretion and subsequent renal outcomes. *Kidney International, 86*(6), 1205–1212. https://doi.org/10.1038/ki.2014.214.

Song, Y., Hsu, Y.-H., Niu, T., Manson, J. E., Buring, J. E., & Liu, S. (2009). Common genetic variants of the ion channel transient receptor potential membrane melastatin 6 and 7 (TRPM6 and TRPM7), magnesium intake, and risk of type 2 diabetes in women. *BMC Medical Genetics, 10*(1), 4. https://doi.org/10.1186/1471-2350-10-4.

Song, J., Knepper, M. A., Verbalis, J. G., & Ecelbarger, C. A. (2003). Increased renal ENaC subunit and sodium transporter abundances in streptozotocin-induced type 1 diabetes. *American Journal of Physiology. Renal Physiology, 285*(6), F1125–F1137. https://doi.org/10.1152/ajprenal.00143.2003.

Song, P., Onishi, A., Koepsell, H., & Vallon, V. (2016). Sodium glucose cotransporter SGLT1 as a therapeutic target in diabetes mellitus. *Expert Opinion on Therapeutic Targets, 20*(9), 1109–1125. https://doi.org/10.1517/14728222.2016.1168808.

Sonneveld, R., van der Vlag, J., Baltissen, M. P., Verkaart, S. A., Wetzels, J. F., Berden, J. H., et al. (2014). Glucose specifically regulates TRPC6 expression in the podocyte in an AngII-dependent manner. *The American Journal of Pathology*, *184*(6), 1715–1726. https://doi.org/10.1016/j.ajpath.2014.02.008.

Spatola, L., Finazzi, S., Angelini, C., Dauriz, M., & Badalamenti, S. (2018). SGLT1 and SGLT1 inhibitors: A role to be assessed in the current clinical practice. *Diabetes Therapy*, *9*(1), 427–430. https://doi.org/10.1007/s13300-017-0342-8.

Spires, D., Ilatovskaya, D. V., Levchenko, V., North, P. E., Geurts, A. M., Palygin, O., et al. (2018). Protective role of Trpc6 knockout in the progression of diabetic kidney disease. *American Journal of Physiology. Renal Physiology*, *315*(4), F1091–F1097. https://doi.org/10.1152/ajprenal.00155.2018.

Staruschenko, A. (2012). Regulation of transport in the connecting tubule and cortical collecting duct. *Comprehensive Physiology*, *2*(2), 1541–1584.

Staruschenko, A. (2017). Hypertension and diabetes mellitus: The chicken and egg problem. *Hypertension*, *69*(5), 787–788. https://doi.org/10.1161/HYPERTENSIONAHA.117.08671.

Staruschenko, A. (2019). TRPC6 in diabetic kidney disease: Good guy or bad guy? *Kidney International*, *95*(2), 256–258. https://doi.org/10.1016/j.kint.2018.10.027.

Sugano, N., Hayashi, K., Hosoya, T., & Yokoo, T. (2013). Mechanistic view of renal protective action of calcium channel blockade. *Current Hypertension Reviews*, *9*(3), 187–192.

Svenningsen, P., Skott, O., & Jensen, B. L. (2012). Proteinuric diseases with sodium retention: Is plasmin the link? *Clinical and Experimental Pharmacology & Physiology*, *39*(1), 117–124. https://doi.org/10.1111/j.1440-1681.2011.05524.x.

Szablewski, L. (2017). Distribution of glucose transporters in renal diseases. *Journal of Biomedical Science*, *24*(1), 64. https://doi.org/10.1186/s12929-017-0371-7.

Tahrani, A. A., Barnett, A. H., & Bailey, C. J. (2013). SGLT inhibitors in management of diabetes. *The Lancet Diabetes and Endocrinology*, *1*(2), 140–151. https://doi.org/10.1016/s2213-8587(13)70050-0.

Tanaka, S., Sugiura, Y., Saito, H., Sugahara, M., Higashijima, Y., Yamaguchi, J., et al. (2018). Sodium-glucose cotransporter 2 inhibition normalizes glucose metabolism and suppresses oxidative stress in the kidneys of diabetic mice. *Kidney International*, *94*(5), 912–925. https://doi.org/10.1016/j.kint.2018.04.025.

Tarnow, L., Grarup, N., Hansen, T., Parving, H. H., & Pedersen, O. (2001). Diabetic microvascular complications are not associated with two polymorphisms in the GLUT-1 and PC-1 genes regulating glucose metabolism in Caucasian type 1 diabetic patients. *Nephrology, Dialysis, Transplantation*, *16*(8), 1653–1656.

Thorens, B., & Mueckler, M. (2010). Glucose transporters in the 21st century. *American Journal of Physiology. Endocrinology and Metabolism*, *298*(2), E141–E145. https://doi.org/10.1152/ajpendo.00712.2009.

Tiwari, S., Nordquist, L., Halagappa, V. K., & Ecelbarger, C. A. (2007). Trafficking of ENaC subunits in response to acute insulin in mouse kidney. *American Journal of Physiology. Renal Physiology*, *293*(1), F178–F185.

Tomilin, V., Mamenko, M., Zaika, O., & Pochynyuk, O. (2016). Role of renal TRP channels in physiology and pathology. *Seminars in Immunopathology*, *38*(3), 371–383. https://doi.org/10.1007/s00281-015-0527-z.

Tonneijck, L., Muskiet, M. H., Smits, M. M., van Bommel, E. J., Heerspink, H. J., van Raalte, D. H., et al. (2017). Glomerular hyperfiltration in diabetes: Mechanisms, clinical significance, and treatment. *Journal of the American Society of Nephrology*, *28*(4), 1023–1039. https://doi.org/10.1681/asn.2016060666.

Tuttle, K. R. (2017). Back to the future: Glomerular hyperfiltration and the diabetic kidney. *Diabetes*, *66*(1), 14–16. https://doi.org/10.2337/dbi16-0056.

Umanath, K., & Lewis, J. B. (2018). Update on diabetic nephropathy: Core curriculum 2018. *American Journal of Kidney Diseases*, *71*(6), 884–895. https://doi.org/10.1053/j.ajkd.2017.10.026.

Unoki, H., Takahashi, A., Kawaguchi, T., Hara, K., Horikoshi, M., Andersen, G., et al. (2008). SNPs in KCNQ1 are associated with susceptibility to type 2 diabetes in East Asian and European populations. *Nature Genetics*, *40*(9), 1098–1102. https://doi.org/10.1038/ng.208.

Unruh, M. L., Pankratz, V. S., Demko, J. E., Ray, E. C., Hughey, R. P., & Kleyman, T. R. (2017). Trial of amiloride in type 2 diabetes with proteinuria. *Kidney International Reports*, *2*(5), 893–904. https://doi.org/10.1016/j.ekir.2017.05.008.

Vallon, V., Blantz, R. C., & Thomson, S. (2003). Glomerular hyperfiltration and the salt paradox in early [corrected] type 1 diabetes mellitus: A tubulo-centric view. *Journal of the American Society of Nephrology*, *14*(2), 530–537.

Vallon, V., & Komers, R. (2011). Pathophysiology of the diabetic kidney. *Comprehensive Physiology*, *1*(3), 1175–1232. https://doi.org/10.1002/cphy.c100049.

Vallon, V., Platt, K. A., Cunard, R., Schroth, J., Whaley, J., Thomson, S. C., et al. (2011). SGLT2 mediates glucose reabsorption in the early proximal tubule. *Journal of the American Society of Nephrology*, *22*(1), 104–112. https://doi.org/10.1681/asn.2010030246.

Vaulont, S., & Kahn, A. (1994). Transcriptional control of metabolic regulation genes by carbohydrates. *The FASEB Journal*, *8*(1), 28–35.

Vazquez, G., Wedel, B. J., Aziz, O., Trebak, M., & Putney, J. W., Jr. (2004). The mammalian TRPC cation channels. *Biochimica et Biophysica Acta*, *1742*(1–3), 21–36. https://doi.org/10.1016/j.bbamcr.2004.08.015.

Wakisaka, M. (2016). Empagliflozin and progression of kidney disease in type 2 diabetes. *The New England Journal of Medicine*, *375*(18), 1799–1800. https://doi.org/10.1056/NEJMc1611290.

Wakisaka, M., Nagao, T., & Yoshinari, M. (2016). Sodium glucose cotransporter 2 (SGLT2) plays as a physiological glucose sensor and regulates cellular contractility in rat mesangial cells. *PLoS One*, *11*(3), e0151585. https://doi.org/10.1371/journal.pone.0151585.

Walder, R. Y., Landau, D., Meyer, P., Shalev, H., Tsolia, M., Borochowitz, Z., et al. (2002). Mutation of TRPM6 causes familial hypomagnesemia with secondary hypocalcemia. *Nature Genetics*, *31*(2), 171–174.

Wang, L., Chang, J.-H., Buckley, A. F., & Spurney, R. F. (2019). Knockout of TRPC6 promotes insulin resistance and exacerbates glomerular injury in Akita mice. *Kidney International*, *95*(2), 321–332.

Wang, Z., do Carmo, J. M., Aberdein, N., Zhou, X., Williams, J. M., da Silva, A. A., et al. (2017). Synergistic interaction of hypertension and diabetes in promoting kidney injury and the role of endoplasmic reticulum stress. *Hypertension*, *69*(5), 879–891. https://doi.org/10.1161/HYPERTENSIONAHA.116.08560.

Wang, Y., Heilig, K., Saunders, T., Minto, A., Deb, D. K., Chang, A., et al. (2010). Transgenic overexpression of GLUT1 in mouse glomeruli produces renal disease resembling diabetic glomerulosclerosis. *American Journal of Physiology. Renal Physiology*, *299*(1), F99–f111. https://doi.org/10.1152/ajprenal.00466.2009.

Wang, X. X., Levi, J., Luo, Y., Myakala, K., Herman-Edelstein, M., Qiu, L., et al. (2017). SGLT2 protein expression is increased in human diabetic nephropathy: SGLT2 protein inhibition decreases renal lipid accumulation, inflammation, and the development of nephropathy in diabetic mice. *The Journal of Biological Chemistry*, *292*(13), 5335–5348. https://doi.org/10.1074/jbc.M117.779520.

Wang, Q., Song, B., Jiang, S., Liang, C., Chen, X., Shi, J., et al. (2015). Hydrogen sulfide prevents advanced glycation end-products induced activation of the epithelial sodium channel. *Oxidative Medicine and Cellular Longevity*, *2015*, 976848. https://doi.org/10.1155/2015/976848.

Wasik, A. A., & Lehtonen, S. (2018). Glucose transporters in diabetic kidney disease—Friends or foes? *Frontiers in Endocrinology (Lausanne), 9*, 155. https://doi.org/10.3389/fendo.2018.00155.

White, J. R., Jr., & Campbell, R. K. (1993). Magnesium and diabetes: A review. *The Annals of Pharmacotherapy, 27*(6), 775–780. https://doi.org/10.1177/106002809302700619.

Wilding, J. P., Woo, V., Soler, N. G., Pahor, A., Sugg, J., Rohwedder, K., et al. (2012). Long-term efficacy of dapagliflozin in patients with type 2 diabetes mellitus receiving high doses of insulin: A randomized trial. *Annals of Internal Medicine, 156*(6), 405–415. https://doi.org/10.7326/0003-4819-156-6-201203200-00003.

Winn, M. P., Conlon, P. J., Lynn, K. L., Farrington, M. K., Creazzo, T., Hawkins, A. F., et al. (2005). A mutation in the TRPC6 cation channel causes familial focal segmental glomerulosclerosis. *Science, 308*(5729), 1801–1804.

Woudenberg-Vrenken, T. E., Bindels, R. J., & Hoenderop, J. G. (2009). The role of transient receptor potential channels in kidney disease. *Nature Reviews. Nephrology, 5*(8), 441–449. https://doi.org/10.1038/nrneph.2009.100.

Wright, E. M., Loo, D. D., & Hirayama, B. A. (2011). Biology of human sodium glucose transporters. *Physiological Reviews, 91*(2), 733–794. https://doi.org/10.1152/physrev.00055.2009.

Wu, K. L., Khan, S., Lakhe-Reddy, S., Wang, L., Jarad, G., Miller, R. T., et al. (2003). Renal tubular epithelial cell apoptosis is associated with caspase cleavage of the NHE1 Na+/H+ exchanger. *American Journal of Physiology. Renal Physiology, 284*(4), F829–F839. https://doi.org/10.1152/ajprenal.00314.2002.

Wulff, H., & Castle, N. A. (2010). Therapeutic potential of KCa3.1 blockers: Recent advances and promising trends. *Expert Review of Clinical Pharmacology, 3*(3), 385–396. https://doi.org/10.1586/ecp.10.11.

Xu, H., Chen, R., & Ghishan, F. K. (2005). Subcloning, localization, and expression of the rat intestinal sodium-hydrogen exchanger isoform 8. *American Journal of Physiology. Gastrointestinal and Liver Physiology, 289*(1), G36–G41. https://doi.org/10.1152/ajpgi.00552.2004.

Yale, J. F., Bakris, G., Cariou, B., Yue, D., David-Neto, E., Xi, L., et al. (2013). Efficacy and safety of canagliflozin in subjects with type 2 diabetes and chronic kidney disease. *Diabetes, Obesity & Metabolism, 15*(5), 463–473. https://doi.org/10.1111/dom.12090.

Yan, N. (2015). Structural biology of the major facilitator superfamily transporters. *Annual Review of Biophysics, 44*, 257–283. https://doi.org/10.1146/annurev-biophys-060414-033901.

Yee, J. (2018). Magnesium: An important orphan. *Advances in Chronic Kidney Disease, 25*(3), 217–221. https://doi.org/10.1053/j.ackd.2018.04.001.

Zambrowicz, B., Freiman, J., Brown, P. M., Frazier, K. S., Turnage, A., Bronner, J., et al. (2012). LX4211, a dual SGLT1/SGLT2 inhibitor, improved glycemic control in patients with type 2 diabetes in a randomized, placebo-controlled trial. *Clinical Pharmacology and Therapeutics, 92*(2), 158–169. https://doi.org/10.1038/clpt.2012.58.

Zeni, L., Norden, A. G. W., Cancarini, G., & Unwin, R. J. (2017). A more tubulocentric view of diabetic kidney disease. *Journal of Nephrology, 30*(6), 701–717. https://doi.org/10.1007/s40620-017-0423-9.

Zhang, H., Ding, J., Fan, Q., & Liu, S. (2009). TRPC6 up-regulation in Ang II-induced podocyte apoptosis might result from ERK activation and NF-kappaB translocation. *Experimental Biology and Medicine (Maywood, N.J.), 234*(9), 1029–1036. https://doi.org/10.3181/0901-rm-11.

Zhang, H., Schin, M., Saha, J., Burke, K., Holzman, L. B., Filipiak, W., et al. (2010). Podocyte-specific overexpression of GLUT1 surprisingly reduces mesangial matrix expansion in diabetic nephropathy in mice. *American Journal of Physiology. Renal Physiology, 299*(1), F91–F98. https://doi.org/10.1152/ajprenal.00021.2010.

Zillich, A. J., Garg, J., Basu, S., Bakris, G. L., & Carter, B. L. (2006). Thiazide diuretics, potassium, and the development of diabetes: A quantitative review. *Hypertension*, *48*(2), 219–224. https://doi.org/10.1161/01.HYP.0000231552.10054.aa.

Zou, H., Zhou, B., & Xu, G. (2017). SGLT2 inhibitors: A novel choice for the combination therapy in diabetic kidney disease. *Cardiovascular Diabetology*, *16*(1), 65. https://doi.org/10.1186/s12933-017-0547-1.